Biostratigraphy in Production and Development Geology

Geological Society Special Publications

Series Editors

A. J. Fleet

R. E. Holdsworth

A. C. Morton

M. S. Stoker

It is recommended that reference to all or part of this book should be made in one of the following ways.

Jones, R. W. & Simmons, M. D. (eds) 1999. *Biostratigraphy in Production and Development Geology*. Geological Society, London, Special Publications, **152**.

Payne, S. N. J., Ewen, D. F. & Bowman, M. J. 1999. The role and value of 'high-impact biostratigraphy' in reservoir appraisal and development. *In*: Jones, R. W. & Simmons, M. D. (eds) *Biostratigraphy in Production and Development Geology*. Geological Society, London, Special Publications, **152**, 5–22.

GEOLOGICAL SOCIETY SPECIAL PUBLICATION NO. 152

Biostratigraphy in Production and Development Geology

EDITED BY

R. W. JONES
BP Exploration, Sunbury-on-Thames, UK

AND

M. D. SIMMONS
University of Aberdeen, UK

1999

Published by

The Geological Society

London

THE GEOLOGICAL SOCIETY

The Society was founded in 1807 as The Geological Society of London and is the oldest geological society in the world. It received its Royal Charter in 1825 for the purpose of 'investigating the mineral structure of the Earth'. The Society is Britain's national society for geology with a membership of around 8500. It has countrywide coverage and approximately 1500 members reside overseas. The Society is responsible for all aspects of the geological sciences including professional matters. The Society has its own publishing house, which produces the Society's international journals, books and maps, and which acts as the European distributor for publications of the American Association of Petroleum Geologists, SEPM and the Geological Society of America.

Fellowship is open to those holding a recognized honours degree in geology or cognate subject and who have at least two years' relevant postgraduate experience, or who have not less than six years' relevant experience in geology or a cognate subject. A Fellow who has not less than five years' relevant postgraduate experience in the practice of geology may apply for validation and, subject to approval, may be able to use the designatory letters C Geol (Chartered Geologist).

Further information about the Society is available from the Membership Manager, The Geological Society, Burlington House, Piccadilly, London W1V 0JU, UK. The Society is a Registered Charity, No. 210161.

Published by The Geological Society from:
The Geological Society Publishing House
Unit 7, Brassmill Enterprise Centre
Brassmill Lane
Bath BA1 3JN
UK
(*Orders*: Tel. 01225 445046
Fax 01225 442836)

First published 1999

British Library Cataloguing in Publication Data
A catalogue record for this book is available from the British Library.

ISBN 1-86239-031-2

Typeset by Aarontype Ltd, Bristol, UK

Printed by Cambridge University Press,
Cambridge, UK

Distributors

USA
AAPG Bookstore
PO Box 979
Tulsa
OK 74101-0979
USA
(*Orders*: Tel. (918) 584-2555
Fax (918) 560-2652)

Australia
Australian Mineral Foundation
63 Conyngham Street
Glenside
South Australia 5065
Australia
(*Orders*: Tel. (08) 379-0444
Fax (08) 379-4634)

India
Affiliated East-West Press PVT Ltd
G-1/16 Ansari Road
New Delhi 110 002
India
(*Orders*: Tel. (11) 327-9113
Fax (11) 326-0538)

Japan
Kanda Book Trading Co.
Cityhouse Tama 204
Tsurumaki 1-3-10
Tama-Shi
Tokyo 0206-0034
Japan
(*Orders*: Tel. (0423) 57-7650
Fax (0423) 57-7651)

Contents

Preface and Introduction

R. W. JONES[1] & M. D. SIMMONS[2]

[1] *BP Exploration, Building 200, Chertsey Road, Sunbury-on-Thames, Middlesex, TW16 7LN, UK*

[2] *Department of Geology and Petroleum Geology, University of Aberdeen, Aberdeen, AB24 3UE, UK*

This book records some of the recent advances that biostratigraphy has made in production and development geology. It serves to illustrate to non-biostratigraphers the potential applicability of biostratigraphy in this arena, and to encourage biostratigraphers to further explore and evaluate this potential.

In the production as opposed to the exploration arena, biostratigraphic and related techniques are employed to address reservoir-scale problems such as detailed correlation (often utilizing local, but nonetheless extremely useful, marker events), interpretation of depositional environment, geometry, connectivity and compartmentalization, and reserves estimation and optimization of recovery. This involves the biostratigrapher working as a member of an integrated multidisciplinary reservoir team alongside sedimentologists, petrophysicists, development geophysicists and geologists and engineers, and being familiar with core, wireline log, seismic and other data. It also involves maintaining a high level of specialist biostratigraphic knowledge and the capability to undertake analyses as appropriate across a wide range of geographies, stratigraphies and reservoir depositional environments and/or to be an 'informed buyer' (and interpreter) of vendor/contractor analytical data.

High-resolution and quantitative biostratigraphy and integrated reservoir description are among the comparatively novel techniques employed. Biostratigraphic steering (bio-steering) of wells at well-site, enabling optimal penetration of reservoir sections, has become an exceedingly important application (and one bringing significant benefits to the operator in terms of saving time and money).

These sorts of applications are likely to become increasingly important in a future in which the oil companies are likely to continue to focus on enhancing recovery from producing fields rather than exploring for new ones.

Introduction

In this volume, case histories of applications of biostratigraphic and related techniques in the North Sea are given first, and case histories of applications in the international arena (Euramerica, Borneo, Venezuela, Nigeria and the Gulf of Mexico) second. The overall representation of different geographies, stratigraphies and reservoir depositional environments is wide, although the North Sea is particularly well represented, reflecting the fact that it was here that many of the techniques, notably bio-steering (in Chalk reservoirs) were first practised.

In the first of the North Sea contributions, **Payne *et al.*** discuss the role and value of 'high-impact' biostratigraphy in reservoir appraisal and development, with worked examples of applications in the Donan, Andrew and Forties Fields in the UK Sector (Late Palaeocene submarine fan reservoirs).

Duxbury *et al.* then discuss the sequence stratigraphic subdivision of the Humber Group (Late Oxfordian to Ryazanian) of the Outer Moray Firth area of the UK Sector, with worked examples of applications in the Tartan, Highlander and Petronella Fields. **Morris *et al.*** continue the Mesozoic theme with a contribution on the micropalaeontological biostratigraphy of the Magnus Sandstone Member (Kimmeridgian–Early Volgian) of the Magnus Field, emphasizing its role in integrated reservoir description and reservoir management (extending production life).

Shipp, Bergen & Sikora and Sikora *et al.* all discuss aspects of the biostratigraphy of Chalk reservoirs in the North Sea (Late Cretaceous). **Shipp** focuses on well-site applications of high-resolution biostratigraphy in bio-steering in Chalk Fields in the Danish Sector, with worked examples of applications in the Dan Field and mention of several others (Harald, Kraka, Gorm,

JONES, R. W. & SIMMONS, M. D. 1999. Preface and Introduction. *In:* JONES, R. W. & SIMMONS, M. D. (eds) *Biostratigraphy in Production and Development Geology*. Geological Society, London, Special Publications, **152**, 1–3.

Roar, Skjold, Svend, Tyra and Valdemar). **Bergen & Sikora** and **Sikora *et al.***, respectively, discuss diachronism and depositional interpretation and the implications thereof for the Chalk Fields of the Norwegian Sector.

Bidgood *et al.* describe the stratigraphic potential of diatoms in the Late Palaeocene–Early Eocene of the North Sea (together with some of the taxonomic problems that require to be addressed before this potential can be fully realized), with an example of an application from the 29/25-1 well (just south of the Auk and Fulmar Fields) in the UK Sector.

Holmes resumes the bio-steering theme with examples of contrasting applications in the Andrew Formation (Late Palaeocene) of the Joanne and Andrew Fields of the UK Sector. In the case of the Andrew Field, an innovative form of microfacies analysis was used to derive interpretations of the depositional environment (whether turbiditic or interturbiditic) and likely extent of mudstones within the reservoir section, and their likely effects on fluid flow (whether barriers or baffles). These interpretations were in turn used to maximize production through appropriate well placement and stand-off from oil–water and gas–oil contacts (using barrier shales as 'umbrellas' to protect wells from gas invasion).

Mangerud *et al.* discuss the high-resolution biostratigraphy and sequence stratigraphy of the Palaeocene succession in the Grane Field in the Norwegian Sector and the implications thereof, including the resolution of heterogeneities in reservoir architecture (and consequences for fluid flow) not revealed by early models based solely on wireline log correlations.

In the last of the North Sea contributions, **Jones** attempts to demonstrate the value of historical micropalaeontological data in integrated reservoir description, with an example from the Forties Field (Late Palaeocene submarine fan reservoir).

In the first of the contributions from the international arena, **McLean & Davies** present an exhaustive discussion on constraints on the application of palynology to the correlation of Euramerican Late Carboniferous clastic reservoirs.

The remainder of the contributions from the international arena deal with Neogene–Pleistogene fluvial, paralic, peri-deltaic and submarine fan clastic reservoirs in low to moderate latitudes (Borneo, Venezuela, Nigeria, Gulf of Mexico).

Simmons *et al.* describe the use of microfossil assemblages (both foraminifera and palynomorphs) to determine the precise depositional setting of reservoir sands in the Neogene successions of northwest Borneo. Outcrop analogues of reservoir sands in a variety of depositional settings (from fluvial, through paralic and perideltaic to submarine fan) have been identified and an empirical observation of the variations in microfossil assemblages made. When applied to the subsurface, this ability to precisely identify depositional setting should allow for optimal production strategies to be applied.

Jones *et al.* discuss the reservoir biostratigraphy of the Pedernales Field in the Eastern Venezuelan Basin (Late Miocene–Early Pliocene peri-deltaic to submarine fan reservoirs), emphasizing its role in integrated (biostratigraphic, sedimentological, seismic, wireline log) reservoir description.

Armentrout *et al.* discuss the integrated high-resolution sequence stratigraphy of the peri-deltaic reservoir of the Oso Field, Nigeria. In this case, high-resolution biostratigraphy contributed significantly to the characterization of the reservoir, and thus indirectly to the infill- and enhanced recovery-drilling programmes based on this characterization and designed to maximize cost-effectiveness in the field development strategy.

Van der Zwan & Brugman describe 'bio-signals' from the EA Field, also in Nigeria. This new high-resolution (essentially climatostratigraphic) tool provided a detailed biostratigraphic subdivision within the existing biozonation and enabled easier correlation of reservoir units across growth-faults and recognition of fault cut-offs.

Finally, **O'Neill *et al.*** discuss uses of applied biostratigraphy in the Gulf of Mexico, with examples of applications from the Bonnie discovery well in Eugene Island Block 95 and from Mars Field in Mississippi Canyon Blocks 763, 806 and 807. In the latter example, biostratigraphy helped to resolve stratigraphic and structural relationships poorly imaged by seismic and thus to appraise reserves estimates.

This volume arose from the Petroleum Group conference on 'Biostratigraphy in Production and Development Geology' held in Aberdeen in June 1997, at which most of the papers in this volume were first presented. This conference was sponsored by BP Exploration Operating Company Ltd, Badley Ashton & Associates Ltd, Chevron Europe, Philips Petroleum Norway, RPS Paleo and StrataData. This volume itself is sponsored by BP.

Sheila Barnette, Mike Charnock, Phil Copestake, Dave Ewen, Nick Holmes, Mike Kaminski, Steve Lowe, Simon Payne, Dave Pocknall, Bob Ravn, Osman Varol and Paul Ventris are thanked for refereeing manuscripts (the remainder of the refereeing was undertaken by the editors).

The staff at the Geological Society and at the Geological Society Publishing House are thanked for helping to see the project through to publication.

Cambridge University Press and the Gulf Coast Section of the Society of Economic Paleontologists and Mineralogists Foundation are thanked for permission to reproduce figures in the Bidgood *et al.* and O'Neill *et al.* papers, respectively.

The role and value of 'high-impact biostratigraphy' in reservoir appraisal and development

S. N. J. PAYNE,[1] D. F. EWEN[1] & M. J. BOWMAN[2]

[1] *BP Exploration Ltd, Farburn Industrial Estate, Dyce, Aberdeen AB21 7PB, UK*
[2] *BP Exploration Technology Ltd, Chertsey Road, Sunbury-on-Thames, Middlesex TW16 7LN, UK*

Abstract: Over recent years changes in the application of biostratigraphy in the reservoir appraisal and development arena have greatly increased the impact and value of the discipline, giving it a central role in integrated reservoir description. These changes include placing emphasis on local field-scale bioevents to erect a reservoir framework of time slices through which reservoir heterogeneity can be modelled and the application of biosteering to maximize reservoir penetration. In addition, palaeoenvironmentally diagnostic benthonic microfacies are used to model the lateral continuity of intra-reservoir mudstones in an attempt to understand their potential as baffles/barriers to fluid flow. The evolution of this cost-effective methodology is discussed by reference to three Palaeocene turbidite reservoirs from the North Sea UK continental shelf (UKCS); the Donan, Forties and Andrew fields.

To understand the role of biostratigraphy in its widest context, it is helpful to step back and consider the role that all geotechnical resources play in the commercial exploitation of hydrocarbons. Geoscience imparts a mechanism for providing answers to, options for, and limiting uncertainty throughout the entire range of the oil exploration and production business. The key business objective into which technology feeds is to ensure company profitability and cash flow whilst maintaining a safe and environmentally sound operation. The latter is of particular significance to biostratigraphers given the suite of potentially harmful chemicals used in processing techniques.

Underlying this high-level objective, any geoscience must provide input into several business drivers throughout the life of an oil field to add value. These include sustaining base production, delivery of new options for renewal or growth, accessing and understanding future opportunities, and managing lifting costs. Geoscience impacts at all stages – so should biostratigraphy. However, to maximize impact a clear understanding of where and how biostratigraphy can deliver the highest possible value to the business in any given project or stage of project is essential. The requirement for biostratigraphy must be focused and business driven, fully understanding why it is undertaken and what it can deliver and not, as has happened too often in the past, out of quasi-academic interest or because 'it has always been done'.

Historical perspective

Biostratigraphy has been utilized in hydrocarbon exploration since the 1890s, its first application being well correlation in Poland (Kaminski *et al.* 1993). Since then it has impacted hydrocarbon exploration globally, but particularly in the Gulf of Mexico, the Middle East, the North Sea and Southeast Asia. For much of this time biostratigraphy has been recognized by the wider geoscience population as a valid tool for dating and correlating sediments in the exploration realm. At the reservoir scale, however, continual improvements in wireline log and seismic technology, such as 3D seismic, have often provided a finer subsurface resolution than could be achieved biostratigraphically.

An inherent conservatism has dogged reservoir-scale biostratigraphy; too often emphasis has been routinely placed on tying the reservoir into regional- or semi-regional-scale biozonations. Indeed, for a biostratigraphic event to be considered valid, a geographically widespread occurrence was generally the prerequisite, with more areally restricted, facies-controlled events downgraded or discarded. More focused, local schemes were seldom utilized outside syn-rift

PAYNE, S. N. J., EWEN, D. F. & BOWMAN, M. J. 1999. The role and value of 'high-impact biostratigraphy' in reservoir appraisal and development. *In:* JONES, R. W. & SIMMONS, M. D. (eds) *Biostratigraphy in Production and Development Geology*. Geological Society, London, Special Publications, **152**, 5–22.

settings in which local taphrogenic control resulted in rapid lateral variations in depositional facies and microfossil assemblages such as in Jurassic syn-rift packages in North Sea half-grabens (for example in the Brae Field; see Riley *et al.* 1989).

Traditional regional biozonations have often provided too coarse a resolution to impact reservoir and production issues. These biozones, as used in industry, are frequently defined by a number of individual bioevents, typically first downhole occurrences (stratigraphic extinctions), the recovery of any one of which can be taken as definitive. Problems can arise when attempting to correlate biozones thus defined. Frequently this approach will engender the correlation of different individual bioevents under the umbrella of one biozone, creating potential for miscorrelation and thereby 'blurring' stratigraphic resolution. The impact of applying basin-scale schemes to individual reservoirs has been minimal, pay zones often being encapsulated within one biozone.

The impact of biostratigraphy was frequently worsened by a lack of integration into broader geoscience (Fig. 1), perhaps in part a result of the difficulty in communicating an interpretative, non-numerate science in an essentially numerate realm, coupled with the relative dearth of oil company biostratigraphers co-located with field groups.

In addition, biostratigraphy (and biostratigraphers) has often been culpable in its perception as an untimely dataset, with biostratigraphic input arriving too late too often to impact materially the post-well review and update geological models. Unfocused or broadly spaced sample programmes have also, on occasions, hindered the potential impact of the discipline. Very often biostratigraphic results have been issued as a *fait accompli* with limited assessment and communication of confidence levels, crucial factors in weighing up the relative merits of a variety of data sources.

A sense that biostratigraphy was an outmoded tool seeped into parts of the geoscience community, with the perception of biostratigraphy as a static, non-progressive science. One comparison neatly illustrates this point: whilst reprocessing of seismic data was often seen as a valid means of boosting subsurface resolution over a reservoir, the reprocessing of material for microfossils, a complementary and comparatively inexpensive exercise to generate higher resolution results, was not always viewed in the same light.

Biostratigraphy has thereby been largely regarded as a tool confined to the exploration arena, the value of biostratigraphy at the reservoir scale often being perceived as low and its real impact negligible.

Recent developments

In recent years there has been an assertion from within the biostratigraphic community that the science has considerably more to offer at the

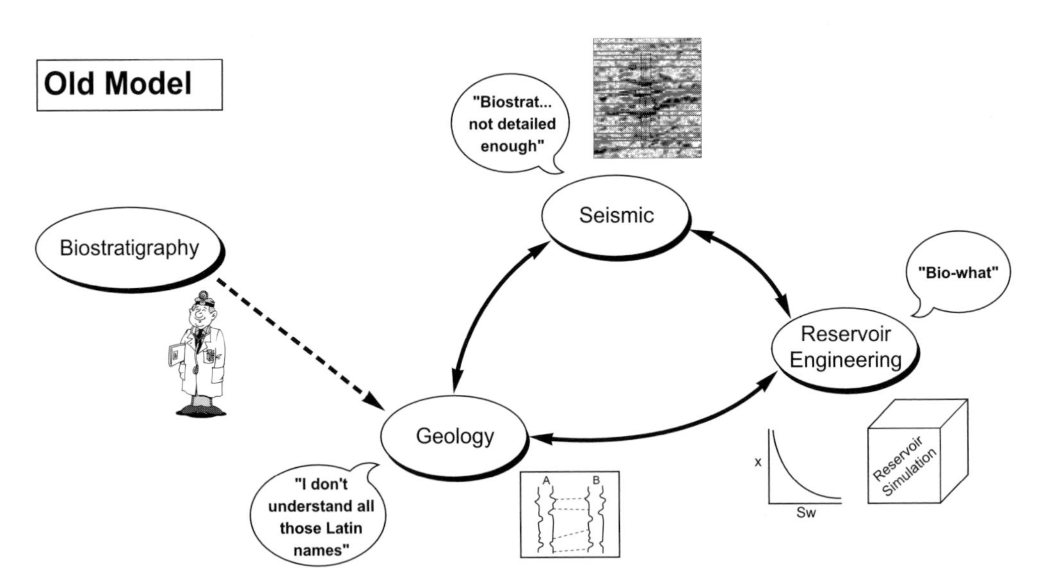

Fig. 1. Past perception of the geoscience community to reservoir biostratigraphy: old model showing a lack of integration with broader geoscience.

reservoir scale and should play a major role in all phases of field life, from exploration through appraisal, development and into production. A more pragmatic, field-focused approach has been adopted by operators and service companies, allowing greater stratigraphic subdivision of reservoirs beyond the control provided by regional biozonation schemes. Much of this impetus has been derived through work in the North Sea basin, a mature province with a large biostratigraphic database. With many fields now in the production or post-plateau stage, the industry has seen a drive to greater resolution in reservoir description in order to access remaining reserves.

High-resolution biostratigraphy

High-resolution biostratigraphy looks to use any repeatable bioevent of potential field-wide chronostratigraphic significance to 'fingerprint' mudstones, which can then be integrated with wireline log data and tied around the field to give a wireline log correlation constrained by biostratigraphy. Central to high-resolution biostratigraphy is thinking at field scale and, judiciously, pushing the data hard but always integrating and iterating with other geoscience disciplines.

Dropping a formal biozonation in favour of the use of a series of finer scale bioevents has been key to the application of this methodology. Whilst it is still possible to relate the local bioevents back to a broader regional biozonation scheme, strong emphasis is placed on 'anything goes' – to develop a localized, field-focused scheme driven by any repeatable data.

The brief period of geological time during which many reservoir intervals were deposited allowed for little genetic evolution to occur in the biota, and thereby the 'traditional' extinctions ('tops') or evolutionary appearances ('bases') may provide too coarse a subdivision. Instead, much reliance is placed on the use of assemblage characteristics and local acme events or influxes in individual species or genera. These acmes are the biological responses to local dynamic and/or physico-chemical changes in the palaeowater mass. In the case of dinocysts, radiolaria, diatoms, calcareous nannofossils and planktonic foraminifera these typically reflect changes in the upper water mass within the photic zone, whilst changes at the sediment–water interface affect the benthonic foraminifera. These changes enabled forms that could flourish under specific water mass regimes to 'bloom' for a short period whilst others might decline. This ephemeral aspect to these 'blooms' renders them potentially viable as local stratigraphic indices.

It might be thought that relying on the biotic impact from environmental changes as opposed to true extinctions/inceptions could result in applying diachronous events and is therefore a flawed ploy. However, within the confines of a limited geographical area such as a field, and with due consideration for palaeogeographic control, the assertion is that these palaeoenvironmental changes are effectively isochronous, allowing a correlatable field-scale 'ecostratigraphy' to be developed. Some of these bioevents appear to be subtle, but when the data are generated, interpreted and integrated with care they may frequently be seen to occur consistently across a field.

If biostratigraphy is to impact reservoir appraisal and development, a number of prerequisites usually need to be met. A focused pilot study of nearby offset wells, using a consistent preparation technique and, ideally, one micropalaeontologist, should be used to generate the field bioevent scheme. Where possible, core and sidewall core material should be used in preference to ditch cuttings, reducing sample contamination through downhole caving. The use of core also allows greater integration with sedimentological and palaeoenvironmental models. The reliability of the bioevents generated therein should be expressed in terms of confidence limits, allowing their value to be more clearly assessed.

High-impact biostratigraphy

The crucial tenet to this 'new' approach of high-impact biostratigraphy is the focus on the maximization of hydrocarbon recovery through the development and use of a robust, local, stratigraphic framework which addresses the key issues of reservoir management and field development. This sharper focus ensures that biostratigraphy, if successful, directly and materially impacts the business and adds real value. The alignment of high-resolution biostratigraphy with the attainment of business goals is what we imply by the term 'high-impact biostratigraphy'. To maximize impact it is vital to use applied biostratigraphy as a fully integrated part of the project (Fig. 2) rather than the previous model where biostratigraphy frequently was viewed as a non-critical path in the subsurface evaluation (Fig. 1).

High-impact biostratigraphy has a key role in the correlation of reservoir units, understanding reservoir architecture, helping to model the connectivity or the 'plumbing' of a reservoir and assisting well placement through well-site biostratigraphy.

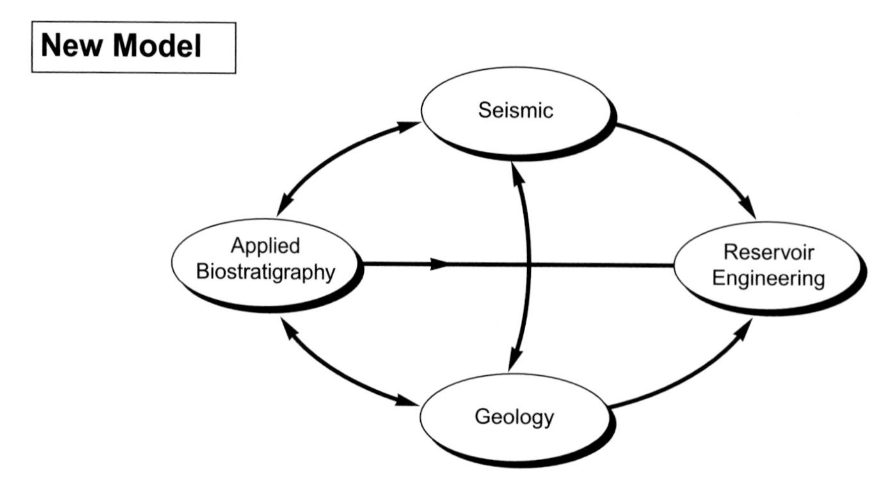

Fig. 2. Reservoir biostratigraphy as part of a fully integrated project: new model showing full interaction with other subsurfaces disciplines.

The use of well-site biostratigraphy has become an increasingly important arena for applying reservoir-scale stratigraphy. It has provided tangible, material benefit during the drilling of high-angle and horizontal wells through its unique ability to discriminate between the individual mudstones that separate and envelop reservoir bodies.

In considering the use of high-impact biostratigraphy, it is important to remember the value derived from capital spend. This is a relatively inexpensive tool with the potential to impact a project out of all proportion to its cost. In the experience of the authors, the fullest biostratigraphic coverage on an individual well, including well-site analysis, has never comprised more than 1% of the full well cost.

The key messages to the delivery of high impact biostratigraphy are:

- keep the focus on understanding the business aim – what questions need to be answered?;
- think field scale and field specific, and push the data hard;
- think 'bioevents' not 'biozonation';
- communicate confidence limits on your data points;
- integrate and iterate with other geoscience;
- realize the 'technology' can work at the well site.

Objectives of reservoir biostratigraphy

Applying these principles, reservoir biostratigraphy can have a profound impact in integrated reservoir characterization and description, from development planning to well placement. As a qualitative tool operating in a numerate environment, biostratigraphy 'sees' reservoir issues differently from many other datasets, adding a greater breadth and 'ground-truth' to models and thereby acting as a key contributor in uncertainty reduction.

The objectives of reservoir biostratigraphy are detailed below.

Framework for integrated reservoir description

- Template for integrating geoscience;
- constrains log picks and calibrates seismic.

Reservoir biostratigraphy provides a field-specific stratigraphic framework, a fundamental building block that forms the template for the integration of other geoscientific elements. It thereby acts as the 'glue', holding the reservoir framework together. The nature of biostratigraphy allows it to be uniquely definitive in recognizing individual mudstones, providing a higher order of constraint on wireline log and seismic interpretation.

Reservoir architecture and heterogeneity

- Time-slice mapping – plot facies changes in time;
- palaeoenvironmental facies – plot facies changes in space;
- lateral/vertical heterogeneity.

Through dividing the reservoir into isochronous time slices, biostratigraphy allows reservoir heterogeneity, in particular the lateral and vertical continuity of the reservoir per time slice, to be understood. Time slice mapping provides the ability to plot facies temporally (vertically) and spatially (laterally). Biofacies analysis, in conjunction with sedimentology, complements this by determining the depositional environment, allowing analogue data to be worked into a field model, impacting the prediction of reservoir architecture.

Reservoir compartmentalization

- Continuity of intra-reservoir mudstones;
- hemipelagic v. interturbidite – barrier or baffle?
- impacts well placement/stand-off from fluid contacts.

The definition of mudstone type through microfacies analysis can impact understanding of stratigraphic reservoir compartmentalization through helping to assess the relative lateral extent of mudstone bodies. Within turbidite reservoir packages, interturbiditic or channel wing mudstones comprise more 'proximal' units and may be notionally modelled as having a relatively lesser areal continuity, thereby acting as less extensive baffles to fluid flow. The more 'distal' hemipelagic mudstones may reflect more profound events in basin evolution such as relative rises in sea level. These are considered more likely to form areally extensive 'blankets' which, if preserved, may act as more profound barriers to reservoir fluid flow (Fig. 3). Integration with sedimentological and pressure (RFT) data to model depositional environment and lateral mudstone continuity is important.

In reservoirs deposited under deep-marine conditions, variations in the benthonic agglutinated foraminiferal 'morphogroups' (Jones & Charnock 1985) can provide a means of characterizing mudstone type. The use of microfacies in understanding reservoir compartmentalization is illustrated with reference to the Andrew Field later in this chapter.

Understanding the lateral continuity of intra-reservoir mudstones may significantly impact the reservoir model and producer well placement. It may be desirable to use the more extensive mudstones as an 'umbrella' to protect the producing well from early gas and/or water breakthrough, or it may be advisable to ensure that laterally extensive mudstones are 'punctured' to obviate their potential as baffles or barriers, reducing the likelihood of leaving pockets of oil unswept.

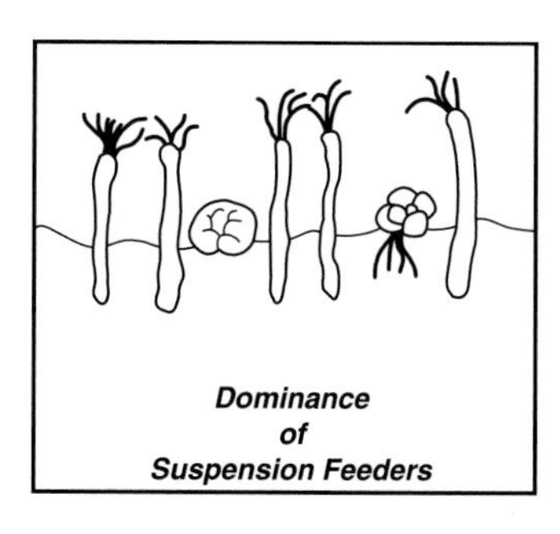

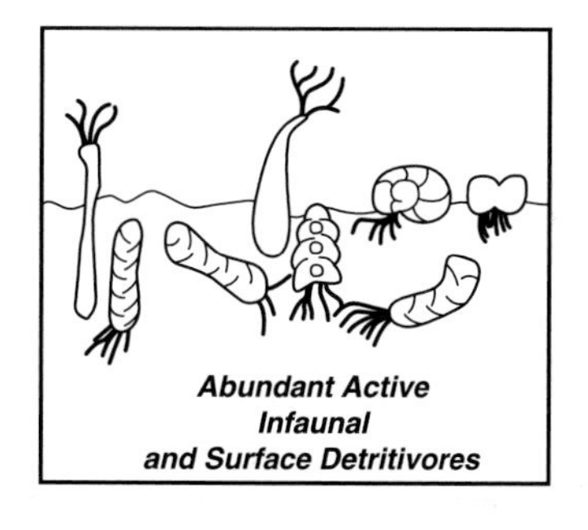

Fig. 3. Microfacies concepts using agglutinated foraminifera – a qualitative means of estimating the lateral extent of mudstone bodies in turbidite reservoirs.

Biosteeering

- 'Real-time' impact on well-site decision-making process;
- maximize reservoir penetration and through this the production index (P.I.) in horizontal and high-angle wells;
- build angle in high-angle wells to enter reservoir at optimal angle/location;
- casing, coring, total depth (TD) decisions;
- fast-track development campaigns – immediate constraint on reservoir model, iteration for sidetrack issues and ensuing wells.

Well-site biostratigraphy on vertical or subvertical exploration and production wells is a long-established tool for the 'real-time' stratigraphic monitoring of drilling, principally used to determine the stratigraphic position of the drill-bit, and to pick coring and casing points and total depths (TD). Over many years this type of well-site biostratigraphy has proved its value in saving drilling costs and continues to be in demand.

The increase in the drilling of horizontal or subhorizontal wells, concomitant with the drive for high-resolution biostratigraphy, has seen a renewed demand for well-site biostratigraphy and the advent of biosteering. Biosteering is intended to maximize reservoir penetration by biostratigraphically 'fingerprinting' the reservoir-enveloping non-pay packages during drilling. Highly detailed local bioevent schemes can be erected in high-angle wells, capable of detecting stratigraphic changes over vertical thicknesses of less than 1 foot in the case of the carbonate fields in the Danish sector of the North Sea, where wells have been biosteered successfully to follow vertical targets of 20 feet or less for several thousand feet (Shipp and Marshall 1995). If the well-bore encounters non-pay, having passed up through the top of the reservoir or down through the base, or passes out of the reservoir due to offset by faults (often of subseismic resolution), high-resolution biostratigraphy gives us a tool to steer the well-bore back into the reservoir. In addition, biosteering can be used in the supra-reservoir interval to monitor and calibrate angle-build in high-angle wells to ensure reservoir entry at the optimal angle. Integration with 'logging while drilling' (LWD) and lithological data is essential in biosteering. However, it is frequently biostratigraphy that provides the greatest resolution.

With many field developments now becoming fast-track campaigns, an additional but significant facet of utilizing well-site biostratigraphy is the provision of a preliminary stratigraphic understanding of the reservoir at, and prior to, TD. This allows more rapid data integration for possible sidetrack decisions and future well planning in a 'back-to-back' drilling schedule.

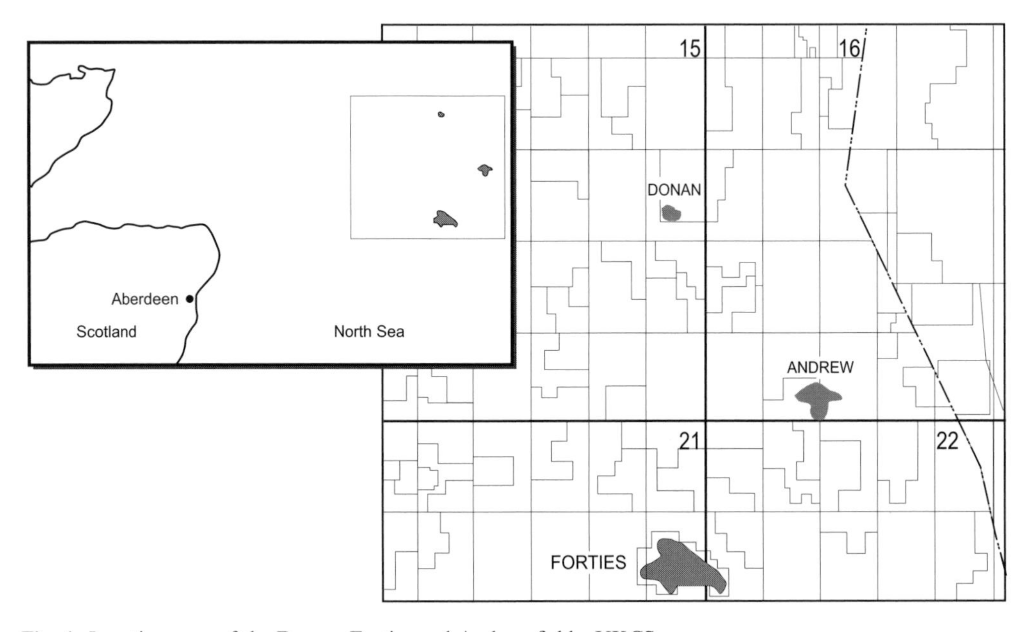

Fig. 4. Location map of the Donan, Forties and Andrew fields, UKCS.

Case histories

The application of high-impact biostratigraphy methodology is discussed by reference to three Palaeocene turbidite reservoirs from the North Sea UKCS; the Donan, Forties and Andrew fields.

Donan Field

The Donan Field reservoir comprises turbiditic sands within the upper part of the Late Palaeocene Lista Formation, deposited towards the northern flank of the Lista submarine fan system. The trapping mechanism is a subtle, low-relief four-way dip closure sitting over a Mesozoic basement high. Initial mean reserves stood at 16.5 mmbls and the field has been in production since April 1992. The field lies in UK Licence Block 15/20a (Fig. 4).

In 1993 there was a drive to increase oil production through the placement of infill producer wells. Central to this was the requirement to understand more fully the reservoir architecture and heterogeneity of this field. Seismic definition over the thin Donan reservoir is relatively poor (thickness of gross reservoir interval is 21.5 m true vertical depth (TVD) at maximum closure) and only top sand could be mapped reliably.

Hitherto it was perceived that this reservoir had a relatively simple 'layer-cake' architecture comprising laterally continuous and connected sands separated by mudstones, with the top sand being isochronous across the field; this model was supported by performance data suggesting excellent and predictable reservoir connectivity (Fig. 5a).

The existing biostratigraphic control was of a broad regional scale which provided insufficient resolution to impact the reservoir model. In the absence of more discriminatory biostratigraphic control, other techniques had been used to subdivide the reservoir. These included unconstrained wireline log correlation and correlating tuff horizons. The latter was particularly fraught with potential error as the tuffs are not definitively ash-fall deposits but are probably reworked from the hinterland and are therefore likely to be diachronous.

With the concept of field-specific high-resolution bioevent stratigraphy gaining credibility, a dinocyst-based pilot study was undertaken. The aim of the study was to discriminate between the intra-reservoir mudstones and thereby assess uncertainty on the lateral continuity of the reservoir sands.

Four acme bioevents were consistently recorded through the Donan reservoir in this study, giving greater resolution for discriminating intra-reservoir shales. From top to bottom these bioevents are:

- first downhole appearance (FDA) common *Achomosphaera alcicornu*;
- FDA super-abundant *Areoligera gippingensis*;
- FDA super-abundant *Spiniferites 'rhomboideus'*;
- FDA abundant *Cordosphaeridium gracile*.

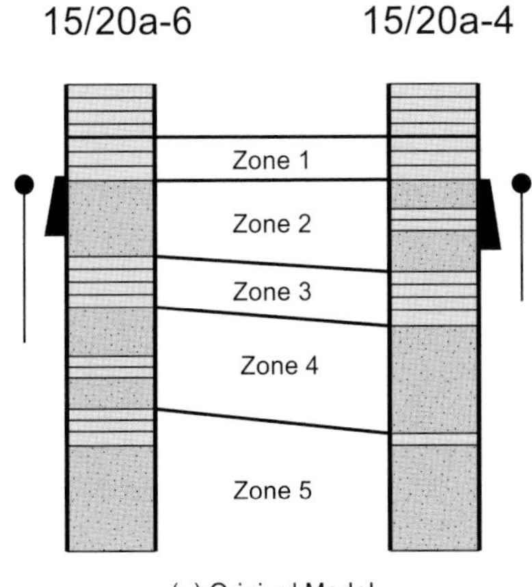

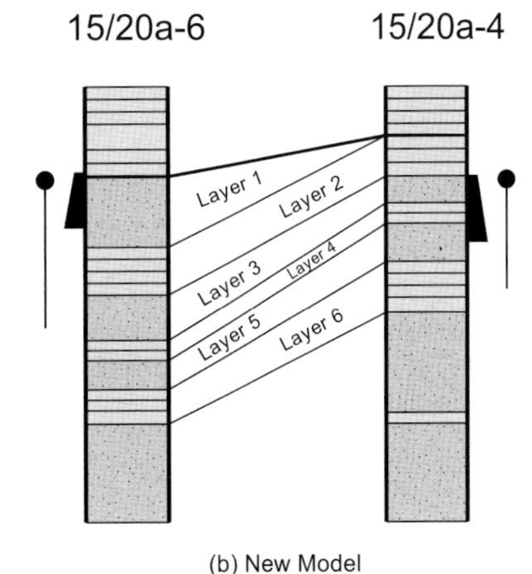

Fig. 5. Donan Field – original model v. new model (**a**) Original model – a 'layer-cake' reservoir beyond biostratigraphic resolution. (**b**) New model – high-resolution biostratigraphy indicates a more heterogenous system.

This enhanced defintion substantially changed the existing reservoir model by suggesting that (Fig. 5b):

- the Donan reservoir comprises a strongly heterogenous system with sand bodies pinching out laterally, often over short distances (hundreds of metres);
- the reservoir interval in the two producers, 15/20a-4 and 15/20a-6, was formed of three separate sand bodies of different ages, separated by mudstone packages. The Layer 1 sand in 15/20a-6 comprised the thickest sands with the maximum stand-off from the oil–water contact (OWC);
- the top reservoir sand is markedly diachronous.

These implications impacted the exploitation strategy for Donan; the revised objectives were to place high-angle wells into Layer 1 in proximity to well 15/20a-6, the thickest discrete reservoir sand package and furthest away from the oil–water contact (OWC). In 1995 a two-well programme was planned around this objective, with biosteering to be utilized to maximize the penetration of sands equivalent to the uppermost reservoir unit in well 15/20a-6 by 'fingerprinting' the reservoir-enveloping mudstone layers. There was an expectation that some of the well paths might be offset by small throw faults of sub-seismic resolution and that biosteering would help indicate the required direction to maintain maximum reservoir penetration.

This technique proved highly successful in the series of wells drilled and consistently proved to have the unique, lithology-independent capability to indicate the location of the drill-bit relative to the reservoir and subdivide the reservoir package, profoundly impacting the drilling decisions. Furthermore, the greater measured depth thickness of reservoir section and mudstones penetrated in this series of high-angle wells (in contrast to the vertical sections analysed in the pilot study) allowed the recognition of additional bioevents during drilling, enhancing the pilot study resolution.

Following the drilling of the two original wells and associated sidetracks (Fig. 6), the results suggest a laterally extensive sand-rich depositional system (pre-abundant *C. gracile*) becoming progressively more laterally restricted and heterolithic through time. Ultimately sand deposition persisted in the western part of the field around wells 15/20b-12X and 15/20a-6, with contemporaneous mudstone deposition occurring in 15/20a-4.

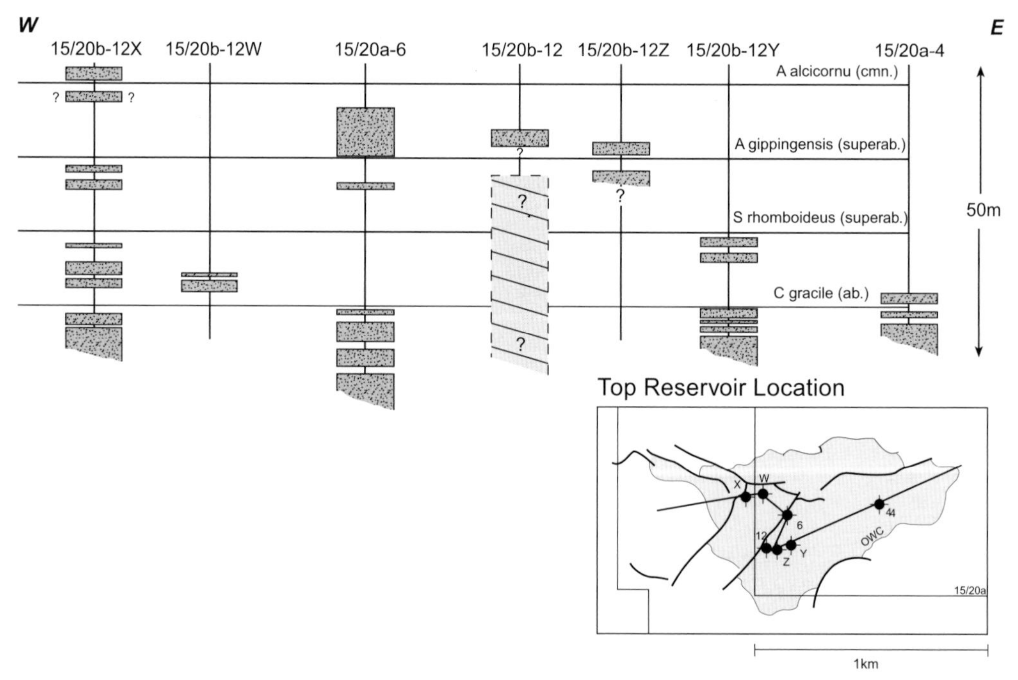

Fig. 6. Donan sand-body chronostratigraphy illustrating time-slice heterogeneity. Heterogeneity in this abandonment facies is greater than suggested in the initial 2 well dataset (15/20a-4 and 6). In high-angle wells the greater measured depth thickness of reservoir and mudstones can allow field biostratigraphy to 'evolve' further at the well site.

Between the biostratigraphically 'fingerprinted' mudstones great lateral heterogeneity is evident, with sand bodies clearly pinching out over very short distances, locally less than 100 m. This is primarily considered to reflect complexity in the abandonment facies of a marginal area of the Lista fan. In addition, the low relief on seafloor topography at time of deposition may have resulted in the deposition of successive sedimentary gravity flows in the topographic lows in the basin floor, away from the positive relief created by the previous turbidite. Syn-depositional faulting may also be a factor in explaining local increases in sand thickness. The heterogeneity encountered is certainly greater than suggested by the pilot study, but the study was crucial in creating an awareness of the probability of the lateral impersistence of individual sand bodies and discarding a 'layer-cake' model.

Despite this complexity, sandstone connectivity as suggested by the strong aquifer drive appears to be extremely good. There is probably a sufficient amount of net sand in the gross reservoir interval to provide connectivity, possibly aided by sand injection. Therefore, despite the discontinuous nature of the sand bodies, connectivity is good and sweep efficiency very high.

High-impact biostratigraphy has had a direct business impact on Donan, providing a model that radically changed the view of reservoir continuity, altered the strategy for the placement of producer wells and thereby has impacted the management of offtake from the field.

Forties Field

The Forties Field (see also Jones 1999) comprises a turbidite package of Late Palaeocene age, part of a large submarine fan system within the basal Sele Formation. It is located predominantly in UK Block 21/10 (Fig. 4). Initial reserves were estimated as 2500 mmbls (Wills 1991) and to date production has exceeded 2350 mmbls.

Oil production commenced in 1975 with the bulk of the production coming from the Main Sand, a complex of stacked progradational lobe and channel sequences of a mid- to lower-fan environment, ranging from 60 to 260 m in thickness. Reservoir pressure decline data are essentially uniform indicating that there is hydrodynamic communication throughout the Main Sand (Carman & Young 1981). This unit is now largely swept of hydrocarbons, with only thin zones of bypassed oil present.

The remaining oil now sits principally within the upper part of the reservoir in the Charlie Sand, a more areally restricted, complex and heterolithic mid-fan channel system with a maximum thickness of 70 m. This sits in pressure isolation from the Main Sand, the two units separated by the Charlie Shale.

A better understanding of the Charlie Sand is essential for maximizing recovery in late field life. Key issues include:

- understanding the heterogeneity and lateral continuity of the Charlie Sand system;
- placement of infill wells – producers and injectors – to enable remaining oil to be more efficiently located and swept;
- calibration and constraint of reprocessed 3D seismic to enable extrapolation between wells and to constrain the new full field model.

A 1987 biozonation of the Forties Field subdivided the reservoir into four biozones, based primarily on autochthonous marine dinocyst abundances, principally of the genus *Apectodinium* (Fig. 7). It was the cornerstone in the early understanding of the field stratigraphy and was adequate to describe early production. This scheme, tied to the seismic, was a major influence in erecting a laterally extensive, largely 'layer-cake' model for the reservoir, implying laterally persistent sands, separated by similarly extensive mudstones (Fig. 8). Whilst dynamic data support this model for the Main Sand, there is increasing conflict between a 'layer-cake' model and dynamic data and the biostratigraphy in the Charlie Sand. The 1987 biozonation has not been able to provide sufficient resolution to help in understanding the heterogeneity of the Charlie Sand.

In 1993, the 1988 vintage 3D seismic dataset was reprocessed and provided a tool offering potentially better resolution than the 1987 biozonation scheme. The reprocessed 3D seismic was used to produce a new reservoir model in part of the Forties Alpha (FA) area, markedly improving the dynamic model and the history match. However, carrying the seismic picks more widely across the field to the Forties Charlie (FC) or Forties Delta (FD) area required the seismic picks to cross-cut the reservoir model based on the 1987 biozonation.The need to employ a higher resolution tool as an independent means of verification for the seismic picks became imperative.

The time of deposition of the Forties fan is associated with environmental stress in the North Sea basin. Microfossil and sedimentological data indicate a stratified, restricted basin with dysaerobic bottom waters. Benthonic microfossils such as agglutinated and calcareous benthonic

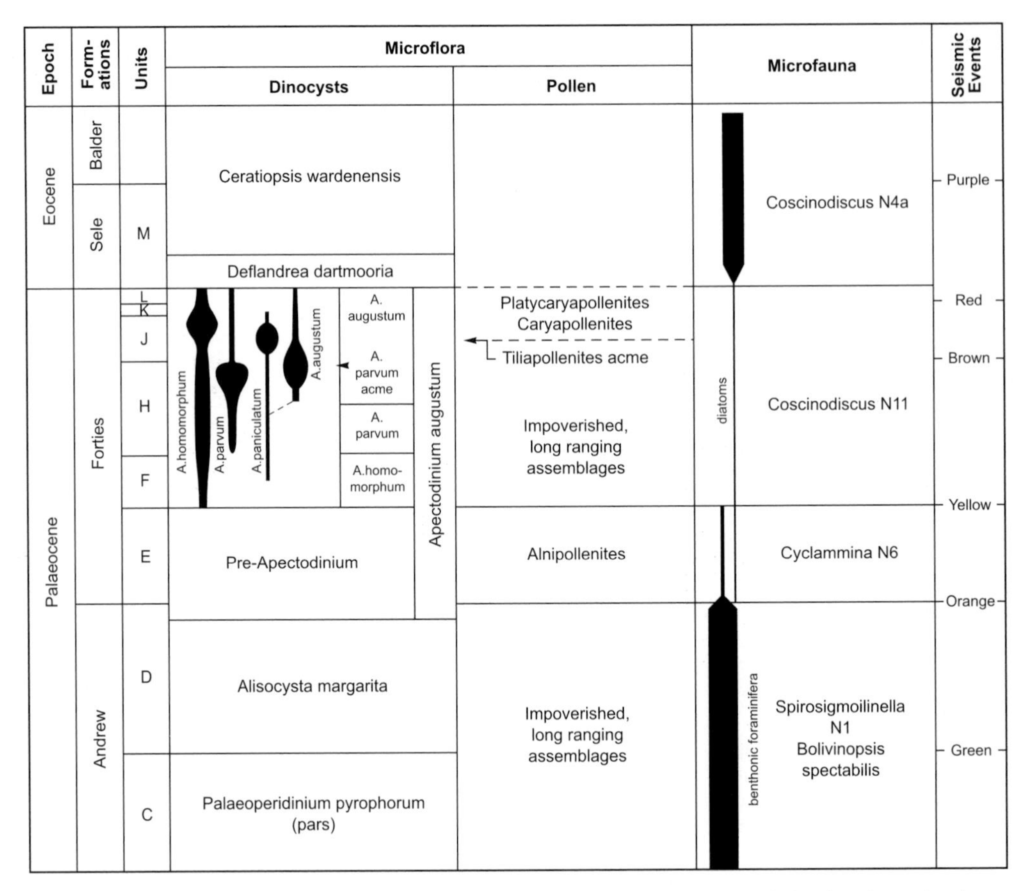

Fig. 7. Original stratigraphic framework for reservoir subdivision in the Forties Field showing the predominance of the use of dinocysts, notably the genus *Apectodinium*.

foraminifera are rare, and the lack of planktonic foraminifera suggests low levels of oxygenation even at higher levels in the water column and/or dissolution. The most significant *in situ* biotic elements comprise low-diversity and high-abundance dinocyst assemblages dominated by an intergrading plexus of *Apectodinium* spp. Allied to this lack of biodiversity is the relatively short period of geological time throughout which the Forties fan was deposited. This allowed for little genetic evolution to occur in an already poorly developed palynoflora, rendering the exclusive use of the autochthonous marine palynoflora for stratigraphic subdivision difficult and ultimately of limited impact.

Schroder (1992) illustrated the potential for subdividing the Forties interval through combining dinocyst and contemporaneously derived spore and pollen bioevents, the latter typically being extremely abundant in the Forties lowstand fan. Utilizing these allochthonous forms brings with it several concerns. First, the transport and deposition of the non-marine palynomorphs over some 200–300 km from the palaeoshoreline would introduce the artefact of hydraulic sorting (influenced by density, morphology and size). Secondly, these forms are often long ranging, with few useable stratigraphic inceptions or extinctions. Thirdly, the fact that many of these bioevents are repeatable through time, given the appropriate palaeoclimatic conditions. Any biovent stratigraphy would thereby reflect an interplay between hinterland ecostratigraphy, palaeoclimate and sea level, the sum of which will have been modified by hydrodynamic processes.

However, in the light of our new philosophy it was felt that a bioevent framework incorporating the most robust bioevents from the 1987 dinocyst biozonation complemented with additional local dinocyst and contemporaneously derived pollen and spores events might provide a more precise bioevent stratigraphy over the Forties Field.

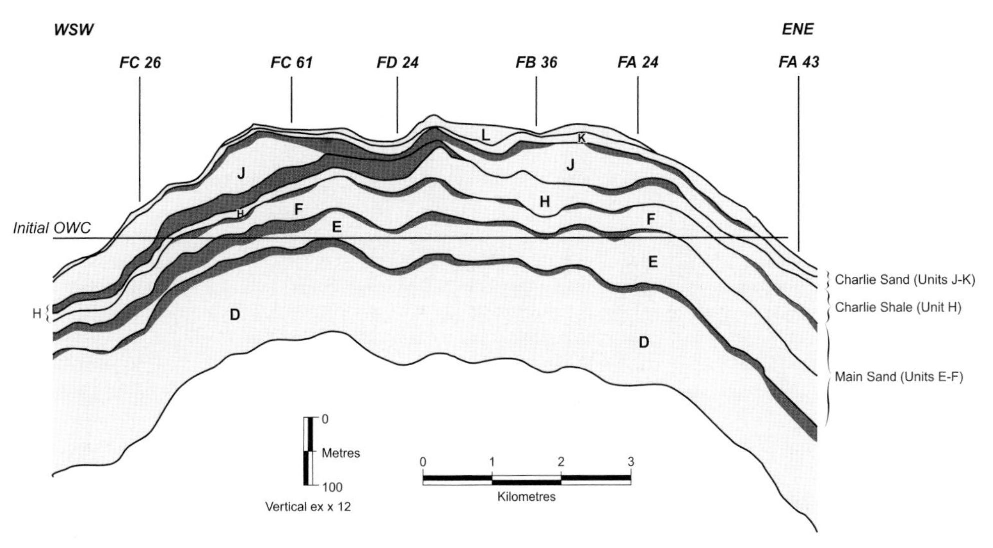

Fig. 8. 1987 Reservoir zonation for the Forties Field showing the lateral persistence of many of the reservoir units, as modelled. Modified from fig. 6 of Wills & Peattie (1990), reproduced with permission from Kluwer Academic.

A pilot study demonstrated that many bioevents appear to have potential for subdividing the reservoir interval in Forties Field. Up to 11 primary correlatable bioevents, together with 10 secondary bioevents, have been used:

- first downhole appearance (FDA) *Apectodinium augustum* – confidence level 100%;
- downhole increase in *Apectodinium* spp. – confidence level 90%;
- FDA frequent *Kallosphaeridium* spp. – confidence level 80%;
- FDA abundant fungal spores and hyphae – confidence level 85%;
- FDA frequent/common *Trudopollis* spp. – confidence level 65%;
- FDA abundant *Inaperturopollenites* spp. > abundant *Deltoidospora* spp. – confidence level 60%;
- FDA frequent/common *Alnipollenites verus* – confidence level 75%;
- FDA frequent/common *Sequoiapollenites* spp. – confidence level 75%;
- FDA common *Osmundacidites/Baculatisporites* spp. – confidence level 70%;
- FDA common *Impletosphaeridium* spp. – confidence level 70%;
- FDA *Alisocysta margarita* – confidence level 100%.

In addition to these primary bioevents, there is a series of secondary bioevents that may be applied, albeit with caution, in addition to or in the absence of the primary bioevents. These include:

- FDA *Glaphyrocysta 'reticulata'*;
- FDA *Interpollis supplingenis*;
- FDA *Labrapollis labraferoides*;
- FDA common/abundant *Deltoidospora* spp.;
- FDA common/abundant *Laevigatosporites* spp.;
- FDA *Pesavis* spp.;
- FDA *Tiliaepollenites 'diktyotus'*;
- last downhole appearance (LDA) *Tiliaepollenites 'diktyotus'*;
- LDA *Apectodinium parvum*;
- LDA consistent *Apectodinium* spp.

The new biostratigraphic framework has been successfully tested, providing the stratigraphic constraint on well-ties which are then used to calibrate the 3D seismic. Through this iteration a new reservoir zonation has been erected. Tying detailed sedimentology into the new reservoir zonation has added a further level of definition to the understanding of reservoir architecture. A key understanding to arise from this work is the confirmation that the Main Sand is very much a 'layer-cake' across the field, but whilst it was immediately succeeded by more areally constrained and channelized deposits of the Alpha–Bravo channel in the eastern area of the field, to the west it shales out rapidly (Fig. 9).

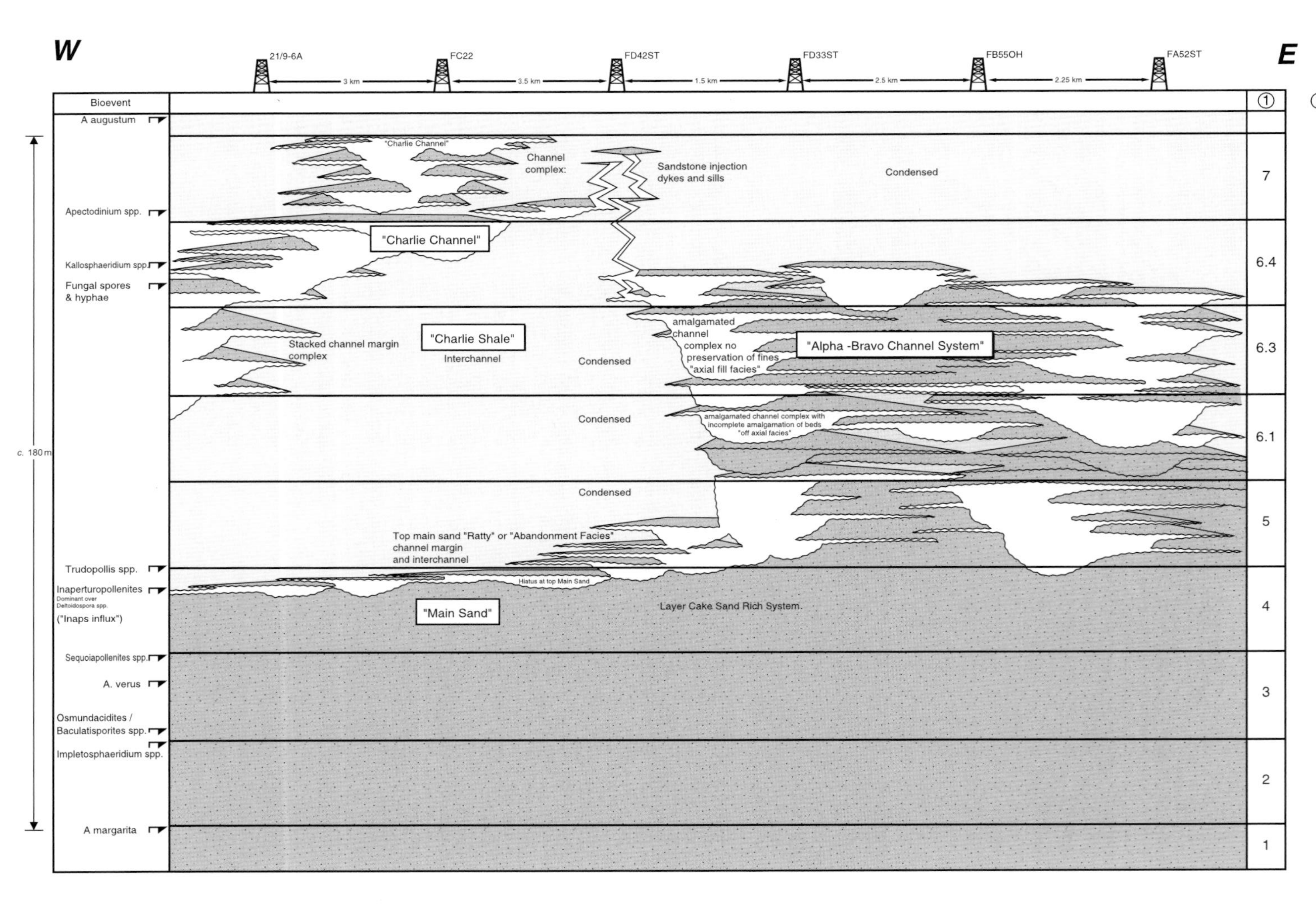
W
E
21/9-6A
FC22
FD42ST
FD33ST
FB55OH
FA52ST
3 km
3.5 km
1.5 km
2.5 km
2.25 km
Bioevent
① Reservoir Zone
A augustum
Apectodinium spp.
Kallosphaeridium spp.
Fungal spores & hyphae
Trudopollis spp.
Inaperturopollenites
Dominant over Deltoidospora spp.
("Inaps influx")
Sequoiapollenites spp.
A. verus
Osmundacidites / Baculatisporites spp.
Impletosphaeridium spp.
A margarita
c. 180 m
"Charlie Channel"
Channel complex:
Sandstone injection dykes and sills
Condensed
"Charlie Channel"
Stacked channel margin complex
"Charlie Shale"
Interchannel
Condensed
amalgamated channel complex no preservation of fines "axial fill facies"
"Alpha -Bravo Channel System"
Condensed
amalgamated channel complex with incomplete amalgamation of beds "off axial facies"
Condensed
Top main sand "Ratty" or "Abandonment Facies" channel margin and interchannel
Hiatus at top Main Sand
"Main Sand"
Layer Cake Sand Rich System.
7
6.4
6.3
6.1
5
4
3
2
1

Additionally, as the Alpha–Bravo channel system wanes, the Charlie channel system became established, the locus of sand deposition having switched in the upper sands across the field. The Charlie channel persists until the Forties fan abandonment facies, in contrast to earlier abandoned Alpha–Bravo channel; top sand is thereby strongly diachronous.

Using the new reservoir model, the heterogeneity of the Charlie Sand can be plotted spatially and temporally, allowing the diachroneity and connectivity of the major sedimentary systems to be better understood and predicted. The stratigraphic resolution over the Forties Field has been increased at least three-fold and frequently provides a finer definition than the seismic. In areas of equivocal seismic, wireline log and dynamic data, biostratigraphy is often the key uncertainty reduction tool, suggesting the 'best-fit' correlation. This is successfully impacting the location of new infill well sites enabling better placement of producer and injector wells.

Andrew Field

The Andrew Field (see also Holmes 1999) lies in UKCS blocks 16/27a and 16/28, and comprises a stacked series of turbidite sandstones, deposited in the central area of the Lista submarine fan system of Late Palaeocene age (Fig. 4). Salt diapirism has exerted a profound influence on Andrew, imposing some penecontemporaneous control on sedimentation as well as providing the hydrocarbon trapping mechanism for the pool through enhancing the four-way dip closure. The field has a thin oil column (58 m) with mean reserves of 118 mmbls and has been developed using horizontal wells. Production commenced in June 1996 and the field is currently producing 64 000 barrels of oil per day.

The value of the Andrew project will be largely controlled by the amount of oil produced from the wells prior to gas and/or water breakthrough. This can be maximized by the optimal placement of the horizontal wells relative to fluid contacts. Fluid movement in the reservoir is controlled by key heterogeneities such as laterally extensive shales and higher permeability sands. High-impact biostratigraphy and sedimentological modelling are important in defining these heterogeneities.

High-resolution biostratigraphy, integrated with wireline log and sedimentological data, impacts Andrew Field depletion by:

- providing a detailed framework in order to model reservoir heterogeneity per time slice;
- using foraminiferal microfacies to understand the uncertainty around the lateral continuity of shales, hence their potential as fluid flow barriers/baffles;
- biosteering horizontal wells by 'fingerprinting' individual shale bodies that separate the reservoir sands.

As a first step a localized bioevent stratigraphy was erected to provide a chronostratigraphic framework. Biostratigraphic data were integrated with wireline logs, sedimentology and seismic to produce a new reservoir model for the field (Fig. 10). The model utilizes a series of palynological and micropalaeontological bioevents to divide the reservoir into seven reservoir zones (A1–B2), equivalent to time slices. Confidence limits are provided for each primary bioevent in order to communicate their perceived reliability. Secondary bioevents are also noted.

The benthonic microfacies concept was then applied in an attempt to model gross mudstone type (hemipelagic or interturbidite/channel wing) and hence potential lateral continuity. In the Andrew Field the more energetic 'proximal' interturbidite mudstones yield a predominance of simple epifaunal suspension-feeding agglutinated foraminifera of low diversity and high-species dominance. These foraminifera can be considered as primary colonizers or 'doomed pioneers', ephemerally establishing themselves during a pause in turbidite deposition. In contrast, 'distal' hemipelagic mudstones typically yield assemblages of high diversity, high abundance and low dominance, characterized by an increased proportion of complex, sophisticated forms including active infaunal and surface detritivores (Grun *et al.* 1964). This 'climax community' is considered to reflect a less 'disturbed' episode of greater duration within a tranquil environment.

Fig. 9. Qualitative chronostratigraphic model for deposition of the Forties reservoir through a E–W section showing the diachroneity of the major sand bodies. Note the widespread and largely 'layer-cake' Main Sand is succeeded by the Alpha–Bravo channel system in the east. Later Alpha–Bravo sand deposition is coeval with the initiation of the Charlie channel system further west, the youngest sands within the Forties resevoir. Modified from fig. 12 of Wills & Peattie (1990), reproduced with permission from Kluwer Academic.

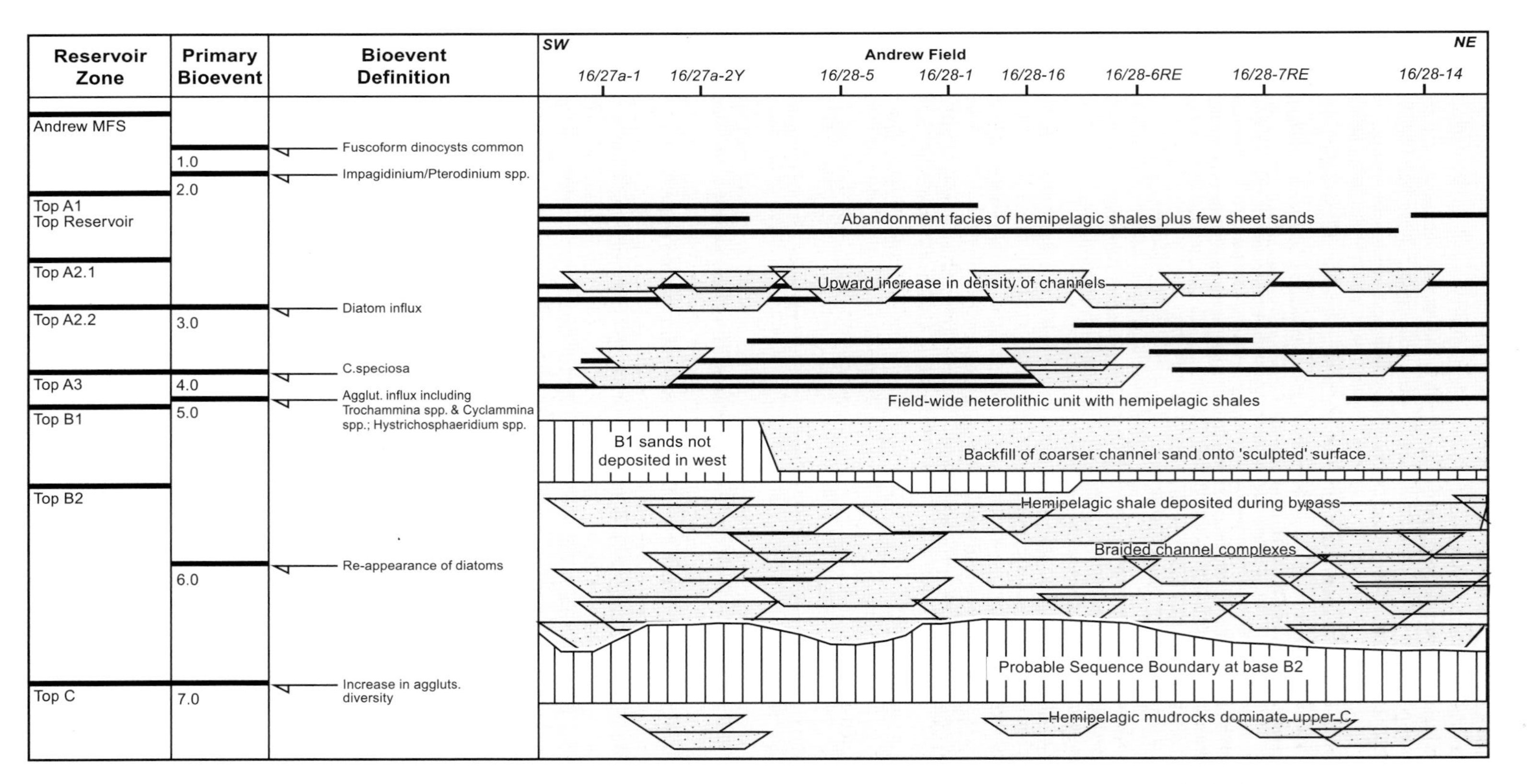

Fig. 10. Andrew Field depositional framework utilizing integrated high-resolution biostratigraphy, sedimentology, wireline logs and seismic.

Unit B1 is the main reservoir unit, a higher permeability, laterally constrained, linear, NW–SE-trending channel sandstone up to 23 m in thickness. It incorporates coarse material up to granule size, and the base of this unit is locally erosive. The overlying Unit A is more heterolithic and the sandstones are typically of a more sheet-like nature. This base of this unit, A3, incorporates mudstones characterized by relatively complex, diverse and abundant agglutinated foraminifera including *Trochammina* spp. and *Cyclammina* spp. (morphogroups C and B3 of Jones & Charnock 1985), locally including *C. amplectens* (Fig. 11). This assemblage is considered to suggest a hemipelagic origin associated with the raising of relative sea level and/or autocyclic fan switching. Therefore, this mudstone is notionally considered to be more likely to form a laterally extensive 'blanket' of potentially field-wide extent. Unit A3 is thereby considered to represent a field-wide, intra-reservoir, correlatable shut-down after the dominantly channelized sytem of Unit B recording a change in sedimentary regime from relatively proximal to relatively distal. This mudstone might be expected to form a significant barrier to reservoir fluid flow if not subsequently eroded or 'punctured'.

In contrast, the other intra-reservoir mudstones on Andrew Field, such as the mudstone that separates Unit A2.2 from A2.1, show a predominance of morphogroup A astrorhizids, simple epifaunal suspension-feeders of low diversity and high-specific dominance. These foraminiferal assemblages are considered to reflect the recolonization of the sediment surface during a break in turbidite deposition (or in a depositional setting away from the immediate locus of turbidite deposition) within a more energetic 'proximal' setting. The integration of morphogroup and sedimentological data allow these sediments to be defined as interturbidite or channel wing mudstones. These may be more localized in distribution, potentially forming less extensive baffles to reservoir fluid flow.

The only mudstones in the gross Andrew reservoir package with foraminiferal assemblages comparable with Unit A3 are seen in the Unit A1 abandonment facies continuing into the 'Andrew Shale'. These intervals are considered to reflect the retrogradation of the Andrew sand system under increasing pelagic and transgressive influence, culminating in the top Lista maximum flooding surface (Fig. 11).

Over the central part of Andrew Field the base of A3 is effectively coincident with the gas–oil

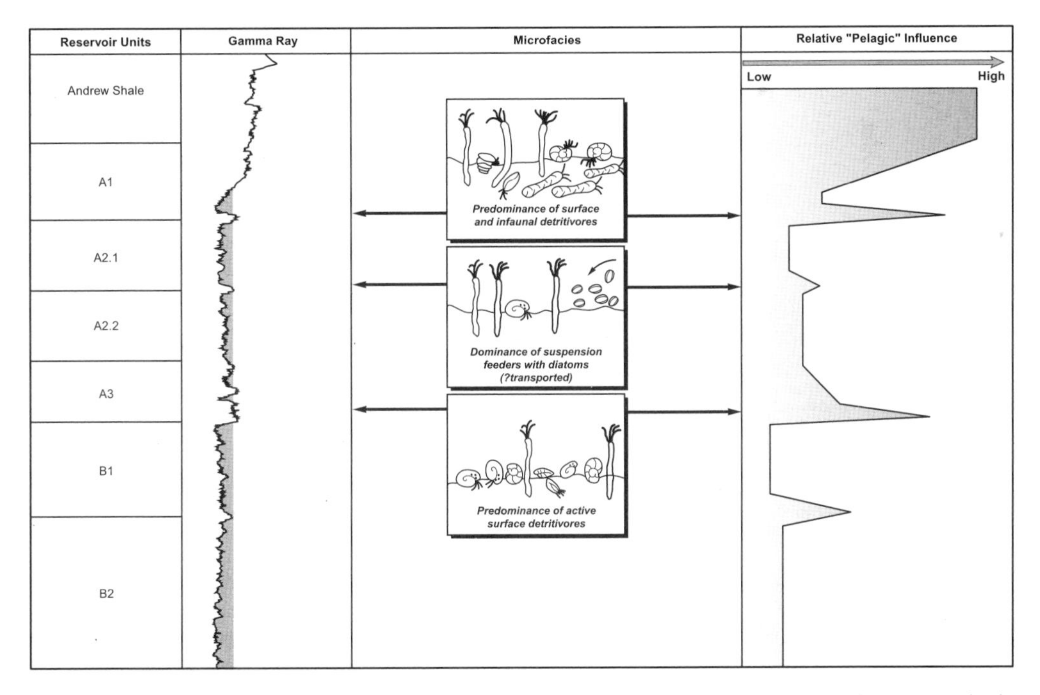

Fig. 11. Andrew Field – microfacies model illustrating the relative complexity of agglutinated foraminifera in the A3 mudstone. The complex, diverse and abundant assemblages seen in Unit the A3 suggest that these mudstones are of hemipelagic origin and may be laterally extensive.

contact (GOC). It is significant that this laterally extensive mudstone runs subparallel to this important fluid contact through much of the field. The oil–water contact (OWC) lies within the B Unit, with the main pay-zone in the upper B1. The juxtaposition of the B1 reservoir unit with the A3 mudstone has proved to be important when considering potential fluid flow into the horizontal producers. An understanding of these contacts in relation to the stratigraphy has profoundly impacted the planning of well paths as it has been considered desirable to use the A3 mudstone as an 'umbrella' to protect the producing well from early gas breakthrough, the prime risk to well productivity. One of the key risks from this model is the presence of local erosional 'holes' in A3 mudstones, juxtaposing Unit A sands on Unit B sands which could provide a ready conduit for gas coning. This risk is reduced by detailed sedimentological facies mapping, tracing the major Unit A channel systems which should be avoided in placing producer wells.

Prior to this understanding, horizontal well paths were planned to penetrate Unit B reservoir at a constant 63% : 37% stand-off between GOC and OWC, maintained through to TD (Fig. 12a). This was deemed optimal to avoid early gas and/or water breakthrough, and was designed to run close to the OWC as water is less mobile than gas.

As a result of the study, wells are now planned to run more medially through the Unit B reservoir at a 50% : 50% stand-off from the GOC and OWC over the crestal part of the field, using the A3 mudstones as an 'umbrella' to protect the well from gas invasion (Fig. 12b). This allows the well path to run at a further distance from the OWC. Away from the crest, on the limb of the structure, the 'umbrella' is lost as the A3 mudstone is necessarily penetrated. Well trajectory in the toe section is dropped to

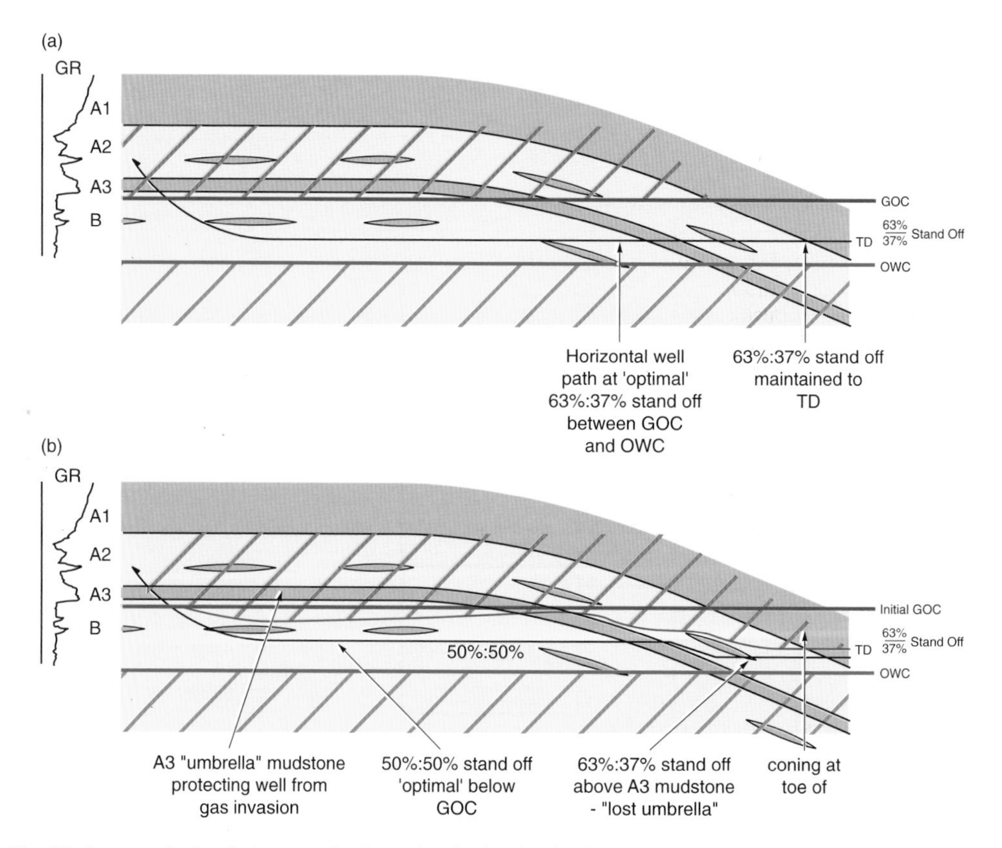

Fig. 12. Impact of microfacies on a horizontal well planning in the Andrew Field. (**a**) Before recognition of the impact of the A3 mudstone – original well path as planned. (**b**) After recognition of the impact of the A3 mudstone – the A3 'umbrella' allows the well path to run closer to the GOC and further from the OWC giving the probability of extended well life and productivity.

provide a conventional 63% : 37% stand-off from GOC in the higher A2 reservoir with the greater risk of gas coning. Having impacted the well-planning stage, well-site biosteering is used to monitor progress through the reservoir interval as a whole in 'real time', picking the key mudstone units, including A3, at which well trajectory is lowered.

Biostratigraphy has materially added value to the Andrew project by providing a mechanism for understanding reservoir heterogeneity, modelling mudstone type to impact horizontal well placement and picking key mudstone horizons at the well site. Through this, individual well life is extended and productivity is enhanced. The acquisition of dynamic production data has now supported the pre-drill model described, with the A3 mudstone acting as the most significant pressure barrier within the reservoir.

The Andrew Field group currently assigns an additional 10 mmbls of recoverable reserves through field life to this biostratigraphic input.

Conclusions

The application of biostratigraphy to reservoir-scale problems has accelerated in recent times due to the increazed resolution being offered by the use of localized bioevents rather than traditional regional biozones. However, to maximize value from biostratigraphy, work has to be focused and pragmatic but, most importantly, must not be used as a stand-alone discipline. Biostratigraphy undertaken from a business-driven perspective rather than out of quasi-academic interest or inertia, provides the potential for high-value, high-impact, but low-cost solutions.

The three examples presented provide an insight into Palaeocene fields in the North Sea at different stages in their production history, with different issues and problems. The biostratigraphic tools used to help in an integrated solution to these problems have, likewise, been very different.

The key lessons learnt are:

- high-impact biostratigraphy is a key component of integrated reservoir description;
- it impacts reservoir description, field development, well-path planning and 'real time' biosteering;
- this low-cost discipline can have a major effect on the business bottom-line.

This paper is written to maintain the impetus of high-impact biostratigraphy, and is intended to keep geoscientists aware of the potential business impact that the many and varied strands of biostratigraphy can offer in the solution of reservoir problems.

The authors wish to acknowledge BP Exploration Ltd for permission to publish. The approval of Donan Field partners Conoco (UK) Ltd, OMV (UK) Ltd and Candecca Resources Ltd, Forties Field partners Shell UK Exploration and Production, Esso Exploration and Production UK Ltd, British Borneo plc, LEPCO plc and Samedan Oil Corporation and Andrew Field partners LASMO plc, MOC Exploration (UK) Ltd, Clyde Petroleum plc and Talisman Energy (UK) Ltd is gratefully acknowledged.

The authors wish to acknowledge the technical input of Aubrey Hewson and Simon Todd (BP Exploration Ltd) and Paul Middleton (formerly BP Exploration Ltd, now Talisman Energy (UK) Ltd), together with Nick Holmes of Ichron Ltd and Neil Campion and Keith Marshall of Robertson Research International Ltd.

Bob Jones and Mike Simmons are thanked for reviewing this manuscript.

Figures 8 and 9 are modified from Wills & Peattie (1990) (figs 6 and 12, respectively) and are published with kind permission from Kluwer Academic.

References

Carman, G. J. & Young, R. 1981. Reservoir geology of the Forties oilfield. *In*: Illing, L. V. & Hobson, G. D. (eds) *Petroleum Geology of the Continental Shelf of North-West Europe*. Heyden, London, 371–391.

Copestake, P. 1993. *Application of micropalaeontology to hydrocarbon exploration in the North Sea basin*. *In*: Jenkins, D. G. (ed.) *Applied Micropalaeontology*. Kluwer, Dordrecht, 93–152.

Grun, W., Lauer, G., Niedermayr, G. & Schnabel, W. 1964. Die Kreide-Tertiar im Wienerwaldflysch bei Hochstrasse (Niederosterreich). *Sonderabdruck aus den Verhanlungen der Geologischen Bundesanstalt*, **2**, 226–283.

Holmes, N. A. 1999. The Andrew Formation and 'biosteering' – different reservoirs, different approaches. *This volume*.

Jones, R. W. 1999. Forties Field (North Sea) revisited: a demonstration of the value of historical micropalaeontological data. *This volume*.

—— & Charnock, M. A. 1985. 'Morphogroups' of agglutinating foraminifera. Their life positions and feeding habits and potential applicability in (paleo)ecological studies. *Revue de Paleobiologie*, **4**, 311–320.

Kaminski, M. A., Geroch, S. & Kaminski, D. 1993. *The Origins of Applied Micropalaeontology: The School of Josef Grzybowski*. Alden Press/Grzybowski Foundation, Oxford.

Riley, L. A., Roberts, M. J. & Connell, E. R. 1989. The application of palynology in the interpretation of Brae Formation stratigraphy and reservoir geology in the South Brae Field area, British North Sea. *In*: Collinson, J. (ed.) *Correlation in Hydrocarbon Exploration*. Graham and Trotman, London, 339–356.

SCHRODER, T. 1992. A palynological zonation for the Paleocene of the North Sea basin. *Journal of Micropalaeontology*, **11,** 113–126.

SIMMONS, M. & LOWE, S. 1996. The future for palaeontology? An industrial perspective. *Geoscientist*, **6**, 14–16.

SHIPP, D. J. & MARSHALL, P. R. 1995. Biostratigraphic steering of horizontal wells. *In*: AL-HUSSEINI, M. I. (ed.) *Geo '94 Middle East Petroleum Geoscience*. Gulf Petrolink, Bahrain, 849–860.

WILLS, J. M. 1991. The Forties Field, Block 21/10, 22/6a, UK North Sea. *In*: ABBOTTS, I. L. (ed.) *United Kingdom Oil and Gas Fields, 25 Years Commemorative Volume*. Geological Society, London, Memoirs, **14**, 301–308.

—— & PEATTIE, D. K. 1990. The Forties Field and the evolution of a reservoir management strategy. *In*: BULLER, A. T. (ed.) *North Sea Oil and Gas Reservoirs – II*. Graham and Trotman, London, 1–23.

Sequence stratigraphic subdivision of the Humber Group in the Outer Moray Firth area (UKCS, North Sea)

S. DUXBURY,[1] D. KADOLSKY[2] & S. JOHANSEN[3]

[1] *Duxbury Stratigraphic Consultants, 4 Coldstone Avenue, Kingswells, Aberdeen AB15 8TT, UK*

[2] *Texaco Ltd, 1 Westferry Circus, Canary Wharf, London E14 4HA, UK*

[3] *Texaco Inc., Exploration and Production Technology Department, 3901 Briarpark, Houston, TX 77042, USA*

Abstract: The Humber Group (Late Oxfordian–Ryazanian) in the Outer Moray Firth has been subdivided into 19 depositional sequences, which are firmly related to a newly defined, consistent biostratigraphic framework. Each sequence has been defined by core sedimentology (where possible), biostratigraphy and electric log patterns. In non-marine to shallow-marine environments (e.g. Piper Formation *sensu lato*), well-defined transgressive and highstand (or regressive) systems tracts are recognized, which are extensively cored in the area due to their importance as reservoirs of major fields. In younger sections, sequence recognition is based on the presence of submarine fans and maximum flooding surfaces; sequences so defined are to some extent supported by the cyclicity of equivalent shallow-marine sand sequences near the basin margins. The sequence stratigraphy approach has clarified the lateral stratigraphic relationships of reservoir sand units within the Tartan Field, and the relationships between the reservoir sand units in the Tartan, Highlander and Petronella fields. The main causes of the observed cyclic sedimentation are interpreted as the complex interplay between sediment supply, subsidence rates and eustasy. Although the relative importance of the last two remains unclear, the sequence stratigraphic approach followed here has allowed rigorous description and consistent interpretation of vertical and lateral facies associations.

Nomenclature

AOM	Amorphous organic matter
AST	Aggrading systems tract
FDO	First downhole occurrence (extinction)
GR	Gamma ray
HST	Highstand systems tract
LDO	Last downhole occurrence (first appearance)
LK	Lower Cretaceous
LKP	Lower Cretaceous palyzone
LST	Lowstand systems tract
MD	Measured depth
MFS	Maximum flooding surface
OMF	Outer Moray Firth
RS	Ravinement surface
SB	Sequence boundary
TST	Transgressive systems tract
TVD	True vertical depth subsea
UJ	Upper Jurassic
UJP	Upper Jurassic palyzone

The Humber Group (Oxfordian–Ryazanian) in the Outer Moray Firth area (OMF) (Fig. 1) is well known as the provider of reservoir and hydrocarbon source rocks for fields such as Piper, Tartan, Claymore, Scott and many others (David 1996). Accounts focusing on its biostratigraphy and lithostratigraphy have been published by Harker *et al.* (1987, 1993), and additional stratigraphic information can be found in descriptions of individual fields, viz.: Chanter: Schmitt (1991); Claymore: Maher & Harker (1987) and Harker *et al.* (1991); Highlander: Whitehead & Pinnock (1991); Hamish: Currie (1996); Ivanhoe: Brealey (1990), Parker (1991) and Currie (1996); Lowlander: McCants & Burley (1996); Petronella: Waddams & Clark (1991); Piper: Williams *et al.* (1975), Maher (1980, 1981), Schmitt & Gordon (1991), Harker (1998); Rob Roy: Brealey (1990), Parker (1991) and Currie (1996); Saltire: Casey *et al.* (1993); Tartan: Coward *et al.* (1991); Telford: Syms *et al.* (in press).

The regional lithostratigraphy is treated in the UKOOA account of the North Sea (Richards *et al.* 1993), and aspects of the regional stratigraphic and sedimentary history are dealt with by Andrews & Brown (1987), Boote & Gustav (1987), Boldy & Brealey (1990), O'Driscoll *et al.*

DUXBURY, S., KADOLSKY, D. & JOHANSEN, S. 1999. Sequence stratigraphic subdivision of the Humber Group in the Outer Moray Firth area (UKCS, North Sea). *In:* JONES, R. W. & SIMMONS, M. D. (eds) *Biostratigraphy in Production and Development Geology*. Geological Society, London, Special Publications, **152**, 23–54.

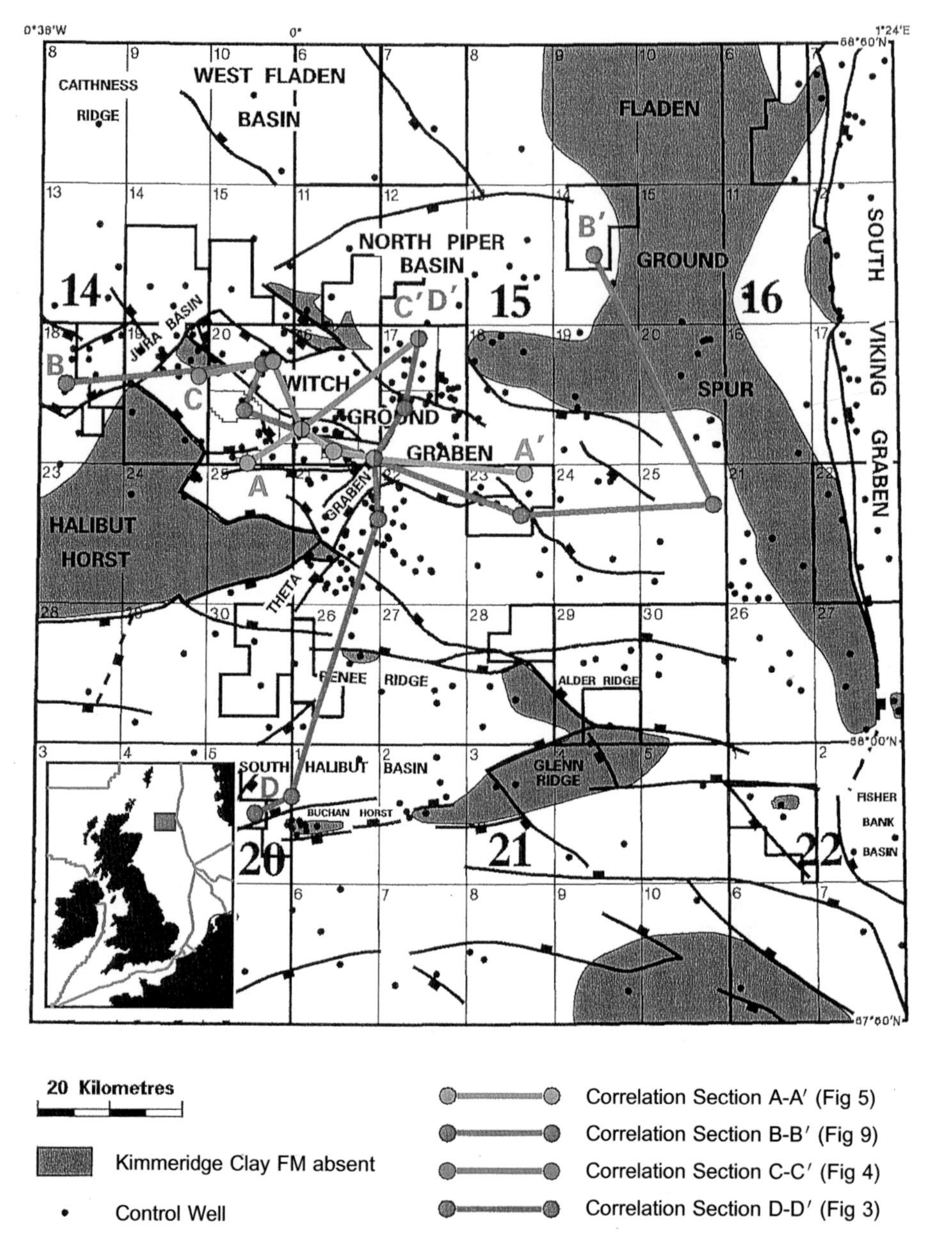

Fig. 1. Location map.

(1990), Rattey & Hayward (1993), Hallsworth *et al.* (1996) and Harker & Rieuf (1996). Published biostratigraphic data are, however, sparse and general, and a sequence stratigraphy interpretation of the entire Humber Group in the OMF has only been proposed by Carruthers *et al.* (1996) and by Harker & Rieuf (1996), while other accounts treat the Inner and Outer Moray Firth sequence stratigraphy only up to the Kimmeridgian (Davies *et al.* 1996; Stephen & Davies 1998). An outline of the sedimentary history of the OMF Humber Group, based on the sequence subdivision presented in this paper, is presented by Kadolsky *et al.* (in press).

Our own studies suggested that a considerable refinement of sequence (or parasequence) definitions and their biostratigraphic calibrations beyond the work published to date is possible; in this paper we wish to describe the sequences identified and the biostratigraphical zonation applied to them, and to show wireline logs of the best-documented examples of these sequences and their facies changes across the OMF basin. We will further demonstrate that the combined approach of sequence stratigraphy and biostratigraphy lends itself both to field-scale subdivisions for the purposes of reservoir modelling and provides a consistent regional framework. This overcomes the limitations of the lithostratigraphical approach, in particular in a syn-rift setting with its inherent major lateral facies changes. The study is a test of the applicability of the Jurassic sequence stratigraphy schemes for the entire North Sea as proposed by Partington *et al.* (1993*a*, *b*); also, it fulfills the demand of Miall (1986, 1992) that the eustatic concept of sequence stratigraphy be separated from a tectonic overprint by documenting the sequences of each basin separately, so that eventually sequences representing eustatic sealevel fluctuations may be separated from localized tectonic effects.

In this paper, S. Duxbury is responsible for the biostratigraphy and characterization of maximum flooding surfaces; D. Kadolsky for sequence identifications and correlations; and S. Johansen for advice on methodical aspects of sequence stratigraphy and particularly on the interpretation of the Piper Formation.

Objectives of the study and data used

In 1994 Texaco geoscientists initiated a regional sequence stratigraphic study of the Humber Group (Oxfordian–Ryazanian) in the Outer Moray Firth (OMF). Major in-house and consultant studies over a 2-year period have allowed the integration of first-hand and third-party data.

The purpose of the study was to:

- establish/confirm the genetic sequences in the Upper Jurassic of the study area;
- apply a consistent and modern biostratigraphic interpretation to the wells on the basis of already available biostratigraphic data (which were of mixed origin, age and quality) and selected new analyses;
- reinterpret the stratigraphy of the existing wells in terms of sequence stratigraphy;
- interpret the palaeogeography of each sequence with particular emphasis on the origin, depositional environment and distribution of reservoir sands;
- assist in the detailed subdivison of fields for reservoir modelling purposes;
- assist in the evaluation of exploration prospects and leads.

In a first phase all wells on Texaco's blocks 14/20 and 15/16, in a second phase all wells on the Texaco-operated blocks 15/23 and 15/29 and their immediate surroundings, and eventually all available well data in the OMF area were studied.

On the four Texaco blocks mentioned above, 78 wells penetrated the Humber Group at least partially. All Humber Group core material was either examined by ourselves or a detailed third-party sedimentological log was studied. All available biostratigraphical reports were reinterpreted, and all cores were resampled for new analyses, thereby enabling a definite correlation between microplankton assemblages, depositional environment and systems tracts to be made.

Outside these four blocks, wireline logs, core and biostratigraphy data of approximately 500 wells with Humber Group penetration were utilized; first-hand core inspections and biostratigraphic work was carried out on selected additional wells.

Interpretation procedure

(A) Biostratigraphy. A1 – available biostratigraphic data were reinterpreted in terms of the dinocyst zonation scheme presented in this paper (Fig. 2; see the Appendix for details).

A2 – in order to optimize the microplankton recovery of new preparations, a palynofloral sample preparation trial was made. Significant quality differences between five contractors were noted and the contractor producing the best results was chosen for new sample preparations.

A3 – new quantitative biostratigraphical analyses were made of key sections of blocks 15/16, 14/20, 15/23 and 15/29 wells, and of several released competitor wells.

(B) Sequence stratigraphy. B1 – almost all core material of all wells on blocks 15/16, 14/20, 15/23 and 15/29, and of several released competitor wells, was re-examined and its sedimentology was re-interpreted in terms of systems tracts.

B2 – sedimentological core logs were plotted against wireline logs and thus the log expressions of sedimentological features were calibrated in a qualitative way.

B3 – in uncored well sections, sequences and their systems tracts were identified by wireline log responses.

B4 – the ages of the sequences were identified from biostratigraphic data, and the sequence stratigraphic surfaces were correlated between wells.

B5 – as a final step, palaeogeographic maps of individual or combined sequences were constructed (Kadolsky *et al.* in press).

Most of the above interpretation steps were iterative.

Sequence stratigraphic framework for the Humber Group (Late Oxfordian–Ryazanian) in the Outer Moray Firth

Basis for sequence recognition

The presence of several well-developed transgressive–regressive cycles in the lower part of the Humber Group combined with good biostratigraphic calibration facilitated first-hand definition of the sequences in the dinocyst zones UJP5–UJP8.2 (i.e. within the Piper Formation *s.l.*, including the Sgiath and 'Heather' formations). In the following interval from UJP 9.1 to LKP4, shoreface sediments are restricted to the basin margins and show less well-pronounced transgressive–regressive cyclicity. Nonetheless, first-hand observations of cored shoreface sands up to UJP 14.2 (Late Volgian) were possible. In parallel, the equivalent basinal facies of dysaerobic black mudstones with mass flow clastics (Kimmeridge Clay Formation) was subdivided on the basis of discrete episodes of clastics emplacement and intervening episodes of clastic starvation, interpreted as maximum flooding surfaces. The frequency and ages of sequences derived from these two facies belts are very similar and are hence combined.

Biostratigraphy

The biostratigraphic calibration of the sequences is based on first-hand observations in the reference wells. Core material was analysed wherever possible.The definition of the dinocyst zones is given in Fig. 2 and described in detail in the Appendix. The dinocyst zones of the Humber Group are denoted by the prefixes UJP or LKP (= Upper Jurassic or Lower Cretaceous palynozone), while in the sequence designations (see below) the 'P' is omitted.

Sequence naming

Our preferred sequence nomenclature is based on the dinocyst zonal age of the MFSs: the sequences, and all component systems tracts and surfaces of a sequence are named after the dinocyst zonal age of the pertinent MFS, even if the entire sequence may extend beyond the dinocyst (sub)zonal age of its MFS. For example, the UJ13.1 sequence is the sequence with a MFS in dinocyst zone UJP13.1, but its boundaries may be outwith that subzone, or their biostratigraphic age may be not precisely identified. Thus, the zonal naming scheme should be considered as a numbering scheme. If more than one MFS lies within one dinocyst zone or subzone, they are distinguished by the addition of a letter to the zonal number (Fig. 2). As an alternative to a numbering scheme, the sequences may be named after a characteristic dinocyst, and such a parallel scheme is proposed here.

The use of ammonite zone names for MFSs has not been followed here, and directly observed palynofloral criteria have been preferred. As noted by Veldkamp *et al.* (1996), the correlation of the MFSs to ammonite zones is at the moment tentative, as direct evidence from core material is very sparse; reference to the actual data is preferable. Nonetheless, for reference purposes, ammonite zone attributions of the dinocyst zones and of the MFSs are quoted, following Partington *et al.* (1993*a*); occasional ammonite recoveries in cores from study wells proved inconclusive due to the poor state of preservation of the material.

Reference wells for the sequences are proposed here in order to highlight a section where their typical attributes are best documented. Figures 6, 7, 8, 10 & 11 depict log responses, sequence boundaries and biostratigraphical data of the most significant reference wells; Figs 5 and 9 show the correlations between the reference wells. Figure 4 shows correlations in the Piper Formation *s.l.* between the Claymore, Lowlander, Tartan and Piper fields, and a N–S correlation section through the OMF basin is represented in Fig. 3. Figure 12 illustrates the Galley Field reservoir stratigraphy.

Description of the sequences

The following account will concentrate on the sequence definitions and their expressions in the reference wells. An account of the Humber Group sedimentary history, developed from sequence identifications in most wells in the OMF, is given elsewhere (Kadolsky *et al.*, in press).

The following discussions include brief descriptions of sequences defined here, followed by their reference sections (including log depths of their MFSs), and by palynofloral characteristics of the MFSs, as an aid to their discrimination.

UJ4 (Polonicum) *sequence*

Alternative name (ammonite nomenclature): *Glosense* sequence.

The oldest sequence of the Humber Group in the OMF, tentatively placed into dinocyst zone UJP4, is identified in the southwest (Buchan Graben, e.g. wells 20/5b-2, 21/1-5, 14/30-3 and 15/26-5 (Fig. 3)) and in the northwest (West Fladen basin, e.g. wells 14/14a-2, 14/15a-5 and 14/18a-10).

A pre-UJ5 sequence is tentatively recognized in well 21/1-5 in the Buchan Graben, on the basis of a report of *Compositosphaeridium polonicum* in the 'Scott Member' of well 21/1-5 by Partington *et al.* (1993*a*, p. 381). The sand in this well is therefore interpreted as the UJ4 HST. Wireline log correlations to the type well of the Scott Sand, 15/21a-15 (Fig. 3), using sequence stratigraphical principles, place the Scott Sand into a younger sequence. We are not aware of any biostratigraphical evidence to support a UJP4 age of the Scott Sand sequence in its type area on blocks 15/21 and 15/22; it is therefore placed in the UJ5 HST and the UJ6a AST.

Harker & Rieuf's (1996) report of a widespread first '*Glosense*' (= UJ4 = *Polonicum*) transgression in the OMF is therefore interpreted to refer mainly to the UJ5 (and 'UJ5a') highstand(s) (see below).

A UJP4 age is provisionally attributed to this sequence in 21/1-5, which is correlated into well 20/5b-2 (Fig. 3), although no UJP4 marker species were identified in the latter. Davies *et al.* (1996) attributed an '*Athleta*/*Lamberti*' age (i.e. Middle–Upper Callovian) to the UJ4 MFS in well 20/5b-2, on the basis of reports of *Wanaea fimbriata*, *C. continuum* and *C. sellwoodi*. Our re-examination of the original palynological slides did not confirm the presence of any of these three species. The samples examined here from 20/5b-2 were entirely ditch cuttings, and the apparent FDO of *S. redcliffense*, the UJP5 marker species, within what is interpreted as the UJ4 sequence is low. No UJP4 marker species were encountered; the observed assemblages are consistent with an UJP5–UJP6 age, which would be consistent with an assemblage of UJP4 age in which the infrequent occurrences of the marker species had not been recorded in the available samples.

In the West Fladen basin, palyzone UJP4 was tentatively identified by the occurrence of *Nannoceratopsis pellucida* in the upper part of the basal fluviatile to coastal plain sediments of the Sgiath Formation. This is only tentatively regarded as evidence for a UJP4 age, as *Compositosphaeridium polonicum* is missing.

The sedimentation begins in 20/5b-2 (Fig. 3) with an AST consisting of a coal-bearing unit 180 feet in thickness, followed by a very-fine-grained non-marine sand with a slight fining-upward tendency. In 21/1-5 there is no sand above the coal. In both wells, a thin UJ4 TST consisting of mudstone leads to a MFS. In 20/5b-2 this is overlain by a thick unit (*c.* 250 feet) of low GR mudstones with a subtle coarsening-upward tendency. In 21/1-5 the coarsening-upward tendency is more pronounced and culminates in a very fine grained sand. Available palynofloral records are sparse, but they suggest that euhaline conditions may only have been reached around the MFS level, with the HST supporting low-diversity dinocyst assemblages and acmes of the freshwater alga *Botryococcus* sp.

Reference section: none proposed as the sequence has only tentatively been identified.

'UJ5a' sequence

Alternative name: none (not previously recognized).

The M1 Shale of the Piper Field reservoir subdivision is tentatively identified as representing the earliest marine transgression in that area. Direct biostratigraphic evidence for the presence of palyzone UJP5 is missing, however. The identification of the Piper Field reservoir units with our sequences is further discussed below. In the extended sections of wells 20/5b-2 and 21/1-5 (Fig. 3) this additional unit can be easily accommodated, although again the biostratigraphical data are not detailed enough to distinguish the UJP5 and UJP6 zones in these wells.

This sequence was initially identified only in Piper Field as being separate from sequence UJ5. Numerous transgressive–regressive episodes have been identified in the Piper Field reservoir, which formed the basis for a detailed and consistently recognizable reservoir zonation (Maher 1980, 1981; Schmitt & Gordon 1991; see well 15/17-4, Fig. 4). These cycles are attributed to the sequences recognized in our study as follows:

- C-Shale: sequence UJ7 highstand ?;
- D-Shale: sequence UJ6b highstand;
- E-Shale: sequence UJ6a highstand;
- I: sequence UJ5 highstand;
- M1-Shale: sequence 'UJ5a' highstand.

Available biostratigraphic data only enable us to date the interval from the D-Shale (inclusive) to the M1-Shale as UJP6–UJP5, so that the attribution to sequences is mainly based on the superposition of flooding events.

The 'UJ5a' sequence marks the onset of an areally widespread sedimentation in the OMF. Typically, the sequence begins with a coal seam (e.g. wells 15/16-T8, 15/16-9, 15/21a-15 and 15/17-9; Figs 3 and 5) which formed on the coastal plain in response to rising groundwater level caused by the rising base level. In well 21/1-5 the superposition of two coal seams attributable to the UJ4 and 'UJ5a' sequences is evident (Fig. 3).

The coal is followed by a laminated mudstone ('Skene Member'). The presence of agglutinating foraminifera and a low-diversity dinocyst assemblage of almost exclusively *Dissiliodinium* sp. CMS suggests slightly brackish conditions; a bay or estuary opening to the south may be envisaged. Most likely the 'UJ5a' transgression invaded this body of fresh to slightly brackish water. Within and around the Theta Graben the lack of sand-grade clastics makes the recognition of this sequence difficult. Only at the fringes of the basin, such as the Piper area, a prograding sandy shoreface system developed. The preceding non-marine Skene Member is absent in type well 15/17-4 (Figs 3 and 4), but is present in other wells in Piper, thus indicating onlap on some topographic relief in this area.

This beach–bar trend extended from Piper to the east of Tartan Field; in Tartan itself the sequence is developed in lagoonal laminated marine mudstone facies which includes the T8 Sand (well 15/16-T8, Fig. 5), interpreted as a flood tidal delta, that is, well 15/16-T8 was situated depositionally up-dip of the beach–bar sandy shoreface which fed the T8 Sand.

Further south, in the depositional centre of the North Buchan basin, the sequence 'UJ5a' is tentatively identified in the expanded muddy sections (Fig. 3) here. Specific biostratigraphical evidence for the presence of zone UJP5 is not available, however. The interval attributed to the sequences 'UJ5a', UJ5, UJ6a and UJ6b is biostratigraphically only identifiable as UJP5–UJ6; the sequence identifications are based on the observation of fining-upward and coarsening-upward trends of the sedimentary column. In practice, the separation between the 'UJ5a' and UJ5 sequences is not always feasible, and the sequence designated as UJ5 in such wells may include an equivalent of the 'UJ5a' sequence.

Tentative reference section: well 15/17-4, MFS at 8715 feet MD (−8636 feet TVD) (Fig. 4).

UJ5 (Redcliffense) *sequence*

Alternative name (ammonite nomenclature): *Serratum* sequence.

This sequence includes a significant transgression which oversteps the area reached by the UJ5a transgression. For example, in the Tartan area (e.g., 15/16-T8, Fig. 5) it oversteps the marine lagoonal sediments of the 'UJ5a' sequence with a well-developed ravinement surface and a highstand in bioturbated sandy offshore mud facies. Clean sands are only rarely developed in the TST, but admixture of fine sand to the marine muds is widespread and is clearly visible in the log responses. In Piper Field (Block 15/17) the I-Shale is interpreted to present the UJ5 highstand. Ammonite evidence was initially taken to put the I-Shale in the *Serratum* ammonite zone (Maher 1980 implicitly), and subsequently re-interpreted as the *Rosenkrantzi* Zone (Maher 1981), respectively as the *Baylei* Zone (Harker & Rieuf 1996).

Reference section: well 15/16b-22, MFS at 12 828 feet MD (−12 727 feet TVD) (Fig. 6).

Alternative reference section: well 15/16-T8, MFS at 13 897 feet MD (−10 244 feet TVD) (Fig. 5).

UJ5 MFS biostratigraphic calibration. The characteristic palynoflora of this flooding event are similar to those for the UJ6a MFS, but with the addition of *Stephanelytron redcliffense*, which has its highest occurrence just above the MFS. In all known instances (wells 15/16b-22, 15/16b-21, 15/16b-20 and 15/16b-T8, and wells in Block 15/21) the presence of *S. redcliffense* is restricted to the marine shales ('Saltire Member') below the Scott Sand, or its equivalent, the Tartan Lower Sand, which are therefore placed in the UJ5 HST and the following UJ6a AST. Peak occurrences of *Dissiliodinium* sp. CMS and *Glossodinium dimorphum* occur below the UJ5 MFS.

UJ6a (Cinctum) *sequence*

Alternative name (ammonite nomenclature): *Regulare* sequence.

In well 15/16b-22 this sequence begins with a sand-rich AST which is equivalent to the larger part of the Scott Sand (= the Lower Tartan Sand). The TST–HST couplet is developed as thoroughly bioturbated argillaceous sands grading into arenaceous mudstone.

On Block 14/20 the UJ6a sequence is the oldest unit of the Humber Group, onlapping the Triassic Smith Bank Formation. It is thin, but widespread and its TST and HST are clearly developed (e.g. well 14/20b-22, Fig. 4), unless the sequence is locally eroded.

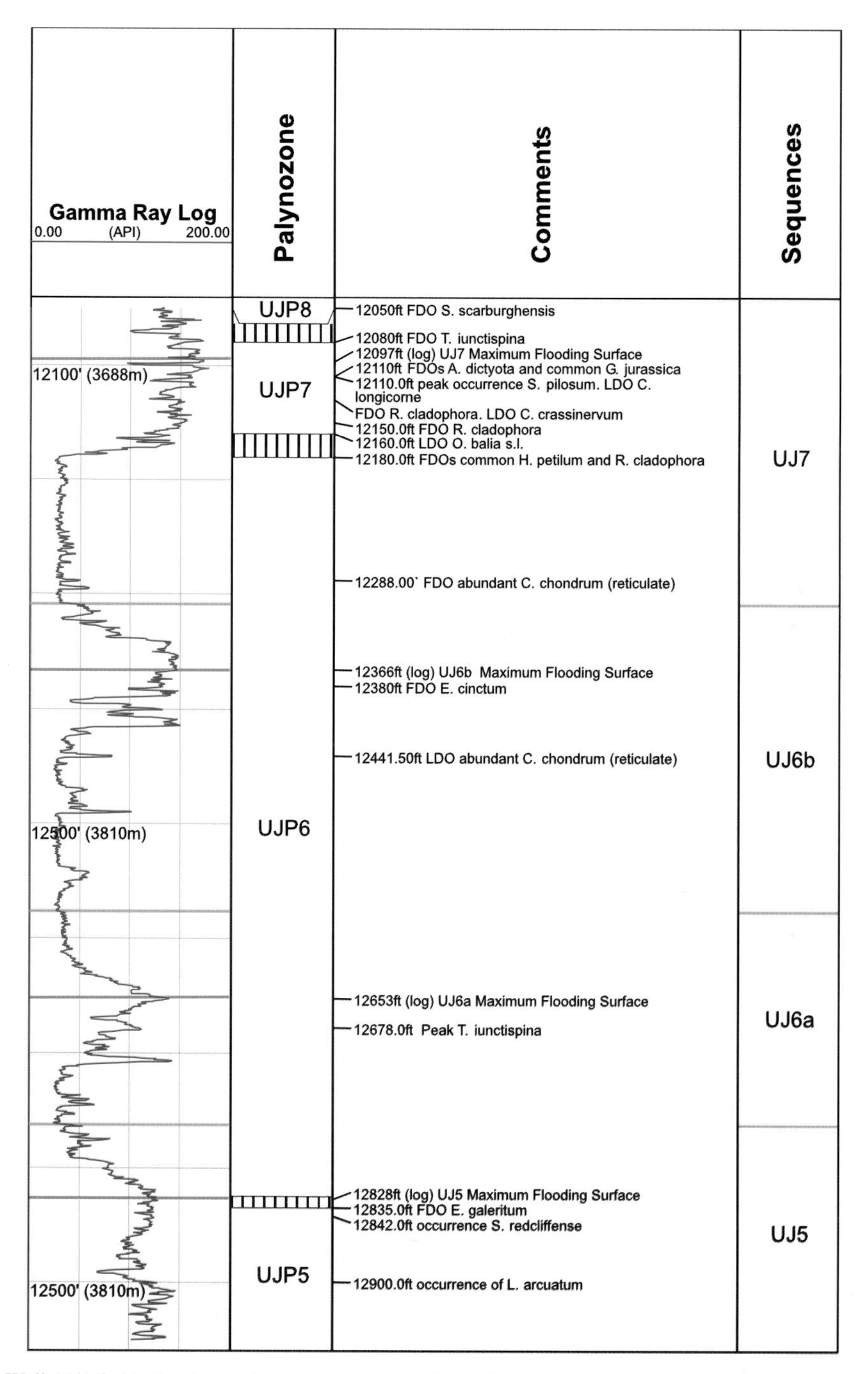

Fig. 6. Well 15/16b-22. GR log and significant biostratigraphical events.

In the Ivanhoe and Rob Roy fields on Block 15/21, there is only one significant highstand (the 'Mid-Shale Unit') between the UJ5 and the UJ7 sequences. The presence of abundant *Cteni-dodinium chondrum* (reticulate) suggests the Mid-Shale may represent the UJ6b highstand. The underlying 'lower reservoir unit' is composed of five coarsening- and cleaning-upward units (Currie 1996). One of them has locally a well-developed argillaceous highstand interval, which is here interpreted as the UJ6a highstand (e.g. well 15/21a-31 (Currie 1996, Fig. 9 labelled as 15/21a-19) and well 15/21a-22). Its absence in adjoining wells is probably caused by erosion by tidal channels of the succeeding AST. The 'Main Piper' reservoir of these fields is therefore the amalgamated UJ6a and UJ6b ASTs. The remaining four coarsening-upward units mentioned by Currie (1996) may be manifestations of smaller scale sea-level fluctuations.

On Block 14/19 the second weakly developed marine ingression below the UJ7 sequence is regarded as the UJ6a highstand (14/19-4, Fig. 4). This is based on some evidence (ostracods) for the presence of UJP5 in the lower Sgiath Formation and indications of ravinement by the overlying UJ7 SB = RS, suggesting removal of the UJ6b sequence as in the Tartan area and on Block 14/20.

Reference section: well 15/16b-22, MFS at 12 653 feet MD (−12554 feet TVD) (Fig. 6).

Alternative reference sections: well 15/16-9, MFS at 12 595 feet MD (−12 496 feet TVD) (Fig. 7).

UJ6a MFS biostratigraphic calibration. The criteria for the recognition of dinocyst zone UJP6 apply to this and the overlying UJ6b highstands. In some wells the UJ6a MFS can be readily distinguished from the UJ6b MFS as it occurs below the FDO of common *Ellipsoidictyum cinctum* and an influx of *Dissiliodinium* sp. CMS. Also, the UJ6a MFS occurs above an influx of *Taeniophora iunctispina*. In well 15/16b-22 the UJ6b highstand is distinguished from the UJ6a highstand by a mass occurrence of *Ctenidodinium chondrum* and *Ctenidodinium chondrum* (reticulate). As these forms are not frequent in the only obvious highstand of UJP6 age occurring in Tartan (the 'Lower Shale'; core samples of 15/16-11, 15/16-12, 15/16-13 and 15/16-16 analysed), the latter is considered to be the UJ6a highstand, with the UJ6b highstand not being developed here.

Some of these distinguishing events may be facies controlled and hence of localized significance only.

UJ6b (Cladophora) *sequence*

Alternative name (ammonite nomenclature): *Rosenkrantzi* sequence.

This sequence begins with a well-developed, sand-rich AST in well 15/16b-22. The highstand is developed as a massive bioturbated shale in 15/16b-22, in which a restricted environment is indicated by the presence of monospecific shell layers (oysters), but the dinocyst assemblages are fully marine and diverse. The HST is developed as the third sand-rich shoreface system of the area.

The UJ6b sequence is well developed in the Theta Graben area, where the thick 'Mid-Shale' is thought to represent the offshore highstand muds. In Tartan there is no fully developed sequence assignable to it, but log correlations suggest that the largest part of the 'Main Sand' represents the UJ6b AST (compare the log motifs in wells 15/16b-22 and 15/16-T8 (Fig. 5), and in 15/16b-20 (Kadolsky *et al.* in press)). In Tartan this AST consists mainly of tidal flat sands and to a lesser degree of fluvial sands; an allochthonous coal seam is present in 15/16-T4 and surroundings. In eastern Tartan wells GR 'spikes' in the upper part represent fine-grained, bioturbated sands which are caused by minor estuarine flooding episodes prior to the principal UJ6b transgression. The absence of this main transgression in the Tartan area, while the precursory estuarine floodings are widespread, suggests removal by erosion. The UJ6b highstand is absent along the northwestern Theta Graben shoulder (e.g. Tartan Field) and on parts of the Highlander–Piper Ridge (Kadolsky *et al.* in press), suggesting removal by graben flank uplift.

A currently unresolved problem of Mid-Shale biostratigraphy is the relatively young ammonite age (*Mutabilis*, *Cymodoce* and *Baylei* zones) reported by Boldy & Brealey (1990) and by Brealey (1990), whereas an age in the *Regulare–Rosenkrantzii* interval would have been expected according to Partington *et al.* (1993*a*). On the basis of available palynofloral data (see also Currie 1996) the Mid-Shale can clearly be assigned to palynozone UJP6. However, Riding & Thomas (1997) report assemblages of UJP6 character as high as basal *Cymodoce* Zone from Staffin Shale Formation outcrops of the Isle of Skye. Their report of the occurrence of *Perisseiasphaeridium pannosum* within the *Baylei* Zone is incompatible with what is known about the range of this species. Further investigations are necessary to clarify these matters.

Reference section: well 15/16b-22, MFS at 12 384 feet MD (−12 288 feet TVD) (Fig. 6).

UJ6b MFS biostratigraphic calibration. See under the preceding UJ6a sequence. This highstand shale is the youngest to contain *E. cinctum*, *E. galeritum* and *S. crystallinum*; but these species occur infrequently and are not always reported from the unit. Some caution must be exercised in using *Ctenidodinium chondrum* (reticulate) floods as a consistent age-indicator for the UJ6b highstand; there is some evidence of palaeoenvironmental control, and therefore diachroneity of its occurrence within the latest Oxfordian–Early Kimmeridgian of this area is possible.

UJ7 (Jurassica) *sequence*

Alternative name (ammonite nomenclature): *Baylei* sequence.

In the Tartan area the TST of this sequence overlies the UJ6b AST; the latter had been identified by correlation to the more complete succession in well 15/16b-22. In 15/16b-22, a UJ7 AST is also present, developed as a massive clean sand deposited probably in a tidal flat environment.

The actual highstand sediment is a bioturbated arenaceous shaly interval termed the 'White Zone' in western Tartan (e.g. 15/16-9), followed by a thick sandy coarsening-upward HST, which together with the overlying UJ8.2a AST forms the '15/16-6 Sand' or 'Six Sand'. A similar facies development is present to the west on Block 14/20, for example, within the PR-1 unit of Lowlander (wells 14/20b-27, 14/20b-17, 14/20b-22 and 14/20b-23, cf. McCants & Burley 1996, Fig. 13), but here the grain size is coarser throughout. To the east of 15/16-9, the sequence was laid down in increasing water depth, leading to a lateral facies change into fine-grained argillaceous sands (15/16-5, 15/16-T2 and 15/16-T8) and further down-dip into offshore muds (15/16b-22). The thinness in 15/16-T8, compared to 15/16b-22 (Fig. 5) was probably caused by greater tectonic subsidence in the 15/16b-22 area and a contemporaneous reversal of the subsidence trend in upthrown Tartan, again suggesting graben shoulder uplift. This is apparent in wells 15/16-1 and 15/16-2Z, where the entire UJ7 highstand, as well as the UJ8.2a sequences, are absent and a very clean lag deposit is present instead.

East of 15/16b-22 in the OMF the topography raises up to the Fladen Ground Spur, and a sand-rich UJ7 Sequence is identified in 15/23-6A (Fig. 5) based on positive evidence for a UJP8.2 age of the overlying sand unit.

Reference section: well 15/16-9, MFS at 12442 feet MD (−12343 feet TVD) (Fig. 7).

Alternative reference: well 15/16b-22, MFS at 12099 feet MD (−12005 feet TVD) (Fig. 6).

UJ7 MFS biostratigraphic calibration. A critical palynofloral feature of this MFS is the cooccurrence of common *Gonyaulacysta jurassica* and consistent *Occisucysta balia* group. Neither of these were recognized as important by Partington *et al.* (1993*a*).

The peak occurrence of *Sentusidinium pilosum* and the LDO of *Cribroperidinium longicorne* are also associated with this MFS, as is the common occurrence of *Chytroeisphaeridia chytroeides*. A peak occurrence of *Systematophora fasciculigera* is typically observed at this level, and the FDO of *Rhynchodiniopsis cladophora* (rare) is below it.

UJ8.2a (*Rare* Pannosum) *sequence*

Alternative name (ammonite nomenclature): *Mutabilis* sequence.

In western Tartan (Figs 4, 5 and 7: well 15/16-9) the AST is a well-developed succession of mainly tidal deposits, often with bimodal grain-size distribution (upper Six Sand). As in the preceding UJ7 HST (lower Six Sand), the depositional dip is eastwards, and as a result a marked eastward grain size decrease, as well as the disappearance or merger of the AST with the TST, is observed (cf. Six Sand in the log panel showing 15/16-9, 6 T5 and T2 by Coward *et al.* 1991, Fig. 6: the AST disappears between T5 and T2; see also 15/16-T8 on Fig. 5 for a strongly condensed Six Sand equivalent in lower shoreface facies). Depositionally up-dip to the west on Block 14/20, the AST is the predominant part of the sequence (e.g. 14/20–15, see Fig. 5 and Lowlander, cf. McCants & Burley (1996): mainly unit PR-2; and Fig. 4 of this paper, well 14/20b-22), where coarse and very coarse grain sizes are typical. The UJ8.2a highstand in 15/16-9 peaks in a thin argillaceous unit, which grades into the coarsening-upward HST part of the 'Hot Sand' reservoir unit. This is a completely bioturbated lower to middle shoreface sand with an elevated GR caused by the presence of 20–30% K-feldspar. The increase in feldspar content began already in the Six Sand, and high feldspar contents are observed in all sands of a UJP8.2 age upwards in the vicinity of the Halibut Horst (Fig. 1), persisting into the Volgian (e.g. Petronella shoreface sands, 14/20-11; Claymore massflow sands, block 14/19; cf. O'Driscoll *et al.* 1990). It clearly indicates that from UJ7 time the Halibut Horst emerged as a positive feature which shed large amounts of clastics into adjacent lows.

In Lowlander well sections, a slight upward decrease in grain size is interpreted as an expression of the UJ8.2a highstand (cf. the PR-3 unit in 14/20b-22 and 14/20b-23; McCants & Burley (1996) fig. 13, and Fig. 4 of this paper), although open marine conditions are not reached. The

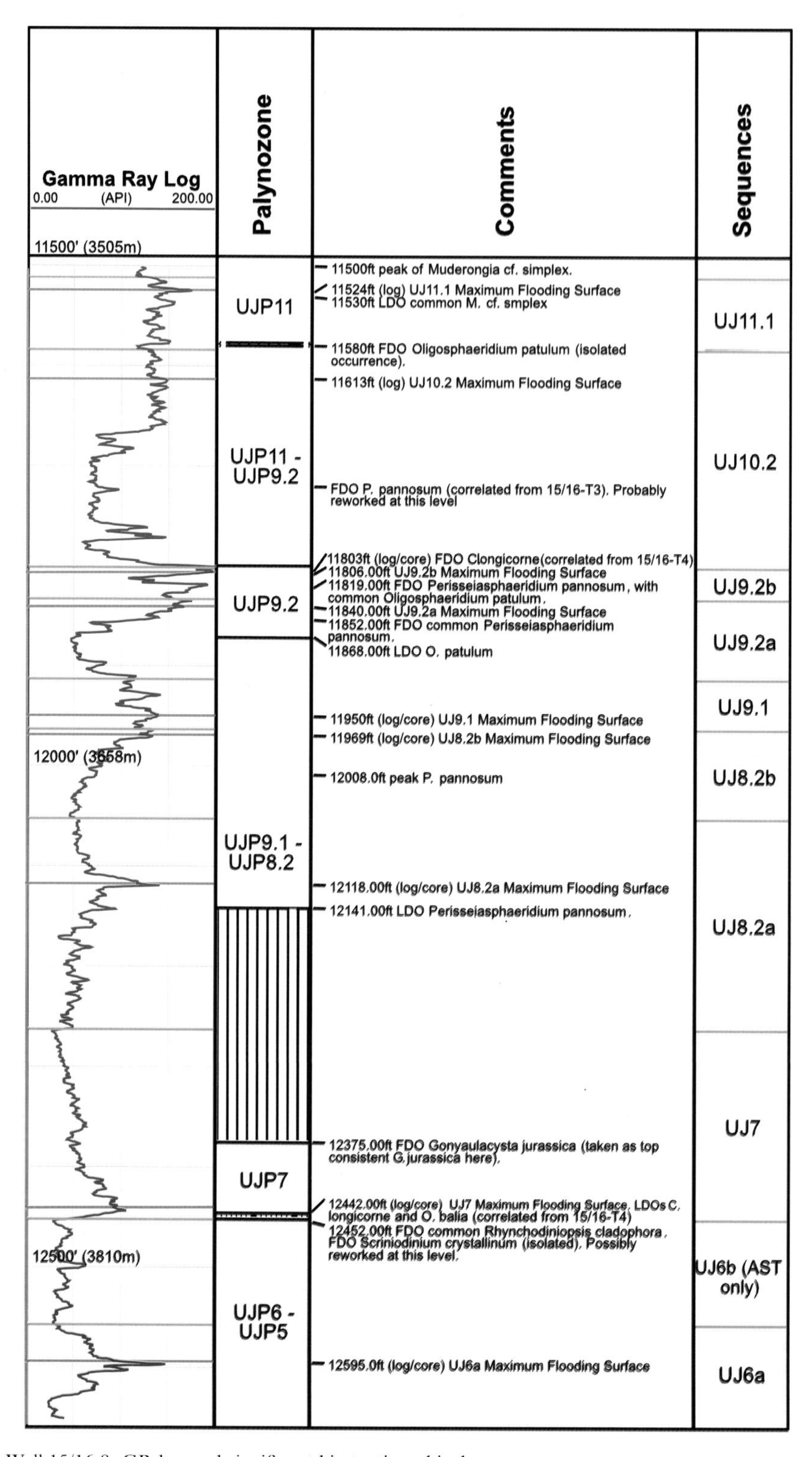

Fig. 7. Well 15/16-9. GR log and significant biostratigraphical events.

rocks are usually palynologically barren, but a rare occurrence of *P. pannosum* or *O. patulum* at this level (14/20b-27, core sample 14 928 feet MD = 12 768 feet TVD) proves an age no older than UJP8.2.

To the east of 15/16-9, the entire sequence becomes rapidly thinner and more argillaceous, as it is deposited at increasing water depth in the Theta Graben. Still further east, in 15/23-6A (Fig. 5), a well-developed sandy TST and HST were identified by the occurrence of *P. pannosum* and *C. longicorne* below the *P. pannosum* peak indicative of the UJ8.2b MFS, testifying to the presence of this sequence on the eastern OMF basin margin.

Reference section: well 15/16-9, MFS at 12 118 feet MD (−12 020 feet TVD) (Fig. 7).

Alternative reference: well 15/23-6A, MFS at 13 202 feet MD (−13 110 feet TVD) (Fig. 5).

UJ8.2a MFS biostratigraphic calibration. The MFS is placed at a high gamma unit immediately above the deepest (rare) occurrence of *Perisseiasphaeridium pannosum*. The age of this horizon is based on the known basal occurrence of *Perisseiasphaeridium pannosum* within the *Aulacostephanoides mutabilis* ammonite zone (Riding & Thomas 1988).

UJ8.2b (*Peak* Pannosum) *sequence*

Alternative name (ammonite nomenclature): *Eudoxus* sequence.

The *Eudoxus* MFS was described in Partington *et al.* (1993*a, b*) as exhibiting 'distinct footwall and hangingwall stratigraphies', and they classed it as a TEMFS (tectonically enhanced maximum flooding surface). To illustrate the differing footwall and hangingwall sedimentation across faults, they show well logs of the Tartan Downthrown wells 15/16-9, 15/16-6, 15/16-T5 and 15/16-T2 (Partington *et al.* 1993*a*, Fig. 4). However, the MFS determined as the '*Eudoxus* MFS' (= UJ8.2b = Peak *Pannosum* MFS) we identify with the UJ9.2b (*Longicorne*) MFS. The displayed interval comprises the UJ8.2a–LK4 sequences, all of which show a gradual thinning to the east. Initially, that is up to the UJ8.2b MFS, this was caused by the eastward gradation of lower shoreface sands (the Hot Sand) into argillaceous sediments deposited at increased water depth. The subsequent sequences are interpreted as an expression of sedimentation on a single rotating fault block which subsided more in the west than in the east.

Transition from the shoreface sands of the Piper (*s.l.*) Formation into the dysaerobic mudstones of the Kimmeridge Clay Formation occurred across a large area of the OMF at the UJ8.2b MFS. However, in the basin centre the transition occurred earlier, and on its margins later (up to the Ryazanian) or not at all. The UJ8.2b sequence is doubtlessly associated with tectonic activity in the study area (Boldy and Brealey 1990), but the rifting activity did not peak in this sequence.

In well 15/16-9 the fining upward part of the Hot Sand culminates in a highstand approximately at the level of the peak occurrence of *P. pannosum*. Westward, in the Highlander and Lowlander fields in Block 14/20, a coarse- to very-coarse-grained AST is interpreted, although biostratigraphical control is very poor. A highstand is missing here, as the overlying shale is already of late UJP9.2 or of UJP10.2 age, indicating a tectonically induced hiatus. This hiatus occurs in a SW–NE-trending area from the Higlander Field in Block 14/20 to the Piper Field in Block 15/17, and is here termed the Highlander–Piper Ridge (see Kadolsky *et al.* in press, for a more detailed account). A similar hiatus was also found in the eastern Tartan area and in well 15/16b-21 to the north-northeast, where it may be related to an uplift of the Theta Graben shoulder simultaneously with rifting inside this graben. In all other areas sedimentation is continuous, and water depth is increased relative to the preceding sequence (Figs 3–5).

Shoreface sands are restricted to the Petronella area (14/20-11, see Fig. 5) and are expected west of 15/23-6A.

Reference section: well 15/16-9, MFS at 11 971 feet MD (−11 873 feet TVD) (Fig. 7).

Alternative reference: well 15/23-6A, MFS at 13 112 feet MD (−13 021 feet TVD) (Fig. 5).

UJ 8.2b biostratigraphic calibration. This MFS is characterized by a major influx of *Perisseiasphaeridium pannosum*, below the highest abundant occurrence of *Geiselodinium paeminosum* (when seen).

Unfortunately, the last species has not been observed in high numbers in the reference wells, although it has been observed commonly above the UJ8.2b (*Eudoxus*) MFS (as described by Partington, *et al.* 1993*b*) elsewhere in the Witch Ground Graben.

UJ9.1 (Crassinervum) *sequence*

Alternative name (ammonite nomenclature): *Autissiodorensis* sequence.

In the study area, this is typically an incospicuous unit. Positive biostratigraphic evidence is usually sparse, but the position above the peak of *P. pannosum* and below the LDO of *Oligosphaeridium patulum* help to identify this sequence. In the Petronella area (14/20-11,

Fig. 5), a clearly fining- and then coarsening-upward shallow-marine sand unit is placed in this sequence, but due to the lack of direct biostratigraphic evidence and lack of core samples this section is not chosen as a reference section.

Tentative reference section: well 15/16-9, MFS at 11 950 feet MD (−11852 feet TVD) (Fig. 7).

UJ9.1 MFS biostratigraphic calibration. The UJ9.1 maximum flooding surface is characterized by the presence of rare *Endoscrinium luridum* and *Cribroperidinium crassinervum*.

UJ9.2a (Patulum) *sequence*

Alternative name (ammonite nomenclature): *Wheatleyensis* sequence.

For a brief discussion of this sequence, see remarks under the UJ9.2b (*Longicorne*) sequence below.

Reference section: well 15/16-9, MFS at 11 840 feet MD (−11 742 feet TVD) (Fig. 7).

UJ9.2a MFS biostratigraphic calibration. Biostratigraphic characteristics displayed by the UJ9.2a and UJ9.2b MFSs are very similar, and common *Oligosphaeridium patulum* and rare *Cribroperidinium longicorne* are present in both cases.

However, the common occurrence of *C. longicorne* is restricted to an interval immediately below the UJ9.2a MFS.

UJ9.2b (Longicorne) *sequence*

Alternative name (ammonite nomenclature): *Hudlestoni* sequence.

The UJP9.2b MFS is the culmination of a very significant flooding event in the Witch Ground Graben, and it is invariably associated with a major high gamma feature.

The UJ9.2b MFS is very close to the UJ9.2a maximum flooding surface, and separation of these events is difficult on palynofloral criteria alone. A safe distinction is only possible where both are observed in superposition. Well-documented development of both sequences occurs in:

- 14/20-11 in shallow-marine sands (Fig. 5);
- 15/16-9 and adjacent wells within the Spiculitic Sands (mass flow sands) (Figs 5 and 7);
- 15/23a-12 in mass flow sands and conglomerates, where two fining-upward turbidite cycles are present within zone UJP9.2 (Fig. 12);
- 15/23-6A in offshore muds: two subtle expressions of clastics-starved condensed shales are seen within subzone UJP9.2 (Fig. 5).

In Tartan the earliest mass flow deposits (Hot Lens B and part of the Spiculitic Sands) are included in the UJ9.2a and UJ9.2b sequences. The earliest significant Galley and Claymore Sands can be attributed to the same sequences. As the Hot Lens B Galley and Claymore Sands have separate origins (Kadolsky *et al.* in prep.), a major tectonic event affecting the base level of possible source areas is evident at this time.

In most of the OMF, the facies of the underlying UJ9.2a sequence continues into the UJ9.2b sequence, but overall water depth increases, as suggested by the increase in shaliness.

A rapid facies change is perceptible in the Lowlander wells, where black shale dated UJP9.2 overlies with a sharp contact non-marine, very coarse sands which are placed in the UJ8.2b AST. The hiatus indicates the presence of a temporary uplift, referred to us the Highlander–Piper Ridge (Kadolsky *et al.* in press). The transgression is placed into UJ9.2b, because only one UJ9.2 highstand can be recognized, and this dates the beginning of Witchground Graben rifting. On the adjacent footwall, a shallow-marine UJ9.2b highstand is preserved only in well 14/20-12. It is a poorly sorted, fine to very coarse sand with abundant wood fragments, lying on similar sands of the UJ8.2b AST, which lack wood fragments and marine plankton. This UJ9.2b sand is probably formed by reworking of the subcrop during the UJ9.2b transgression, when the footwall of the Witchground Graben was just reached by the sea level. At the hiatus between the UJ8.2b AST and the UJ9.2b TST in Lowlander sand units have been removed by erosion, as suggested by the presence of remnants of sand with increased GR, but poor sedimentological and biostratigraphical constraints in wells 14/20b-27, 14/20b-23 and 14/20b-17 (PR-7 Unit of McCants & Burley 1996, Fig. 13). The UJ9.2b TST in 14/20b-20 also contains *C. crassinervum* (UJP 9.1) and *G. paeminosum* (UJP8.2), which are considered reworked.

Reference section: well 15/16-9, MFS at 11 806 feet MD (−11 708 feet TVD) (Fig. 7).

UJ9.2b MFS biostratigraphic calibration. See remarks under UJ9.2a (above)

UJ10.2 (Telaspinosum) *sequence*

Alternative name (ammonite nomenclature): *Rotunda* sequence.

In Tartan (15/16-9, T8) this sequence begins with a more substantial mass flow sand, the ‘Hot Lens A’. In Claymore and in Galley the deposition of mass flow sands and conglomerates (in South Galley), which began in UJ9.2a, continues.

Only in Petronella are shallow-marine sands known, from a short core in 14/20-11. As these sands are completely structureless, their depositional environment is not recognizable, but the marine microflora contains frequent *Cyclopsiella* sp., a brackish and shallow-marine alga. Therefore, and because of the lack of a consistent grain-size trend, an UJ10.2 AST is interpreted here. Positive UJP10 age evidence of the sand is provided by the occurrence of *K. telaspinosum*, *Senoniasphaera jurassica* and *Cribroperidinium gigas*. The overlying black shale supports a UJP10 age with the additional presence of *O. patulum*.

In the northern and southern footwalls of the Witchground Graben, for example in wells 14/20-15 and 14/20b-29, respectively, black shales of the UJ10.2 highstand lie with a sharp contact on the coarse and very-coarse-grained UJ8.2b AST sands. This highstand dates the flooding of the Witchground Graben shoulders and hence the disappearance of the Highlander–Piper Ridge (Kadolsky *et al.* in press).

Reference section: well 15/16-9, MFS at 11613 feet MD (−11 515 feet TVD) (Fig. 7).

Alternative reference: 14/20-11, MFS at 7728 feet MD (−7645 feet TVD) (Fig. 5).

UJ10.2 MFS biostratigraphic calibration. The presence of *O. patulum* above *C. longicorne* and below consistent occurrences of *Muderongia* cf. *simplex* usually identifies this sequence. The oldest occurrences of *Cribroperidinium gigas* and *Muderongia* cf. *simplex* appear to be within the *Pavlovia rotunda* ammonite zone, and these can be useful markers for the UJP10.2 MFS, although they are very rare at that level.

The deepest consistent occurrence of *Tanyosphaeridium* cf. *variecalamum* is below the UJP10.2 MFS, although this species can range (very rarely) into the Early Volgian, zone UJP10.1.

UJ11.1 (Granulosum) *sequence*

Alternative name (ammonite nomenclature): *Fittoni* sequence.

The UJ11.1 MFS was described by Partington *et al.* (1993*a*, *b*) as a 'tectonically enhanced maximum flooding surface', situated between the Claymore and Galley depositional systems. This statement is, however, incorrect as far as the Galley Sands are concerned: in both systems the onset of minor mass flow sands is in sequence UJ9.1, and the onset of the principal sands is in sequence UJ9.2a. The Galley system persisted into sequence UJ13.1, however, while the Claymore system terminated within UJ11.1.

In the study area, shale sections show a marked reduction in gamma log response above the UJ11.1 highstand.

In Tartan (15/16-9, T8) there is no more mass flow sand in this or any younger sequence, but fluctuations in the shaliness serve to differentiate further sequences. The UJ11.1 highstand is widely developed as a double (or triple) GR peak, suggesting uniform conditions through those parts of the basin which were not affected by mass flow sand deposition. The thin, but massive 'Hot Lens Equivalent' of Highlander is attributed to this sequence, whereas the Hot Lens A of Tartan Field, of similar log motifs, is a part of the preceding UJ10.2 sequence (cf. well 14/20-15, Figs 8 and 9, and well 15/16-9, Figs 5 and 7).

Reference section: well 15/16-9, MFS at 11 524 feet MD (−11 426 feet TVD) (Fig. 7).

Alternative reference: well 14/20-15, MFS at 10 044 feet MD (−9340 feet TVD) (Fig. 8).

UJ11.1 MFS biostratigraphic calibration. The UJ11.1 MFS is identified by the criteria as described in the Appendix for the dinocyst subzone UJP11.1, particularly the common occurrence of *Muderongia* cf. *simplex* just below its acme. The highest occurrence of *Oligosphaeridium patulum* probably is immediately above this MFS, as suggested by Partington *et al.* (1993*b*), but it is very rare and sporadic in its distribution at that level, and cannot be used to consistently recognize this MFS.

UJ12 (Muderongia) *sequence*

Alternative name (ammonite nomenclature): *Okusensis* sequence.

In shallow-marine environments, a highstand of UJP12 age is interpreted for the basal transgressive units in wells 15/25b-1A and 15/14a-3 (Fig. 5). It is barren in 15/25b-1A but overlain by a transgressive unit of proven UJP13.1 age. In 15/14a-3 (Fig. 10) a bioturbated sand of proven UJP11.2 age onlaps on Triassic rocks. This is interpreted as a truncated remnant of the UJ12 TST. The overlying sequence, which has a pebble lag at its base, is dated UJP13.1.

In deep-marine sections, the UJP12 age of the condensed mudstones following the termination of UJP11.2–UJP12 mass flow sand deposition is

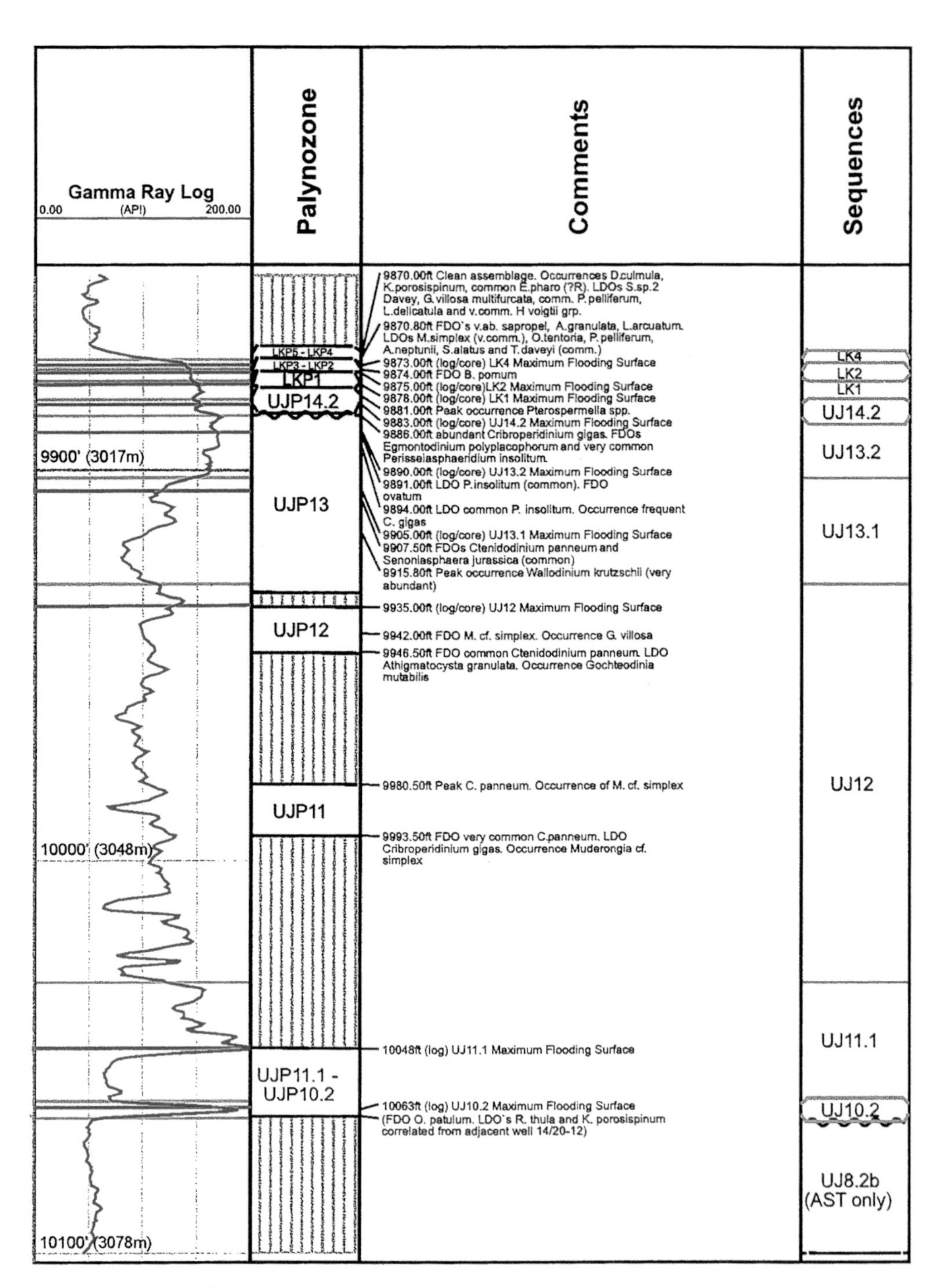

Fig. 8. Well 14/20-15. GR log and significant biostratigraphical events.

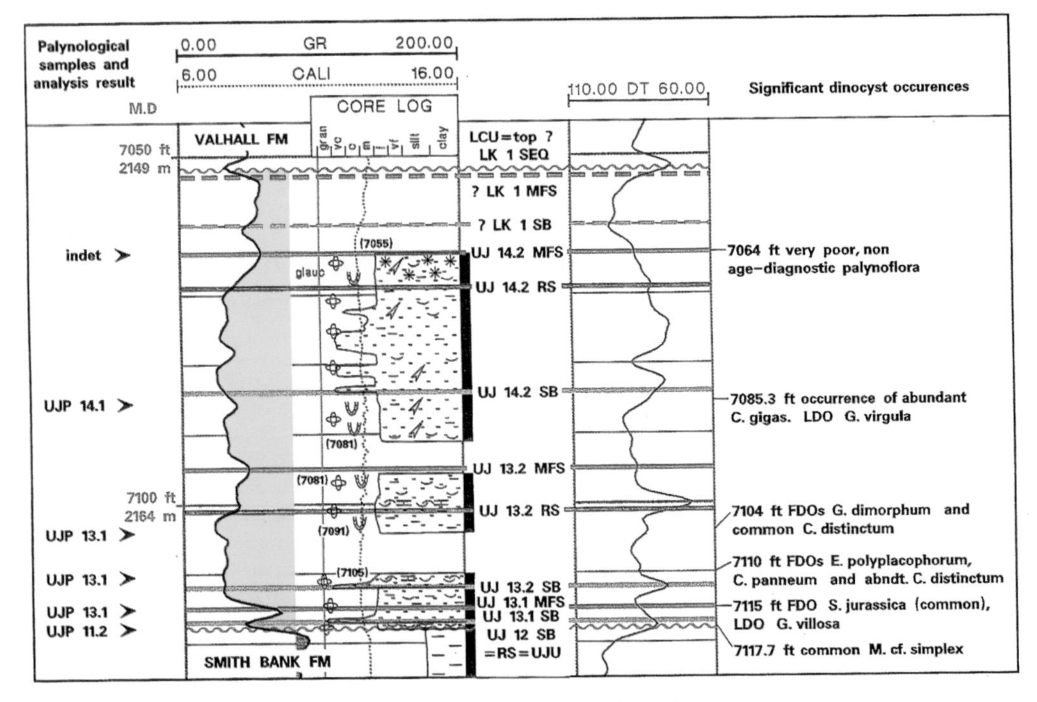

Fig. 10. Well 15/14a-3. Well logs, core log and significant biostratigraphic events.

proven in 14/20-11 and in 14/20-15 (Figs 5, 8 and 9). In the Galley area, mass flow sands and conglomerates continued to be deposited.

Reference section: well 14/20-15, MFS at 9934 feet MD (−9230 feet TVD) (Fig. 8).

UJ12 MFS biostratigraphic calibration. This MFS, which in well 14/20-15 is close to the top of zone UJP12, has been recognized widely in the Central North Sea area. The highest occurrence of *Muderongia* cf. *simplex* (rare) approximates to this level.

The FDO of *Gochteodinia mutabilis* is within the unit, and the LDO of 'primitive' *G. villosa* occurs near the top of UJP12. *Gochteodinia mutabilis* is very rare near the top of its range, and there is a possibility that there is a direct transition from it to *G. villosa*, within UJP12.

UJ13.1 (Arcuatum) *sequence*

Alternative name (ammonite nomenclature): *Anguiformis* sequence.

This sequence has been positively identified in shallow-marine settings in 15/25b-1A and 15/14a-3 (Figs 9 and 10). In well 15/25b-1A it begins with an AST, which consists of very poorly sorted pebbly quartz sand with large intact bivalve shells. The combination of high transport energy, lack of winnowing, marine influence and preservation of fragile shells may be explained as a storm deposit on a beach. This is followed by a muddy bioturbated sand in which the proportion of coarse and very coarse quartz grains gradually diminishes upwards; entrainment from below by burrowing is envisaged. This grades into a fairly uniform very-fine-grained burrowed lower shoreface sand which constitutes the HST of the sequence.

In well 15/14a-3 the sequence lacks the AST and begins with a pebbly lag deposit; fine- to very-fine-grained, burrowed fining- and then coarsening-upward sands constitute the entire sequence.

In the offshore realm, the highstand event terminates the deposition of the massive Galley Sands (cf. 15/23a-12), as well as the youngest sand-bearing mass flow deposits in 14/20-15 and the 'Kimm Silt' fine-grained clastics in the Claymore area. The UJ13.1 MFS is therefore the culmination of a rapid marine flooding event, of major significance in the Witch Ground Graben.

Remarkable is the low GR response of the highstand shales in 14/20-15, where the absence of any sand laminae is evidenced by the core as well as by the sonic, neutron and density log responses. The highstand shales are truncated in 15/16-9 (Fig. 7), as there is no MFS candidate of a UJP13.1 age, and adjacent wells show the intercalation of additional section.

Reference section: well 14/20-15, MFS at 9905 feet MD (−9202 feet TVD) (Fig. 8).

Alternative reference sections: well 15/23a-12, MFS at 12 844 feet MD (−12 767 feet TVD) (Fig. 9); well 15/14a-3, MFS at 7115 feet MD (−7007 feet TVD) (Fig. 10); well 15/25b-1A, MFS at 9079 feet MD (−8998 feet TVD) (Fig. 9).

UJ13.1 MFS biostratigraphic calibration. This MFS occurs immediately below the highest occurrence of *Senoniasphaera jurassica* and is often characterized by an influx of *Leptodinium arcuatum*. The deepest occurrences of *Kleihriasphaeridium fasciatum* group, rare *Dingodinium spinosum* and common *Perisseiasphaeridium insolitum* are immediately below it. An influx of *Cyclonephelium distinctum* is invariably observed at about this level.

UJ13.2 (Insolitum) *sequence*

Alternative name (ammonite nomenclature): *Oppressus* sequence.

The UJ13.2 sequence is again developed in shoreface facies in wells 15/14a-3 and 15/25b-1A. In the latter, a fining- and coarsening-upward couplet is suggested by the fluctuations of the GR and neutron/density logs. In well 15/14a-3 the sequence begins with a shell concentration as a lag deposit, followed by a burrowed fine-grained sand with a subtle fining- and coarsening-upward couplet (Fig. 10).

Offshore, no significant mass flow deposits are recognized within this sequence. A marked GR increase commences in the UJ13.2 sequence and persists into the Ryazanian, unless the shaliness is 'diluted' by mass flow sands. The increase in GR corresponds to an increase in shaliness as shown by the other lithology logs.

The UJ13.2 MFS does not appear to have been differentiated by Partington *et al.* (1993*a*, *b*), although it often shows a distinctive high gamma signature, is widespread in the Central North Sea and is of consistent palynofloral character.

Reference section: well 14/20-15, MFS at 9890 feet MD (−9187 feet TVD) (Fig. 8).

Alternative references: well 15/23a-12, 12 834 feet MD (−12 757 feet TVD) (Fig. 9).

UJ13.2 MFS biostratigraphic calibration. This MFS is marked by an influx of the chorate dinocyst *Perisseiasphaeridium insolitum*. The acme of this species is effectively restricted to this MFS, and can therefore be easily diluted in ditch cuttings.

In this case, the presence of abundant *Cribroperidinium gigas* can be used as a valuable secondary indicator.

UJ14.2 (Spinosum) *sequence*

Alternative name (ammonite nomenclature): *Lamplughi* sequence.

This sequence is again developed in shallow-marine facies in wells 15/14a-3 and 15/25b-1A. In 15/14a-3 *c.* 20 feet of bioturbated bimodal sands are interpreted as an AST of tidal sediments, as channelized transport in a fully marine environment at shallow depth is envisaged. This is overlain by a well-sorted glauconitic fine sand, for which a greater water depth is assumed; the most argillaceous parts of the section, that is the inferred UJ14.2 MFS and HST, are above the cored section, however. In 15/25b-1A *c.* 60 feet of very gently fining-upward fine-grained bioturbated sands, deposited in a lower shoreface environment, represent the TST of this sequence. The highstand part of it is above the cored section.

Elsewhere, deep-water muds and local mass flow sands were deposited. Notable as a reservoir is the Dirk Sand in South Galley (15/23a-12, Fig. 9), of which the more significant lower part belongs to this sequence. Distal turbidites are also present in the Lowlander area (e.g. 14/20b-23, Fig. 9).

Reference section: well 14/20b-23, MFS at 12 269 feet MD (−12 147 feet TVD) (Fig. 11).

Alternative references: well 14/20-15, MFS at 9983 feet MD (−9180 feet TVD) (Fig. 8); well 15/23a-12, MFS at 12 703 feet MD (−12 626 feet TVD) (Fig. 9).

UJ14.2 MFS biostratigraphic calibration. Refer to the subzonal definition in the Appendix. The MFS is known to approximate the Jurassic–Cretaceous boundary, as indicated in Partington *et al.* (1993*b*). It is placed within the *Subcraspedites lamplughi* ammonite zone here rather than in the *Subcraspedites preplicomphalus* zone, as by those authors.

LK1 (Compta) *sequence*

Alternative name (ammonite nomenclature): *Runctoni* sequence.

Shallow-marine sands of this sequence are not proven in the wells studied, but may possibly be present in well 15/20-2 below a section of probable LKP2–3 age (see below).

The basinal facies is similar to the facies of the preceding UJ14.2 sequence, that is dysaerobic offshore muds with local mass flow sands in South Galley and in the Lowlander area.

The LK1 MFS is usually close to the preceding UJ14.2 MFS, and these may have been interpreted as a single MFS elsewhere. The extended section in the Lowlander area (e.g. 14/20b-23) clearly shows two separable highstand shales with intercalated distal mass flow sands; the two highstand events can be traced throughout the OMF.

Reference section: well 14/20b-23, MFS at 12 243 feet MD (−12 121 feet TVD) (Fig. 11).

Alternative reference: well 15/16-9, MFS at 11 354 feet MD (−11 256 feet TVD) (Fig. 7).

LK1 MFS biostratigraphic calibration. This MFS occurs at a level of very abundant *Canningia compta*, immediately above the extinction of *Egmontodinium expiratum, Cribroperidinium gigas* (rare), *Leptodinium arcuatum* and *L. eumorphum.*

LK2 (Thula) *sequence*

Alternative name (ammonite nomenclature): *Kochi* sequence.

This sequence is represented in most wells by offshore muds. Locally it can be condensed due to sediment starvation on a submarine high (compare wells 14/20-15 and 14/20b-23 as examples of footwall and hangingwall developments, Fig. 9). In the Lowlander area distal mass flow turbidites continue to be deposited after the LK1 highstand shales.

In well 15/20-2, high on the flanks of the Fladen Ground Spur, 211 feet of very-fine- to coarse-grained sands are intercalated between Devonian (?) and Barremian sediments. Two SWCs in the uppermost quarter of the section yielded dinocysts, including *Endoscrinium pharo, Oligosphaeridium diluculum, Daveya boresphaera* and *Batioladinium pomum* or *radiculatum*. This assemblage would indicate an LKP2–LKP3 age, and suggest the presence of the LK2 and LK1 sequences in a shoreface sand facies. Late Volgian shoreface sands have been encountered further down-dip in 15/25b-1A, where they are overlain by undated condensed shales of presumed Ryazanian age. These observations suggest that the highest relative sea-level rise in the Kimmeridge Clay Formation was achieved in the Ryazanian.

Reference section: well 14/20b-23, MFS at 12 166 feet MD (−12 044 feet TVD) (Fig. 11).

LK2 MFS biostratigraphic calibration. See zone definitions in the Appendix. In their description of the *Kochi* MFS, Partington *et al.* (1993*b*) state that it occurs 'below the FDA of common *Rotosphaeropsis thula*'. However, this species occurs only rarely at this level in the study area.

LK4 (Simplex) *sequence*

Alternative name (ammonite nomenclature): *Stenomphalus* sequence.

This sequence is similar to the underlying ones, represented by often condensed mudstone sections. Only in the Lowlander area distal mass flow sands are present, and a separate mass flow sand unit was encountered in well 14/18b-12 (Fig. 9).

The nature of the contact to the overlying Valhall Formation has been extensively discussed by Rattey & Hayward (1993). According to these authors the boundary marks 'the final failure of the North Sea Jurassic rift basin with the transfer of all crustal extension out onto the Atlantic margin'. It is conceivable that such shift in extensional tectonics resulted in hydrographical changes that brought about the sudden facies change from the dark, organic-rich, dysaerobic, almost non-calcareous Kimmeridge Clay Formation to the light-coloured, organic-poor, aerobic (usually bioturbated) and very calcareous Valhall Formation. Tectonic activity in the OMF did not cease, however, and many of the present-day structural traps were clearly affected by Early Cretaceous faulting; Early Cretaceous mass flow sands are also significant, for example in the Scapa and Britannia fields. Locally there is evidence of Early Cretaceous basin inversion, particularly in well 15/16b-22, where an originally very thick graben fill was severely truncated down into the UJ9.2b sequence (Fig. 5).

Rattey & Hayward (1993) argue that sedimentation across the Kimmeridge Clay–Valhall Formation boundary was continuous, but often condensed, or locally absent due to extreme sediment starvation in a pelagic setting. Continuous sedimentation (at least within the resolution of biostratigraphy and sedimentology) has been observed in most wells in structurally low settings; our data are insufficient to ascertain whether this was also the case in structurally high settings where both Upper Jurassic and Lower Cretaceous sections are very thin.

Reference section: well 14/20b-23, MFS at 12 089 feet MD (−11 967 feet TVD) (Fig. 11).

Alternative reference: well 14/18b-12, MFS 12 feet below top Kimmeridge Clay Formation (Fig. 9).

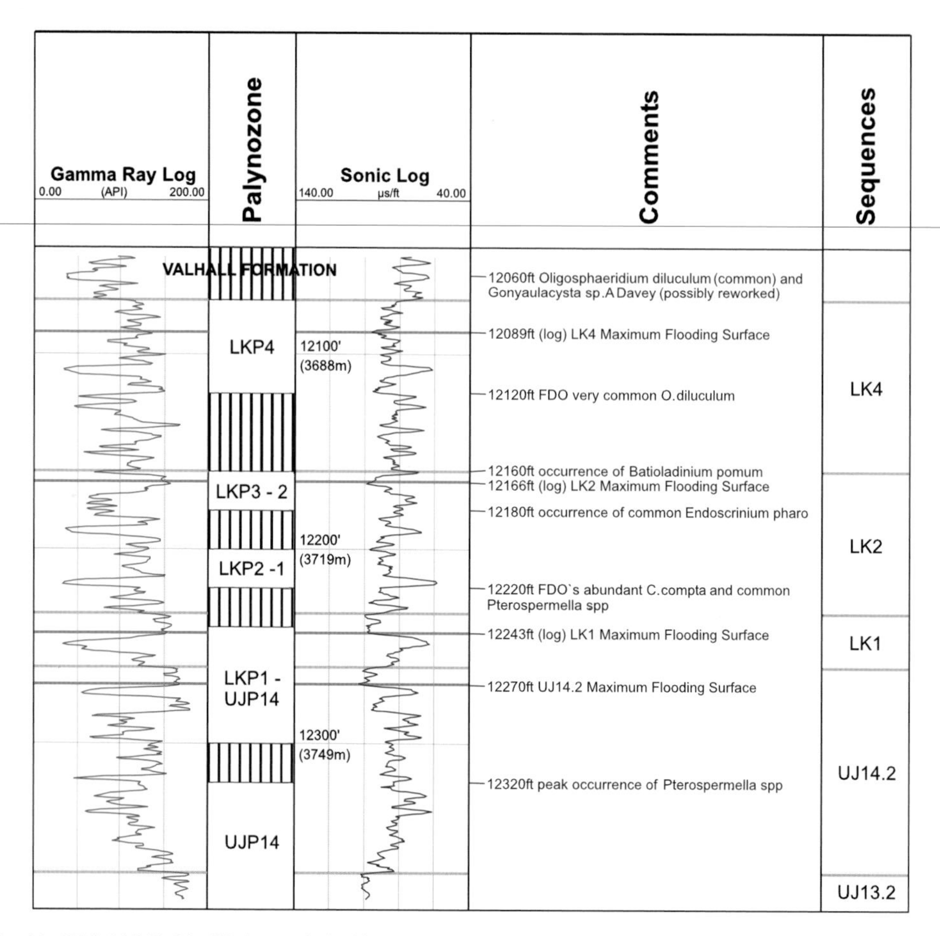

Fig. 11. Well 14/20b-23. GR log and significant biostratigraphical events.

LK4 MFS biostratigraphic calibration. See zonal definitions in the Appendix. The MFS is close to the base of the Valhall Formation, into which its marker taxa may often be reworked. Biostratigraphic constraint can be difficult without knowledge of the log break between the Valhall and Kimmeridge Clay Formations, but the incoming of amorphous organic matter may be used as a secondary criterion to recognize sequence LK4 and the presence of the Kimmeridge Clay Formation.

Application example: the Galley Field (Fig. 12)

Introduction

The reservoir zonation and correlation of many important fields in the OMF has been established prior to the emergence of sequence stratigraphy concepts. As examples, the current reservoir zonations of the Tartan and Piper fields are shown in Fig. 10. Although the definition of the reservoir layers in these fields pre-dates the present study, many boundaries chosen equate or approximate a boundary identified by the application of sequence stratigraphy. By contrast, in the Galley Field, sequence stratigraphy and the biostratigraphic zonation scheme described in this paper were applied from the beginning of field appraisal studies. Biostratigraphy proved to be essential in reservoir correlation and hence in imposing constraints on the sedimentological reservoir model. Figure 12 illustrates the wells discussed below.

Overview of Galley Field

The Galley Field consists of several oil accumulations, of which the two most significant ones are

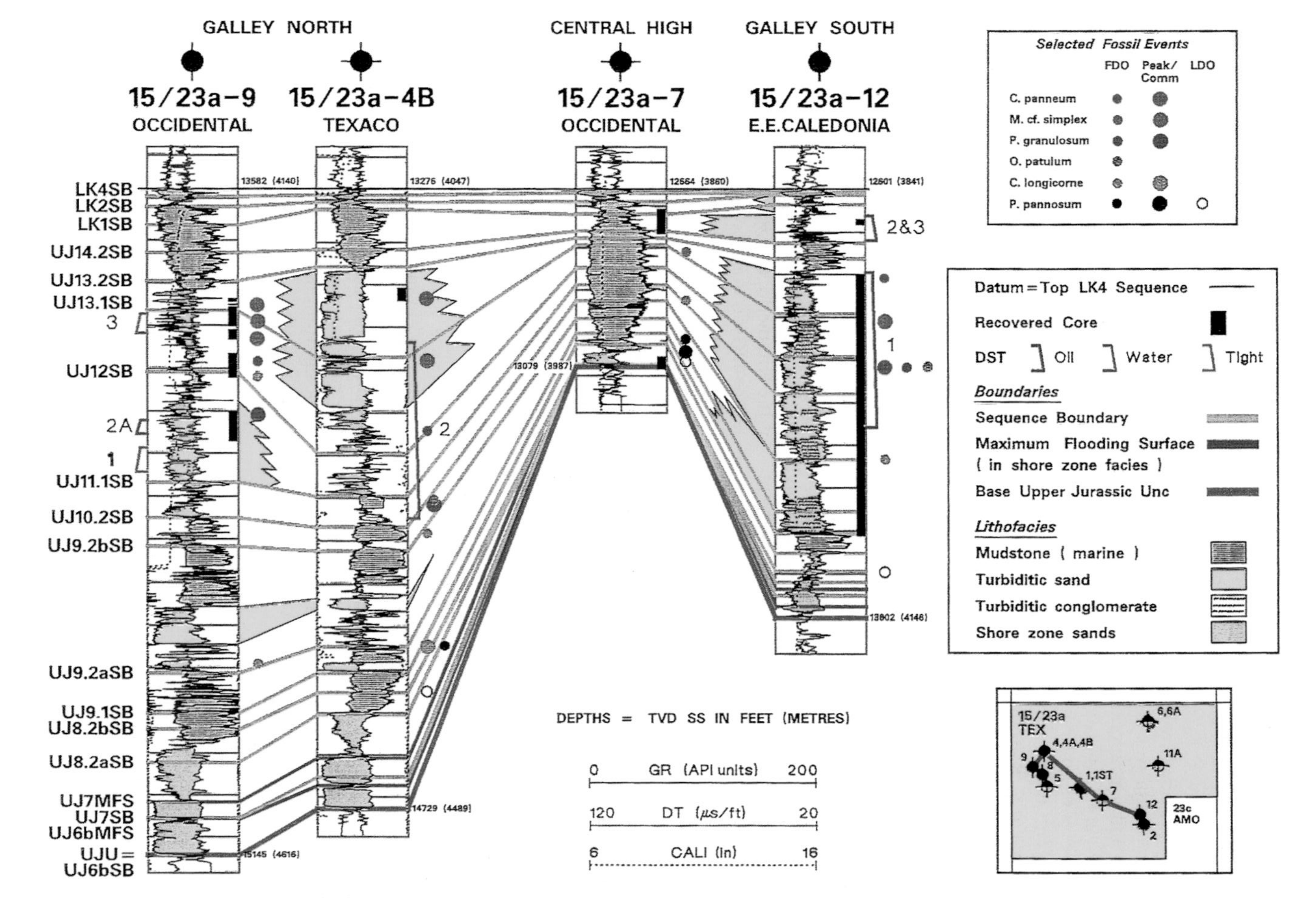

Fig. 12. Field reservoir correlation: Galley Field.

termed Galley South and Galley North. In both of them Humber Group mass flow clastics form the reservoir. As the Galley Field accumulations are small and are being developed with few wells, a thorough understanding of the sedimentary architecture is particularly important for optimum well placement. An overview of Galley is given, followed by a presentation of crucial evidence provided by biostratigraphy.

Galley South was discovered by well 15/23-2 in 1976 and was appraised by well 15/23a-12 in 1991. Figure 12 illustrates the latter, as it has extensive core coverage. Producible hydrocarbons occur in the Dirk Sand Member (Late Volgian–Early Ryazanian) and in the Galley Sand Member (Early–Middle Volgian). The Dirk Sand comprises low-density turbidites, mainly channelized sands of limited extent. The Galley Sand Member of South Galley is an almost 600 feet thick unit of stacked high-density turbidite sands and conglomerates with very few low-density turbidites or shales. The majority of the pebble- to boulder-sized clasts are composed of basalts derived from the Pentland Formation; well-rounded quartz pebbles, mudstones of the Kimmeridge Clay Formation, Piper Formation and Pentland Formation, and shell debris (belemnites and bivalves) do also occur, together with sandstones of Carboniferous age. Zechstein carbonate pebbles are a rare component.

The Central Galley High lies between Galley South and Galley North. It was drilled by wells 15/23-1 (1974) and 15/23a-7 (1983), and the latter is illustrated. Galley Sands are completely absent, although the somewhat sparse biostratigraphic data indicate the Kimmeridge Clay Formation is complete from the Ryazanian to the Late Kimmeridgian. A Dirk Sand equivalent in overbank facies ('tigerstripe' sands) is, however developed in 15/23a-7, and below the Kimmeridge Clay Formation is a clean mass flow sand of UJP8.2 age lying on a 7-foot remnant of very calcareous Piper Formation sands which contain an uncharacteristic brackish microflora of indeterminate age.

Galley North was discovered by well 15/23-4B in 1977 and appraised by 15/23a-8 in 1983 and 15/23a-9 in 1984. Well 15/23a-8 is not considered here because its principal reservoir is very similar to the one in 15/23-4B: a massive clean channel sand in the youngest Galley Member Sequence UJ13.1. Well 15/23a-9 proved the highest Galley Sands to be wet, but an underlying sand body to be oil-bearing.

The Galley North sands differ from the Galley South sands by an almost complete lack of pebble-sized components; the rare pebbles observed are quartz and sandstone clasts. The sands reflect discrete episodes of high- and low-density turbidite channel sands alternating with sandy overbank deposits and silty mudstones. A separate fan system is invoked for Galley North, which was separated by the Central Galley High acting as a submarine bypass area.

The biostratigraphic studies conducted on these wells were designed to address, *inter alia*, the following questions.

Can biostratigraphy aid in the subdivision and correlation of the Galley Sandstone Member in the Galley South area? With the extensive core coverage, the analysed samples were free of cavings. The analysis showed the Volgian marker species and species associations appearing in the same succession as in the equivalent mudstone facies, although the FDOs of some species were higher than normal. This suggests the absence of a (within the resolution of biostratigraphy) significant hiatus and the presence of some reworking. The origin of the dinocysts within the clastics was probably complex: (1) mudstones of the Kimmeridge Clay Formation deposited in the channels in quiet periods between individual flows and scoured by them; this is supported by a separate analysis of individual mudstone clasts which demonstrated that some of them were indeed Kimmeridge Clay Formation material up to the age derived from composite samples (see item (3) below); (2) the channel walls may have contributed somewhat older Kimmeridge Clay material; (3) contemporaneous shoreface sands in which dinocysts were incorporated and which sourced the Galley Sands. Palynological preparations of Upper Jurassic shoreface sands elsewhere in the OMF demonstrated that they usually contain marine dinoflagellates; a significant portion of the Galley Sands is derived from shoreface sands, as suggested by the sometimes frequent occurrence of calcareous shells; and (4) the infiltration of turbidite deposits by pelagic material settling out of suspension may have contributed, although in this case infiltrated clay minerals should also be expected, but their quantities are negligible.

The actual subdivision of the South Galley sands is based on the observation of three main 'fining-upward' units, each of which may be subdivided into less distinctive subunits. The fining-upward takes place in the pebble fraction and in the thickness of associated individual beds, leading to pebble-free clean sand intervals. In well 15/23a-12 (Fig. 12) the bulk content of volcanic pebbles can be gauged from the GR log which shows the increased activity of Pentland Formation volcaniclastic material, and from the sonic which responds to the pervasive calcite cement typically associated with the conglomeratic intervals. The resulting six units were then attributed to dinocyst zones and to the sequences as recognized elsewhere in the OMF.

Is the Galley South fan system filling a single incised canyon? The very proximal facies of the depositional system of South Galley and its thickness suggested it might represent the fill of a deeply incised canyon. In this case a signifiant biostratigraphic hiatus should be observed at the base of the turbidite section, and young ages should be observed from its bottom upward. However, this was definitely not the case; instead, the clastics section showed only a limited amount of assemblage mixing, and only a few Kimmeridge Clay clasts were present. This suggests that the amount of incision was actually limited, and would have rarely exceeded the thickness of two dinocyst zones.

What fossil evidence is there about the origin of the clastic material? While the derivation of the basaltic pebbles of South Galley from the Middle Jurassic Pentland Formation is obvious, the origin of associated dark mudstone clasts was not. Five individual mudstone pebbles were analysed from the UJ10.2 sequence of well 15/23a-12 for palynology: (1) black moderately calcareous micaceous and carbonaceous mudstone: UJP10.2; (2) calcareous siltstone: UJP8.2; (3) and (4) silty limestone and conglomerate intraclast: Middle–Late Jurassic, indeterminate (non-marine). The most likely origin is the non-marine Middle Jurassic Pentland Formation; the limestone clast contained also Late Permian palynomorphs, suggesting erosion of Permian Zechstein strata during its deposition; and (5) dark grey micritic limestone: Late Permian.

A UJP10.2 age for the sequence was, prior to this clast analysis, interpolated from palynological evidence recovered from composite samples in over- and underlying units. Thus, the youngest clasts appear to be nearly contemporaneous with the turbidites.

Further, in composite samples Carboniferous miospores were frequently encountered. This may either point to contemporaneous erosion of Early Carboniferous formations, or redeposition of shoreface sands containing them. Carboniferous miospores are encountered in, for example, Piper Sands, throughout the OMF.

The source area and the extent of the Galley submarine fan system are treated by Kadolsky *et al.* (in press). Essentially, an origin on the E–W-trending 15/18–15/19 High to the north is assumed, as this is the nearest area with contemporaneous outcrops of all the component clasts encountered.

Is there any difference in age between the Galley Sands in the Galley South and North areas? Sand deposition in South and North Galley was essential contemporaneous from sequence UJ9.2a until UJ13.1. On the Central Galley High mud deposition occurred in the same interval.

Is there a stratigraphic trapping element in the anomalous hydrocarbon distribution between wells 15/23-4B and 15/23-9? The reservoir in 15/23a-4B consists of 320 feet gross of 'blocky' clean channel sands attributed to the UJ13.1 and UJ12 sequences by direct palynological evidence. These sequences were identified in well 15/23a-9 in a thin levée facies which flowed water on test. These sands are below the lowest proven oil in 15/23-4B; pressure data suggest the same overpressure regime, and hence their position below the oil–water contact (OWC) as well as outside the #4B channel appears plausible.

These water-bearing minor sands are underlain by a thicker and oil-bearing sand in 15/23a-9 dated as belonging in the UJ11.1 sequence. The equivalent interval in 15/23-4B is thin and contains only a few levée sands of low porosity, which are water-wet although being structurally higher than the oil-bearing sands in 15/23a-9. This indicates that levée sands are not necessarily in hydraulic continuity with the pertinent channel sands, that is stratigraphic trapping of hydrocarbons in a mass flow channel complex may occur.

Similarly, the water level seen in the UJ13.1–UJ12 levée sands of 15/23a-9 may not be a constraint on the OWC in the equivalent 15/23-4B channel sands, although in this case hydraulic continuity is more likely as indicated by the good agreement of pressure data, and the much better porosity in the levée sands of 15/23b-9.

What is the likelihood of aquifer support to the Galley North and Galley South accumulations? For Galley North, it follows from the preceding paragraph that good hydraulic connectivity would be restricted to deposits in individual channels, and hence a significant aquifer support is not considered likely. Vertical flow is unlikely due to the presence of significant argillaceous horizons.

In Galley South, however, argillaceous interbeds are negligible. Pressure measurements suggest vertical connectivity exists, although the strongly cemented conglomeratic intervals are expected to be barriers to vertical flow during production. Biostratigraphic analysis of well 15/23b-10 (not figured) 5 km to the south demonstrates that all sand-bearing sequences present in South Galley extend in clean sand facies in similar thickness, at least to this well. The inferred continuity of the South Galley mass flow sequences to the south, supported by the pressure data in 15/23b-10, suggest a high probability of aquifer support for Galley South production.

Discussion

Exploration applications: a general comment

An obvious advantage of sequence stratigraphy is the provision of a methodical framework to group laterally and vertically different facies associations (the systems tracts) into units (sequences) bounded by isochronous surfaces (maximum flooding surfaces and sequence boundaries).

The exact contemporaneity of an MFS is usually well below the resolution of any dating method and can, strictly speaking, currently not be proven, as Miall (1992) correctly pointed out. We worked therefore with the hypothesis that the SB (where conformable) and the MFS are chronostratigraphical surfaces within the study area. This lead to no apparent contradictions or inconsistencies in our interpretations; in addition, the working hypothesis of contemporaneous highstand events facilitates interpretations where, in the absence of any theoretical framework, no consistent interpretation would have been possible. This holds particularly true for sections with poor or no constraining biostratigraphical and sedimentological information.

The utility of sequence stratigraphy is well demonstrable in the OMF, where strong depositional gradients and resulting lateral facies changes are the norm. Tectonic overprint is clearly very important, and sediment thicknesses and log motifs often change abruptly across faults. In Figs 4 and 5 we give our view of the sequence stratigraphical relationships of the Claymore, Lowlander, Tartan, Scott and Piper field shoreface reservoirs, and the Highlander Field sequence identification is shown in Fig. 9. Sequences may therefore be regarded as natural units for purposes of palaeogeographical reconstruction with its obvious application in exploration, to model the origin and distribution of reservoir rocks. The results of our study, using the same dataset, are published elsewhere (Kadolsky *et al.* in press).

The usefulness or otherwise of sequence stratigraphy for reservoir prediction in exploration has been viewed differently by various authors, with Partington *et al.* (1993*b*) and Veldkamp *et al.* (1996) providing some polarity. We offer the following observations.

The sequence subdivision of the Humber Group proposed here is based on well control and is far beyond the seismic resolution achieved in the area. Compared to a sediment package for which 3D seismic provides a good resolution of internal sediment architecture, a severe constraint is imposed on extrapolations beyond areas of dense well control. Therefore the accuracy of reservoir rock predictions in lightly drilled areas cannot be expected to be high, although the analysis of the sedimentary record should provide constraints and point towards the most likely facies developments. The most important value of sequence stratigraphy to exploration models is the provision of a unifying stratigraphical nomenclature for approximately chronostratigraphic units, and a consistent methodology to pick the boundaries of these units even where fundamental facies changes occur.

A hierarchy of Humber Group sequences?

In the foregoing, 19 sequences are described in the interval from the Late Oxfordian to the Late Ryazanian. Using the timetable of Gradstein *et al.* (1994), and assuming that the base of the Humber Group in the OMF coincides with the Middle–Upper Oxfordian boundary (156.2 Ma), and its top with the top Berriasian (137 Ma), the average duration of each sequence would approximate 1.0 Ma. Vail *et al.* (1977) propose a hierarchy of sequences with three interregionally recognizable levels; the duration of their third-order sequences is given as 1–10 Ma. The duration of the 19 sequences described here falls in the high-frequency end of this range; we attribute this to the greater level of detailed study in a more restricted area. Some of these 19 'sequences' are probably local events rather than of global significance. This question can only be addressed by comparative studies of additional basins which is beyond the scope of this study.

An attempt was made to recognize higher order sequences in the sequences in the present study. To this end, the areal onlap of the 19 sequences in the OMF basin was mapped (Kadolsky *et al.* in press). This showed a steady onlap of successive sequences on to the eastern and western basin margins, that is to the Fladen Ground Spur in the east and the combination of Halibut Horst and Caithness Ridge in the west. The northern and southern basin margins were outside the study area. The culmination of onlap occurred in the Ryazanian. Thus, only one single low-frequency event can be identified throughout the entire Humber Group, implying an overall transgressive regime throughout.

Three significant tectonic reorganizations have been recognized in the Humber Group in the OMF (Kadolsky *et al.* in press), at the UJ5a SB, UJ7 SB and UJ9.2a/b SB. These events involved changes in the orientation of rifting activity and the emergence of highs as sand sources. While a subdivision on this basis would be possible, it would amount to localized tectonostratigraphy, and, hence lithostratigraphy.

Comparison with previously published sequence stratigraphy schemes in the study area

Partington *et al.* (1993*a*, *b*). The scheme of Partington *et al.* (1993*a*, *b*) (Fig. 2) combined observations from the entire North Sea. It does not appear to rely on detailed sedimentological observations on cores, but mainly on the interpretation of log motifs, in combination with biostratigraphy and broad facies, and palaeogeography considerations. Most of the flooding surfaces of Partington *et al.* (1993*a*, *b*) have been confirmed here, and we have added several others. Those not directly confirmed may equate to MFSs described here, but which have been placed at a different stratigraphical level. Further studies of wells outside the OMF are required to decide whether the additional sequences are present regionally.

Partington *et al.* (1993*b*) distinguished 'tectonically enhanced maximum flooding surfaces' (TEMFS) from other 'maximum flooding surfaces' (MFS). TEMFSs were characterized as 'the most obvious marine condensed intervals in exploration wells. They drown footwalls, switch off deposition of coarse clastics in both the basin centre and marginal areas, and lead to major reorganization of both the basinal and shelfal paleogeographies'. These authors recognized three TEMFSs in the North Sea: 'Eudoxus' = UJ8.2b MFS; 'Fittoni' = UJ11.1 MFS; and 'Anguiformis' = UJ13.1 or 13.2 MFS.

As discussed above, in the OMF we found insufficient evidence to consider these three highstands to be more important than the others. Moreover, in the OMF the most significant tectonic 'reorganization' occurred approximately within the UJ9.2a–9.2b sequences, when the orientation of rifting changed from the NE–SW-trending Theta Graben trend to the WNW–ESE-trending Witchground Graben trend. It is to be expected that tectonic processes occur at different times in different basins, creating TEMFSs of different ages. For example, the often quoted 'Eudoxus' (UJ8.2b) flooding event does not stand out in the OMF as more significant than those above or below it. The correlation sections of Figs 4, 5 and 9 are hung on this level to illustrate this point. Possibly in less detailed analyses the transition from the Piper (*s.l.*) sands to the Kimmeridge Clay Formation has been assigned a UJ8.2b (or '*Eudoxus*') age, while in reality it may occur in any sequence between UJ8.2a and UJ9.2b.

Carruthers et al. *(1996)*. These authors subdivided the Humber Group of the Central Graben and the southern OMF into six genetic sequences bounded by MFSs. The earliest one, LJ1 (comprising the Early and Middle Oxfordian), was not recognized by us in the OMF and is not considered in the following discussion. In the interval from UJP4 to LK4 (Late Oxfordian–Late Ryazanian) five principal sequences, and hence MFSs and three additional MFSs of lower rank, are recognized. These nine highstands compare to our 19. The basis for the proposed hierarchy seems to be a distinction between flooding events thought to be recognizable in the entire North Sea, as opposed to 'higher frequency flooding events'. More detailed data are desirable to confirm the regional presence of the six flooding events recognized by Carruthers *et al.*, and the apparent absence outside the OMF of the additional events observed by us. A possible cause of the recognition of a low number of sequences is the strong lateral variability of the sedimentary expression of a sequence which may only locally be well developed, and rather inconspicuous elsewhere, and hence may not be identified as a sequence in its own right. Examples are UJ7 in wells 15/16-T8 and 15/23-6A, and in many wells in Block 15/21; and UJ8.2a in wells 14/20-11 and 15/16-T8. A sequence may be removed by erosion over large areas, for example UJ6b in blocks 15/16, 14/20 and 14/19. Such phenomena can be recognized in areas with dense well spacing, but in regional studies different flooding events of similar age may be mistaken for one and the same if only one of them is well developed.

Carruthers *et al.* (1996, Fig. 2) recognize an underlying low-frequency trend in the relative sea-level curve with two highstands of similar magnitude in the Early Volgian and in the Ryazanian. As discussed above, the overall transgressive regime in the OMF peaked only within the Ryazanian; the inferred Mid-Volgian regression is either a phenomenon observed outside the OMF, or an incorrect conclusion from the reduced GR readings in the Middle Volgian.

The onlap behaviour clearly shows that in the OMF basin the sequences following the UJ9.2a and UJ9.2b sequences transgress successively further on the margins of the basin, that is in each successive sequence the relative sea-level increased further until its maximum in the Ryazanian. The onlap pattern of the Central Graben has not been documented.

A possible reason for the postulated Middle Volgian regression may be the widely observed lowering of the gamma ray log readings. In the Middle Volgian (UJ10.2–UJ13.1 sequences) sandy mass flow deposits are widespread, of which the finest fractions (very fine sand, silt and clay) spread throughout the basins and 'dilute' the shaliness. This can, however, only have contributed to the phenomenon, as in the Early Kimmeridgian UJ9.2a and 9.2b sequences with their high GR shales and sandy turbidites are also widespread in the OMF. The upper part of the Middle Volgian low GR interval, in sequence UJ13.1, is composed of sand and silt-free low GR shales, as shown by the core recovered in well 14/20-15; the UJ11.1 and UJ12 shales in this well with more elevated GR readings are associated with distal sandy turbidites. Thus, the causes for the lowering of the GR in the UJ11.1–UJ13.1 sequences are complex and do not necessarily include a relative sea-level fall.

Harker & Rieuf (1996). Harker & Rieuf (1996) recognize six MFSs in the Humber Group, pre-UJ4 MFSs not included, compared to Carruther *et al.*'s six main MFSs and our 19. Their genetic sequences are similar to those recognized by Carruthers *et al.* (1996), although the boundaries differ in many cases. The comments made above about the difficulties to find any objective criterion to establish a hierarchical grouping of sea-level fluctuations apply here as well.

A number of detailed comments need to be made.

A 'base H4' (= '*glosense*' = UJ4) flooding event is described as the first marine transgression in the OMF. The underlying Skene Member (coal and non-marine shales) is placed in the preceding 'H3' sequence which begins with the Middle Oxfordian UJ2 (= '*densiplicatum*') MFS. However, the first regional marine transgression occurs in the Saltire Member with a highstand consistently dated as UJP5. Only in the North Buchan Graben and in the West Fladen basin is it underlain by a UJ4 sequence (Fig. 3). The non-marine Skene Member is therefore younger than the UJ4 highstand.

The 'base H5' (= '*baylei*') flooding event in well 15/17-4 is correlated with the UJ5 transgression, in wells 15/17-16 and 15/21a-15 with the UJ6a transgression, and in well 14/19-4 with the UJ7 transgression (Harker & Rieuf 1996, Fig. 17b). The term I-Shale was proposed for the UJ5 highstand shales in the Piper Field and should be restricted to this field unless a correlation into an outside area is proven. The conventional tie of the *Baylei* ammonite zone is with the UJP7 dinocyst zone. According to Boldy & Brealey (1990) and Brealey (1990) the Rob Roy Mid-Shale yielded ammonites of the *Baylei*, *Cymodoce* and *Mutabilis* zones; from the Piper I-shale suggested, however, the much older *Serratum* zone (Maher 1980). The basis for the later reinterpretations of the I-Shale ammonite age as *Rosenkrantzi* zone (Maher 1981) resp. *Baylei* Zone (Harker & Rieuf 1996) is unknown. The dinocyst assemblages of the I-Shale and the Mid-Shale are very similar and certainly fit into the UJ5–UJ6b sequences, which may be difficult to distinguish biostratigraphically. The 'Baylei' MFS in Claymore (well 14/19-4; Harker & Rieuf 1996: Figs 14b and 17b; here Fig. 4) we identify, however, with the UJ7 MFS. Although the UJ7 sequence is an important MFS in the OMF, it is not recognized as separate from the UJ5 and UJ6 MFSs by Harker & Rieuf (1996).

The Glamis Sand (well 16/21-6; Harker & Rieuf 1996, fig. 22) is placed in the H8 sequence, that is post-UJ13.2 MFS. However, *Oligosphaeridium patulum* was reported from its upper part and in the base of the overlying Kimmeridge Clay Formation, giving a pre-UJ10.2 highstand age for the upper part of the sands, if *in situ*. The error may have been caused by Partington *et al.* (1993*b*, fig. 12) showing a post-UJ13.1 MFS age for the sands in nearby well 16/21-9. This may still be correct, as shoreface sands of up to UJP14.2 age have been encountered elsewhere on the flank of the Fladen Ground Spur, and topographic relief may well have caused the deposition of shoreface sands of such a young age at the 16/21-9 location, coeval with condensed shales depositionally down-dip in the Glamis Field area.

The Kimmeridge Clay cover of Piper Field well 15/17–4 is shown to be of 'H8' (= UJ13.2–MFS to LK4 MFS) age. However, the UJ10.2–UJ12 sequences are biostratigraphically identified within the Kimmeridge Clay Formation. The base of the latter is biostratigraphically poorly constrained, but the absence of *Cribroperidinium longicorne* suggests absence of zones UJP9.2–UJP8.2, and an isolated record of *Perisseiasphaeridium pannosum*, if correctly identified, is therefore thought to represent reworking. A hiatus between the Kimm Clay Formation and the Piper sands is thus present, but of a more limited extent than Harker & Rieuf suggest.

Conclusions

- The Humber Group in the OMF (Fig. 1) has been divided into 19 sequences. The sequences were identified on the basis of sedimentological criteria observed in well sections. They are characterized by dinocyst biostratigraphy (scheme in Fig. 2 and well examples in Figs 3–7), and exemplified by reference wells (Figs 8 and 9) and additional examples (Figs 10–12).
- The oldest sequence, which is restricted to the southwestern and northwestern OMF, is UJ4 (*Polonicum*) of Late Oxfordian age (Fig. 11).
- Relative water depth increased throughout the Humber Group, culminating in the Ryazanian; evidence includes the onlap pattern of the western and eastern margins of the OMF basin, and the facies development within individual well sections.
- A natural grouping of the 19 sequences into higher order sequences is not apparent on the basis of water depth development and onlap pattern. On the basis of tectonostratigraphic developments significant reorganizations of depositional centres occurred during sequences UJ5, UJ7, UJ9.2a and UJ9.2b.
- In the offshore mud facies of the Kimmeridge Clay Formation intervals of higher and lower GR, each comprising several sequences, can be distinguished which appear to be age constant. The causes for this variation are not well understood, but lowering of the GR is not necessarily an indication of a lower sea level.
- In the younger part of the Humber Group the frequency of sequences defined in deep-water facies (alternation of mass flow clastics and condensed shale sections) agrees with the frequency of sequences recognized in shoreface settings.
- The utility of sequence stratigraphy is mainly in the provision of a methodical framework to recognize and and interpret the spatial arrangements of facies associations, the systems tracts. From this follows the recognition of time surfaces and their correlation from one well to another. This becomes essential in situations where depositional gradients cause strong facies variation. Biostratigraphy is indispensable to identify the age of the sequences and hence to correlate them. In development geology, the most applicable aspect is the reservoir rock subdivision and the spatial arrangement of facies associations. In exploration, a sequence or group of sequences is the natural unit for palaeogeographical reconstructions as it is a time-bound rock unit. Prediction of reservoir rock distribution for exploration purposes is possible; the reliability of such predictions depends on well density and seismic resolution, as well as a realistic concept of sediment origin, quantity and transport pattern.
- In the deep-water clastics facies of the Galley Field reservoir, important aspects of reservoir architecture and provenance have been clarified with the support of biostratigraphy.

This study was only possible by building on the work of many geoscientists who have worked this area over the past 25 or so years. In particular, we wish to mention the unpublished biostratigraphic work of K.-H. Georgi and J. Lund (then Deutsche Texaco AG, Wietze, Germany) prior to 1988, and N. Clark's support for and comments on the Galley Field section. We thank Texaco's partners Amerada Hess Limited for permitting publication of information on the unreleased well 15/16b-22, and likewise Monument and OMV for 14/18b-12, and Texaco's management for permission to publish this study. The interpretations expressed in this article are the opinions of the authors.

Appendix: dinoflagellate zonation of the Oxfordian–Ryazanian

The Duxbury Stratigraphic Consultants palynofloral scheme is used in the present study, and this is illustrated in Fig. 2. All defining events are the result of first-hand observation by the author (S. Duxbury). A definition of each Late Oxfordian–Ryazanian zone and subzone is given below. Zonal definitions are included to the base of the Late Oxfordian, although in the OMF no first-hand or third-party evidence was observed for any sediments older than palyzone UJP4.

Zone UJP1

Age: Early Oxfordian, *Cordatum–Mariae* ammonite zones.

Definition: the interval from the highest occurrence of *Wanaea fimbriata* to the highest occurrences of *Energlynia acollaris*, *Meiourogonyaulax caytonensis* group and abundant *Mendicodinium groenlandicum* (top Callovian markers).

Comments: this zone can be further subdivided into two subzones.

Subzone UJP1.1. Definition: the interval from the highest occurrence of common *Surculosphaeridium*? *vestitum* to the highest occurrences of *Energlynia acollaris*, *Meiourogonyaulax caytonensis* group and abundant *Mendicodinium groenlandicum* (top Callovian markers).

Comments: this is the oldest Upper Jurassic unit. *Leptodinium mirabile* has its lowest occurrence within it.

Subzone UJP1.2. Definition: the interval from the highest occurrence of *Wanaea fimbriata* to the highest occurrence of common *Surculosphaeridium*? *vestitum*.

Comments: this subzone is restricted to the *Cordatum* ammonite zone.

Zone UJP2

Age: Middle Oxfordian, *Densiplicatum* ammonite zone.

Definition: the highest occurrence of *Liesburgia scarburghensis* to the highest occurrence of *Wanaea fimbriata*.

Comments: this very thin biozone is defined by the extinction events of two very distinctive species. In addition, the deepest occurrence of *Glossodinium dimorphum* occurs within this zone.

Zone UJP3

Age: Middle Oxfordian, *Tenuiserratum–Densiplicatum* ammonite zones.

Definition: the interval from the highest occurrence of *Gonyaulacysta jurassica longicornis* and/or *Rigaudella aemula* (rare) to the highest occurrence of *Liesburgia scarburghensis*.

Comments: This zone can be further subdivided into two subzones.

Subzone UJP3.1. Definition: The interval from the highest occurrence of common *Rigaudella aemula* to the highest occurrence of *Liesburgia scarburghensis*.

Subzone UJP3.2. Definition: The interval from the highest occurrence of *Gonyaulacysta jurassica longicornis* and/or *Rigaudella aemula* (rare) to the highest occurrence of common *Rigaudella aemula*.

Comments: a further defining event for the top of this subzone is the highest occurrence of *Chytroeisphaeridia cerastes*. However, this species can be rare near the top of its range.

Stephanelytron redcliffense can be common in this subzone.

Zone UJP4

Age: Late Oxfordian, *Glosense* ammonite zone.

Definition: the interval from the highest occurrence of *Compositosphaeridium polonicum* to the highest occurrence of *Gonyaulacysta jurassica longicornis* and/or *Rigaudella aemula* (rare).

Comments: the highest occurrence of *Nannoceratopsis pellucida* is within this zone, although this is essentially a Middle Oxfordian and older taxon.

Zone UJP5

Age: Late Oxfordian, *Regulare–Glosense* ammonite zones.

Definition: the interval from the highest occurrence of *Stephanelytron redcliffense* to the highest occurrence of *Compositosphaeridium polonicum*.

Comments: The highest occurrence of *Stephanelytron redcliffense* has been observed in the present study, but it was very rare.

Peak occurrences of *Dissiliodinium* sp. CMS and *Glossodinium dimorphum* occur near the base of this unit.

Zone UJP6

Age: Late Oxfordian, *Rosenkrantzi–Regulare* ammonite zones.

Definition: the interval from the FDO of common *Hystrichosphaeridium petilum* and *Rhynchodiniopsis cladophora* to the FDO of *Stephanelytron redcliffense*.

Comments: the top of this unit is defined on the highset common occurrences of *H. petilum* and *R. cladophora*. However, the first appears to be localized as the common occurrence of *H. petilum* has also been observed in the Kimmeridgian elsewhere.

A very distinctive event which occurs near the top of UJP6 is the FDO of *Ellipsoidictyum cinctum*. This approximates to the FDO of common *Scriniodinium crystallinum*, and to a marked increase in *Gonyaulacysta jurassica*.

The highest common occurrence of *E. cinctum* is also within this biozone, and corresponds to an influx of *Dissiliodinium* sp.CMS.

An influx of *Taeniophora iunctispina* typically occurs near the base of this unit.

The deepest occurrence of *Occisucysta balia* group is within this biozone, but above the UJ6a MFS.

Zone UJP7

Age: Middle–Early Kimmeridgian, *Cymodoce–Baylei* ammonite zones.

Definition: the interval from the FDO of common *Gonyaulacysta jurassica* to the FDOs of common *Hystrichosphaeridium petilum* and *Rhynchodiniopsis cladophora*.

Comments: although *Gonyaulacysta jurassica* has been recorded considerably higher than this interval by several authors, these records tend to be isolated. The common occurrence of this species has not been observed above UJP7 here.

Further events of stratigraphic significance here include the FDO (rare) of *Taeniophora iunctispina* and the highest common occurrence of *Leptodinium mirabile* at the top of this unit.

The FDO of common *Chytroeisphaeridia chytroeides* and the peak occurrence of *Sentusidinium pilosum* are within this zone

The LDO of *Cribroperidinium longicorne* has been described elsewhere within the *Cymodoce* ammonite zone. In the present study, however, this species has been recorded (very rare) within UJP7, as deep as the UJ7 maximum flooding surface, which is thought to be within the *Baylei* ammonite zone. The deepest consistent occurrence of *Occisucysta balia* group is at the base of this unit.

Rare *Rhynchodiniopsis cladophora* occur in the lowest part of this zone, although this is essentially an older taxon.

Zone UJP8

Age: Late–Middle Kimmeridgian, *Eudoxus–Cymodoce* ammonite zones.

Definition: the interval from the highest occurrence of consistently common *Perisseiasphaeridium pannosum* to the highest common occurrence of *Gonyaulacysta jurassica*.

Comments: this zone can be further subdivided into two subzones.

Subzone UJP8.1. Definition: the interval from the highest occurrence of *Stephanelytron scarburghense* to the highest common occurrence of *Gonyaulacysta jurassica*.

Comments: the FDO of *Taeniophora iunctispina* provides an alternative indicator for the base of this unit, although it is very rare at that level.

Subzone UJP8.2. Definition: the interval from the highest occurrence of consistently common *Perisseiasphaeridium pannosum* to the highest occurrence of *Stephanelytron scarburghense*.

Comments: a major influx of *Perisseiasphaeridium pannosum* occurs within the *Eudoxus* ammonite zone, and the consistently common to abundant occurrence of this species is apparently restricted to sediments of that age in this area.

The deepest occurrence of *P. pannosum* is within UJP8.2.

The abundant occurrence of *Geiselodinium paeminosum* occurs at the top of UJP8.2 elsewhere. However, in this area, this does not provide a consistent marker, because of general rarity, and because of the condensed nature of the Late Kimmeridgian section.

The FDO of *Leptodinium mirabile* is within UJP8.2, and this appears to coincide with the highest consistent occurrence of *Gonyaulacysta jurassica* in the Witch Ground Graben.

The LDO of *Perisseiasphaeridium pannosum* (rare) is immediately below these events. A further event within this subzone is the extinction level of *Aldorfia dictyota*, in the lower part of the unit.

Zone UJP9

Age: Early Volgian–Late Kimmeridgian, *Hudlestoni–Autissiodorensis* ammonite zones.

Definition: the interval from the highest occurrence of *Cribroperidinium longicorne* to the highest occurrence of consistently common *Perisseiasphaeridium pannosum*.

Comments: this zone can be further subdivided into two subzones.

Subzone UJP9.1. Definition: the interval from the highest occurrences of *Endoscrinium luridum* and/or *Cribroperidinium crassinervum* to the highest occurrence of consistently common *Perisseiasphaeridium pannosum*.

Comments: both of the species defining the top of this unit can be very rare at the tops of their ranges, and it is sometimes difficult to differentiate this subzone on palynofloral criteria alone.

Subzone UJP9.2. Definition: the interval from the highest occurrence of *Cribroperidinium longicorne* to the highest occurrences of *Endoscrinium luridum* and/or *Cribroperidinium crassinervum*.

Comments: the consistently common occurrence of *Cribroperidinium longicorne* is restricted to this unit, and this has been described from the *Scitulus* and *Wheatleyensis* ammonite zones elsewhere. However, common *C. longicorne* can also be observed in exceptional samples within the latest Kimmeridgian, palyzone UJP9.1.

Perisseiasphaeridium pannosum has its highest occurrence within UJP9.2, and there is a minor influx of this species towards the base of this unit. Poor preservation often causes difficulty in distinguishing this species from the very similar *Oligosphaeridium patulum*.

The highest common occurrence of *Systematophora fasciculigera* is also an intra-UJP9.2 event.

The LDOs of several species are observed near the top of this subzone, including *Kleithriasphaeridium porosispinum*, *K. telaspinosum*, *Rotosphaeropsis thula* and *Pseudomuderongia* spp.

In addition, the LDOs of *Senoniasphaera jurassica* and *Oligosphaeridium patulum* are in the lower part of UJP9.2, with the deepest common occurrences of those species also providing valuable intra-UJP9.2 markers.

Zone UJP10

Age: Middle–Early Volgian, *Rotunda–Hudlestoni* ammonite zones.

Definition: The interval from the FDO of *Kleithriasphaeridium telaspinosum* and/or common *Prolixosphaeridium granulosum* to the FDO of *Cribroperidinium longicorne*.

Comments: this zone can be further subdivided into two subzones.

Subzone UJP10.1. Definition: the interval from the FDO of common *Oligosphaeridium patulum* to the FDO of *Cribroperidinium longicorne*.

Comments: the LDO of *Tanyosphaeridium* cf. *variecalamum* is within this subzone, although its occurrence is sporadic below UJP10.2.

This subzone tends to be thin in this area, and the FDO of common *O. patulum* often occurs with *Cribroperidinium longicorne*, at the top of UJP9.

Subzone UJP10.2. Definition: the interval from the FDOs of *Kleithriasphaeridium telaspinosum* and/or common *Prolixosphaeridium granulosum* to the FDO of common *Oligosphaeridium patulum.*

Comments: *Oligosphaeridium patulum* can be observed consistently to the top of this subzone, although it becomes increasingly rare towards the top of its range, and in samples of variable quality it is sometimes difficult to pick out.

An influx of *Leptodinium arcuatum* occurs immediately below the top of this unit, and the FDOs of consistent *Scriniodinium inritibilum* and consistent *Sirmiodiniopsis frisia* are within this subzone.

The LDO of *Muderongia* cf. *simplex* (very rare) is within the uppermost part of this interval, and the deepest consistent occurrence of *Tanyosphaeridium* cf. *variecalamum* is within the lower part of UJP10.2.

Zone UJP11

Age: Middle Volgian, *Okusensis–Rotunda* ammonite zones.

Definition: the interval from the FDO of common *Muderongia* cf. *simplex* to the FDO of *Kleithriasphaeridium telaspinosum* and/or common *Prolixosphaeridium granulosum.*

Comments: this zone can be further subdivided into two subzones.

Subzone UJP11.1. Definition: the interval from the FDO of *Prolixosphaeridium granulosum* to the FDO of *Kleithriasphaeridium telaspinosum* and/or common *Prolixosphaeridium granulosum.*

Comments: the upper part of this subzone contains common to abundant *Muderongia* cf. *simplex*, and the deepest common occurrence of this taxon is within UJP11.1.

The peak occurrence of *M.* cf. *simplex* is within this unit, at approximately the same level as the FDOs of *Rhynchodiniopsis 'machaera'* and *Pseudomuderongia* spp., close to the UJ11.1 MFS.

An influx of *Cribroperidinium gigas* has been observed within palyzone UJP11.1, although this species is rare throughout the lower part of the unit, and absent below.

Common to abundant *Cyclonephelium distinctum* are recorded to the base of UJP11.1, but this species is rare below this unit.

Rare *Oligosphaeridium patulum* are occasionally observed here, but this species is consistently present only to the top of UJP10.

Subzone UJP11.2. Definition: the interval from the FDO of very common *Muderongia* cf. *simplex* to the FDO of *Prolixosphaeridium granulosum.*

Comments: there is a marked downhole increase in ceratioid dinocysts at the top of this interval, and both *Muderongia* cf. *simplex* and *Senoniasphaera jurassica* can be abundant throughout.

Also, the peak occurrence of *Ctenidodinium panneum* and the deepest common occurrence of *Wallodinium krutzschii* are within this subzone.

Zone UJP12

Age: Middle Volgian, *Okusensis–Glaucolithus* ammonite zones.

Definition: the interval from the FDO of *Muderongia* cf. *simplex* to the FDO of common *Muderongia* cf. *simplex.*

Comments: there is a marked downhole increase in numbers of *Ctenidodinium panneum* within this zone.

Zone UJP13

Age: Middle Volgian, *Oppressus–Kerberus* ammonite zones.

Definition: the interval from the FDO of *Egmontodinium polyplacophorum* to the FDO of *Muderongia* cf. *simplex.*

Comments: This zone can be further subdivided into three subzones.

Subzone UJP13.1. Definition: the interval from the FDO of *Senoniasphaera jurassica* to the FDO of *Muderongia* cf. *simplex.*

Comments: the top of this subzone corresponds to the FDO of common *Leptodinium arcuatum.* A major influx of *Cyclonephelium distinctum* is often observed near the top of this unit, associated with the markedly lower gamma response section ('Kimmeridgian Silt') normally present.

Wallodinium krutzschii has its peak abundance near the top of this subzone, and the FDO of common *Senoniasphaera jurassica* is within it.

The deepest occurrences of several taxa have been recorded near the top of this subzone, including *Kleithriasphaeridium fasciatum* group, *Isthmocystis* cf. *distincta* and *Dingodinium spinosum* (rare).

The deepest common occurrence of *P. insolitum* is also near the top of UJP13.1, and this species is effectively absent below that level.

Subzone UJP13.2. Definition: the interval from the FDO of *Ctenidodinium panneum* and/or *Glossodinium dimorphum* to the FDO of *Senoniasphaera jurassica.*

Comments: in the study area, the UJ13.2 marine flooding event is a marked feature (see below) and it is associated with the peak occurrence (abundant) of *Perisseiasphaeridium insolitum.* In the absence of the key subzonal taxa, this flood can be used as an alternative diagnostic feature.

The deepest occurrence of *Gochteodinia virgula* group is within this subzone.

Subzone UJP13.3. Definition: the interval from the FDO of *Egmontodinium polyplacophorum* to the FDO of *Ctenidodinium panneum* and/or *Glossodinium dimorphum.*

Comments: the top of the Middle Volgian is sometimes difficult to place, because of the rarity of *E. polyplacophorum* near the top of its range. However, the FDO of common *Perisseiasphaeridium insolitum* provides a useful marker within this subzone.

In addition, the deepest occurrence of common *Gochteodinia villosa* is within UJP13.3. Specimens recorded in this and younger sections are larger than those below, usually with longer spines.

Zone UJP14

Age: Late Volgian, *Lamplughi–Primitivus* ammonite zones.

Definition: the interval from the highest occurrence of *Egmontodinium expiratum* to the highest occurrence of *Egmontodinium polyplacophorum*.

Comments: this zone can be further subdivided into two subzones.

Subzone UJP14.1. Definition: the interval from the FDO of common *Cribroperidinium gigas* to the FDO of *Egmontodinium polyplacophorum*.

Comments: this earliest Late Volgian subzone is usually thin in the study area, and it is difficult to resolve, particularly in widely spaced ditch cuttings samples.

Subzone UJP14.2. Definition: the interval from the highest occurrence of *Egmontodinium expiratum* to the highest common occurrence of *Cribroperidinium gigas*.

Comments: the highest occurrences of *Cribroperidinium gigas* (rare), *Leptodinium arcuatum* and *L. eumorphum* are observed near the top of this unit. However, they are very rare at that level, and difficult to pick out consistently.

Although *Pterospermella* spp. is abundant in the overlying biozone, the peak occurrence of this genus is within the upper part of UJP14.2. The FDO of common *Wallodinium krutzschii* is at approximately the same level.

The highest occurrence of *Gochteodinia virgula* group (including that species and *Gochteodinia* sp. 1 Davey, 1982) is within this subzone.

In addition, the deepest occurrences of common *Stiphrosphaeridium dictyophorum* and common *Stephanelytron cretaceum* are near the top of this unit, and the deepest common occurrences of *Endoscrinium pharo* and *Dingodinium spinosum* are near its base.

Zone LKP1

Age: Early Ryazanian, *Runctoni* ammonite zone.

Definition: the interval from the highest occurrence of abundant *Pterospermella* spp. to the highest occurrence of *Egmontodinium expiratum*.

Comments: this zone is the oldest Cretaceous biozone, and it shows characteristics transitional between 'typically Jurassic' and 'typically Cretaceous' assemblages.

The lower part of this unit is characterized by the highest very abundant occurrence of *Canningia compta*.

The deepest occurrence of *Daveya boresphaera* (= *Gonyaulacysta* sp. A Davey, 1979) is within this unit.

Zone LKP2

Age: Early Ryazanian, *Kochi–Runctoni* ammonite zones.

Definition: the interval from the highest occurrence of *Rotosphaeropsis thula* to the highest occurrence of abundant *Pterospermella* spp.

Comments: the deepest occurrence of *Batioladinium radiculatum*, and the deepest abundant occurrence of *Oligosphaeridium diluculum* occur near the top of this unit.

Zone LKP3

Age: Late Ryazanian, *Icenii* ammonite zone.

Definition: the interval from the FDO of *Batioladinium pomum* to the FDO of *Rotosphaeropsis thula*.

Comments: the highest occurrence of *Batioladinium radiculatum* appears to be within this unit.

Zone LKP4

Age: Late Ryazanian, *Stenomphalus* ammonite zone.

Definition: the interval from the highest occurrence of abundant amorphous organic matter (AOM) to the highest occurrence of *Batioladinium pomum*.

Comments: there is a marked palynofloral change at the top of this zone, with clean, diverse assemblages above being replaced by more restricted palynofloras, including abundant AOM and common to abundant gymnosperm pollen.

The FDO of *Daveya boresphaera* (= *Gonyaulacysta* sp. A Davey, 1979) is near the top of this zone, and the deepest occurrence of *Muderongia simplex* is near its base.

This zone represents the highest unit of the Kimmeridge Clay Formation. In the study area, reworking of the upper Kimmeridge Clay Formation into the Valhall Formation is common, and the true top of this unit is sometimes difficult to pick out on biostratigraphic criteria alone.

Also, the uppermost Kimmeridge Clay Formation tends to be very condensed, and discrimination of individual biozones may therefore be difficult without densely spaced core or sidewall core samples.

Zone LKP5

Age: Late Ryazanian, *Albidum* and *Stenomphalus* ammonite zones.

Definition: the interval from the highest occurrence of *Dingodinium spinosum* to the highest occurrence of abundant AOM.

Comments: taxa which became extinct at or near the top of the Ryazanian include *Kleithriasphaeridium porosispinum* and *Dichadogonyaulax culmula*, and these may be used as alternative markers for the top of this zone, in the absence of *Dingodinium spinosum*.

Several taxa typical of the Cromer Knoll Group Valhall Formation have their inceptions within this biozone. These include *Ctenidodinium elegantulum*, *Phoberocysta neocomica*, *Lagenorhytis delicatula*, *Diacanthum hollisterii*, *Achomosphaera neptunii* group, *Heslertonia heslertonensis*, *Hystrichodinium voigtii*, *Pseudoceratium pelliferum* and *Tehamadinium daveyi*.

This zone is essentially the post-Kimmeridge Clay Formation part of the Ryazanian, and it is typically very thin and difficult to resolve, particularly where only ditch cuttings samples are available.

References

ANDREWS, I. J. & BROWN, S. 1987. Stratigraphic evolution of the Jurassic, Moray Firth. *In*: BROOKS, J. & GLENNIE, K. (eds) *Petroleum Geology of North West Europe*. Graham & Trotman, London, 785–795.

BOLDY, S. A. R. & BREALEY, S. 1990. Timing, nature and sedimentary result of Jurassic tectonism in the Outer Moray Firth. *In*: HARDMAN, R. F. P. & BROOKS, J. (eds) *Tectonic Events Responsible for Britain's Oil and Gas Reserves*. Geological Society, London, Special Publications, **55**, 259–279.

BOOTE, D. R. D. & GUSTAV, S. H. 1987. Evolving depositional systems within an active rift, Witch Ground Graben, North Sea. *In*: BROOKS, J. & GLENNIE, K. (eds) *Petroleum Geology of North West Europe*. Graham & Trotman, London, 819–833.

BREALEY, S. 1990. *The Late Oxfordian to Kimmeridgian History of the Rob Roy and Ivanhoe Fields, Outer Moray Firth*. PhD thesis, University College, London.

CARRUTHERS, A., MCKIE, T., PRICE, J., DYER, R., WILLIAMS, G. & WATSON, P. 1996. The application of sequence stratigraphy to the understanding of Late Jurassic turbidite plays in the Central North Sea, UKCS. *In*: HURST, A., JOHNSON, H., BURLEY, S. D., CANHAM, A. C. & MACKERTICH, D. S. (eds) *Geology of the Humber Group: Central Graben and Moray Firth, UKCS*. Geological Society, London, Special Publications, **114**, 29–45.

CASEY, B. J., ROMANI, R. S. & SCHMITT, R. H. 1993. Appraisal geology of the Saltire Field, Witch Ground Graben, North Sea. *In*: PARKER, J. R. (ed.) *Petroleum Geology of Northwest Europe: Proceedings of the 4th Conference*. The Geological Society, London, 507–517.

COWARD, R. N., CLARK, N. M. & PINNOCK, S. J. 1991. The Tartan Field, Block 15/16, UK North Sea. *In*: ABBOTTS, I. L. (ed.) *U.K. Oil and Gas Fields, 25 Years Commemorative Volume*. Geological Society, London, Memoirs **14,** 377–384.

CURRIE, S. 1996. The development of the Ivanhoe, Rob Roy and Hamish Fields, Block 15/21A, UK North Sea. *In*: HURST, A., JOHNSON, H., BURLEY, S. D., CANHAM, A. C. & MACKERTICH, D. S. (eds) *Geology of the Humber Group: Central Graben and Moray Firth, UKCS*. Geological Society, London, Special Publications, **114,** 329–342.

DAVIES, R. J., STEPHEN, K. J. & UNDERHILL, J. R. 1996. A re-evaluation of Middle and Upper Jurassic stratigraphy and the flooding history of the Moray Firth Rift System, North Sea. *In*: HURST, A., JOHNSON, H., BURLEY, S. D., CANHAM, A. C. & MACKERTICH, D. S. (eds) *Geology of the Humber Group: Central Graben and Moray Firth, UKCS*. Geological Society, London, Special Publications, **114,** 81–108.

DAVEY, R. J. 1979. The stratigraphic distribution of dinocysts in the Portlandian (latest Jurassic) to Barremian (Early Cretaceous) of Northwest Europe. *American Association of Stratigraphic Palynologists, Contribution Series*, **5B**, 49–81.

——1982. Dinocyst stratigraphy of the latest Jurassic to Early Cretaceous of the Haldager No. 1 borehole, Denmark. *Danmarks Geologiske Undersgelse, Serie B*, **6**, 1–57.

DAVIES, R. J., O'DONNELL, D., BENTHAM, P. N., GIBSON, J. P. C., CURRY, M. R. & DUNAY, R. E. in press. The origin and genesis of major Jurassic unconformities within the triple junction area of the North Sea, UK. *In*: FLEET, A. J. & BOLDY, S. A. R. (eds) *Petroleum Geology of Northwest Europe: Proceedings of the 5th Conference*. Geological Society, London.

DAVID, M. J. 1996. History of hydrocarbon exploration in the Moray Firth. *In*: HURST, A., JOHNSON, H., BURLEY, S. D., CANHAM, A. C. & MACKERTICH, D. S. (eds) *Geology of the Humber Group: Central Graben and Moray Firth, UKCS*. Geological Society, London, Special Publications, **114**, 47–80.

GRADSTEIN, F. M., AGTERBERG, F. P., OGG, J. G., HARDENBOL, J., VAN VEEN, P., THIERRY, J. & HUANG, Z. 1994. A Mesozoic timescale. *Journal of Geophysical Research* **99**, 24 051–24 074.

HALLSWORTH, C., MORTON, A. C. & DORÉ, G. 1996. Contrasting mineralogy of Upper Jurassic sandstones in the Outer Moray Firth, North Sea: implications for the evolution of sediment dispersal patterns. *In*: HURST, A., JOHNSON, H., BURLEY, S. D., CANHAM, A. C. & MACKERTICH, D. S. (eds) *Geology of the Humber Group: Central Graben and Moray Firth, UKCS*. Geological Society, London, Special Publications, **114**, 131–144.

HAQ, B. U., HARDENBOL, J. & VAIL, P. R. 1988. Mesozoic and Cenozoic chronostratigraphy and eustatic cycles. *In*: WILGUS, C. K., HASTINGS, B. S., KENDALL, C. G. S. C., POSAMENTIER, H. W., ROSS, C. A. & VAN WAGONER, J. C. (eds), *Sea-level Changes: An Integrated Approach*. Society of Economic Paleontologists and Mineralogists, Special Publication **42**, 71–108.

HARKER, S. 1998. The palingenesy of the Piper oil field, UK North Sea. *Petroleum Geoscience*, **4**, 271–286.

HARKER, S. D., GREEN, S. C. H. & ROMANI, R. S. 1991. The Claymore Field, Block 14/19, UK North Sea. *In*: ABBOTTS, I. L. (ed.) *U.K. Oil and Gas Fields, 25 Years Commemorative Volume*. Geological Society, London, Memoirs, **14**, 269–278.

——, GUSTAV, S. H. & RILEY, L. A. 1987. Triassic to Cenomanian stratigraphy of the Witch Ground Graben. *In*: BROOKS, J. & GLENNIE, K. (eds) *Petroleum Geology of Northwest Europe*. Graham & Trotman, London, 809–818.

——, MANTEL, K. A., MORTON, D. J. & RILEY, L. A. 1993. The stratigraphy of Oxfordian–Kimmeridgian (Late Jurassic) reservoir sandstones in the Witch Ground Graben, United Kingdom North Sea. *American Association of Petroleum Geologists Bulletin* **77**, 1693–1709.

—— & RIEUF, M. 1996. Genetic stratigraphy and sandstone distribution of the Moray Firth Humber Group (Upper Jurassic). *In*: HURST, A., JOHNSON, H., BURLEY, S. D., CANHAM, A. C. & MACKERTICH, D. S. (eds) *Geology of the Humber Group: Central Graben and Moray Firth, UKCS*. Geological Society, London, Special Publications **114**, 109–130.

KADOLSKY, D., JOHANSEN, S. J. & DUXBURY, S. in press. Sequence stratigraphy and sedimentary history of the Humber Group (Late Jurassic–Ryazanian) in the Outer Moray Firth (UKCS, North Sea). *In*: FLEET, A. J. & BOLDY, S. A. R. (eds) *Petroleum Geology of Northwest Europe: Proceedings of the 5th Conference*. Geological Society, London.

MAHER, C. 1980. Piper oil field. *In*: HALBOUTY, M. T. (ed.) *Giant Oil and Gas Fields of the Decade 1968–1978*. American Association of Petroleum Geologists Memoir, **30**, 131–172.

——1981. The Piper oilfield. *In*: ILLING, L. V. & HOBSON, G. D. (eds) *Petroleum Geology of the Continental Shelf of North-West Europe*. Heyden, London, 358–370.

—— & HARKER, S. D. 1987. Claymore oil field. *In*: BROOKS, J. & GLENNIE, K. (eds) *Petroleum Geology of North West Europe*. Graham & Trotman, London, 835–845.

MCCANTS, C. Y. & BURLEY, S. 1996. Reservoir architecture and diagenesis in downthrown fault block plays: the Lowlander Prospect of Block 14/20, Witch Ground Graben, Outer Moray Firth, UK North Sea. *In*: HURST, A., JOHNSON, H., BURLEY, S. D., CANHAM, A. C. & MACKERTICH, D. S. (eds) *Geology of the Humber Group: Central Graben and Moray Firth, UKCS*. Geological Society, London, Special Publications, **114**, 251–285.

MIALL, A. D. 1986. Eustatic sea level changes interpreted from seismic stratigraphy: a critique of the methodology with particular reference to the North Sea Jurassic record. *American Association of Petroleum Geologists Bulletin*, **70**, 131–137.

——1992. Exxon global cycle chart: an event for every occasion? *Geology*, **20**, 787–790.

O'DRISCOLL, D., HINDLE, A. D. & LONG, D. C. 1990. The structural controls on Upper Jurassic and Lower Cretaceous reservoir sandstones in the Witch Ground Graben, UK North Sea. *In*: HARDMAN, R. F. P. & BROOKS, J. (eds) *Tectonic Events Responsible for Britain's Oil and Gas Reserves*. Geological Society, London, Special Publications, **55**, 299–323.

PARKER, R. H. 1991. The Ivanhoe and Rob Roy Fields, Block 15/21a-b, UK North Sea. *In*: Abbotts, I. L. (ed.) *U.K. Oil and Gas Fields, 25 Years Commemorative Volume*. Geological Society, London, Memoirs, **14**, 331–338.

PARTINGTON, M. A., COPESTAKE, P., MITCHENER, B. C. & UNDERHILL, J. R. 1993*a*. Biostratigraphic calibration of genetic stratigraphic sequences in the Jurassic–lowermost Cretaceous (Hettangian to Ryazanian) of the North Sea and adjacent areas. *In*: PARKER, J. R. (ed.) *Petroleum Geology of Northwest Europe: Proceedings of the 4th Conference*. Geological Society, London, 371–386.

——, MITCHENER, B. C., MILTON, N. J. & FRASER, A. J. 1993*b*. Genetic sequence stratigraphy for the North Sea Late Jurassic and Early Cretaceous: distribution and prediction of Kimmeridgian–Late Ryazanian reservoirs in the North Sea and adjacent areas. *In*: PARKER, J. R. (ed.) *Petroleum Geology of Northwest Europe: Proceedings of the 4th Conference*. Geological Society, London, 347–370.

RATTEY, R. P. & HAYWARD, A. P. 1993. Sequence stratigraphy of a failed rift system: the Middle Jurassic to Early Cretaceous basin evolution of the Central and Northern North Sea. *In*: PARKER, J. R. (ed.) *Petroleum Geology of Northwest Europe: Proceedings of the 4th Conference*. Geological Society, London, 215–249.

RICHARDS, P. C., LOTT, G. K., JOHNSON, H., KNOX, R. W. O'B. & RIDING, J. B. 1993. *Lithostratigraphic Nomenclature of the U.K. North Sea. 3. Jurassic of the Central and Northern North Sea*. British Geological Survey and UKOOA, Keyworth, Nottingham.

RIDING, J. B. & THOMAS, J. E. 1988. Dinoflagellate cyst stratigraphy of the Kimmeridge Clay (Upper Jurassic) from the Dorset coast, Southern England. *Palynology*, **12**, 65–88.

—— & ——1997. Marine palynomorphs from the Staffin Bay and Staffin Shale formations (Middle–Upper Jurassic) of the Trotternish Peninsula, NW Skye. *Scottish Journal of Geology*, **33**, 59–74.

SCHMITT, H. R. H. 1991. The Chanter Field, Block 15/17, UK North Sea. *In*: ABBOTTS, I. L. (ed.) *U.K. Oil and Gas Fields, 25 Years Commemorative Volume*. Geological Society, London, Memoirs, **14**, 261–268.

—— & GORDON, A. F. 1991. The Piper Field, Block 15/17, UK North Sea. *In*: ABBOTTS, I. L. (ed.) *U.K. Oil and Gas Fields, 25 Years Commemorative Volume*. Geological Society, London, Memoirs, **14**, 361–368.

STEPHEN, K. J. & DAVIES, R. J. 1998. Documentation of Jurassic sedimentary cycles from the Moray Firth Basin, UK North Sea. *In*: DE GRACIANSKY, P. C., HARDENBOL, J., JACQUIN, T., FARLEY, M. & VAIL, P. R. (eds) *Mesozoic and Cenozoic Sequence Stratigraphy of European Basins*. Society of Economic Paleontologists and Mineralogists, Special Publication, **60**, 485–510.

SYMS, R. M., SAVORY, D. F., WARD, C. J., EBDON, C. C., AQUILLINA, P. M. & SMITH, M. A. in press. Integrating sequence stratigraphy in field development and reservoir management – the Telford Field. *In*: FLEET, A. J. & BOLDY, S. A. R. (eds) *Petroleum Geology of Northwest Europe: Proceedings of the 5th Conference*. Geological Society, London.

VAIL, P. R., MITCHUM, R. M., JR. & THOMPSON, S., III 1977. Seismic stratigraphy and global changes of sea level, part 4: global cycles of relative changes of sea level. *In*: *Seismic Stratigraphy – Applications to Hydrocarbon Exploration*. American Association of Petroleum Geologists Memoir, **26,** 83–97.

VELDKAMP, J. J., GAILLARD, M. G., JONKERS, H. A. & LEVELL, B. K. 1996. A Kimmeridgian time-slice through the Humber Group of the central North Sea: a test of sequence stratigraphic methods. *In*: HURST, A., JOHNSON, H., BURLEY, S. D., CANHAM, A. C. & MACKERTICH, D. S. (eds) *Geology of the Humber Group: Central Graben and Moray Firth, UKCS*. Geological Society, London, Special Publications, **114**, 1–28.

WADDAMS, P. & CLARK, N. M. 1991. The Petronella Field, Block 14/20b, UK North Sea. *In*: ABBOTTS, I. L. (ed.) *U.K. Oil and Gas Fields, 25 Years Commemorative Volume*. Geological Society, London, Memoirs, **14**, 353–360.

WHITEHEAD, M. & PINNOCK, S. J. 1991. The Highlander Field, Block 14/20b, UK North Sea. *In*: ABBOTTS, I. L. (ed.) *U.K. Oil and Gas Fields, 25 Years Commemorative Volume*. Geological Society, London, Memoirs, **14,** 323–329.

WILLIAMS, J., CONNER, D. & PETERSEN, K. E. 1975. Piper oil field, North Sea: fault block structure with Upper Jurassic beach/bar reservoir sands. *AAPG Bulletin*, **59**, 1581–1601.

Micropalaeontological biostratigraphy of the Magnus Sandstone Member (Kimmeridgian–Early Volgian), Magnus Field, UK North Sea

P. H. MORRIS,[1] S. N. J. PAYNE[2] & D. P. J. RICHARDS[2]

[1] *MicroStrat Services, Southcott, Buckland Brewer, Bideford, North Devon EX39 5LU, UK*
[2] *BP Exploration, Farburn Industrial Estate, Dyce, Aberdeen AB21 7PB, UK*

Abstract: The micropalaeontological biostratigraphy of the Magnus Sandstone Member of Kimmeridgian–Early Volgian age is described for the first time using data from 27 Magnus Field wells (blocks 211/12a and 211/7a). Despite extensive re-sedimentation within the Magnus Sandstone Member turbidite reservoir, a consistent sequence of nine bioevents is recognized field-wide. These include both regionally extensive bioevents and localized extinction and acme bioevents. The microfaunas are documented and key taxa, which include numerous radiolarian species, are described and figured.

The bioevent scheme provides the basis for the chronostratigraphic and genetic sequence stratigraphic subdivision (*sensu* Galloway) of the Magnus Sandstone Member. It allows the recognition and biostratigraphic calibration of three maximum flooding surfaces (MFS) which in ascending stratigraphic order comprise: the pre-Magnus Sandstone Member top J62–basal J63 '*Eudoxus*' of Partington *et al.* (upper limit of the 'Lower Kimmeridge Clay Formation'); the intra-reservoir top J63–basal J64 '*Autissiodorensis*' ('B Shale'); and the supra-reservoir top J64–basal J66 '*Hudlestoni*' (upper limit of the 'H Unit').

The erection of a field-specific biostratigraphic framework that characterizes and discriminates between the individual shales associated with the Magnus Sandstone Member has impacted on reservoir modelling, primarily in the area of fluid pressure distribution, where shales such as the 'B Shale' and those of the 'F Unit' form significant intra-reservoir pressure barriers, placing sands in pressure isolation. Using a bioevent-constrained well correlation in the central field area this pressure compartmentalization is illustrated with reference to repeat formation tester tool (RFT) pressure depletion data.

At a broader scale, the biostratigraphically-defined 'time slices' are used in conjunction with three-dimensional (3D) seismic data to model Magnus Sandstone Member submarine fan evolution, which is interpreted in terms of three main depositional phases: (1) a latest Kimmeridgian (J63) single submarine fan lobe with a depocentre in the mid-field, sourced from the northwest and displaying laterally confined, fault-controlled canyon-fill; (2) Early Volgian (J64 lower) bilobate accretion with two depocentres forming either side of the N–S-trending Brent High; and (3) late Early Volgian (J64 upper) as before but with the southward migration of the northern lobe depocentre, with fault confinement.

Micropalaeontology has also been used at the well site in the Magnus Field where it has influenced drilling operations in 'real time'. The primary role to date has been the precise picking of casing points above high-pressured or low-pressured sands (e.g. 'A Unit') and calibration of supra-reservoir angle-build in deviated wells.

It is concluded that the application of biostratigraphy to integrated reservoir description has considerably enhanced our knowledge of submarine fan heterogeneity in Magnus, which has in turn led to more effective reservoir management and an extended production life for the field.

The Magnus Field is the most northerly producing oil field in the UK sector of the North Sea, located 160 km northeast of the Shetland Isles in licence blocks 211/12a and 211/7a (Fig. 1). Oil production, which began in 1983, is mainly from the Magnus Sandstone Member (De'Ath & Schuylemann 1981; Shepherd *et al.* 1991) with secondary production from the deeper 'Lower Kimmeridge Clay Formation' (Fig. 2). Original oil in place was estimated at 1.65 billion barrels with recoverable reserves calculated at 0.66 billion barrels, giving the field 'Giant' status (see Shepherd *et al.* 1991 for details of field development and production history). Current production (1997) stands at around 70 000 barrels per day, having declined from a peak plateau production rate of 160 000 barrels per day in 1990.

Morris, P. H., Payne, S. N. J. & Richards, D. P. J. 1999. Micropalaeontological biostratigraphy of the Magnus Sandstone Member (Kimmeridgian–Early Volgian), Magnus Field, UK North Sea. *In:* Jones, R. W. & Simmons, M. D. (eds) *Biostratigraphy in Production and Development Geology*. Geological Society, London, Special Publications, **152**, 55–73.

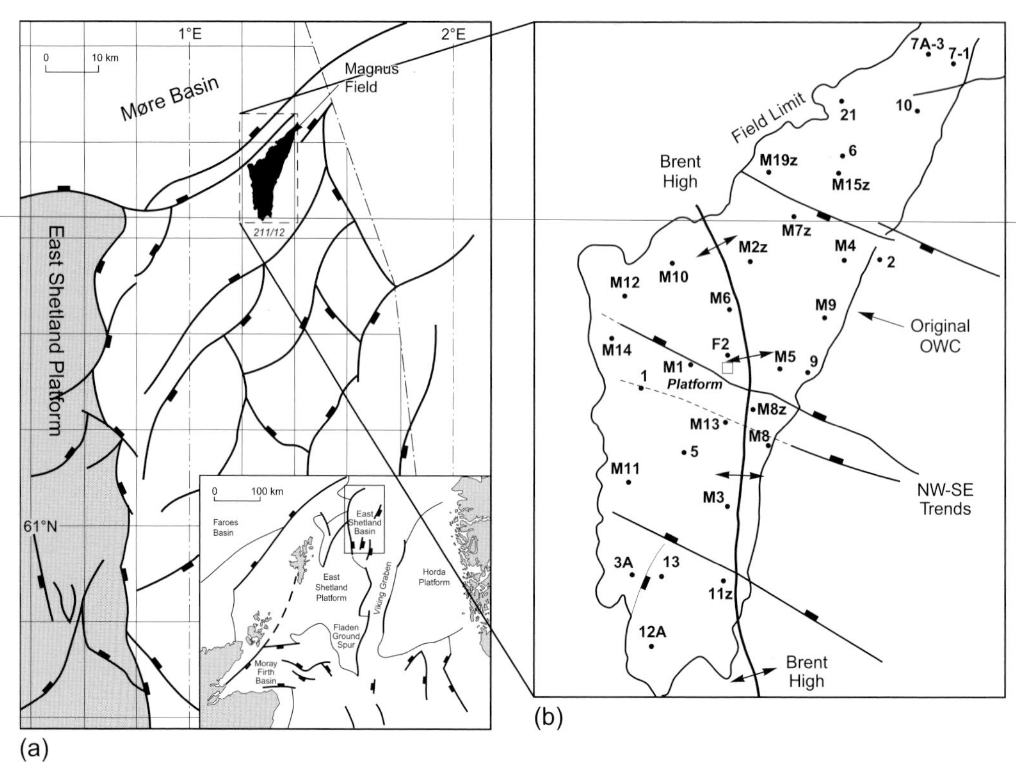

Fig. 1. Location map – Magnus Field, UKCS North Sea, blocks 211/7a and 211/12a. (**a**) Regional structural setting. (**b**) Study wells database.

At an early stage in the field's development, sedimentological analysis of cores taken from the Magnus Sandstone Member revealed the characteristic association of rock types and sedimentary structures of a submarine fan environment of deposition (De'Ath & Schuylemann 1981; Shepherd *et al.* 1991). Initially biostratigraphic characterization and correlation within the reservoir sequence proved problematic due to: (1) extensive penecontemporaneous reworking of palynomorphs which tends to mask *in situ* elements; (2) the apparent low diversity of radiolarian microfaunas in much of the Magnus Sandstone Member and the long stratigraphic ranges displayed by individual taxa; (3) the short period of geological time – *c.* 3 Ma. (Partington *et al.* 1993*a*) – through which the reservoir was deposited, allowing for little apparent evolution in the microfaunas, with few evolutionary inception and extinction events; and (4) facies dependency exhibited by other microfaunas, notably agglutinated foraminifera. Consequently, despite the necessity for a high-resolution stratigraphic subdivision during the mid-1980s, the role of biostratigraphy was relegated, whilst seismic, wireline log, RFT and sedimentological evaluation continued to be applied rigorously.

New advances in North Sea Late Jurassic micropalaeontological (especially radiolarian) biostratigraphy (Dyer & Copestake 1989) during the late 1980s coincided with the prospect of decline in Magnus Field oil production and led to a renewed interest in the field's biostratigraphy. The main object of this paper is to describe the micropalaeontological biozonation of the Magnus Sandstone Member which was principally erected during this period of field re-evaluation: it aims also to highlight the potential of radiolaria in Late Jurassic production and exploration biostratigraphy. The distribution and chronostratigraphic implications of the microfaunas are also considered briefly in the light of recent advances in Jurassic sequence stratigraphy (e.g. Partington *et al.* 1993*a*, *b*).

Structural framework and general stratigraphy, Magnus Field

The Magnus Field has a complex structural history which has been influenced by: (1) basement structural grain (Caledonide) and Middle–Late

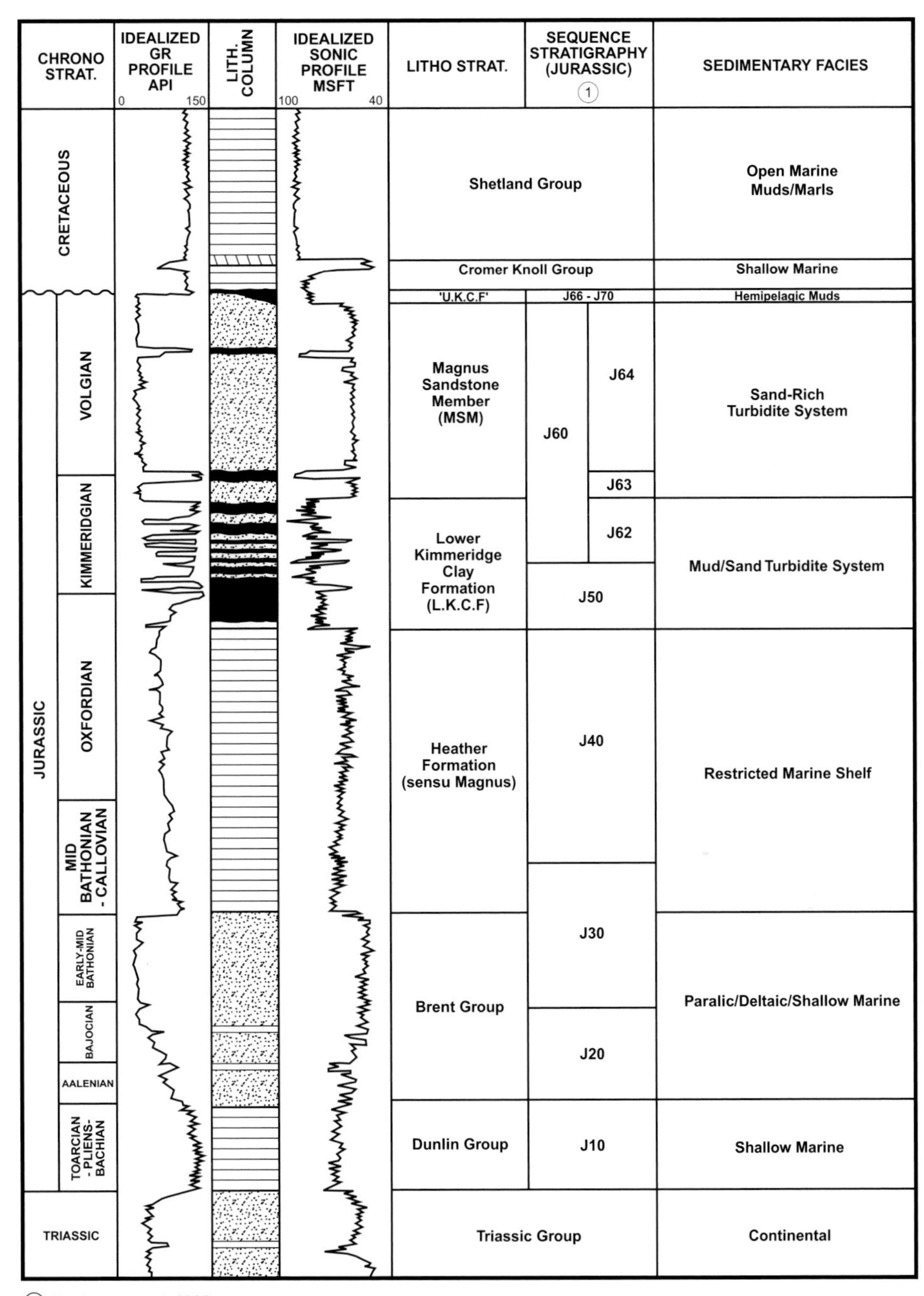

Fig. 2. Magnus Field Mesozoic stratigraphy.

Jurassic rifting trends of the Viking Graben which produced large-scale N–S-trending fault styles typical of the Brent Province (Fig. 1a); and (2) NE–SW faulting activated during the Early Cretaceous opening of the Møre basin and superimposed on older trends (Fig. 1: see Shepherd *et al.* 1991). Recent 3D seismic mapping of the field has defined the presence of a series of localized WNW–ESE-trending normal faults and the N–S aligned fault-activated Brent High

(Fig. 1b): all of these structures appear to have been intermittently active during the Late Jurassic, and have influenced Magnus submarine fan deposition and thus reservoir geometry (see below).

The generalized Mesozoic stratigraphy for the Magnus Field area is presented in Fig. 2. With the exception of the submarine fan development in the Kimmeridge Clay Formation, Middle–Late Jurassic stratigraphy is typical of the East Shetland basin–Viking Graben area. The main reservoir sequence comprises the Magnus Sandstone Member, which is dated as Kimmeridgian–Early Volgian. Owing to the easterly tilting of fault blocks within the field, progressive westerly truncation of the 'Upper Kimmeridge Clay Formation' and Magnus Sandstone Member is seen, such that at crestal locations Early–Late Cretaceous overstep is evident (see Shepherd *et al.* 1991). The clay-rich lithologies of the 'Upper Kimmeridge Clay Formation' and Shetland Group provide an effective seal across the field. The trapping mechanism in Magnus is provided by stratigraphic pinchout and erosional unconfomity.

Stratigraphic subdivision of the Magnus Sandstone Member

Prior to the biostratigraphic re-evaluation of the Magnus Sandstone Member reservoir, subdivision was based on wireline log, RFT and 3D seismic data as summarized in Shepherd *et al.* (1991). At this stage in reservoir evaluation a four-fold subdivision was employed (Shepherd *et al.* 1991). A new, nine-fold reservoir subdivision was generated in 1995 (Fig. 4), reflecting the acquisition of additional well data, a new 3D seismic survey and greater biostratigraphic characterization (see below).

The microfaunal bioevents associated with the Magnus Sandstone Member allow a broad chronostratigraphic subdivision that can be tied to the genetic sequence stratigraphic scheme of Partington *et al.* (1993*a*). This complies with the concept of Galloway (1989*a*, *b*) in subdividing stratigraphic successions into depositional episodes or sequences bounded by maximum flooding surfaces (MFS). In accordance with Partington *et al.* (1993*a*), each MFS defines the base of the overlying stratigraphic sequence. The MFS may represent considerable periods of geological time and equate to condensed sections through the latest Kimmeridgian to the Early Volgian.

The paucity of key palynofloral biostratigraphic indices, compounded by the frequently reworked nature of much of the palynofloral assemblages in the Magnus Sandstone Member, entails that this subdivision is not tied into the more precise Kimmeridgian–Volgian palynostratigraphic biozonation, and hence is applied with caution. The MFS are discussed below in ascending stratigraphic order:

- the pre-Magnus Sandstone Member top J62–base J63 intra-Kimmeridgian '*Eudoxus*' MFS (upper limit of the 'Lower Kimmeridge Clay Formation'). This is defined by bioevents M7a, a downhole numerical decrease in *Trochammina* cf. *lathetica,* and M7, a re-influx of pyritized specimens of *Rhaxella* sp. 1. Both bioevents are considered to be of local stratigraphic value only but appear to occur within biozone MJ20 of Partington *et al.* (1993*a*);
- the intra-lower Magnus Sandstone Member top J63–basal J64 latest Kimmeridgian '*Autissiodorensis*' MFS, represented by the topmost 'B Shale'. This is micropalaeontologically defined as the top of biozone MJ20c (Partington *et al.* 1993*a*). A downhole increase in sponge spicules (bioevent M5b) and radiolaria, particularly *Praeconocaryomma* (?) sp. 2 Dyer & Copestake, both pyritized and siliceous, characterize this flooding surface in Magnus Field. However this bioevent is of local stratigraphic value only;
- the post-reservoir top J64–base J66 intra-Early Volgian, '*Hudlestoni*' MFS, topmost 'H Unit'. This is defined by bioevent M1a, which marks the top of abundant of diverse nasselarian and spumellarian radiolarian assemblages including *Praeconocaryomma hexagona*, *Spongodiscus* sp. 4 Dyer & Copestake and *Orbiculiforma mclaughlini*, together with a marked downhole increase in numbers of the agglutinated foraminifer *Trochammina* cf. *lathetica*.

One additional shale prone unit, the 'F Unit' occurs within J64. It separates 'F' and 'E Unit' sands from 'G Unit' sands (Fig. 4). The 'F Unit' is generally heterolithic, variably comprising shales, siltstones and fine sandstones, and is interpreted as a widespread interturbidite deposit, possibly associated with a higher order flooding surface. The prominence of spumellarian over nassellarian radiolaria supports this, the latter being characteristic of deeper water, hemipelagic shales in Magnus (e.g. 'H Unit'). This may correlate with one of the additional regional maximum flooding surfaces now recognized in the Early Volgian (e.g. 'base J64b MFS; Copestake & Partington pers. comm.)

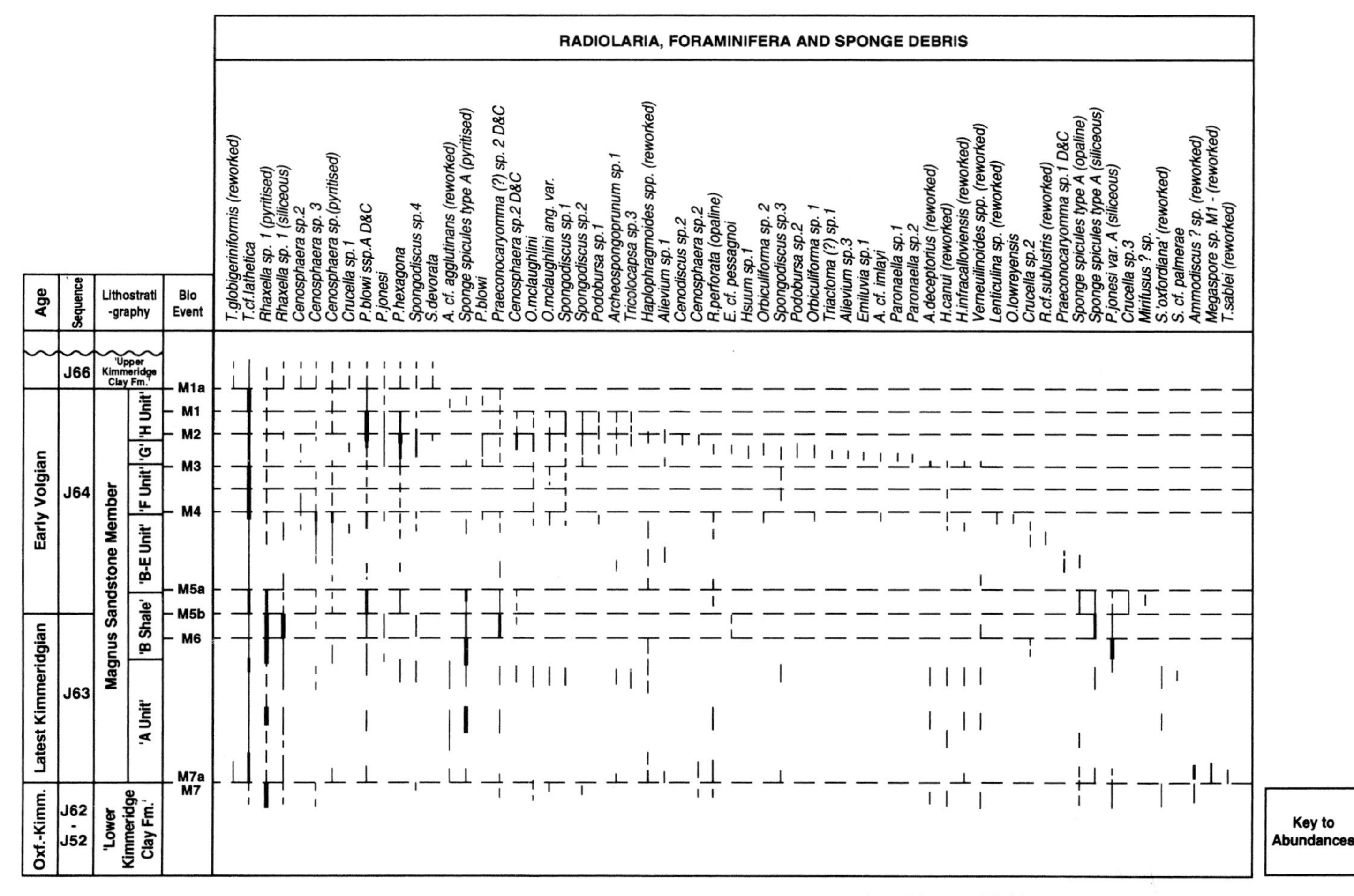

Fig. 3. Generalized stratigraphic distribution of microfaunas through the Magnus Sandstone Member, Magnus Field.

Micropalaeontology of the Magnus Sandstone Member

Micropalaeontological methods and materials

The micropalaeontological evaluation of the Magnus Sandstone Member was based on the analysis of conventional core and ditch cuttings collected from 27 Magnus Field wells (see Fig. 1 for well locations). Sampling points in individual wells were closely tied to wireline logs and sedimentological core descriptions which highlighted slumped, faulted and reworked horizons. Where possible cores were sampled at less than 1 m intervals and cuttings less than 6 m.

Conventional micropalaeontological processing methods were employed, using hydrogen peroxide for sample dissaggregation of core chips and cuttings. Sample residues were picked in the usual manner, but with special emphasis on the the fine-mesh fractions (125 and 63 μm) to optimize radiolarian recovery. Semi-quantitative analysis of picked samples was undertaken which provided species relative abundance trends for all study wells.

General microfaunal character

The generalized stratigraphic distribution of Magnus Sandstone Member microfaunas is presented in Fig. 3. Microfaunas which are clearly not reworked from older formations (e.g. the Heather Formation) fall into three categories: (1) agglutinated foraminifera, essentially a monospecific fauna of *Trochammina* cf. *lathetica*; (2) sponge associated microfossils, comprising *Rhaxella* sp. 1 and sponge spicules – preservation type differentiated into opaline, siliceous and pyrite; and (3) radiolaria – preservation type differentiated into siliceous and pyrite.

The first two groups comprise few species, although these are numerically important and are often dominant at a number of levels within the reservoir (e.g. 'B Shale'). In contrast radiolarian assemblages are more diverse, particularly in the upper reservoir section, with over 40 species/subspecies identified. Radiolarian species present fall mainly into two taxonomic groups: spumellarian (e.g. *Cenosphaera*, *Praeconocaryomma*, *Orbiculiforma*) and nasselarian (e.g. *Parvicingula*, *Stichocapsa*), and appropriately display a northern Boreal character (*sensu* Pessagno *et al.* 1984). Many of the taxa present are seen widely throughout the Kimmeridge Clay Formation and equivalent units (Farsund–Draupne formations) of the Central North Sea (Dyer & Copestake 1989). In Magnus, however, a greater diversity is seen with many undescribed species in evidence, which may reflect the northward increase in Boreal oceanic influence in Late Jurassic times.

The modern and ancient ecological preferences for oceanic, flysch-associated environments of deposition exhibited by the radiolarian group supports the sedimentological evidence for a submarine fan setting for the deposition of the Magnus Sandstone Member.

Definition of bioevents within the Magnus Sandstone Member

The main advance in Magnus Sandstone Member biostratigraphy involved the revision of in-house radiolarian taxonomy which provided a basis for a more objective speciation and a better understanding of species ranges and abundance trends within the reservoir sequence. Unfortunately, owing to the paucity of publications on Late Jurassic Boreal radiolaria, many of the taxa described remain informal although this has not diminished their value for reservoir correlation.

A summary of the key microfaunal bioevents is shown in relation to the Magnus Sandstone Member lithostratigraphy in Fig. 4. A description of these biovents is presented below.

M1a bioevent. Recognition of the M1a microfaunal event is based on a marked downhole increase in *Trochammina* cf. *lathetica* which consistently occurs above the main influx of radiolaria within the 'H Unit'. The top of the M1a event coincides with the log break for the top 'H Unit', which is characterized by a lower gamma ray–higher sonic velocity response, compared to the overlying 'Upper Kimmeridge Clay Formation' (Fig. 4). From Fig. 4 it can be seen that *Trochammina* cf. *lathetica* is normally present in all interlobe mudstone units. In addition to *T.* cf. *lathetica*, low numbers of radiolaria range into the uppermost 'H Unit', with *Parvicingula blowi* ssp. A Dyer & Copestake and *Parvicingula jonesi* (siliceous and pyritized) occurring commonly.

The M1a event has been defined in the majority of wells where a complete 'H Unit' sequence is preserved. As with underlying events, however, the reliance on cuttings has meant that recognition can be adversely affected by caving. Furthermore, the *T.* cf. *lathetica* assemblage clearly constitutes a facies fauna whose distribution must be closely linked to substrate conditions and/or lithology type. However, the consistent occurrence of M1a above the radiolarian M1 event suggests that the event is of correlative value within the Magnus Field, indeed this bioevent may be of regional significance, characteristizing the intra- Early Volgian J64 sequence of Partington *et al.* (1993*a*) (see below).

M1 bioevent. The M1 event marks the top prominent occurrence of radiolarian assemblages with the appearance of dominant *Parvicingula blowi* spp. A Dyer & Copestake and common *Spongodiscus* spp. (Figs 3 and 4). Other radiolarian species making their appearance include consistent *Parvicingula jonesi* (pyrite and siliceous preservation), consistent *Praeconocaryomma hexagona* and rare *Cenosphaera* sp. 2 Dyer & Copestake, *Orbiculiforma mclaughlini*, *O. mclaughlini* angular variety, *Archeospongoprunum* spp. and *Tricolocapsa* sp. 3. *Trochammina* cf. *lathetica* is numerically subordinate to radiolaria in assemblages marking this event. The assemblages appears to correlate with the top of the MJ21 biozone, of Early Volgian age, although the index taxon *Orbiculiforma lowreyensis* is not in evidence in Magnus (see below).

The M1 event generally occurs in the mid to upper part of the 'H Unit' (Fig. 4) where this is preserved beneath the base Cretaceous unconformity.

M2 bioevent. The M2 event is defined on the top range of prominent *Praeconocaryomma hexagona* and common or consistent *Cenosphaera* sp. 2 Dyer & Copestake, *Orbiculiforma mclaughlini* and *O. mclaughlini* angular variety. Numerous other radiolarian taxa occur at this level (Fig. 3) with the event marking the penetration of extremely diverse radiolarian assemblages.

The event occurs in the mid to lower part of the 'H Unit' (Fig. 4). Diverse radiolarian assemblages continue down into the 'G Unit' with a faunal turnover at or near the top of the 'F Unit'.

The M2 bioevent is a highly reliable marker which can be tied regionally to the intra-Early Volgian, top J64–base J66 '*Hudlestoni*' maximum flooding surface (see below).

M3 bioevent. The M3 event is picked on the reappearance of consistently predominant *Trochammina* cf. *lathetica*. A reduction in diverse radiolarian assemblages is also a notable feature which can be related to a major reduction in hemipelagic sedimentation. The M3 event is associated with the mid to upper 'F Unit' mudstone (Fig. 4) which is interpreted as a widespread interturbidite deposit. Facies control on this bioevent is apparent as it can also occur within interturbiditic mudstones in the overlying 'G Unit'.

Whilst the M3 event has been defined in the majority of study wells, in many instances the pick is questionable due to truncation of the upper reservoir section beneath the 'H Unit', Cromer Knoll Group or Shetland Group. In such cases, recognition of the associated and underlying M4 event is required to confidently define the 'F Unit' mudstone.

M4 bioevent. The event is marked by the appearance of low-diversity radiolarian assemblages comprising common to prominent pyritized *Cenosphaera* sp. 3 and *Cenosphaera* sp. (smooth). *Trochammina* cf. *lathetica* continues to be a major component of assemblages of this event. Rarely the appearance of *Orbiculiforma lowreyensis* is associated, this species displaying its regional extinction within the Early Volgian.

The M4 event is associated with mudstones from the mid to lower 'F Unit' (Fig. 4). As with the M3 event, definition of this event can be affected by truncation of the upper reservoir section.

M5a bioevent. A reappearance of pyritized radiolarian assemblages together with pyritized and opaline–siliceous *Rhaxella* sp.1 and sponge spicules type A is seen in many study wells immediately above the 'B Shale', at a basal 'B Unit' level or within the uppermost 'B Shale' (Fig. 4). This event is likely to represent the reworking of the 'B Shale' and older Kimmeridge Clay Formation sediments.

Radiolarian taxa occurring at this level include *Parvicingula blowi* ssp. A Dyer & Copestake, *P. jonesi*, *Praeconocaryomma hexagona* with the appearance of *Crucella* sp. 3 and *Mirifusus* sp. (both rare). The index taxon *Parvicingula jonesi* var. A (siliceous preservation) also occurs rarely.

M5b bioevent. The M5b event marks a downhole shift from radiolarian to sponge spiculite-dominated assemblages composed of prominent pyritized and siliceous *Rhaxella* sp. 1 with sponge spicules type A. The former species (pyritized) is a regional marker for the latest Kimmeridgian, top J63–base J64, '*Autissiodorensis*' maximum flooding surface (Fig. 4; see below), denoted the 'B Shale' in the Magnus Field, and as such is a key marker. Low-diversity radiolarian assemblages are associated, characterized by common *Praeconocaryomma*(?) sp. 2 Dyer & Copestake.

M6 bioevent. The M6 event is defined by the top range of prominent *Parvicingula jonesi* var. A

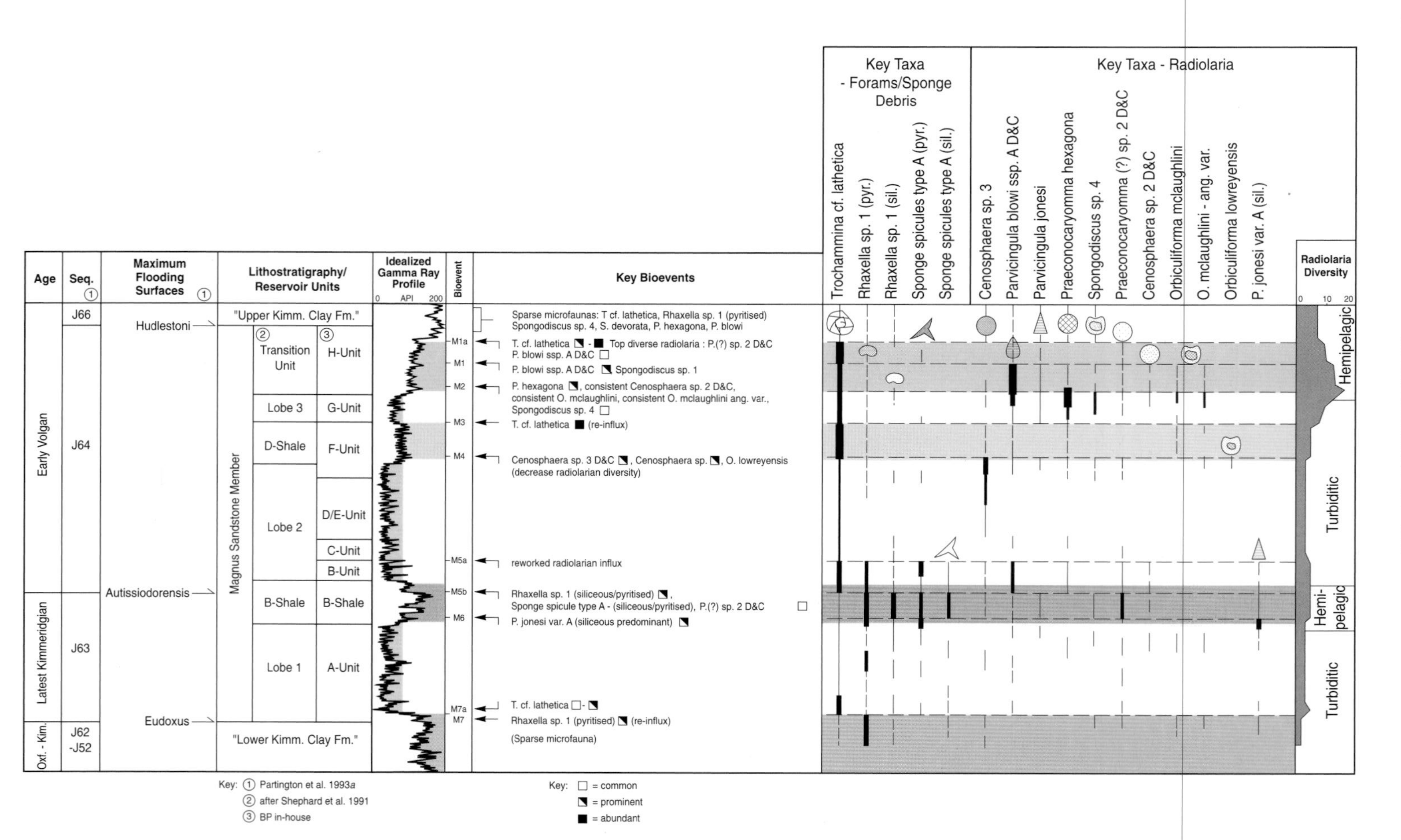

Fig. 4. Magnus Sandstone Member reservoir chronostratigraphy and key microfaunal bioevents.

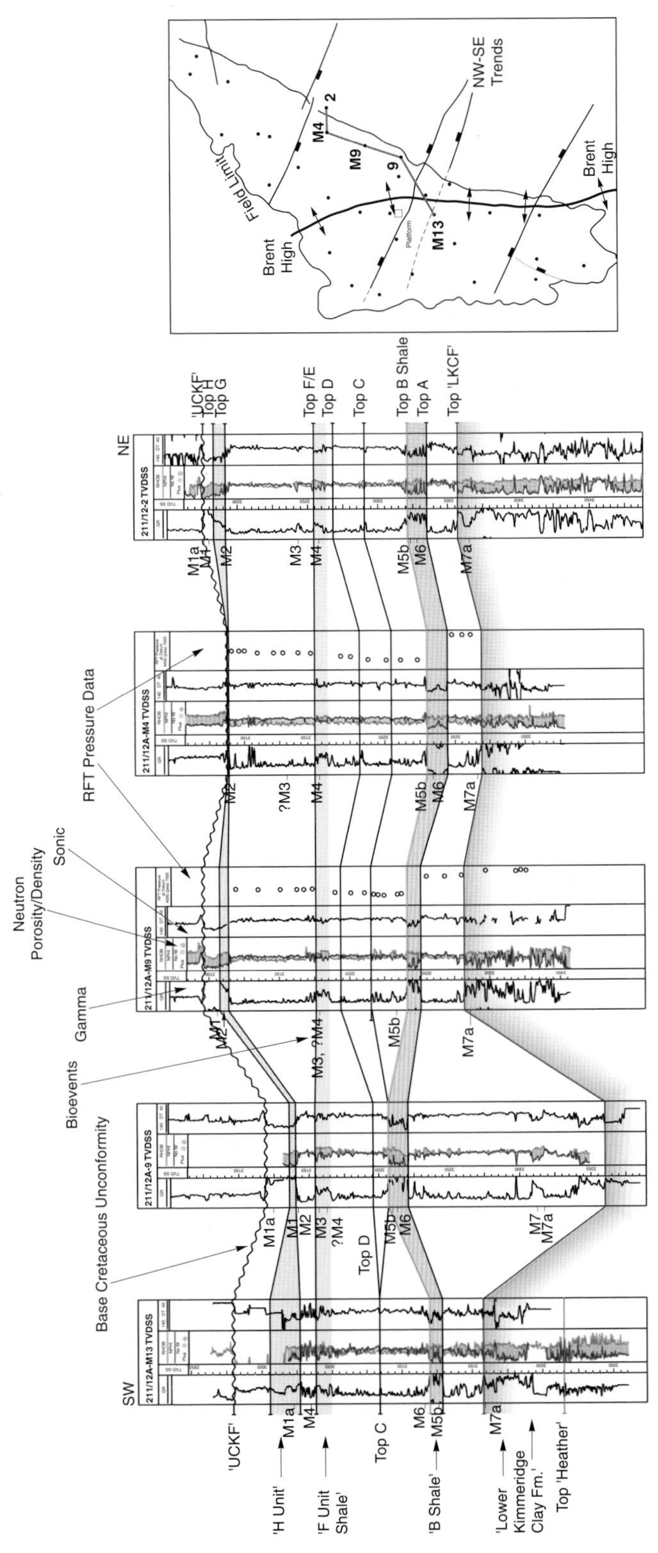

Fig. 5. Magnus Sandstone Member – bioevent-constrained correlation in the central field area.

with siliceous preservation. This event may be associated with the J64 maximum flooding surface which resulted in the switchover to hemipelagic from turbiditic deposition throughout the fan system (see below). *Rhaxella* sp. 1/sponge spicules do, however, continue to occur abundantly, ranging throughout the 'B Shale' (Fig. 4). The M6 event is associated with the mid to lower portion of the 'B Shale' (Fig. 4).

M7a bioevent. The event is defined on the base common to prominent occurrence of *Trochammina* cf. *lathetica,* with high numbers of this taxon being associated with the onset of Magnus Sandstone Member submarine fan deposition. Definition of this event can, however, be adversely affected by cavings which can 'depress' its occurrence when working with cuttings.

M7 bioevent. The M7 event is defined on the reappearance of common to prominent pyritized *Rhaxella* sp. 1 with minor siliceous spiculite (Fig. 4). The event appears to coincide with the top 'Lower Kimmeridge Clay Formation' log break (Fig. 4). Problems in defining the event relate to caving from the overlying 'B Shale' especially where this onlaps the 'Lower Kimmeridge Clay Formation'.

Effects of gravity flow processes on microfossil distribution

Reworking of sediments, markedly evident from the palynofloras, is also expressed microfaunally in the sporadic occurrence of Heather Formation agglutinated foraminifera (e.g. *Ammobaculites deceptorius*, *Haplophragmoides infracalloviensis*) and 'Lower Kimmeridge Clay' radiolaria (Fig. 3). Some patterns of reworking can be defined from the distribution of these microfaunas and reworked palynomorphs. For example, 'A Unit' sandstones of latest Kimmeridgian age consistently display 'Lower Heather Formation' reworking whereas Early Volgian 'B–F' and 'G Unit' sandstones contain mixed Heather and 'Lower Kimmeridge Clay Formation' material. This distinction may relate to a point source, focused canyon-fill origin for sandstones of the 'A Unit' whereas the younger lobes may have been sourced from multiple feeder systems emanating from the North Shetland Platform (see subsection on 'Reservoir geometry and submarine fan evolution').

Sedimentology highlights reworking within the Magnus Sandstone Member interlobe mudstones as evident from the presence of slump structures, debris flow deposits, intraclasts and conglomerates (De'Ath & Schuylemann 1981; Shepherd *et al.* 1991). Even where sequences of undisturbed laminated mudstones exist (e.g. 'B Shale' and 'F Unit') these may be interpreted as low-density turbidites as opposed to hemipelagic sediments (e.g. as in Shepherd *et al.* 1991). The extent to which radiolarian occurrences in these mudstones resulted directly from hemipelagic sedimentation is debatable: other events, however (e.g. M5b spiculite), clearly involved downslope transportation as component taxa have a shallow-marine shelfal origin.

Regardless of the depositional mechanism, a high degree of consistency exists in the sequential occurrence of bioevents, and their association with discrete lithostratigraphic units within the field, providing a reliable stratigraphic framework for reservoir correlation.

Role and impact of biostratigraphy in Magnus Sandstone Member reservoir modelling

Biostratigraphy is a key tool in the reservoir description of the Magnus Sandstone Member, acting as the template for integrating seismic, wireline log and RFT data. In equivocal situations in which no definitive shale to shale correlation is apparent from log or seismic data, biostratigraphy often provides an answer, or adds to the weight of evidence supporting one particular model. It is thereby a valuable 'uncertainty reduction' tool.

The most fundamental role of biostratigraphy involves the erection of a stratigraphic framework which, whilst being drawn from broader regional biozonation schemes, is of higher resolution and is essentially field-specific. An assumption is made that regionally diachronous bioevents may be isochronous on a local scale and may be used with caution; likewise allochthonous bioevents, reflecting phases of sediment reworking, may also be viable on the field scale.

A series of correlatable bioevents are thereby used to characterize individually the key reservoir-enveloping and intra-reservoir shales. These shales often denote flooding surfaces that result in the synchronous cut-off of coarse clastic supply to the Magnus fan system, and their recognition allows the reservoir to be subdivided into collections of sand packages bracketed by discrete shale units. Through this an enhanced understanding of temporal and spatial reservoir heterogeneity has been gained in addition to greater clarity in the modelling of the distribution and continuity of pressure barrier shales.

At the well site biostratigraphic monitoring of the reservoir sequence has proved to be a valuable aid to operations where accuracy is required in casing-point determinations and geosteering. In the former it has often been necessary to case-off within the 'B Shale' prior to drilling on into the high-pressured 'A Unit' sandstone (see below). In such situations a well-site biostratigrapher is deployed to determine the sequence of bioevents down to M5b. Thus, together with the 'measurement while drilling' (MWD) logs, reliable determination of the top 'B Shale' can be achieved. To optimize production from depleted reservoir units high-angle drilling has been increasingly applied in the Magnus Field, and it is here also that well-site biostratigraphy assists operations through the stratigraphic calibration of supra-reservoir angle-build to ensure optimal reservoir entry and ongoing monitoring within a discrete reservoir unit.

Fluid pressure distribution

Figure 5 illustrates a bioevent-constrained correlation in the down-dip area of the centre of the field using five wells, including development wells 211/12a-M9 and 211/12a-M4, which show RFT pressure data through the reservoir. These data highlight the vertical changes in formation pressure depletion across the 'B Shale' and 'F Unit' shale. It can be seen that the 'B Shale' in particular forms a major pressure barrier with the underlying 'A Unit' sand, displaying 1500–2000 psi less depletion. Locally the 'F Unit' shale can also form a significant pressure barrier. Its greater heterogeneity and less well-defined biostratigraphic characterization can render its recognition and prediction difficult.

Mapping the distribution of shale-bounded pressure compartments has proven vital to the location of infill water injector and producer wells for the effective 'sweeping' of the Magnus reservoir. The role of wellsite biostratigraphy can be visualized where there is an operational requirement to pick a casing point in the upper 'B Shale', prior to drilling into 'A Unit' sandstones, which can be at a higher pressure (as in this case) depending on the degree of compartmentalization and the phase of field development.

Reservoir geometry and submarine fan evolution

The accurate definition and correlation of reservoir-encasing shales has provided a means for mapping sandstone distribution on a time-slice basis as the deposition of sandy fan lobes is likely to have been geologically instantaneous when compared to the main interlobe shales. The availability of 3D seismic-derived structural maps provides further constraint, and together an accurate model for submarine fan evolution can be constructed (Fig. 6a–c).

Figure 6a illustrates 'A Unit' sandstone distribution in the latest Kimmeridgian, J63 sequence, which is constrained by study well and 3D seismic data. Compared with the overlying Magnus Sandstone Member reservoir units, the 'A Unit' is genetically distinct in displaying a single central depocentre with a possible fault-controlled feeder system trending NW–SE across the Brent High. The geometry of the 'A Unit', most clearly seen in 211/12a-9 in the axis of 'A Unit' deposition (Fig. 5), is consistent with the development of a single submarine fan lobe, sourced from the northwest and showing partial fault confinement. As such it is modelled here as a laterally confined canyon-fill (Fig. 6a).

Termination of 'A Unit' deposition is likely to have occurred in the latest Kimmeridgian and can be tied regionally to the J64 maximum flooding surface event (Fig. 4; see above): this resulted in the deposition of transgressive 'B Shale' mudstones field-wide. The subsequent resumption of sand deposition of the 'B–F Units' in the Early Volgian followed a markedly different pattern (Figs 5 and 6b). Two depocentres are evident positioned either side of the Brent High, suggesting deflection of deposition around this structure. Log correlation within these sand units in the southwest field sector indicates older infill of topographic lows and onlap onto the Brent High and to the west.

Termination of sand supply resulted in the widespread interturbidite deposition of the 'F Unit' shale. Lateral depositional variation in the succeeding 'G Unit' sandstone is difficult to assess due to Early–mid Cretaceous erosion/reworking of the upper reservoir sequence as evident from the decreasing thickness of this unit, east to west throughout the field (Fig. 6c). Nevertheless, one depocentre remains 'preserved' in the northeast with the remnants of a second, as before, positioned west of the Brent High, indicating that the influence of this structure persisted until the termination of submarine fan deposition in Magnus.

Shutting down of clastic supply to the fan system appears closely linked to the '*Hudlestoni*' regional flooding event within the Early Volgian (base J66 MFS of Partington *et al.* 1993*a*), which led to a field-wide, fining-upward sequence, culminating in the hemipelagic deposition of the

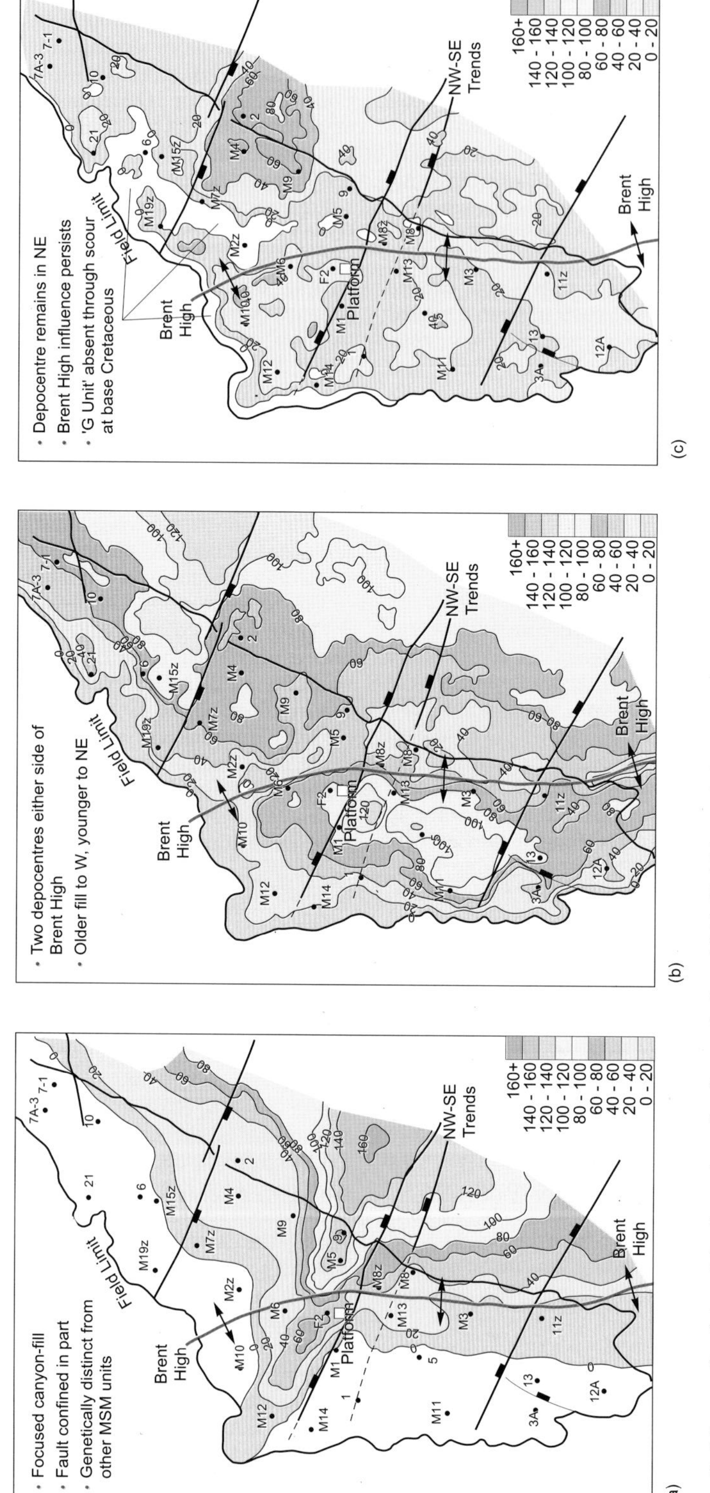

Fig. 6. Magnus Sandstone Member submarine fan evolution: **(a)** 'A Unit' isopach, latest Kimmeridgian (J63 part); **(b)** 'B–F Units' isopach, Early Volgian (J64 lower); and **(c)** 'G Unit' isopachs, Early Volgian (J64 upper).

'H Unit' mudstone and overlying 'Upper Kimmeridge Clay Formation'.

Discussion and conclusions

The stratigraphic evaluation of the Magnus Sandstone Member in the Magnus Field has highlighted both the difficulties and the solutions of working on a complex submarine fan system. The problems encountered such as high net to gross, rapid deposition and reworking are attendant to some degree in all such reservoirs, and Magnus exhibits all these facets particularly well. The arrival at a workable bioevent scheme underpins first the importance of a rigorous taxonomic approach to applied biostratigraphy, drawing on academic research advances where appropriate, as for example in the case of radiolarian speciation. Secondly, it shows that even facies-controlled and resedimented elements such as *Trochammina* cf. *lathetica*, *Rhaxella* and sponge spicules can have stratigraphic value at a local scale and should not be ignored. Having arrived at a workable scheme, the limitations should be borne in mind when applied to incomplete or poorly sampled well sections. In Magnus the weaknesses in the dataset are obviated through the use of biostratigraphy as an 'uncertainity reduction tool', where the definition of the 'F Unit' shale in a well, for example, also takes account of wireline log, RFT and seismic data.

Using the above approach to production and development biostratigraphy in Magnus, it can be concluded that micropalaeontology has proven to be a vital stratigraphic tool, with the microfaunas, especially radiolarian taxa, displaying sensitivity to both regional (tectono-eustatic) and localized palaeoecological controls. The high diversity of radiolarian assemblages in the upper reservoir section can be linked to palaeolatitude suggesting that the group has significant stratigraphic potential in other high-latitude, Late Jurassic, deep-marine prospects.

The microfaunal bioevents determined within the Magnus Sandstone Member allow a broad chronostratigraphic division with 'A Unit' sandstones and the 'B Shale' being assigned to the latest Kimmeridgian, J63 sequence and the remaining 'B–H Units' ascribable to the Early Volgian, J64 sequence. The biostratigraphy of the 'B Shale' and topmost 'H Unit', and their transgressive stratigraphic character in the field, indicate that these units represent the J64 '*Autissiodorensis*' and J66 '*Hudlestoni*' maximum flooding surfaces, respectively: cessation of sand supply to the Magnus submarine fan system corresponds, therefore, to times of maximum transgression.

In characterizing all the reservoir-encasing shale units the biostratigraphy has had significant impact on reservoir modelling through the more effective correlation of pressure barriers (i.e. 'B Shale' and 'F Unit' shale) and through the differentiation of reservoir sand lobes into time-slice isopachs, which when combined with 3D seismic structural mapping provide an accurate insight into the evolution of the Magnus submarine fan system. Thus, the main reservoir fairways can be defined for the more effective placement of infill producer and water injector wells.

The future challenges in late field life will undoubtedly involve biostratigraphy in further increasing the stratigraphic resolution within the Magnus Sandstone Member reservoir: this may lead to the finer definition of heterogeneities within the existing pay zone, and more predictive models of the dimensions of shale bodies, further enhancing the placement of producer/injector wells and increasing sweep efficiency.

The authors are indebted to BP Exploration Co. Ltd, and Magnus Field partners Agip UK, Nippon Oil UK Ltd, Petrobras (UK) Ltd and Talisman Energy (UK) Ltd, for permission to publish this paper. Thanks are also due to the BP Magnus Asset Sub-Surface Team for technical advice and provision of field data. David Ewen and Dr Bob Jones, both of BP, and Dr Phil Copestake, IEDS Ltd, are thanked for constructive criticism of the manuscript.

The biostratigraphic analyses referred to in this paper were originally undertaken by Peter Morris whilst at GeoStrat Ltd; preparation of samples and technical discussion and support at that time is duly acknowledged.

Finally, thin-section and SEM photography of species included in this paper were undertaken in the Department of Geology and Petroleum Geology, University of Aberdeen under the supervision of Dr Mike Simmons who is thanked for his time and organizational skills.

Appendix: notes on key microfaunal species

a: Radiolaria – Spumellariina

Orbiculiforma mclaughlini Pessagno 1977 (Fig. 7p and q; Fig. 8a) – this species ranges throughout the Kimmeridgian–Late Volgian of the North Sea (Dyer & Copestake 1989). In Magnus this species ranges throughout the Magnus Sandstone Member, occurring more abundantly in the 'H Unit'.

Orbiculiforma mclaughlini Pessagno 1977, angular variety (Fig. 7m–o) – this variety is distinguished form *O. mclaughlini* in possessing a more robust test with a pronounced angular margin. In the Magnus

Fig. 7. Radiolaria (SEM micrographs): Early Volgian ('H Unit'). (**a**) *Parvicingula blowi* Pessagno ssp. A Dyer & Copestake. Well 211/12-2, 3227.75 m. (**b**) *Parvicingula blowi* Pessagno ssp. A Dyer & Copestake. Well 211/12a-M1, 3279.85 m. (**c**) *Parvicingula blowi* Pessagno. Well 211/12a-M1, 3279.85 m. (**d**) *Parvicingula jonesi* Pessagno. Juvenile specimen. Well 211/12-5 3040 m. (**e**) *Parvicingula jonesi* Pessagno. Well 211/12a-9, 3195.50 m. (**f**) *Praeconocaryomma hexagona* (Rust). Well 211/12-2, 3227.75 m. (**g**) *Praeconocaryomma hexagona* (Rust). Well 211/12-2, 3227.75 m. (**h**) *Praeconocaryomma hexagona* (Rust). Corroded specimen revealing inner pore framework. Well 211/12-2, 3227.75 m. (**i**) *Cenosphaera* sp. 2 Dyer & Copestake. Well 211/12-2, 3227.75 m. (**j**) *Cenosphaera* sp. 2 Dyer & Copestake. Well 211/12a-M1, 3279.85 m. (**k**) *Spongodiscus* sp. 4 Dyer & Copestake. Well 211/12a-M8, 3645 m. (**l**) *Spongodiscus* sp. 4 Dyer & Copestake. Well 211/12-5, 3050 m. (**m**) *Orbiculiforma mclaughlini* Pessagno, angular variety. Well 211/12a-M1, 3279.85 m. (**n**) *Orbiculiforma mclaughlini* Pessagno, angular variety. Well 211/12-2, 3236.73 m. (**o**) *Orbiculiforma mclaughlini* Pessagno, angular variety, edge view. Well 211/12a-M8, 3645 m. (**p**) *Orbiculiforma mclaughlini* Pessagno. Well 211/12a-M8, 3645 m. (**q**) *Orbiculiforma mclaughlini* Pessagno, edge view. Well 211/12a-M8, 3645 m.

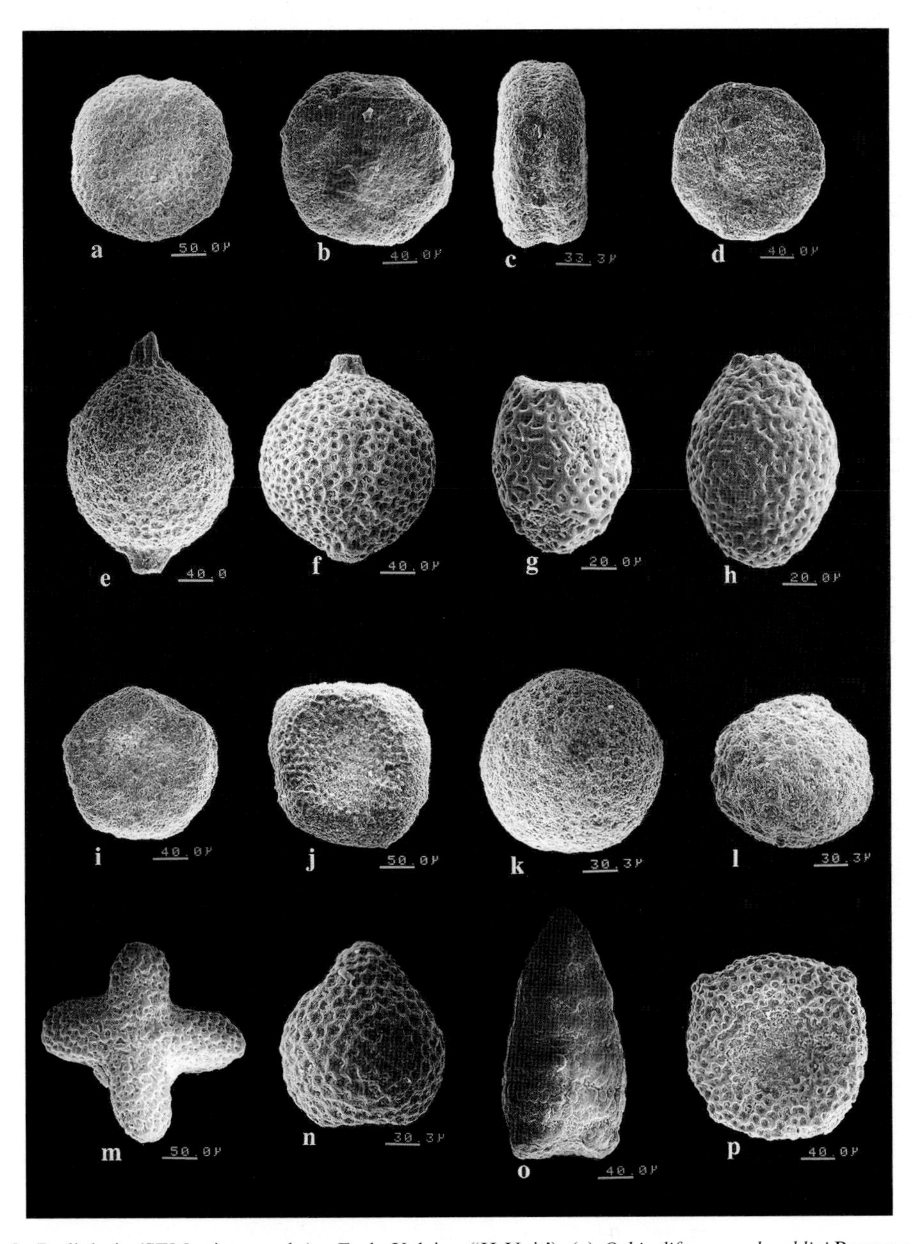

Fig. 8. Radiolaria (SEM micrographs) – Early Volgian ('H Unit'). (**a**) *Orbiculiforma mclaughlini* Pessagno. Well 211/12a-M1, 3283.48 m. (**b**) *Spongodiscus* sp. 3 Dyer & Copestake. Well 211/12a-M8, 3645 m. (**c**) *Spongodiscus* sp. 3 Dyer & Copestake, edge view. Well 211/12a-M8, 3645 m. (**d**) *Spongodiscus* sp. 2 Dyer & Copestake. Well 211/12-5, 3050 m. (**e**) *Archaeospongoprunum* sp. 2. Showing broken polar spines. Well 211/12-2, 3234.45 m. (**f**) *Archaeospongoprunum* sp. 2. Showing broken polar spines. Well 211/12-2, 3234.45 m. (**g**) *Archaeospongoprunum* sp. 1. Well 211/12-2, 3236.73 m. (**h**) *Archaeospongoprunum* sp.1. Well 211/12-2, 3236.73 m. (**i**) *Triactoma*(?) sp. 1. Well 211/12a-M8, 3645 m. (**j**) *Emiluvia* cf. *pessagnoi*. Well 211/12-2, 3227.25 m. (**k**) *Cenosphaera* sp. 2. Well 211/12a-M1, 3283.48 m. (**l**) *Cenosphaera* sp. 1. Well 211/12a-M1, 3283.48 m. (**m**) *Crucella* sp. 1. Well 211/12-2, 3236.73 m. (**n**) *Tricolocapsa* sp. 3. Well 211/12a-M8, 3645 m. (**o**) *Stichocapsa devorata* (Rust), internal cast. Well 211/12-2, 3213 m. (**p**) *Orbiculiforma*(?) sp. 2. Well 211/12-2, 3227.75 m.

Sandstone Member this variant is clearly associated with *O. mclaughlini*, occurring most abundantly in the 'H Unit'.

Orbiculiforma lowreyensis Pessagno 1977 – only a few poorly preserved specimens were recorded in Magnus where it occurs rarely in the lower 'F Unit' mudstones.

Orbiculiforma(?) sp. 2 (Fig. 8p) – test discoidal with central cavity occupying about one half of the test diameter. Test with coarse polygonal framework. Four spine bases are preserved distinguishing this species although no circumferential notch is seen. The species occurs rarely in the upper Magnus Sandstone Member.

Spongodiscus sp. 2. Dyer & Copestake 1989 (Fig. 8d) – this species is distinguished in possessing a flattened test with a thin, rounded margin in edge view. Where preserved, it possesses a fine, spongy mesh framework. This species is mostly restricted to 'H Unit' mudstones in Magnus.

Spongodiscus sp. 3 Dyer & Copestake 1989 (Fig. 8b and c) – this spongodiscid is distinguished by its thick, straight margin in edge view. The species occurs sporadically through the Magnus Sandstone Member, but is recorded commonly in the lower 'H Unit'.

Spongodiscus sp. 4 Dyer & Copestake 1989 (Fig. 7k and l). – this spongodiscid is distinguished by its shallow circumferential notch and circumferential grooves. The species occurs commonly in the lower 'H Unit': it is also recorded from the top 'Lower Kimmeridge Clay Formation'.

Praeconocaryomma hexagona (Rust 1898) (Fig. 7f–h) – this is a distinctive species which is distinguished by the presence of prominent mammae which are regularly spaced in a hexagonal pattern, with interconnecting bars. Problems in speciation can arise, however, where progressive corrosion removes part of the outer medullary layer (e.g. Fig. 7h). The abundant occurrence of *P. hexagona* in the 'H Unit' is consisent with an age no younger than Early Vogian, although in North America it ranges up into the Ryazanian equivalent (Pessagno 1977).

Praeconocaryomma(?) sp. 2 Dyer & Copestake 1989 (Fig. 9j–l) – this species is relatively small (diameter approximately 150 μm) and is distinguished by the presence of numerous, small regularly spaced mammae which give it a 'pimply' appearance when viewed under the stereomicroscope. The species ranges throughout the Magnus Sandstone Member, with an acme occurrence in the 'B Shale', making it a useful field marker.

Cenosphaera sp. 2 Dyer & Copestake 1989 (Fig. 7i and j) – a distinctive speces of *Cenosphaera* possessing a coarse elliptical pore frame with nodes at the points of intersection. This species occurs sporadically through the mid–upper Magnus Sandstone Member, although it can occur commonly in the 'H Unit'. The species is generally considered to be an intra-Kimmeridgian marker so that its occurrence in the Early Volgian of Magnus field extends its range, if the species is *in situ*.

Cenosphaera sp. 3 (Fig. 9a–c) – this species is found mainly pyritized and is distinguished by its small size (diameter approximately 95 μm) and in possessing an extremely fine polygonal pore frame, giving a reticulate appearance under reflected light. The species is known as *Cenosphaera 'reticulata'* or *C. 'microreticulata'* within the oil industry. This radiolarian species ranges throughout the Magnus Sandstone Member with its acme occurrence in lower 'F Unit' shale making it a useful field marker.

Cenosphaera sp. 1 (Fig. 8l) – this species is distinguished by its coarse, open meshwork of hexagonal pore frames, although in the present material these are poorly preserved. The species occurs rarely through the upper Magnus Sandstone Member and top 'Lower Kimmeridge Clay Formation'.

Cenosphaera sp. 2 (Fig. 8k) – this species is distinguished from *Cenosphaera* sp. 1 in possessing a finer meshwork of hexagonal pore frames. This species occurs frequently in the lower 'H Unit' and 'F Unit' shale.

Archaeospongoprunum sp. 1 (Fig. 8g and h) – this species displays an inflated, ellipsoidal test with an irregular, coarse pore frame having elliptical to elongate pores. The polar spines are not preserved in the Magnus material. The species occurs sporadically through the Magnus Sandstone Member.

Archaeospongoprunum sp. 2 (Fig. 8e and f) – this form is distinguished by its inflated, subspherical test with polygonal pore frame with circular to elliptical pores. Polar spines are partially preserved in the present material, showing a tetraradiate cross-section with prominent ridges and groves. The species occurs sporadically through the Magnus Sandstone Member.

Triactoma(?) sp. 1 (Fig. 8i) – the taxonomic affinities of this species are questionable due to poor preservation. Distinguished by its pentagonal test outline. The species occurs rarely in the upper reservoir 'G Unit'.

Emiluvia cf. *pessagnoi* Foreman 1973 (Fig. 8j) – this species is distinguished by its square test with a fine rectangular pore frame possessing nodes at the points of intersection. Spines are not in evidence in the Magnus material. The species occurs rarely in the 'B Shale' and 'G Unit'.

Crucella sp. 1 (Fig. 8m) – test distinctly cruciform with trigonal pore frame enclosing circular to elliptical pores. Spines not in evidence in the Magnus material. The species occurs sporadically in the upper Magnus Sandstone Member – 'Upper Kimmeridge Clay' Formation.

b: Radiolaria – Nassellariina

Parvicingula blowi Pessagno 1977 (Fig. 7c) – distinguished from other species of *Parvicingula* by the development of massive, irregular, polygonal pore frames which extend from the cephal chamber down to the early post-abdominal chamber which associates with an absence of circumferential ridges in this segment. Variation in its preservation can make differentiation from *P. blowi* ssp. A Dyer & Copestake difficult in some instances (see below). This species is consistently present in the upper Magnus Sandstone Member, 'F' and 'H Unit' shales.

Parvicingula blowi Pessagno ssp. A Dyer & Copestake 1989 (Fig. 7a and b) – this subspecies is distinguished by the presence of massive, regular polygonal pore frames on the cephalic, thoracic and abdominal chambers, and absence of circumferential ridges on

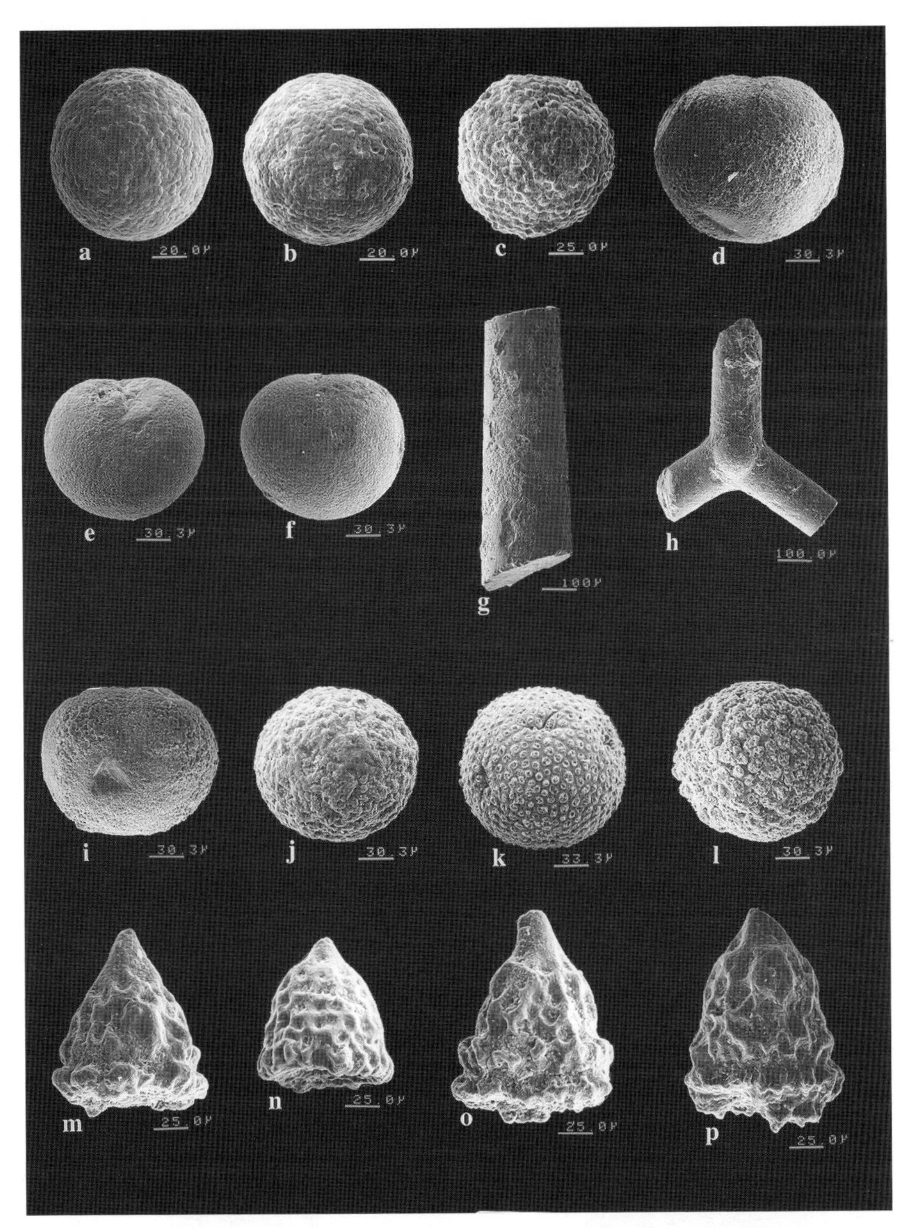

Fig. 9. Radiolaria & Sponge microfossils (SEM micrographs) – Early Volgian and Kimmeridgian ('F Unit' and 'B Shale'). (**a**) *Cenosphaera* sp. 3. Well 211/12-2, 3208 m. (**b**) *Cenosphaera* sp. 3. Well 211/12-2, 3208 m. (**c**) *Cenosphaera* sp. 3. Well 211/12-2, 3208 m. (**d**) *Rhaxella* sp. 1. Well 211/12a-M8, 3774 m. (**e**) *Rhaxella* sp. 1. Well 211/12a-M8, 3774 m. (**f**) *Rhaxella* sp. 1. Well 211/12a-M8, 3774 m. (**g**) Sponge spicule type A (broken). Well 211/12a-M8, 3774 m. (**h**) Sponge spicule type A (broken). Well 211/12a-M1, 3266.00 m. (**i**) *Rhaxella* sp. 1. Well 211/12a-M8, 3774 m. (**j**) *Praeconocaryomma*(?) sp. 2 Dyer & Copestake, corroded specimen. Well 211/12-2, 3236.73 m. (**k**) *Praeconocaryomma*(?) sp. 2 Dyer & Copestake. Well 211/12-3, 3048 m. (**l**) *Praeconocaryomma*(?) sp. 2 Dyer & Copestake. Well 211/12-3, 3048 m. (**m**) *Parvicingula jonesi* Pessagno var. A. Well 211/12-2, 3202.15 m. (**n**) *Parvicingula jonesi* Pessagno var. A. Well 211/12-2, 3202.15 m. (**o**) *Parvicingula jonesi* Pessagno var. A. Well 211/12-2, 3202.15 m. (**p**) *Parvicingula jonesi* Pessagno var. A. Well 211/12-2, 3202.15 m.

the latter. In the present material a range of 'variation' is evident with many specimens appearing to lack circumferential ridges on the post-abdominal chambers (Fig. 7b). This may, however, be attributable to corrosion of the outer pore frame which raises doubts as to validity of subspeciation in some assemblages. This subspecies occurs throughout the Magnus Sandstone Member, occurring abundantly through the 'G–H Units'.

Parvicingula jonesi Pessagno 1977 (Figs 7d and e) – distinguished from *P. blowi* by the presence of a smaller pore framework on cephalic and thoracic chambers, and the separation of the abdominal and post-abdominal chambers by well-developed circumferential ridges: the test is also slimmer with the absence of a tapered distal portion. *P. jonesi* occurs consistently through the 'B Shale' and 'H Unit'.

Parvicingula jonesi Pessagno 1977 var. A (Figs. 9m–p) – this variant is distinguished by its small size (length no greater than 160 μm), with the development of only one or two post-abdominal chambers with circumferential ridges. This form probably represents stunted or dwarf variants of *P. jonesi,* reflecting local environmental conditions. This variant has a restricted range, from 'Lower Kimmeridge Clay Formation' to the 'B Shale' with its acme occurrence in the lower 'B Shale' defining the field-wide M6 event.

Stichocapsa devorata Rust 1885 (Fig. 8o) – this species is poorly represented in Magnus and is found mainly as internal casts, as figured. The species is recorded from the 'H Unit' and overlying 'Upper Kimmeridge Clay' Formation.

Tricolocapsa sp. 3 (Fig. 8n) – test consisting of a trapezoidal thorax and inflated, spherical abdominal chamber (cephal chamber not in evidence). Both chambers possessing a coarse, hexagonal pore frame. In outline the chambers are poorly defined and appear to merge which is a distinguishing feature (e.g. as compared to *Tricolocapsa* sp. 1 Dyer & Copestake). The species occurs rarely through the 'G' and 'H Units'

c: Sponge-associated microfossils

Rhaxella sp. 1 (Fig. 9d–f and i) – this informal species, which is widely known in the industry as

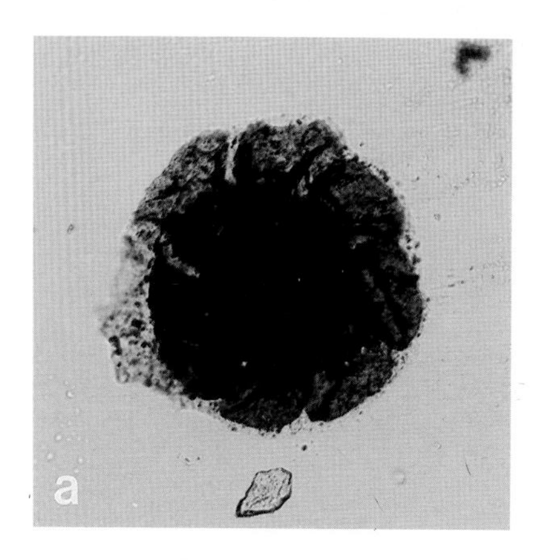

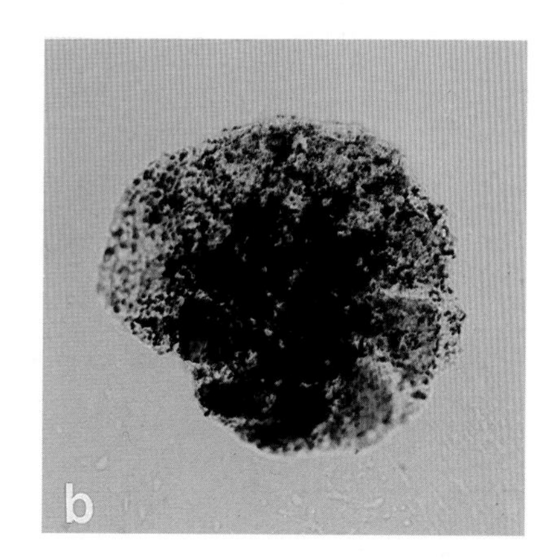

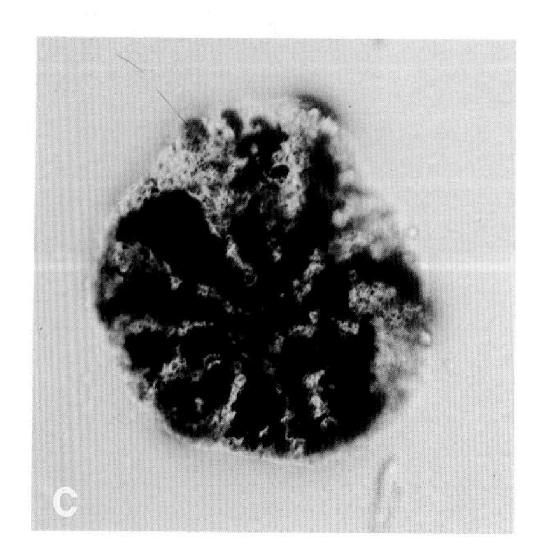

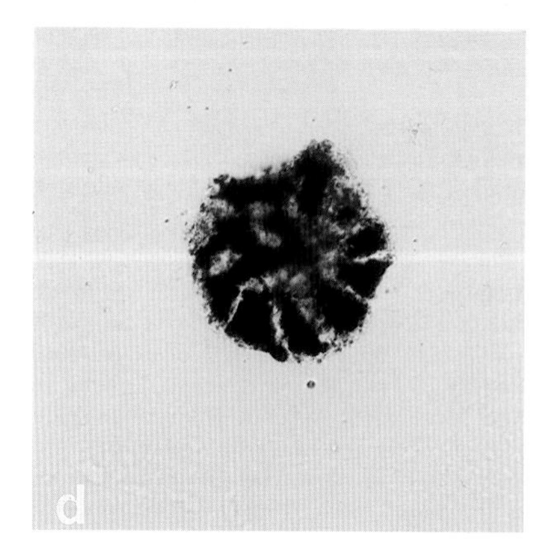

Fig. 10. Foraminifera (transmitted light, all ×450). (**a, b**) *Trochammina* cf. *lathetica* Loeblich & Tappan, showing internal organic wall lining. Well 211/12-2, 3291.85 m. (**c, d**) *Trochammina* cf. *lathetica* Loeblich & Tappan, pyritized specimens. Well 211/12a-M1, 3150.13 m.

Rhaxella 'perforata', is characterized by its opaline, siliceous or pyritized preservation, reniform outline and extremely fine perforated structure which produces a fine reticulate surface ornament. Two distinct morphotypes are evident in Magnus, consistent with its occurrence elsewhere in the Kimmeridge Clay Formation: (1) large, opaline specimens (often altered to silica), associated with sponge spicule type A (as figured) and representing a turbiditic, transported shallow-shelf assemblage; and (2) smaller pyritized forms associated with hemipelagic shales. The latter forms may have been 'floated-out' into deeper waters and thus deposited as part of the hemipelagic sediment load. Both morphotypes occur abundantly through the 'B Unit' and 'B Shale' with acme occurrences defining the M5a and M5b bioevents. Penetration of the top of the 'Lower Kimmeridgian Clay Formation' is also marked by an increase in pyritized *Rhaxella* sp. 1.

Sponge spicule type A (Fig. 9g and h) – material is highly fragmented with specimens usually consisting of smooth, cylindrical or tapered spicules (e.g. Fig. 9g), which may be opaline, siliceous or pyrite in composition. Rarely tri-radiate spicules are preserved with a node evident at the point of intersection (Fig. 9h). Sponge spicules occur sporadically throughout the Magnus Sandstone Member with significant increase in numbers in interturbidite deposits of the 'B Unit' and 'B Shale' marking the M5a and M5b events.

d: Foraminifera – Textulariina

Trochammina cf. *lathetica* Loeblich & Tappan 1950 (Figs. 10a–d) – this trochamminid is distinguished by its finely agglutinated wall with prominent chitinous inner lining, which appears dark-reddish brown due to iron staining. Chambers are arranged in three or four whorls with between eight and twelve chambers per whorl. The sutures are strongly curved backwards with chambers appearing petaloid. Tests are mostly compacted (obscuring details of aperture) and are highly variable in size, with mature specimens ranging up to 400 μm in diameter.

This species is widely known in the industry and is informally designated *T.* cf. *lathetica*. However, *Trochammina lathetica* Loeblich & Tappan is flatter, having fewer chambers per whorl (between five and six) with these being more elongate and trapezoidal in dorsal view. Therefore, the current designation, which is maintained here for consistency, is undoubtly misleading. Some affinities appear to exist between *Trochammina* cf. *lathetica* and *T. septentrionalis* Scharovskaja, described from the Siberian Late Jurassic by Dain (1972).

Trochammina cf. *lathetica* is present thoughout the Magnus Sandstone Member with acme occurrences in the 'F' and uppermost 'H Unit' mudstones.

References

DAIN, L. G. 1972. Foraminifera of Upper Jurassic deposits of Western Siberia. *Translations of the VNIGRI*, **317**, 1–466.

DE'ATH, N. G. & SCHUYLEMAN, S. F. 1981. The geology of the Magnus oilfield. *In*: *Proceeedings of the Second Conference of Petroleum Geology of the Continental Shelf of NW Europe*. Institute of Petroleum Geology, London, 342–351.

DYER, R. & COPESTAKE, P. 1989. A review of Late Jurassic to earliest Cretaceous radiolaria and their biostratigraphic potential to petroleum exploration in the North Sea. *In*: BATTEN, D. & KEEN, M. (eds) *Northwest European Micropalaeontology and Palynology*. Ellis Horwood, Chichester, 214–235.

GALLOWAY, W. E. 1989*a*. Genetic stratigraphic sequences in basin analysis. I: architecture and genesis of flooding-surface bounded depositional units. *American Association of Petroleum Geologists Bulletin*, **69**, 125–142.

——1989*b*. Genetic stratigraphic sequences in basin analysis. II: application to Northwest Gulf of Mexico Cenozoic Basin. *American Association of Petroleum Geologists Bulletin*, **73**, 143–154.

PARTINGTON, M. A., COPESTAKE, P., MITCHENER, B. C. & UNDERHILL, J. R. 1993*a*. Biostratigraphic calibration of genetic stratigraphic sequences in the Jurassic–lowermost Cretaceous (Hettangian to Ryazanian) of the North Sea and adjacent areas. *In*: PARKER, J. R. (ed.) *Proceedings or the Fourth Conference of Petroleum Geology of Northwest Europe*. Geological Society, London, 371–386.

——, MITCHENER, B. C., MILTON, N. J. & FRASER, A. J. 1993*b*. Genetic sequence stratigraphy for the North Sea Late Jurassic and Early Cretaceous: distribution and prediction of Kimmeridgian–late Ryazanian reservoirs in the North Sea and adjacent areas. *In*: PARKER, J. R. (ed.) *Proceedings of the Fourth Conference of Petroleum Geology of Northwest Europe*. Geological Society, London, 347–370.

PESSAGNO, E. A. 1977. Upper Jurassic Radiolaria and radiolarian biostratigraphy of the Californian Coast Ranges. *Micropaleontology*, **23**, 56–113.

——, BLOME, C. D. & LONGORIA, J. F. 1984. A revised radiolarian zonation of the Upper Jurassic of western North America. *Bulletin of American Paleontology*, **87**, 1–51.

SHEPHERD, M., KEARNEY, C. & MILNE, J. H. 1991. Magnus Field. *In*: BEAUMONT, E. A. & FOSTER, N. H. (compilers) *Structural Traps II – Traps Associated with Tectonic Faulting. AAPG Treatise of Petroleum Geology: Atlas of Oil and Gas Fields of the World*. American Association of Petroleum Geologists, Tulsa, Oklahoma, 95–125.

Well-site biostratigraphy of Danish horizontal wells

D. J. SHIPP

Robertson Research International Limited, Llanrhos, Llandudno, Gwynedd LL30 1SA, UK

Abstract: The number of horizontal and subhorizontal wells drilled has significantly increased over the last decade, especially during the last five years, and the use of horizontal wells in oil-field production and development is now routine. In 1987 Maersk Olie and Gas AS drilled the first horizontal well on the Dan Field. Since that time biostratigraphers at Robertson Research International Limited have been involved in Maersk's pioneering work on the development of steering techniques for horizontal wells in the Danish sector of the North Sea. Well-site biostratigraphy has been successfully used to aid in the steering of these horizontal wells.

The background studies leading to the erection of highly detailed local biozonal schemes capable of detecting changes over vertical distances as little as 1 foot are described. The logistics of how well-site biostratigraphy is carried out at the well site is explained. In addition to Dan, well-site biostratigraphy has been successfully applied to development wells from several other Danish North Sea carbonate fields operated by Maersk Olie og Gas AS on behalf of the Danish Underground Consortium (a joint venture between Shell, Texaco and AP Møller). These fields include Harald, Kraka, Gorm, Roar, Skjold, Svend, Tyra and Valdemar. Over 100 wells from Danish fields have to date been effectively steered using high-resolution biostratigraphy. Vertical targets of 10–20 feet or less have been successfully followed over distances of several thousand feet.

The author and his colleagues have been fortunate to have been associated with the Danish-based company Maersk Olie og Gas AS which has drilled over 100 horizontal wells to date, and has been in the forefront of the application of this new technology in the North Sea. As a result of this association it has been possible to develop the required techniques for well-site biostratigraphy to assist in the steering of the horizontal wells.

Geological setting

The Danish sector of the North Sea contains a number of carbonate fields (Fig. 1). Most of these, namely Dan, Gorm, Harald East, Kraka, Roar, Skjold, Svend and Tyra, have reservoirs in Danian or Maastrichtian chalks (Anderson & Doyle 1990). The exception is the large Valdemar Field where the reservoir occurs in Aptian and Barremian limestones or marls of the Sola and Tuxen Formations (Ineson 1993) (Figs 2 and 3).

Origin of horizontal well-site biostratigraphy in Denmark

The use of biostratigraphy as a tool to aid in the geosteering of horizontal wells in Denmark was developed by Maersk Olie og Gas AS on the Dan Field (Fig. 4) where the target is within the upper Maastrichtian. On the early Dan Field horizontal development wells, well-site biostratigraphy was originally used only to identify the top of the Maastrichtian. It was quickly realized, however, that without well-site biostratigraphy it was possible to re-enter the Danian, which overlies the reservoir, while drilling the horizontal part of the well. A need for well-site biostratigraphy to monitor the drilling of the horizontal section was thus identified.

This was first undertaken on MFA-13 where the end of the build-section is stratigraphically the deepest part of the well. Biostratigraphic events identified on the build-section could then be recognized again as the horizontal section moved slowly back up stratigraphy due to the gentle dip of the beds away from the centre of

SHIPP, D. J. 1999. Well-site biostratigraphy of Danish horizontal wells. *In:* JONES, R. W. & SIMMONS, M. D. (eds) *Biostratigraphy in Production and Development Geology*. Geological Society, London, Special Publications, **152**, 75–84.

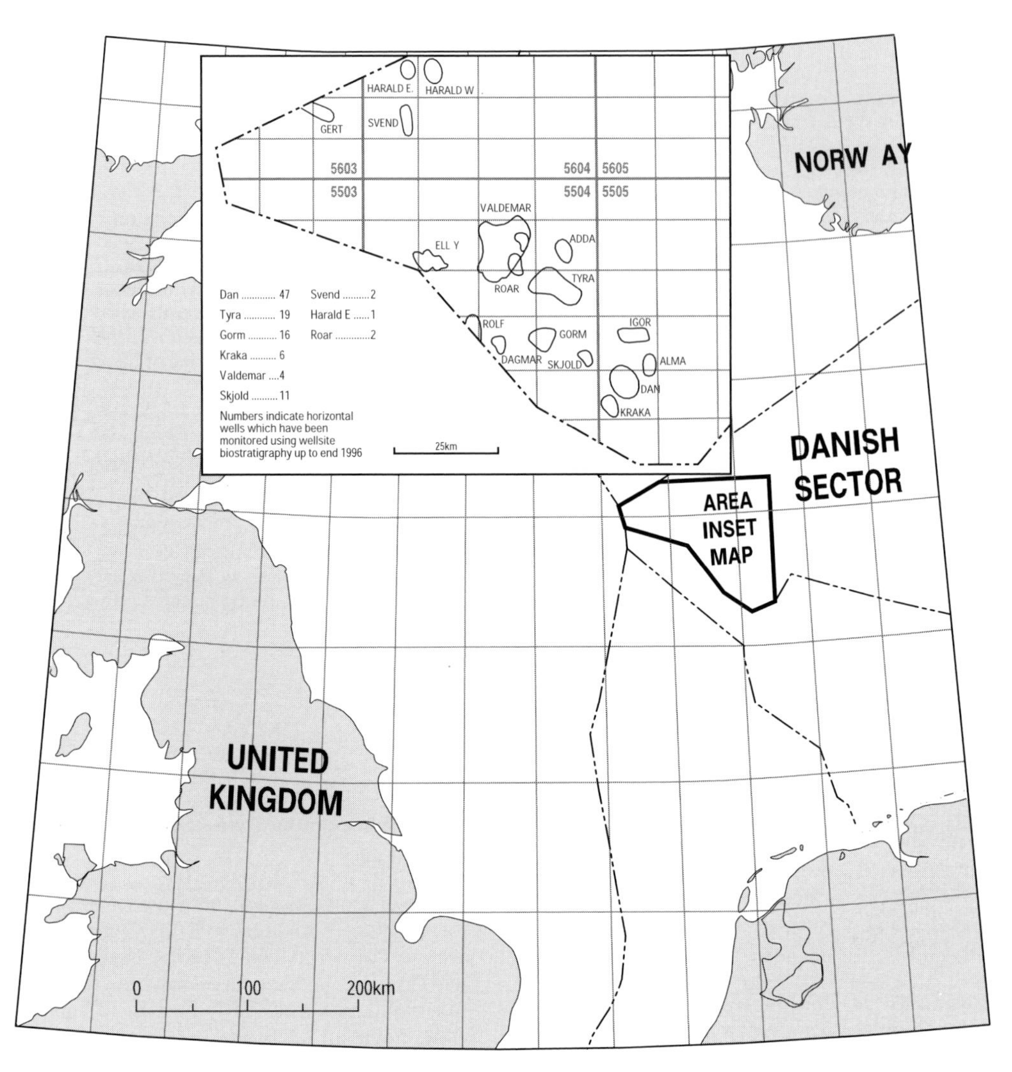

Fig. 1. Location map showing position of Danish Central Graben relative to North Sea and (inset) position of Dansk Undergrounds Consortium Danish Fields together with a list of horizontal wells monitored by biostratigraphy on each field up to the end of 1996 (modified from Shipp & Marshall 1994).

the field (Fig. 5). The crude biostratigraphic zonation of MK1A and MK1B was established while drilling MFA-13 and, although poor, the four data points noted on Fig. 6 were sufficient to monitor progress towards the base of the Danian and permit adjustments to be made to the well path to stay within the Maastrichtian.

High-resolution biostratigraphy

In order to develop a zonal scheme with the degree of refinement required to aid in the steering of a horizontal well, a re-thinking of the way zonations are established was required. Traditionally donations for the Upper Cretaceous were based on ditch cuttings analyses from routine exploration wells, where a distance of 30 feet between samples was the rule. Thus, the uppermost Maastrichtian zone on the existing zonation scheme was 50–60 feet thick (MK1 on Fig. 6).

Traditional zonal schemes based mainly on first and last downhole appearances, and relying primarily on evolutionary changes, do not provide sufficiently close-spaced data for biosteering. A new method that provided a greater number of recognizable events over a given vertical distance was required. Thus, high-resolution

AGE			FIELDS	LITHOSTRATIGRAPHIC UNITS		LITHOLOGY
TERTIARY	PALAEOCENE			ROGALAND GROUP		
TERTIARY	PALAEOCENE	DANIAN	DAN, GORM, TYRA, KRAKA, SKJOLD, HARALD SVEND ROAR	CHALK GROUP	EKOFISK FORMATION	
CRETACEOUS	UPPER	MAASTRICHTIAN		CHALK GROUP	TOR FORMATION	
CRETACEOUS	UPPER			CHALK GROUP		
CRETACEOUS	LOWER	APTIAN AND BARREMIAN	VALDEMAR	CROMER KNOLL GROUP	SOLA AND TUXEN FORMATIONS	
CRETACEOUS	LOWER			CROMER KNOLL GROUP		
JURASSIC	UPPER					

Fig. 2. Simplified stratigraphy for the Danish Central Graben carbonate fields showing age, lithostratigraphic units and lithology for the various fields (modified from Fine *et al.* 1992).

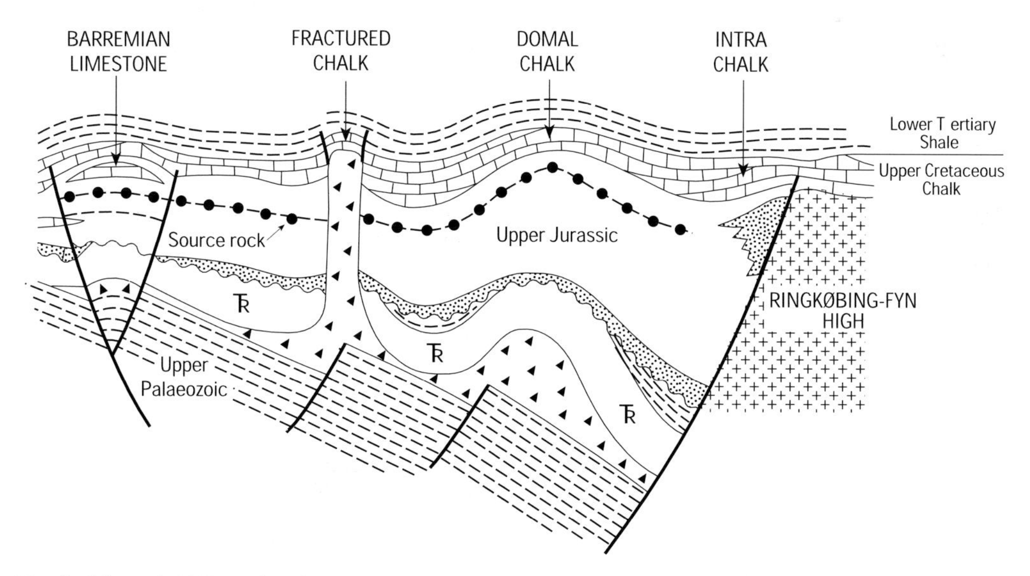

Fig. 3. Schematic E–W section through the Danish Central Graben showing geological setting and trap types of carbonate fields discussed (modified from Anderson & Doyle 1990).

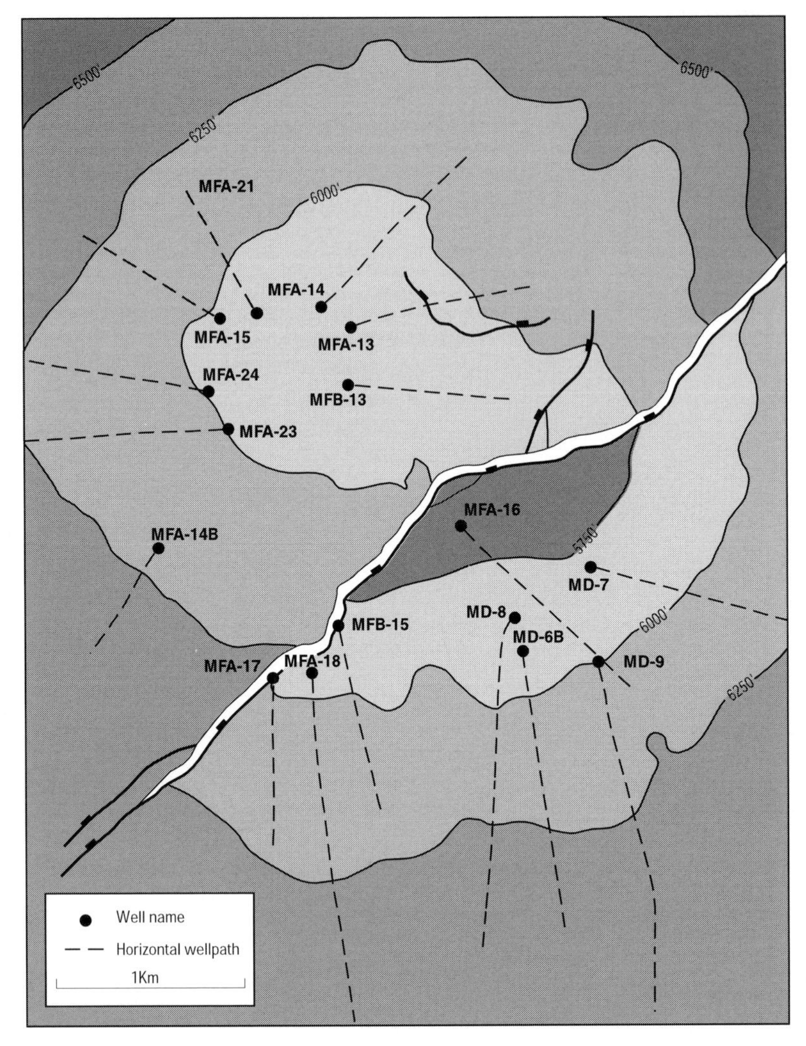

Fig. 4. Contour map of the Dan Field showing horizontal wells drilled up to January 1992. Note wells MFA-13 and MFA-18. Contours are top Maastrichtian, Tor Formation (after Fine *et al.* 1992).

biostratigraphy was developed where counts are made of all the fossils recorded to establish a refined local event stratigraphy. A high-resolution approach reveals changes in the overall composition of the microfossil assemblages which can produce much more refined zonations than presence or absence criteria and total ranges. The zonal schemes are based on local assemblage, partial range and acme biozones, and principally reflect changes in the environment rather than evolution. It might be thought that relying on environmental changes could cause problems with diachronous events. Within the confines of a single field, however, this generally presents no problem, and events can be readily correlated across the field while some can also be correlated between fields.

The first stage in the development of a zonal scheme for a particular field is the analysis of samples from offset wells. These are normally straight holes. Cores and sidewall cores provide the best database as ditch cuttings are often contaminated with cavings. Sample spacing is ideally at 10–5 feet across the target horizon, but studies on samples as close as 1 foot have been undertaken.

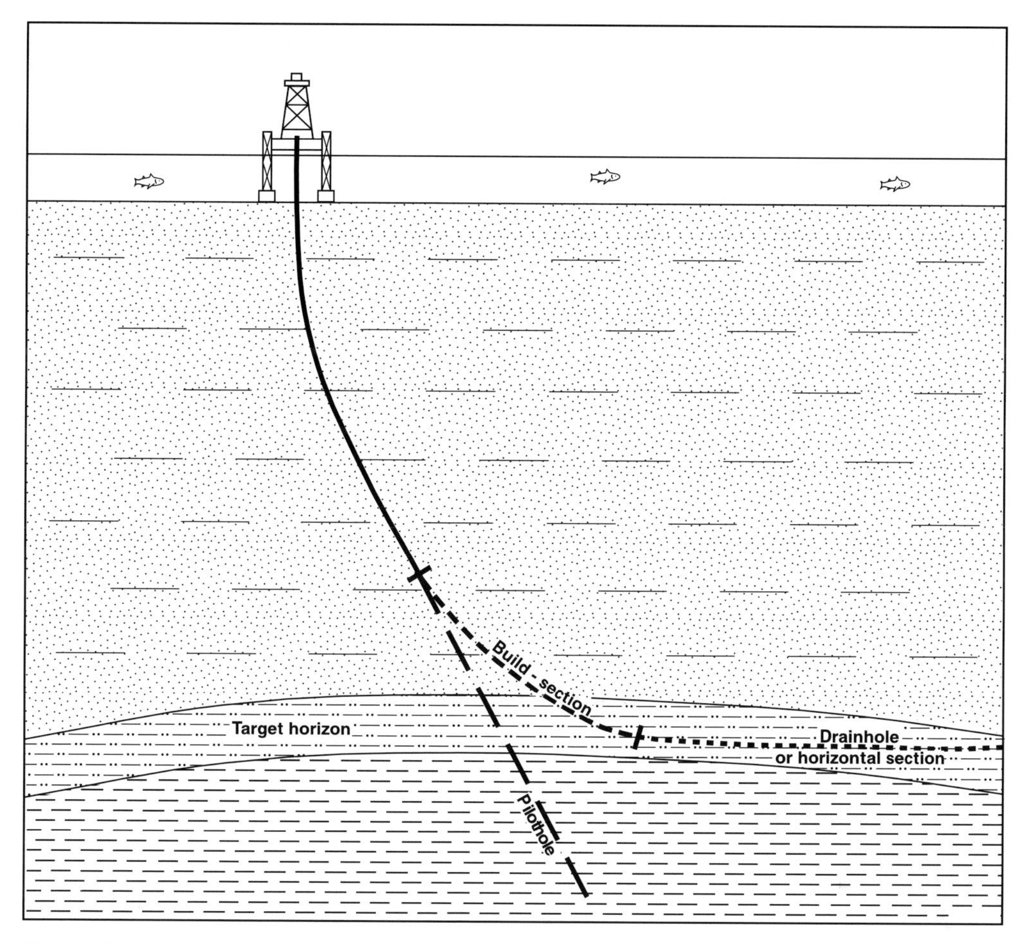

Fig. 5. Diagram of a horizontal well showing the pilot hole, build-section and horizontal or drainhole section. Structure of target horizon is as the Tor Formation on the Dan Field where the end of the build-section is stratigraphically the oldest part of the horizontal well.

A pilot study may include analyses for both micropalaeontology (foraminifera, ostracods, radiolaria and macrofossil debris) and nannofossils (coccoliths and nannoconids). In general, Maastrichtian targets can be adequately zoned by micropalaeontology while in the Danian a combination of both micropalaeontology and nannofossils proves most useful. This is also true of the Lower Cretaceous. A simplified version of a typical biostratigraphic chart showing both nannofossil and micropalaeontological zones is shown in Fig. 7.

Analyses are usually carried out in the laboratory, and a preliminary zonal scheme erected before undertaking any work at the well site. It is possible, however, to erect a zonal scheme on samples from a pilot hole drilled prior to the horizontal well. This was the case on the first horizontal well on the Danian Kraka Field where a preliminary zonal scheme was initially established from core pieces collected from the pilot hole and immediately used when the horizontal section was drilled a few days later. It should be noted that microfossil recovery from horizontal well ditch cuttings samples is generally very good with little contamination by cavings. Consequently, zonal schemes can be significantly improved at the well site especially as a 30-foot spacing between samples may only represent a difference of 1 foot in a vertical sense.

As a result of further work on the Dan Field, including the monitoring of close to 50 horizontal

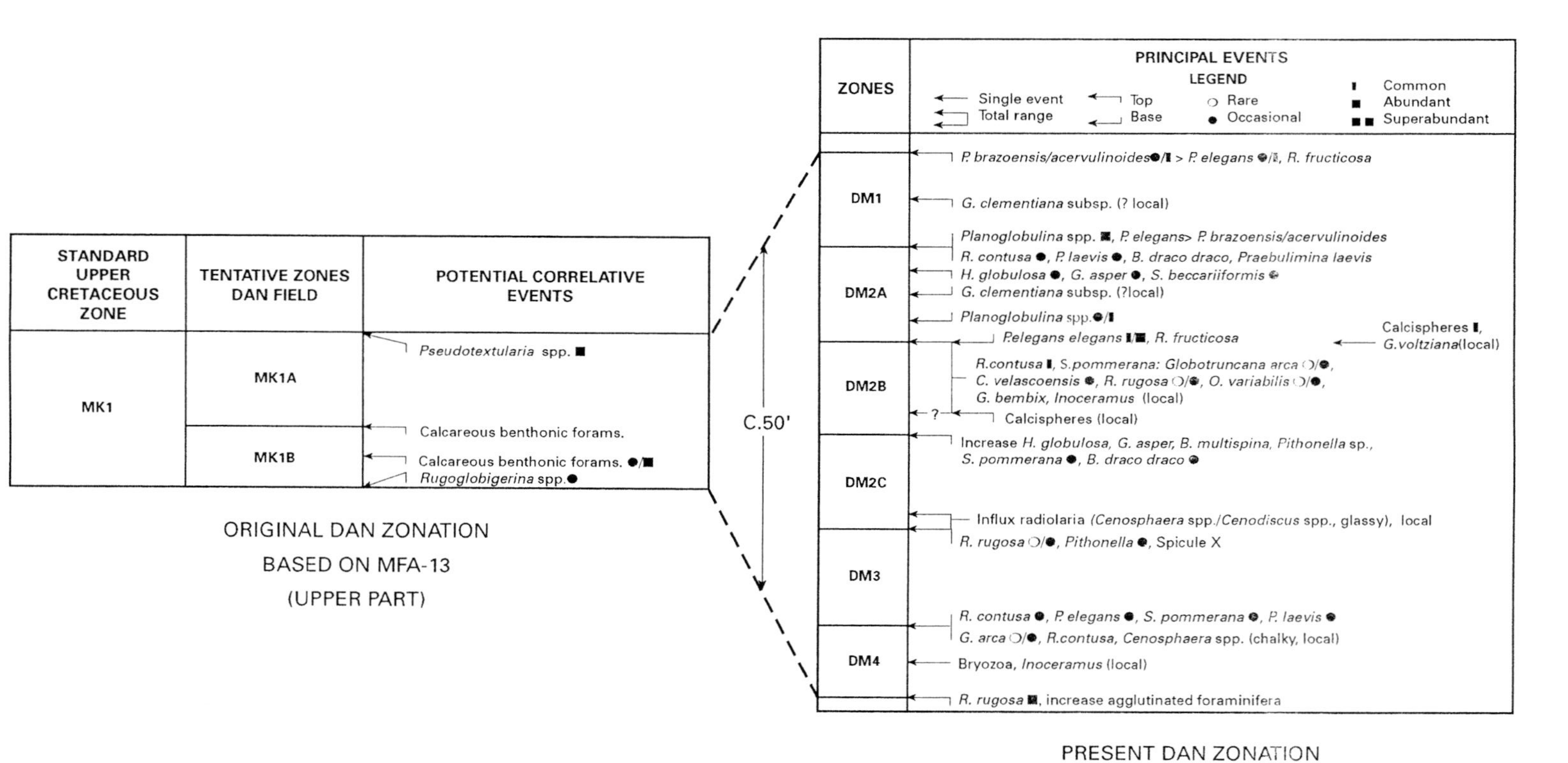

Fig. 6. Dan Field micropalaeontological zonal schemes shoving (left) the early scheme based on MFA-13 together with standard Upper Cretaceous zone at that time and (right) the present-day scheme for the equivalent section (reproduced with permission from Maersk Olie og Gas AS).

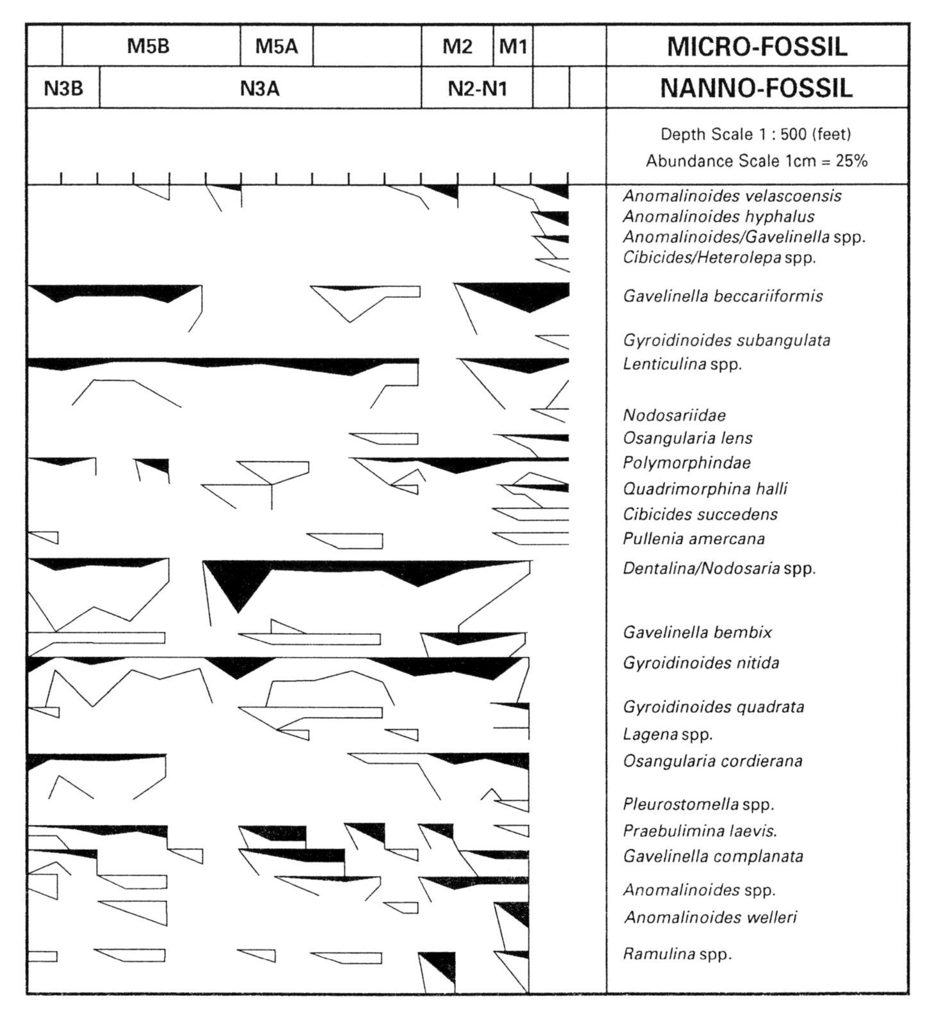

Fig. 7. Simplified biostratigraphic chart from the upper part of the Maastrichtian of Gorm Field. Micropalaeontological and nannofossils zones are shown on the left of the chart, while occurrences and abundances of foraminifera are shown to the right. A similar chart would also be produced for the nannofossils (after Shipp & Marshall 1994).

wells, the present-day Dan zonation (as shown in Fig. 6) with a significantly increased number of micropalaeontological data points has been developed. Individual zones are approximately 10 feet thick and can often be subdivided. Approximately 20 data points are now recognized in place of the four used on MFA-13 and the original single uppermost Maastrichtian zone of the traditional zonal scheme (Fig. 6). As a result of such developments it soon proved possible to monitor wells through the Maastrichtian target zone over a distance of some 6000 feet (Fig. 8). The more recent MFB-2E well, drilled in 1996, was monitored over a horizontal distance of 11 420 feet, at the time a North Sea record.

Even greater refinement has been achieved over part of the Valdemar section, as shown in Fig. 9 which represents part of the zonation covering a 50-foot vertical section. A pilot study to establish the zonation was originally carried out on samples at 5-foot spacings through a core from a nearby well. This resulted in the micropalaeontology and nannofossil zones shown as

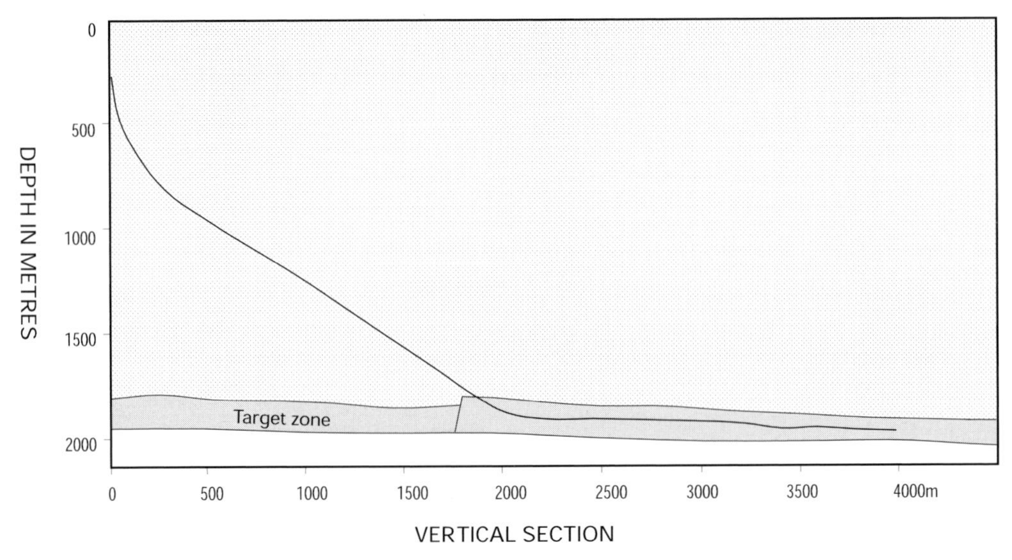

Fig. 8. Dan MFA-18 well showing the trajectory of a horizontal section maintained within the target zone (after Shipp & Marshall 1994).

M zones and N zones. During the drilling of the Valdemar-2 horizontal well, additional events were identified over this part of the section. As a result additional analyses were carried out at 1-foot spacings with the improvement in the zonation indicated by the MV zones and NV zones, where 16 micropalaeontological zones or subzones and 14 nannofossil zones and subzones are recognized over the 50-foot section studied. Some of these zones are only 1 foot in thickness.

At the well site

Mobilization for a well-site assignment can be very rapid as all the equipment required by the micropalaeontologist or nannofossil worker can be hand-carried to the rig.

Foraminiferal microscope work is usually carried out in the logging unit where facilities for washing and drying samples are readily available. Nannofossil work is also ideally undertaken in the logging unit, although occasionally, because of the higher magnification used, excess vibration from the drill floor and shakers may necessitate setting up the microscope elsewhere.

The lag time from the cutting of a sample to its arrival at the shakers can be 1 h or more, and this reinforces the need for a rapid response to any changes. Preparation time takes approximately 5 min with a further 5–10 min required for analysis in the case of both nannofossils and micropalaeontology. The lag time from the cutting of the sample to completion of analysis can therefore be upwards of 1 h, and results are often 100 feet behind the bit depth. However, this is generally within acceptable working, limits, bearing in mind that it is possible to slow down the rate of penetration or to circulate bottom-up samples when necessary.

The biostratigrapher usually tries to analyse every sample collected on the build-section, which can be as close as every 10 feet, to reaffirm the zonal scheme and to identify the correct zonal position at the start of the horizontal section. Once the trajectory is horizontal the sample spacing is usually opened out to 20, 40 or 60 feet depending on the rate of penetration.

It is important for the biostratigrapher to be aware of the angle at which the well is being drilled, and to establish the relationship to the dip of the beds as this can affect the number of fossils recovered. In Fig. 10 a thin bed containing large numbers of a planktonic marker is illustrated. If the well is being drilled along this bed then the marker fossil will appear in abundant numbers as in the first case (A). If the well path cuts the bed at an angle then fewer specimens of the marker will appear in the sample (B), and if it cuts the bed at right angles only rare specimens will be recovered (C). It is also important to be aware whether a zone

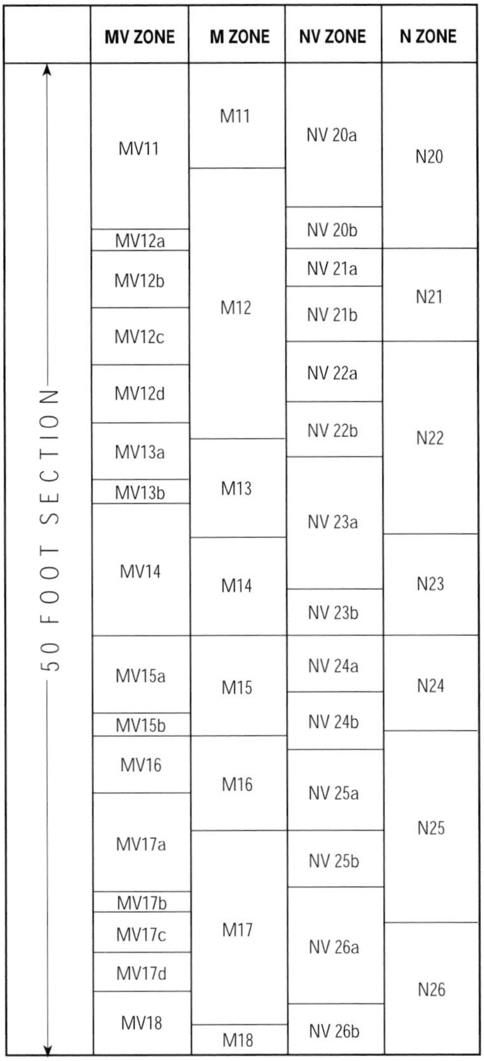

50 FOOT SECTION	MV ZONE	M ZONE	NV ZONE	N ZONE
	MV11	M11	NV 20a	N20
	MV12a	M12	NV 20b	N21
	MV12b	M13	NV 21a	N22
	MV12c	M14	NV 21b	N23
	MV12d	M15	NV 22a	N24
	MV13a	M16	NV 22b	N25
	MV13b	M17	NV 23a	N26
	MV14	M18	NV 23b	
	MV15a		NV 24a	
	MV15b		NV 24b	
	MV16		NV 25a	
	MV17a		NV 25b	
	MV17b		NV 26a	
	MV17c		NV 26b	
	MV17d			
	MV18			

Fig. 9. Part of the Valdemar Field zonation over 50 feet of section showing an increased refinement from M (micropalaeontological) and N (nannofossil) zones, where core samples had been analysed at 5-foot spacings, to MV and NV zones, where core samples were subsequently analysed at 1-foot spacings (after Shipp & Marshall 1994).

is being entered from above or below, as the criteria for identifying the zone are likely to be different in each case. For such reasons experience is vital for the successful interpretation of the zonal schemes.

It must be stressed that high-resolution biostratigraphy is only one of several tools available that together make it possible to geosteer a horizontal well. These other tools include 'logging while drilling' (LWD) or 'measurement while drilling' (MWD), hydrocarbon show evaluation and three-dimensional (3D) seismic (Jeppesen 1994). The well-site geologist, biostratigrapher, MWD or LWD operator, directional driller and company man all form essential elements of the well-site team, which is necessary for the success of any well.

On earlier horizontal wells in the Danish sector a single biostratigrapher carried out the work, but as knowledge of the fields improved so the degree of accuracy required increased. On the Dan Field, for example, the original requirement to stay within the Maastrichtian has changed to more specific requirements such as to follow a particular subzone that correlates to the best reservoir characteristics. In addition, improvements in bit performance has meant that bit trips, which give the palaeontologist an opportunity to rest, are not required so often. Consequently, it was recognized that two biostratigraphers, working 12-h shifts, were required for full 24-h monitoring. Water-injection wells drilled deeper in the reservoir may still only have one biostratigrapher on board, however, as continuous monitoring is less vital.

Case histories

Examples illustrating the contribution made by wellsite biostratigraphy to the drilling of horizontal wells in the Danish sector can be found in the following publications: Anderson *et al.* (1990), Kruse (1991), Fine *et al.* (1992), Jeppesen (1994) and Shipp & Marshall (1994).

The authors wish to thank the management of Robertson Research International Limited, the companies in the Dansk Undergrounds Consortium (Shell Olie og Gasudvinding Danmark BV, Texaco Denmark Inc. and A. P. Møller) and operator Maersk Olie og Gas AS for their consent to publish this paper and for permission to include some of their data and figures. Some parts of this paper were initially presented at GEO '94 and appeared in *Selected Middle East Papers from GEO '94* and we are grateful to the organizers of that conference for permission to include such details in the present updated paper.

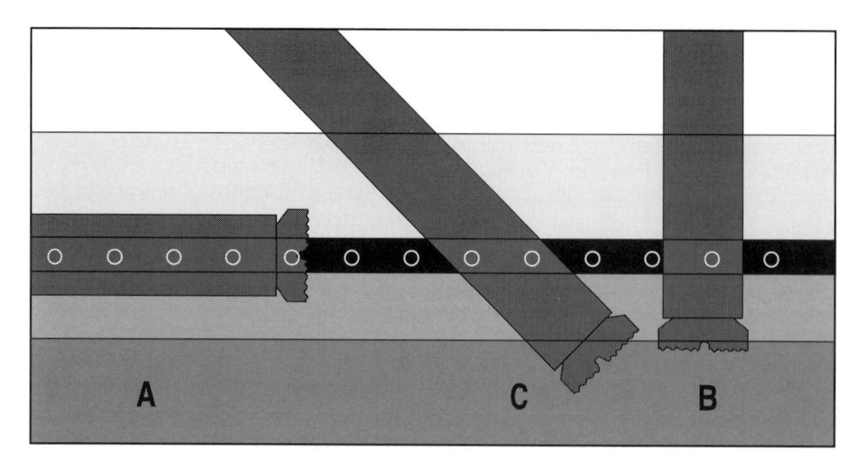

Fig. 10. Diagram to illustrate the different number of microfossils, such as planktonic foraminifera, recovered from a thin bed depending on the angular relationship of the bed to the well path. (**A**) When the well path parallels the bed microfossils are common. (**B**) When the well path cuts the beds at 45° microfossils are occasional. (**C**) When the well path cuts the beds at a right angle microfossils are rare.

References

ANDERSON, C. & DOYLE, C. 1990. Review of hydrocarbon exploration and production in Denmark. *First Break*, **8**(5), 15–165.

ANDERSON, S. A., CONLIN, J. M., FJELDGAARD, K. & HANSEN, S. A. 1990. Exploiting reservoirs and horizontal wells: the Maersk experience. *Schlumberger Oilfield Review*, **2**(3), 11–21.

FINE, S., YUSAS, M. R. & JØRGENSEN, L. N. 1992. Geological aspects of horizontal drilling in chalks from the Danish Sector of the North Sea. *In*: PARKER, J. (ed.) *Petroleum Geology of North West Europe: Proceedings of the 4th Conference*. Geological Society, London, 1483–1490.

INESON, J. R. 1993. The Lower Cretaceous Chalk Play in the Danish Central Trough. *In*: PARKER, J. (ed.) *Petroleum Geology of North West Europe: Proceedings of the 4th Conference*. Geological Society, London, 175–183.

JEPPESEN, M. W. 1994. Geological steering of horizontal wells in chalk reservoirs: examples from the Danish North Sea. *Bulletin of the Geological Society of Denmark*, **41**, 138–144.

KRUSE, I. 1991. Petroleum geoscience – Danish view. *First Break*, **9**(3), 95–106.

SHIPP, D. J. & MARSHALL, P. R. 1994. Biostratigraphic steering of horizontal wells. *In*: HUSSEINI, M. I. (ed.) *The Middle East Petroleum Geosciences. Selected Middle East Papers from the Middle East Geoscience Conference, Bahrain*, Volume 2, 849–860.

Microfossil diachronism in southern Norwegian North Sea chalks: Valhall and Hod fields

J. A. BERGEN & P. J. SIKORA

Amoco Exploration and Production Company, P.O. Box 3092, Houston, TX 77253, USA

Abstract: An integrated late Cretaceous calcareous microfossil framework for the southern Norwegian chalks is presented based on detailed study of 13 wells (many with core) and circum-North Sea outcrops, in addition to many cosmopolitan events recognized in reference sections in Tunisia, North America and Europe. Previous biozonations for the North Sea chalks: (1) lack the biostratigraphic resolution needed in production geology; (2) are largely founded on benthic foraminifera and acme events; (3) fail to recognize numerous stratigraphic discontinuities within the North Sea chalks; and (4) are poorly calibrated. Application of this new integrated framework cautions towards the use of certain acme and benthic foraminifera events as chronostratigraphic horizons both within the basin and from the basin into circum-North Sea outcrops. An additional complicating factor is palaeogeographic segregation between low and high latitudes, which intensified during the middle Campanian and became problematic within the North Sea basin itself during the late Maastrichtian. A final consideration is redeposition, which can lead to erroneous correlations between wells, but may also affect reservoir quality. Established North Sea biozonations have not recognized this primary palaeoenvironmental signal and massive redeposition across major lithostratigraphic units, which has resulted in the correlation of diachronous facies rather than isochronous horizons. This could lead to complications in operational applications, such as biosteering.

Chalks are often ideal for biostratigraphic study because the sediment itself is composed largely of microfossils and nannofossils, and was deposited in a deep-water setting. Nevertheless, the chalks of the Norwegian southern North Sea basin are far from ideal. Remobilization and redeposition have long been suspected as causes for creating porosity within these chalks. Combined with diagenetic effects that can reduce recovery and preservation within reservoir section, both the confidence and precision in biostratigraphic correlations can be seriously compromised.

Nannofossils and foraminifera have carried the bulk of operational biostratigraphic work in the North Sea chalks. Planktic foraminifera and nannofossils are important in global correlation, but their use within the North Sea basin is somewhat limited because a number of index species have Tethyan affinities. Cosmopolitan zonation schemes are further limited by their reliance on first occurrence events. North Sea biozonations have relied more on local events based on acmes and benthic foraminifera. However, both acmes and benthic foraminifera are environmentally controlled and, therefore, may be diachronous. The benthic foraminifera stratigraphic ranges were originally defined in circum-North Sea outcrop sections deposited in much shallower palaeoenvironments, whereas acme events have been established in well-to-well correlations within the basin. Their diachronism has never been tested because a high-resolution biozonation based on planktic microfossils has not been established for the North Sea chalks. In the current study, nannofossils have provided this unifying factor and have shown that many of the acme (nannofossil and foraminifera) and benthic foraminifera events are slightly to highly diachronous. In addition, palaeobiogeographic differentiation and redeposition may complicate the correlation of well sections outside the current production area in the Valhall and Hod fields, especially within the main reservoir in the Tor Formation.

Lithostratigraphy

Description of the Cenomanian–Danian Shetland Group off the north coast of Scotland (Deegan & Scull 1977) was expanded to include the formations of the Chalk Group in the southern Norwegian sector of the North Sea by Isaksen & Tonstad (1989). Isaksen & Tonstad (1989) included five formations (Fig. 1) within the Shetland Group: (1) the Cenomanian Hidra

Bergen, J. A. & Sikora, P. J. 1999. Microfossil diachronism in southern Norwegian North Sea chalks: Valhall and Hod fields. *In:* Jones, R. W. & Simmons, M. D. (eds) *Biostratigraphy in Production and Development Geology*. Geological Society, London, Special Publications, **152**, 85–111.

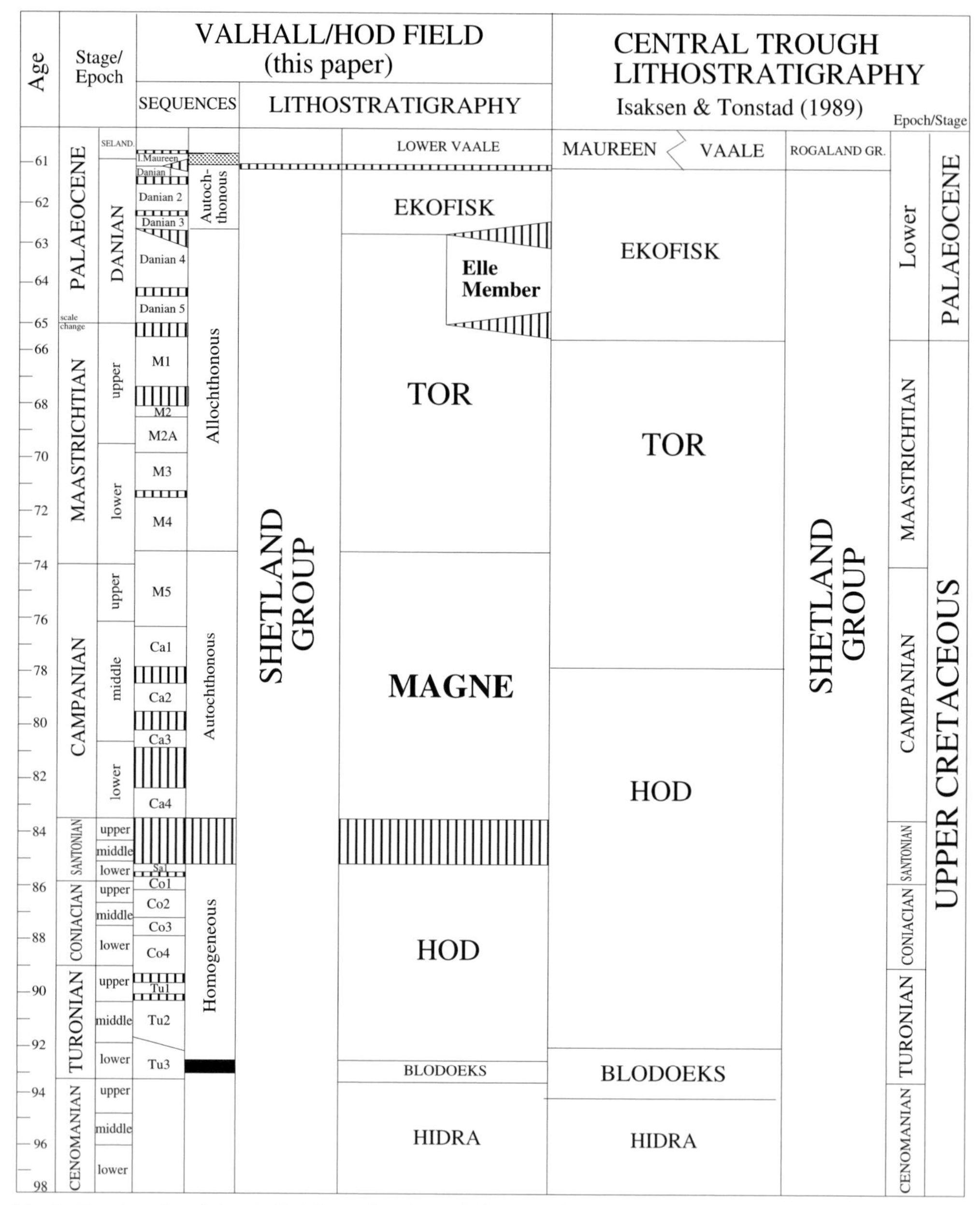

Fig. 1. Stratigraphy of the chalk, Norwegian Central Graben with new lithostratigraphic units in bold.

Formation; (2) the uppermost Cenomanian–lowermost Turonian Bloedoks Formation, which is identified by its high gamma-ray response and is considered an equivalent to the Plenus Marl in outcrop; (3) the Turonian–lower Campanian Hod Formation; (4) the upper Campanian–Maastrichtian Tor Formation; and (5) the Danian Ekofisk Formation. The current paper is focused on the post-Cenomanian Upper Cretaceous, which typically includes the Hod and Tor formations. The highly pelagic Ekofisk Formation is extremely thin within the Valhall Megaclosure and its microfossil biostratigraphy is not discussed.

In a broad sense, we recognize four major depositional regimes in the chalks overlying the Bloedoks Formation within the Valhall and Hod fields, including 23 lower-order units defined by stratigraphic discontinuities (Fig. 1). These four major depositional regimes include: (1) the Turonian-lower Santonian Hod Formation characterized by high pelagic depositional rates over the entire area; (2) Campanian–lowermost Maastrichtian deep-water, autochthonous deposition off-structure with structural crests mantled by submarine hardgrounds, herein informally termed the Magne Formation; (3) the Maastrichtian–lower Danian Tor Formation, characterized by its low gamma-ray signature and largely allochthonous lithologies; and (4) the middle–upper Danian Ekofisk Formation, which forms a thin pelagic drape (1–10 m) over the area. Of particular importance within the Valhall Megaclosure is the recognition of lower Danian section characterized by a low gamma-ray response (Tor Formation) and massive upper Maastrichtian redeposition, previously unrecognized and informally referred herein to the Elle Member of the Tor Formation (Fig. 1). The formal descriptions of the Magne Formation and the Elle Member of the Tor Formation await determination of their lateral extent.

Methodology and biostratigraphy

The current results are based on the study of 13 wells in the Valhall Megaclosure (Valhall–Hod fields) and a reference well for the Ekofisk Formation in the Tor Field (Fig. 2). The wells were selected for their available core material, being more densely sampled in the main reservoir interval of the Tor Formation. In addition, regional trends in event diachronism are partially based upon analysis of a number of wells to the north of the main study area (Block 2/5 and northern Block 2/8).

Graphic correlation methodology (Shaw 1964; Mann & Lane 1989) was used to develop a new biostratigraphical model for the Turonian–Maastrichtian chalks of the Valhall and Hod fields. Critical to this model were the global Amoco Composite Standard Database and its localization through examination of circum-North Sea outcrops and core material. Circum-North Sea outcrops examined for the current study include: (1) the Coniacian–Turonian boundary interval of the Culver Cliff section, Isle of Wight, UK; (2) the middle Coniacian–middle Campanian portion of the Whitecliff section, Isle of Wight, UK; and (3) an upper Campanian–Maastrichtian composite section in the type area of the Maastrichtian in the Netherlands. For the Turonian, core material from two wells in the Valhall Field, including a reference well for the Hod Formation (Isaksen & Tonstad 1989), served as reference material.

Planktic foraminifera are very rare and sporadic in surrounding outcrop sections, thus making them unsuitable for outcrop-to-basin correlations. Our analyses of outcrop sections confirmed the accuracy and great utility of the established benthic foraminiferal zonation in outcrop (Hart *et al.* 1989). However, the majority of benthic indices used in the North Sea basin zonation of King *et al.* (1989) proved ineffective in basin-to-outcrop correlations either because they were absent or extremely rare and sporadic in occurrence in the basin (e.g. *Bolivinoides decoratus*, *Gavelinella usakensis*, *Stensioina granulata levis*) or they were highly diachronous between the basin and the shallower-water outcrop sections (e.g. *Tritaxia capitosa*, *Stensioina exsculpta exsculpta*). This necessitated the construction of an entirely new microfossil zonation for the basin, against which the calcareous nannofossils were used to calibrate between the outcrop and basinal succession.

The integrated calcareous microfossil biostratigraphy for the Turonian–Maastrichtian of the Valhall Megaclosure (Fig. 3) stresses the use of first and last appearance data points of planktonic organisms, many of which can be used for correlations outside the North Sea basin and northwestern Europe. The nannofossil and microfossil biozonations for the Turonian–Maastrichtian chalks of the southern Norwegian sector are described separately in the Appendices; no formal description of an integrated biozonation is attempted. Real (palaeoenvironments and palaeogeography) and apparent (redeposition) diachronism encountered in the chalks of the Valhall and Hod fields are discussed in the following three sections.

Diachronism: palaeoenvironments

Underlying the integrated calcareous microfossil biozonation scheme is the temporal framework provided by composite standard methodology. The resulting biostratigraphic model has defined 18 stratigraphic packages (Fig. 1) delineated by biostratigraphic discontinuities for the Hod–Tor formations within the Valhall Megaclosure. The diachronism of acme and benthic foraminifera events relative to these established schemes and the chronostratigraphic framework are discussed in the following two subsections.

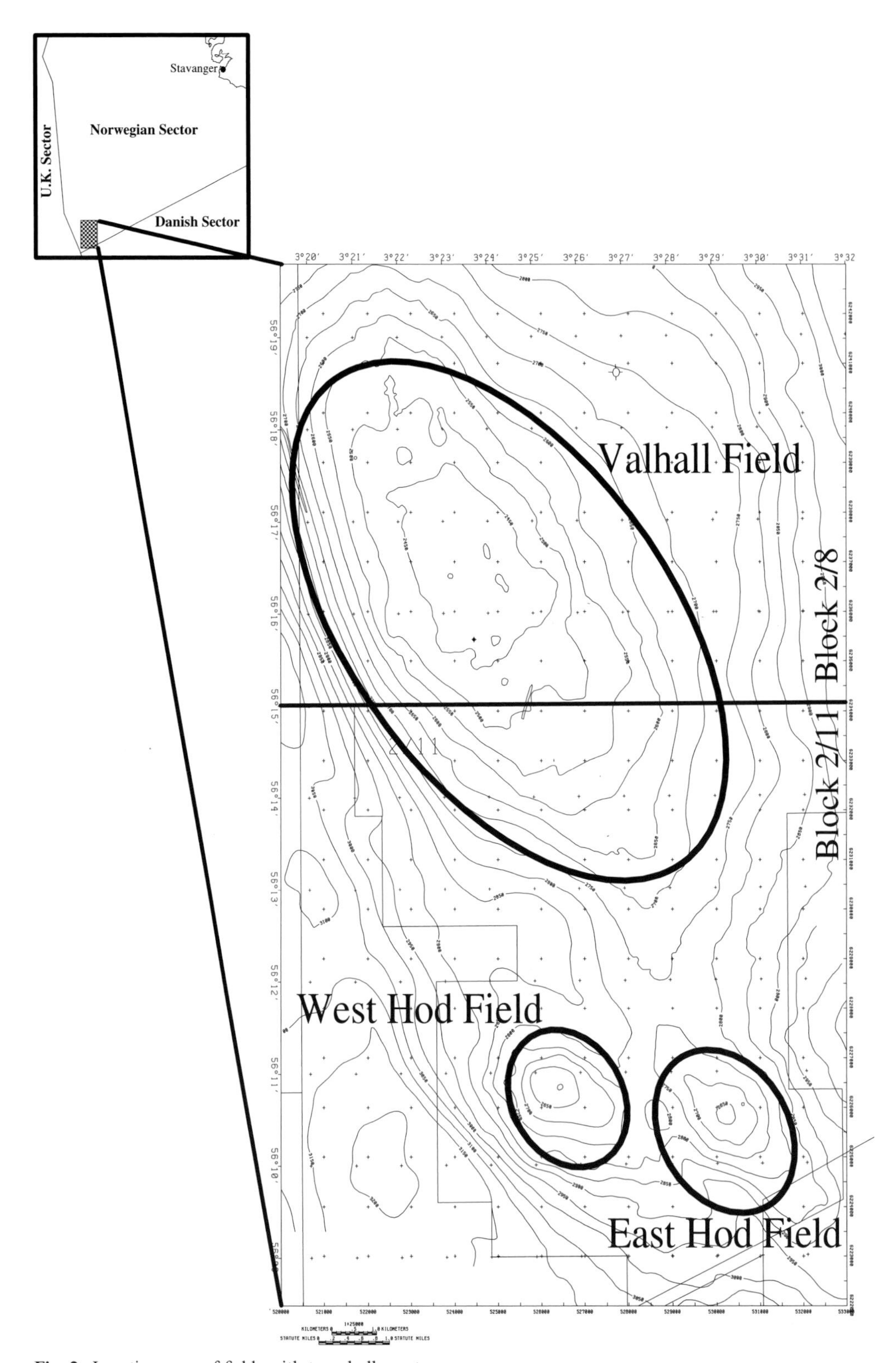

Fig. 2. Location map of fields with top chalk contours.

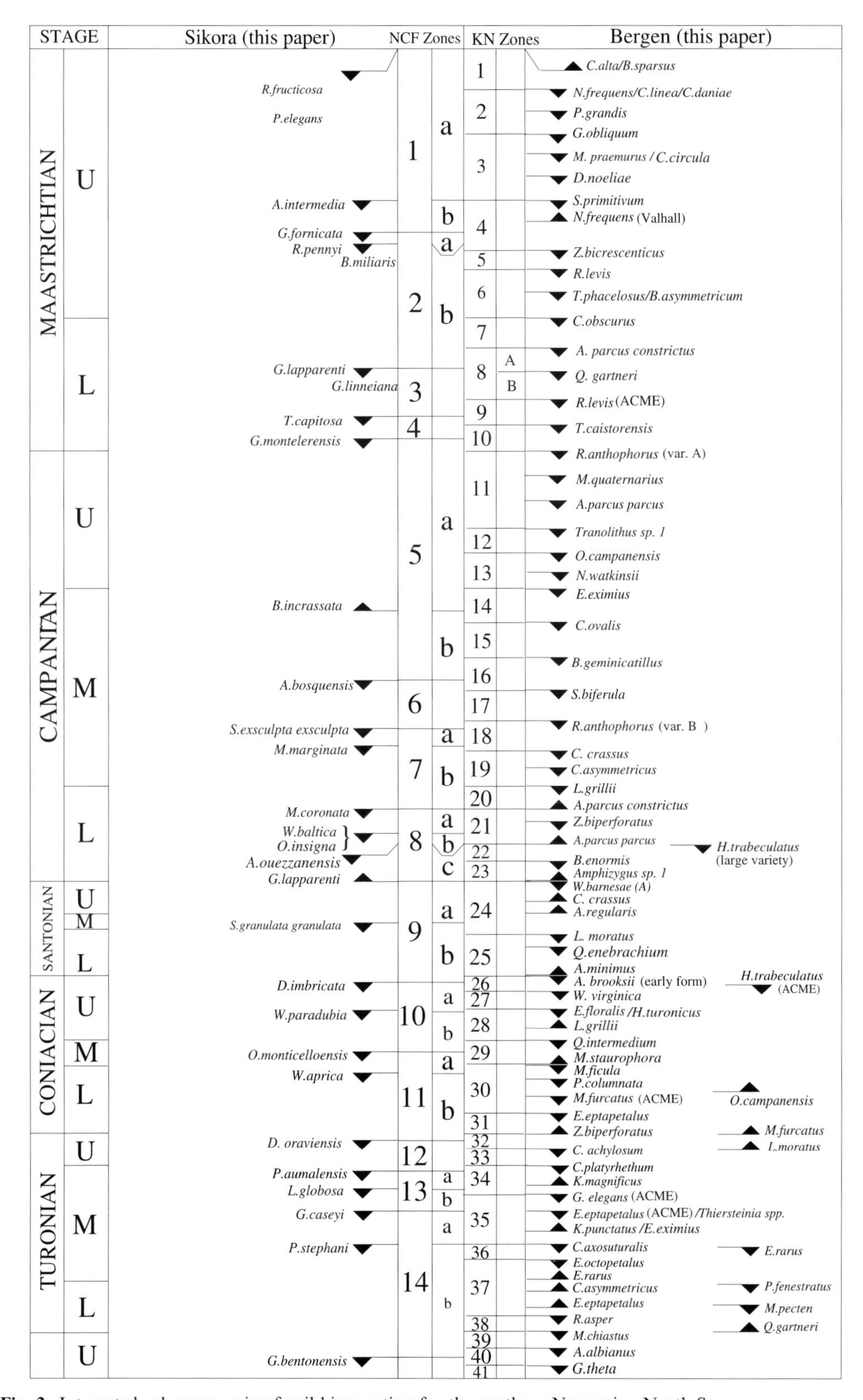

Fig. 3. Integrated calcareous microfossil biozonation for the southern Norwegian North Sea.

Acme events

Calcareous nannofossils. Local nannofossil acmes events are used rather extensively in North Sea biozonations, especially within the Hod Formation (Mortimer 1987) and the Ekofisk Formation (Varol 1989). Their quick determination makes them practical in operational work, but several factors caution towards their rigorous application, especially in production geology.

Five nannofossil acme events (Fig. 4) are used in routine operation work in the Turonian–lower Campanian. The most obvious problem with these events is their calibration, as all five have conveniently marked stage and substage boundaries (Fig. 4). Their definition in terms of abundance can also be difficult in actual application, specifically the three Coniacian–Turonian events (Fig. 4). In the Valhall Megaclosure these three taxa often do not reach their defined abundances (common) or are extremely rare and sporadic in a number of well sections. More often, these acmes are expressed as shifts to higher abundances (single specimens to rare, rare to few) or from sporadic to consistent occurrences. All three taxa also present taxonomic questions. *Helicolithus valhallensis* is invalid as it has never been formally described. Thus, this acme could represent either one or both valid species, *Helicolithus compactus* and *Helicolithus turonicus* (Fig. 4). The *Helicolithus trabeculatus* acme could represent a number of *Helicolithus* species with diagonal cross-structures. Herein, forms greater than 7.2 μm in length are ascribed to this acme. The early Turonian acme of *Lithastrinus* (Mortimer 1987) is confusing because the genus did not appear until the late Turonian. Within the Valhall Megaclosure, this event is better recognized by the highest consistent occurrence of *Eprolithus eptapetalus*.

The diachronism of these local events has never been suspected because they occur in consistent order between well sections (Fig. 4) and operational resolution within the Hod and Magne formations is low. This low resolution is surprising as this interval is characterized by high abundances, good preservation and the general absence of redeposition. The *Eprolithus* acme (*Lithastrinus* of Mortimer 1987) appears isochronous, although it could not be recognized in off-structure wells (Fig. 4). Conversely, the *Helicolithus trabeculatus* acme occurs within the upper Coniacian section in the Hod Field and in off-structure wells (not the terminal Coniacian), and has no utility on-structure in the Valhall Field (probably due to truncation). The most problematic late Cretaceous acme event is that of *Helicolithus turonicus/compactus* (*Helicolithus valhallensis* of Mortimer 1987). It has been used to define the terminal Turonian, yet occurs in lower–middle Coniacian sections in all but two wells examined for this study. Two acmes in *Watznaueria barnesae* have been used to date the tops of the lower Campanian and lower Santonian (Mortimer 1987). A general abundance decrease of this species from the Santonian–Maastrichtian occurs in the southern Norwegian North Sea chalks, until the species disappears immediately below the top of the Cretaceous. The shift down-section to common *Watznaueria barnesae* at the top of the lower Campanian has utility in the basin because it is associated with an intra-Campanian unconformity off-structure and in the Hod Field (Fig. 4), but could not be precisely determined in the more complete section onshore at Whitecliff Bay. The top of abundant *Watznaueria barnesae* is associated with the Santonian–Campanian boundary in the Whitecliff section, but so far has no utility in the basin because upper Santonian section has not been detected. Caution is exercised towards using these events because of massive deposition that can occur near the base of the Magne megasequence (Fig. 4).

Causal mechanisms for these acme events are speculative, although decreasing temperatures from post-Cenomanian time (Jenkyns *et al.* 1994) may be the overriding factor. In the Southern Ocean, decreased abundances of *Watznaueria barnesae* during the late Campanian–Maastrichtian were associated with increased endemism of austral taxa and have been explained by cooling (Watkins *et al.* 1995). A general progressive decline in the abundance of *Watznaueria barnesae* from Campanian–Maastrichtian occurs within the Valhall Megaclosure. *Eprolithus* and *Helicolithus* may also be temperature-dependent. Both genera appeared during a period of rising sea level (late early–late Aptian), when the strong Tethyan-boreal provinciality that had existed since the Tithonian began to breakdown. Thus, abundance decreases and extinctions within these genera in the North Sea basin during the Turonian–early Campanian could be associated with overall late Cretaceous cooling. Another factor could be fertility. An acme event of short duration, such as the early Coniacian acme of *Marthasterites furcatus*, would be better explained by changes in surface-water fertility.

Planktic foraminifera. North Sea contractor biozonations utilize various foraminifera abundance events, mainly those of planktic taxa. Such

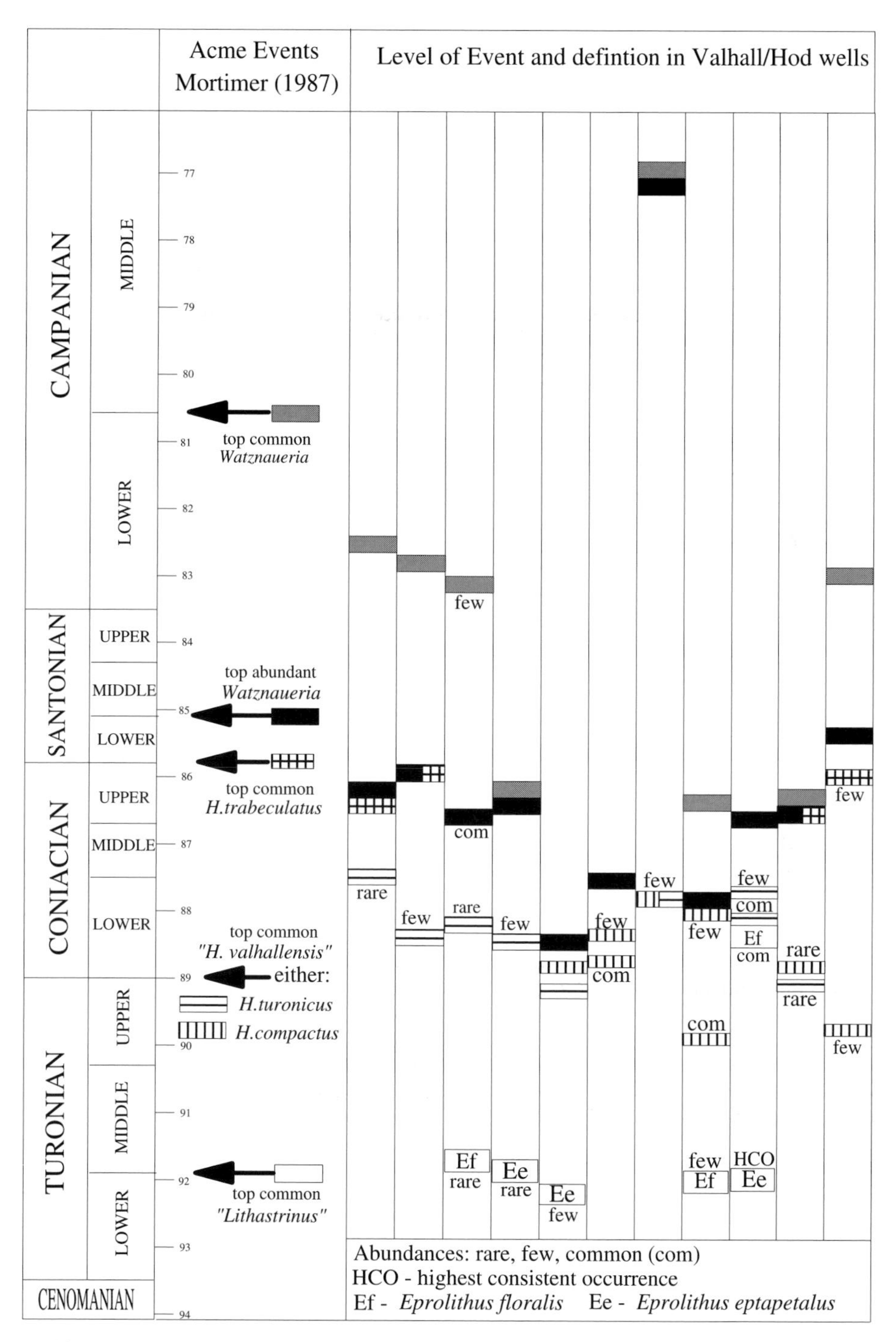

Fig. 4. Chronostratigraphy of Valhall Megaclosure wells and diachronism of Turonian–Campanian nannofossil acmes.

influxes are also noted as subsidiary zonal criteria by King *et al.* (1989). The most commonly used abundance events involve the planktic foraminifer *Rugoglobigerina rugosa* in the Maastrichtian Tor Formation, and multiple influxes of *Marginotruncana marginata* primarily in the Coniacian portion of the Hod Formation. Such events can be extremely useful in directing a well bit during horizontal drilling. However, abundance events are less useful in regional correlation within the chalk for two reasons: (1) palaeoenvironmental control; and (2) sorting in allochthonous chalks.

Marginotruncana marginata is extremely abundant in the upper Turonian–Coniacian portion of the Hod Formation. Two distinct pelagic microfossil assemblages are recovered from the Hod Formation, indicative of two oceanographic water masses (Sikora *et al.* 1999): (1) a deeper-water assemblage dominated by keeled planktic species (such as *M. marginata*) and diverse radiolarians, and most frequently found off-structure; and (2) a shallower-water pelagic assemblage dominated by non-keeled planktic species (mainly *Whiteinella* spp.) and low-diversity radiolarians, which was prevalent along the Valhall–Hod anticline. Nevertheless, there are multiple layers across the Valhall–Hod anticline which contain the deeper-water assemblage, probably indicative of discreet flooding events. Thus, *M. marginata* is generally common from the top of the Hod Formation and continuing down-section in off-structure wells, whereas it may not be common on-structure until lower in the section. Consequently, the first downhole occurrence of abundant *M. marginata* in crestal and off-structure wells is unlikely to be an isochronous datum. Correlation of on-structure wells based upon *M. marginata* events is also hazardous, as crestal Valhall was erosionally truncated during uplift in the latest Coniacian and Santonian. The amount of truncation on top of the Hod Formation varies from the central crestal area to the upper flank. Correlation across the crest based upon *M. marginata* abundance events may therefore result in the correlation of an older flooding deposit from a more deeply truncated central crestal site with a younger flooding event from a less truncated flank section. The first downhole occurrence of common *M. marginata* relative to the chronostratigraphic framework (Fig. 5) demonstrates the very diachronous nature of this event.

Rugoglobigerina rugosa is a species of wide environmental tolerances, and is one of the most common planktic species recovered from the Tor Formation. However, the central crestal area of the Valhall anticline was marked by very shallow-water environments throughout much of the Maastrichtian and, except for discreet flooding events, planktic foraminifera were absent (Sikora *et al.* 1999). Off-structure, the Tor Formation is often characterized by thick, stacked sequences of distal mudflows and turbidites containing well-sorted, tiny foraminifera of the genera *Globigerinelloides* and *Heterohelix*. *R. rugosa* is a relatively large planktic foraminifer and has been sorted out of these deposits. However, occasional coarser-grained debris flows reach these localities transporting larger bioclasts into the basin, including common *R. rugosa*. Thus, the first downhole occurrence of common or abundant *R. rugosa* in different wells may mark a discreet flooding event over the central crestal zone, a typical pelagic assemblage on the structure flank or a debris flow deposit in the basin. It would be pure coincidence if these various manifestations of the *R. rugosa* abundance 'event' represented the same stratigraphic level, and the regional chronostratigraphy indicates considerable diachronism for the event (Fig. 5). In off-structure wells, this diachronism is small. This was a time marked by widespread movement of coarse-grained debris flows into the basin. On-structure in wells, diachronism is much more pronounced, ranging over 2 Ma, or not occurring at all due to palaeoecological exclusion in very shallow-water palaeoenvironments.

Benthic foraminifera

Although the stratigraphic ranges of many planktic and benthic foraminifera are different in the basin than in outcrop, several have utility in well correlations throughout the Valhall Megaclosure. Nevertheless, failure to recognize their diachronism between different palaeobathymetric regimes can lead to erroneous age determinations and correlations. Some biostratigraphic indices have proved to be significantly diachronous not only between outcrop and basin, but within the basin as well. The most prominent example of such a misleading index is *Stensioina granulata polonica*.

The benthic foraminifera *Stensioina granulata polonica* has a very narrow stratigraphic range of lower and middle Santonian in outcrop (Hart *et al.* 1989). In well sections in the Valhall Megaclosure, a total stratigraphic range of upper Coniacian–middle Campanian has been determined for this species, although it is very erratic in stratigraphic and lateral distribution. A Campanian highest occurrence for this species (including the highest common occurrence) is very well constrained by both nannofossil and

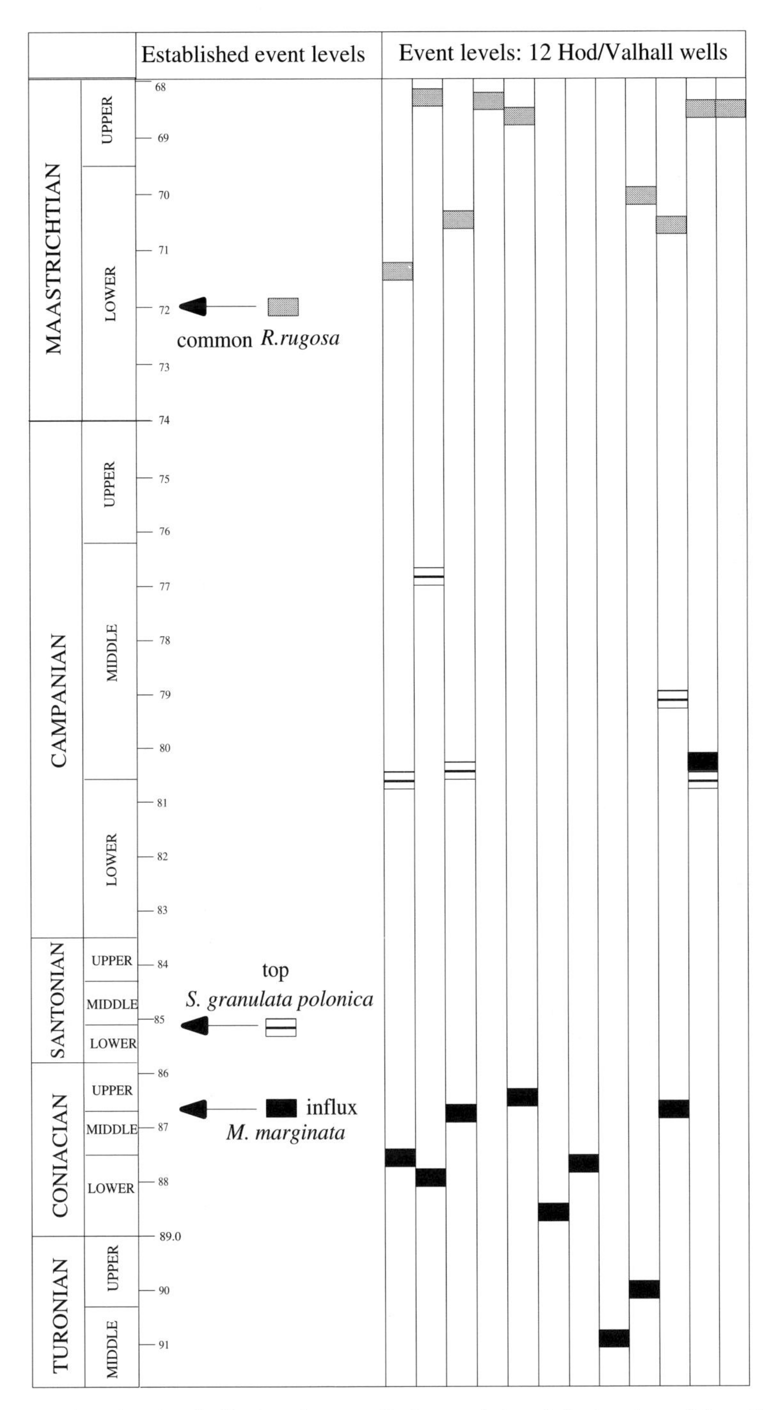

Fig. 5. Chronostratigraphy of Valhall Megaclosure wells (rectangles) and diachronism of three Turonian–Maastrichtian microfossil acmes (arrows at left show accepted biovent levels).

microfossil successions (Fig. 3) in three off-structure wells and in both wells examined from the Hod Field. Its highest occurrence is also highly diachronous amongst these five wells (Figs 5 & 6).

Species of the genus *Stensioina* are deep-water indicators. The occurrence of *S. granulata polonica* in outcrop is probably indicative of a discreet flooding event and does not constitute the entire stratigraphic range of the species. Moving into the basin, however, the stratigraphic range of the species expands in progressively deeper-water facies. However, the distribution of *S. granulata polonica* is discontinuous even within the basin, indicating strong facies dependency. In the Valhall Megaclosure, *S. granulata polonica*

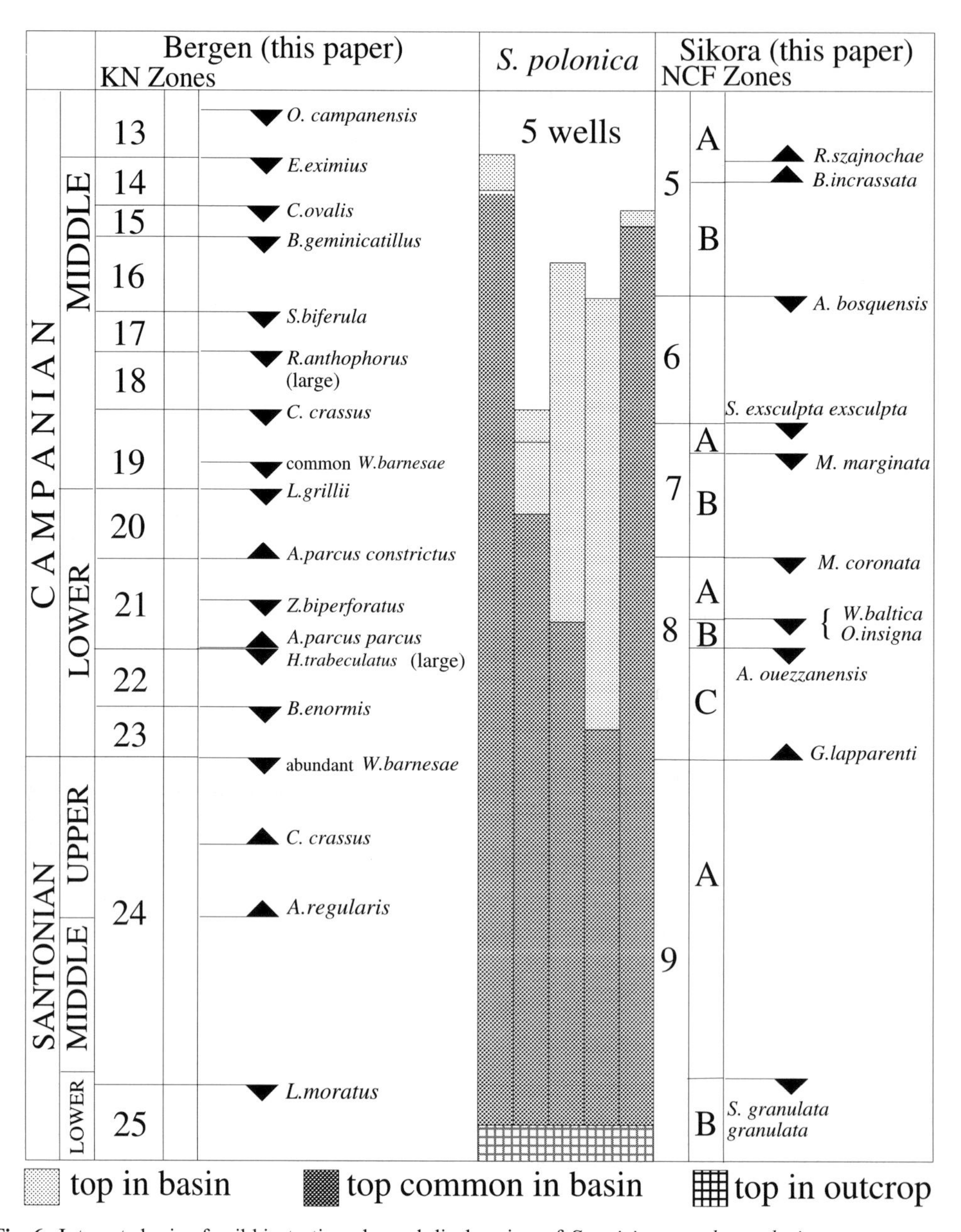

Fig. 6. Integrated microfossil biostratigraphy and diachronism of *Stensioina granulata polonica*.

is characteristic of carbonaceous, pyritic, often laminated chalks signifying low-oxygen palaeoenvironments or, at the very least, depositional environments with a high organic influx. Such palaeoenvironments are highly discontinuous, both geographically and stratigraphically, and tied to local tectonics and palaeoceanographic conditions (Sikora *et al.* 1999). In basinal facies *S. granulata polonica* does not meet the requirements of a good biostratigraphic index. Unfortunately, the species has long been used as a 'golden spike' in most North Sea biozonations, which have assumed that the lower Santonian highest occurrence of the species in outcrop can be applied in the basin. This has exaggerated the amount of Santonian section in the basin at the expense of the Campanian and allowed a major Campanian–Coniacian regional erosional unconformity to go unrecognized. It also has resulted in serious errors in well-to-well correlation, and questions whether existing biozonations are truly integrated amongst various microfossil groups.

Redeposition

Redeposition must be considered in any age determinations and correlations within the North Sea chalks. In the current study, the large amount of core material has been essential in identifying redeposition and its affect on well-to-well correlations. In general, episodic redeposition of varying degrees corresponds to the major lithological units recognized in the Valhall Megaclosure (Fig. 1). Redeposition within the highly pelagic Hod Formation (Turonian–lower Santonian) is obvious because it is limited to a few specimens in isolated samples, and biostratigraphic resolution is very high within this unit. The rich Hod assemblages can be problematic in age determinations when remobilized, especially into the base of the overlying Campanian–lowermost Maastrichtian Magne Formation (Fig. 4). Minor amounts of redeposition also were observed at the base of the Tor Formation in the lower Maastrichtian. Taxa redeposited from the Hod Formation are obvious in these assemblages, whereas rare specimens redeposited from the underlying Magne Formation can complicate age determinations.

Recognition of redeposition is more crucial in the upper Maastrichtian–Palaeocene because it is associated with the main reservoir sections throughout the North Sea chalks. During the late Maastrichtian, redeposition is one possible explanation for the apparent diachronism of first appearance data points between the Valhall Megaclosure and well sections to the north and east (see the section on 'Palaeogeography'). Remobilization of upper Maastrichtian chalks into the Danian is a complex problem throughout the southern Norwegian sector of the North Sea. However, a unique situation may exist in the Valhall Megaclosure and is associated with the Cretaceous–Tertiary boundary and the contact between the Tor and Ekofisk formations.

The contact between the Tor and Ekofisk formations is identified on logs by the sharp increase in the gamma ray response at the base of the Ekofisk Formation (Isaksen & Tonstad 1989). The Ekofisk Formation is very thin throughout most the Valhall Megaclosure (1–10 m), but expands considerably in northern chalks fields (Eldfisk, Ekofisk and southeast Tor fields). Within the Valhall Megaclosure, the Ekofisk Formation is often cored along with the main reservoir of the upper Tor Formation. Cores from this interval were densely sampled in the current study (one or more samples per m).

The first well examined for this study yielded unprecedented results about the contact between the Tor and Ekofisk formations, and its relationship to the Cretaceous–Tertiary boundary (Fig. 7). In this well a succession of early Danian foraminifera and nannofossils was observed in the upper 3.75 m of the Tor Formation, which was otherwise dominated throughout by massive redeposition of upper Maastrichtian nannofossils (99%). In contrast, only about 15% of recovered planktic foraminifera from this 5-m interval were reworked Cretaceous species. The co-occurrence of the planktic foraminifera *Parvularugoglobigerina eugubina* and very small specimens of *P. pseudobulloides* (+2.75 m in Fig. 7) indicate an age close to the Pα–P1a zonal boundary (Berggren *et al.* 1995). The next sample down-section +1.75 m) is marked by a low-diversity assemblage of very abundant, minute specimens of *Globigerinelloides messinae* and *Heterohelix globulosa*, probably representing an earliest Danian assemblage of Cretaceous survivor species (zone P0). The subjacent sample (+0.8 m) nearly barren of foraminifera. An influx of Tethyan planktic foraminifera typical of the latest Maastrichtian in the North Sea basin is first observed down-section at −1.75 m, including *Pseudotextularia elegans*, *Pseudoguembelina excolata* and *Contusotruncana contusa*. The base of the nannofossil zone NP1 (Martini 1971) is well constrained by the lowest occurrences of three nannofossils (Fig. 7) with its lower boundary placed at the lowest occurrence of *Neocrepidolithus dirimosus* at 0 m. The NP1–NP2 zonal boundary (Martini 1971) also falls within the Tor Formation in this well and is

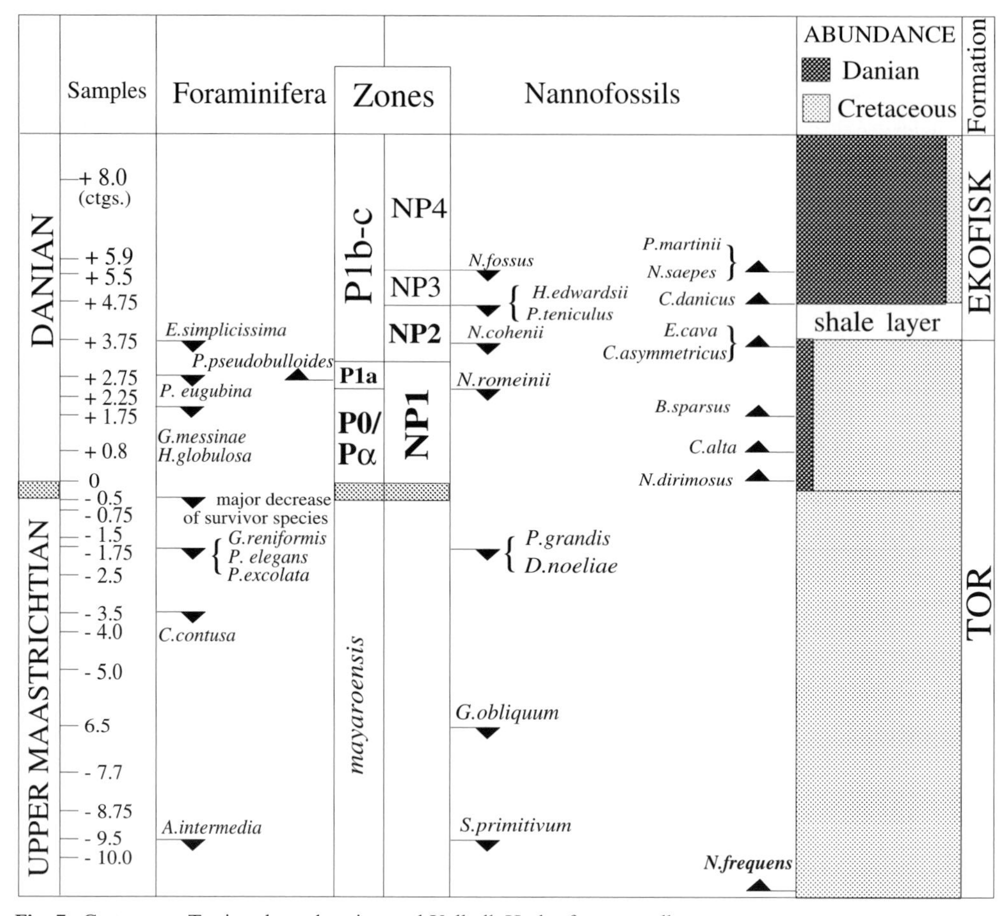

Fig. 7. Cretaceous–Tertiary boundary interval Valhall–Hod reference well.

constrained by the highest occurrence of *Neobiscutum romeinii* at +2.25 m and the lowest occurrences of *Ericsonia cava* and *Cruciplacolithus asymmetricus* at +3.75 m.

Biostratigraphic analyses have identified the Danian section at the top of the Tor Formation in all other wells examined, and this unit has been mapped throughout the Valhall Megaclosure; it is informally termed the Elle Member of the Tor Formation (Fig. 1). The youngest nannofossil assemblages recovered from this unit contain *Chiasmolithus danicus*, which marks the base of zone NP3 (Martini 1971). Nannofossil assemblages in all samples throughout this unit are characterized by massive redeposition from the upper Maastrichtian (nearly 100%), which has obscured previous recognition of this unit as Danian. However, remobilized upper Maastrichtian foraminifera shows trends across the Valhall Megaclosure related to structure and facies.

In the early late Maastrichtian, mudflows and relatively fine-grained debris flows dominated the Valhall region from the structure crest into the basin (Sikora *et al.* 1999). However, by the latest Maastrichtian and continuing into the early Danian (Elle Member of the Tor Formation) shoaling and winnowing of sediment dominated across the central crest feeding coarser-grained, debris flows down the flank. The shedding of coarser-grained bioclasts intensified in the early Danian with erosion of the central crestal zone, whereupon allochthonous sediments contained a very high percentage of reworked Cretaceous material. The amount of reworked material varies with the amount of sorting and transportation which has occurred. Lower Danian chalks from the crestal area are marked by nearly 100% reworked Maastrichtian foraminifera and other microfossils. As mentioned, the amount of redeposited upper

Maastrichtian nannofossils remains nearly constant across the main flank and into the basin. However, the amount of microfossil reworking decreases down the structural gradient to reach a minimum value of about 10% in fine-grained mudflow deposits on the lower flank and into the basin. The larger reworked microfossils were thus substantially sorted out during transportation from the central crestal zone into the basin.

Palaeogeography

Nannofossils

Late Cretaceous palaeoceanographic trends in nannofossil assemblages from the western northern hemisphere are reflected in published results from the Southern Ocean (Wise 1988; Pospichal & Wise 1990; Watkins 1992; Watkins *et al.* 1995). Largely cosmopolitan assemblages were established over a large portion of the western northern hemisphere (North Sea basin, southern Britain, Germany, Gulf of Mexico, Western Interior basin, southern Europe and Tunisia) during the Cenomanian–early Campanian. Nannofossil assemblages recovered from the Southern Ocean, although low in diversity, also consisted largely of cosmopolitan species during this time. However, abundance decreases and extinctions within the genera *Helicolithus* (Coniacian) and *Eprolithus* (Turonian) in the North Sea basin during this time may be an early signal for post-Cenomanian cooling.

A substantial disconformity separates lower and upper Campanian strata throughout the Southern Ocean. An increase in the number and proportion of high-latitude species during the late Campanian–early Maastrichtian in the Southern Ocean has been attributed to the development of a separate (cooler) surface water mass (Watkins *et al.* 1995). Accompanying these changes was the abundance decline of *Watznaueria barnesae*, which was first observed in higher latitude sites and then progressed towards the equator (Watkins *et al.* 1995). In the northern hemisphere, where the stratigraphic sections are more complete, evidence indicates an earlier breakdown of cosmopolitan assemblages. In the North Sea basin *Watznaueria barnesae* is an abundant constituent of Turonian–Santonian assemblages. This species gradually decreases (except across disconformities) in abundance until its disappearance immediately prior to the terminal Cretaceous, and is never common after the early Campanian. Additional evidence for an earlier breakdown in cosmopolitan assemblages and general cooling are the low-latitude markers species *Bukrylithus hayii* and *Marthasterites furcatus*, which are almost entirely excluded from the lower Campanian in northwestern Europe. Conversely, *Saepiovirgata biferula* and *Bifidilithus geminicatillus* are present in both the North Sea basin and the northern Western Interior basin (South Dakota), but have not been observed in lower–middle Campanian sections in the Gulf of Mexico and Tunisia. As in the Southern Ocean, provincialism appears to have intensified during the late Campanian–earliest Maastrichtian. In the northern hemisphere, several species present in lower latitude sections (Tunisia and southern France) during this time are absent from the North Sea basin or are extremely rare in surrounding outcrops. These include important marker taxa such as *Uniplanarius gothicus*, *Uniplanarius trifidus*, *Gorkaea obliquclausus* and *Bukrylithus magnus*. Conversely, certain taxa present in northwestern Europe (*Orastrum campanensis*, *Tortolithus caistorensis* and *Reinhardtites claviclaviformis*) are absent in sections in southwest France and Tunisia. Other longer-ranging taxa (e.g. *Amphizygus brooksii* and *Amphizygus minimus*) having late Campanian extinctions in lower latitudes have been excluded from upper Campanian deposits in the North Sea basin.

The middle early–early late Maastrichtian is characterized by a number of nannofossil extinctions (*Quadrum gartneri*, *Aspidolithus parcus constrictus*, *Tranolithus phacelosus*, *Reinhardtites levis*). A decrease in endemism is suggested by the fact this succession of nannofossil extinctions (zones KN6–KN8 in Appendix A) can be recognized in lower latitude sections (Gulf of Mexico, southern France, Tunisia). In the Southern Ocean an abundance increase of *Watznaueria barnesae* and decline in high-latitude taxa in more equatorial sites during the mid-Maastrichtian has been used to infer a slight moderation in climate (Watkins *et al.* 1995).

Strong latitudinal segregation during the late Maastrichtian is demonstrated by comparisons of nannofossil assemblages recovered from the North Sea basin and southern high latitudes (Watkins 1992; Watkins *et al.* 1995) to those in circum-Mediterranean sections, such as El Kef (Tunisia). Important marker taxa that appeared during the late Maastrichtian in low latitudes (*Micula prinsii*, *Micula murus* and *Ceratolithoides kamptneri*) are absent in both the North Sea basin and southern high latitudes. Conversely, as many as 20 nannofossil species did not survive the late Maastrichtian in the North Sea basin and this is also evident to some degree (e.g. *Prediscosphaera grandis* and *Gartnerago obliquum*) in southern high latitudes (ODP Site 750, Watkins 1992).

Of particular importance is the appearance of *Nephrolithus frequens*, which is ubiquitous in northern and southern high latitudes. Mortimer (1987) erected the late Maastrichtian *Zygodiscus spiralis* interval range zone (NK3) for the southern Norwegian and Danish chalks, which is defined as the interval between the lowest occurrence of *Nephrolithus frequens* and highest occurrence of *Gartnerago obliquum*. In the Valhall Megaclosure the stratigraphic ranges of these two species overlap (Fig. 8). In wells to the north and east of the Valhall Megaclosure *Nephrolithus frequens* has an apparent older first occurrence, as this species co-occurs with such taxa as *Tranolithus phacelosus* and *Reinhardtites levis*. In the southern high latitudes evidence indicates that *Nephrolithus frequens* migrated towards the equator during the late Maastrichtian, but co-occurs with *Reinhardtites levis* in older sections at higher latitude sites (Pospichal & Wise 1990; Watkins *et al.* 1995). Watkins *et al.* (1995) inferred a cooler late Maastrichtian based on the equatorial migration of *Nephrolithus frequens*, along with the continued decline in Austral flora. In the North Sea this would correspond to its migration south and westward into the Valhall Megaclosure. An alternate explanation to the diachronism of *Nephrolithus frequens* would be redeposition of mid-Maastrichtian chalk, especially considering the sharp difference and short distances between the Valhall Megaclosure and sites to the east and north. Examination of cored upper Maastrichtian sections in the North Sea outside the southern Norwegian sector and sampling for magnetostratigraphy could help resolve these uncertainties.

There is some evidence of a brief warming pulse near the Cretaceous–Tertiary boundary in the Southern Ocean, as increased abundances of *Watznaueria barnesae* exhibited a marked poleward migration near the terminal Cretaceous (Watkins *et al.* 1995). No corresponding increase in this species was observed near the terminal Cretaceous in the southern Norwegian sector of the North Sea. However, three nannofossil species not present in the terminal Cretaceous in the North Sea (zone KN1 in Appendix A) could indicate a very brief warming. *Nephrolithus frequens* and *Cribrosphaerella daniae* are

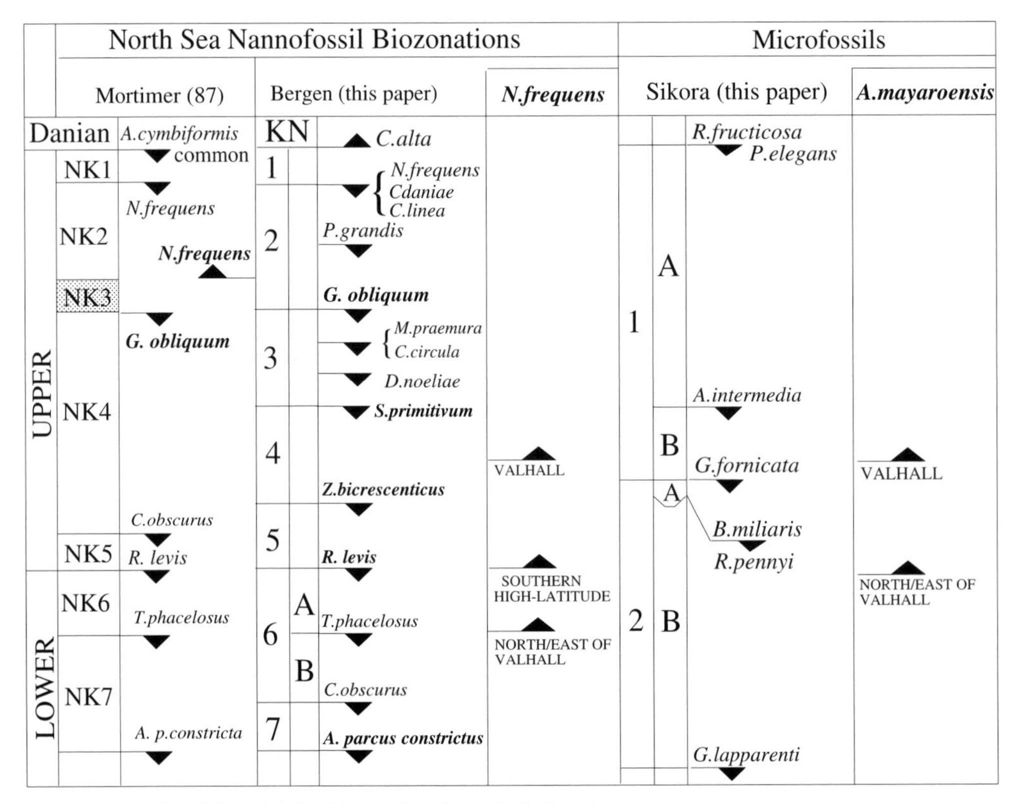

Fig. 8. Mid- to late Maastrichtian biostratigraphy and diachronism.

bipolar species, whereas *Cribrosphaerella ehrenbergii* (taxonomic synonym of *C. linea*) has been noted to have a preference for high southern latitudes (Wise 1988; Pospichal & Wise 1990; Watkins 1992).

Microfossils

Cenomanian–middle Turonian section in the North Sea basin contains planktic foraminifera with marked Tethyan affinities. This includes *Rotalipora* spp. in the Cenomanian Hidra Formation, and a diverse *Praeglobotruncana* assemblage and rare *Helvetoglobotruncana helvetica* in the lower and lower middle Turonian. The Tethyan elements of the planktic foraminifera assemblage began to decrease in the late middle Turonian. Biofacies analysis indicates that the late middle Turonian also was the time of initial uplift of the Valhall anticline (Sikora *et al.* 1999). Thus, the decline of purely Tethyan species at this time indicates a tectonic control on species distribution, probably a poorer connection with the open marine basin.

Planktic foraminifera from the upper middle Turonian–lower middle Campanian chalk section are often dominated by taxa of boreal aspect, such as species of *Whiteinella* and *Hedbergella*. Nevertheless, the presence of often abundant *Archaeoglobigerina cretacea* and species of *Marginotruncana* indicate the presence of an intermediate, rather than fully boreal, water mass. Furthermore, minor Tethyan influences near the Coniacian–Turonian boundary and within the upper Coniacian are indicated by rare occurrences of *Dicarinella* species (mainly *D. imbricata* and, much more rarely, *D. primitiva*). The lower Campanian marks the highest stratigraphic occurrence of this intermediate planktic foraminifera assemblage.

Upper middle Campanian–lower Maastrichtian sections contain evidence of pronounced cooling, marked by planktic foraminifera assemblages dominated by non-keeled taxa, such as species of *Rugoglobigerina*, *Globigerinelloides* and non-tethyan heterohelicids. A minor warming during the middle early Maastrichtian may be indicated by the appearance of rare *Globotruncana* species (mainly *G. lapparenti* and *G. bulloides*), which corresponds to the earliest nannofossil evidence for climatic equability (zone KN8).

The uppermost lower–upper Maastrichtian is marked by frequent changes in the composition of planktic foraminifera assemblages. The uppermost lower Maastrichtian is marked by a return of the boreal assemblage, which characterized most of the middle Campanian–early Maastrichtian. However, the lower upper Maastrichtian chalks are characterized by an influx of relatively Tethyan species of *Globotruncanella*, first appearing in southern Valhall sites about 69 Ma BP and reaching more northern sites (Block 2/5) at approximately 68 Ma BP. This corresponds to a time (nannofossil zones KN6–KN7) when *Nephrolithus frequens* was present only in wells to the north and east of the Valhall Megaclosure. In the Southern Ocean, a similar poleward migration of planktic foraminifers was associated with nannofossil assemblage changes indicating climatic moderation near the early–late Maastrichtian boundary (Huber & Watkins 1992; Watkins *et al.* 1995). Later in the Maastrichtian (approximately 66.7 Ma BP), *N. frequens* extended across the southern Norwegian North Sea and *Globotruncanella* had completely withdrawn. Thus, an early late Maastrichtian warming followed by cooling during the middle late Maastrichtian is indicated.

However, evidence contradictory of this middle late Maastrichtian cooling as defined by nannofossils is provided by another influx of Tethyan planktic foraminifera. This middle late Maastrichtian Tethyan influx in the North Sea has been long known (e.g. Koch 1977) and has been a widely used biostratigraphic datum. Principally, it involves Tethyan heterohelicids, such as *Pseudotextularia elegans*, *Racemiguembelina fructicosa* and *Pseudoguembelina excolata*, but also includes rare to frequent globotruncanids such as *Contusotruncana contusa* and *Abathomphalus mayaroensis*. Most of these species have much longer cosmopolitan stratigraphic ranges, the influx in the North Sea indicative of a local palaeoceanographic change. However, the present study indicates that the appearance of this Tethyan fauna is highly erratic and diachronous across the region. The oldest occurrences of this Tethyan fauna occurs in basinal localities to the north of Valhall. First occurrences of the assemblage occur much later on structure (e.g. the Balder Ridge) or do not occur at all (e.g. on the Valhall anticline). Such a pattern would suggest that the Tethyan fauna is indicative of a deeper water mass, and its influx in the latest Maastrichtian marks improved deep-water connections with the open marine basin as has been previously postulated (Schonfeld & Burnett 1991). However, this pattern would also suggest a transgression of structure during the latest Maastrichtian, whereas biofacies analysis indicates a regressive section (Sikora *et al.* 1999). This contradiction, together with nannofossil evidence of climactic cooling, indicates that further study is needed to fully resolve the palaeoceanographic implications of this Tethyan incursion, as well as the possibility of redeposition.

Conclusions

Detailed examination of post-Cenomanian Upper Cretaceous chalks in the southern Norwegian Sea has produced an integrated, high-resolution calcareous microfossil biozonation and identified numerous stratigraphic discontinuities within these Upper Cretaceous chalks. The stratigraphic units bounded by these discontinuities form the basic unit for mapping and their integration a necessary component of a robust geological model. A high-resolution chronostratigraphy is necessary for operations (biosteering), but also fundamental to understanding facies distributions, especially within reservoir units.

A strong palaeoenvironmental signal has been recorded in the microfossil and nannofossil assemblages within these North Sea chalks, which has been magnified by their high palaeolatitude. Because of this, caution must be exercised in the application of acme and benthic foraminifera events in correlation, both within the basin and from basin to outcrop. Tectonic movements and growth of the Valhall anticline from middle Turonian to Danian time is another important control on the distribution of microfossil assemblages, especially in redeposition across major lithostratigraphic boundaries and the downslope movement and sorting of assemblages during the Maastrichtian–early Danian. Cooling during the post-Cenomanian late Cretaceous is evident in Turonian–early Campanian assemblages recovered from these chalks, although assemblages were generally cosmopolitan during this time. Palaeogeographic segregation from lower latitudes first occurred during the late early Campanian and these trends mirror those observed in the Southern Ocean until the terminal Maastrichtian. Additional study of Campanian–Maastrichtian chalks of the North Sea, including isotope geochemistry and magnetostratigraphy, is needed to address palaeogeographic parameters and redeposition and their effect on chronostratigraphic correlations.

We would like to thank Amoco Norway Oil Company and its partners (Amerada, Elf, Enterprise), as well as Amoco Exploration and Production Technology (Houston), for the opportunity to publish these results. Special thanks are extended to Cathy Farmer (Amoco, Houston) for initiating this work, as well as current members of the Chalk Exploration and Production Team at ANOC for their continued support. Comments and suggestions by Aase Moe of the Norwegian Petroleum Directorate have been especially helpful in directing our research in the southern Norwegian chalks. We would also like to thank David K. Watkins (University of Nebraska) and Dr Osman Varol (Varol Research) for their helpful suggestions and modifications of the original manuscript in their reviews.

Appendices

The biozonations described herein stress the sequences of calcareous nannofossil and microfossil events. The time scale used is that of Gradstein *et al.* (1995). The results of the second International Symposium on Cretaceous Stage Boundaries, held in Brussels during September of 1995, were published by Rawson *et al.* (1996). Concrete recommendations resulting from this symposium included subdivisions of the Upper Cretaceous stages (i.e. two- or three-fold), as well as potential boundary stratotype sections and/or fossil criteria for both stage and substage boundaries (Fig. 9). Calibration of the North Sea chalk biozonations proposed herein to Upper Cretaceous stage and substage definitions is preliminary and based mainly on nannofossil correlations, but includes both unpublished and published data from recommended global stratotypes and/or correlation to recommended fossil criterion for these boundaries in circum-North Sea outcrops (Fig. 9). Scaling of these events in geological time will undergo significant revision with enhanced correlations to: (1) the $^{40}Ar/^{39}Ar$ radiometric ages of Obradovich (1993); (2) Campanian–Maastrichtian magnetostratigraphy; and (3) orbital periodicities identified in marine stratigraphic records. A final step is the correlation of the current chronostratigraphic framework for the southern Norwegian chalks to sequence stratigraphy (e.g. Gale 1996).

Zones and subzones are described and numbered from the top of the Cretaceous down-section. Abbreviations are used for the first downhole occurrence (FDO) and lowest downhole occurrence (LDO) in both the nannofossil and microfossil biozonations.

Appendix A: Calcareous nannofossil biostratigraphy

Four cosmopolitan Upper Cretaceous nannofossil biozonations have been published over the past two decades (Fig. 10). The CC zones of Sissingh (1977, 1978) are based largely on low- to middle-latitude outcrop sections, whereas the NC zones of Roth (1978) are founded on deep-sea sections. Perch-Nielsen (1985) refined the biozonation of Sissingh (1977, 1978), whereas Bralower *et al.* (1995) modified the biozonation of Roth (1978). Upper Cretaceous nannofossil biozonations of circum-North Sea outcrop sections have excluded the Maastrichtian, but are based on study of the Cenomanian–mid-Campanian of southern England (Crux 1982) and the German Campanian–lowermost Maastrichtian (Burnett 1991). Application

STAGES	Substage	CRETACEOUS STRATIGRAPHIC SYMPOSIUM - BRUSSELS 1995		NORTH SEA CALIBRATIONS	
		AMM - ammonite CRIN - crinoid INOC - inoceramid NN - nannofossil		PF - planktonic foram HO - highest occurrence LO - lowest occurrence	
		BOUNDARY EVENT	BOUNDARY STRATOTYPE	BOUNDARY EVENT - NORTH SEA	REFERENCE SECTION/ BOUNDARY EVENT
Maastrichtian	U	iridium event/boundary clay	El Kef, Tunisia		El Kef, Tunisia Brazos River, Texas
	L	no decision	no decision - Zumaya, NE Spain probable	HCO *Calculites obscurus* NN	approximation
71.3 Brussels		LO *Pachydiscus neubergicus* AMM between HO *G.obliqueclausus* and HO *Q.gothicum/Q.trifidum* NN	Tercis Quarry, France @ 116.9m	base Maastrichtian at 71.3Ma and Tercis is much younger than North Sea usage	
74.0 N.Sea Campanian	U	Campanian substages no formal proposals - 3-fold subdivision favored as used in the Western Interior Basin, United States		HO *Reinhardtites anthophorus* NN Bralower *et al.* (1995) used ~74 Ma for HO *Globotruncanita calcarata*	Mississippi, USA & El Kef, Tunisia HO *Globotruncanita calcarata* PF co-eval in low-latitudes with HO *Reinhardtites anthophorus* NN
	M			HO *Eiffellithus eximius* NN (bar angle 0-5 degrees)	mid portion magnetochron 33N: Bottacione, Monechi & Thierstein, 1985 Site 530A, Stradner & Steinmetz, 1984
	L			HO *Lithastrinus grillii* NN	just below magnetochron 33N/33R: DCH-1 Core - Mississippi, USA & Bralower *et al.* (1995)
83.5		HO *Marsupites testudinarius* CRIN	Waxahachie, Texas or Seaford Head, Sussex, UK	LO *Amphizygus sp. 1* NN	Whitecliff, Isle Wight, UK Waxahachie, Waco, Texas HO *Marsupites testudinarius* CRIN
Santonian	U	? LO *Uintacrinus socialis* CRIN	no decision	LO *Actinozygus regularis* NN	Whitecliff, Isle Wight, UK LO *Uintacrinus socialis* CRIN
	M	? LO *Cordiceramus cordiformis* or ? HO *C. undulatoplicatus* INOC	no decision	HCO *Watznaueria porta* NN	Whitecliff, Isle Wight, UK LO *Cordiceramus cordiformis* INOC
	L				
85.8		LO *Cladoceramus undulatoplicatus* INOC	Olazagutia Quarry, Spain; Seaford Head, England; or 10-mile Creek, Dallas, Texas	HO *Amphizygus brooksii* (early form) NN	Whitecliff, Isle Wight, UK LO *Cladoceramus* INOC
Coniacian	U	LO *Magadiceramus subquadratus* INOC	Dallas-Ft. Worth area, Texas or Seaford Head, southern England	HO *Quadrum intermedium* NN	Whitecliff, Isle Wight, UK LO *Platyceramus* INOC
	M	LO *Volviceramus koenini* INOC	Dallas-Ft. Worth area, Texas or Seaford Head, southern England	LO *Micula staurophora* NN	Whitecliff, Isle Wight, UK - top HG in *Micraster cortestudinarium* Zone
	L				
89.0		LO *Cremnoceramus rotundatus* INOC	Salzgitter-Salder Quarry, Lower Saxony, N.Germany within base of Bed 45	LO *Reinhardtites biperforatus* NN	Culver Cliff, Isle Wight ,UK - base *Micraster cortestudinarium* Zone
Turonian	U	no decision - either LO *Romaniceras deverianum* or *S. neptuni* AMM	no decision	HO *Chiastozygus platyrhethum* NN	North Sea - middle/upper Turonian unconformity
	M	LO *Collignoniceras woollgari* AMM	Rock Canyon Anticline, W of Pueblo, Colo., base of Bed 120 in Bridge Ck. Mbr.	between HO *Percivalia fenestratus* NN & LO *Eprolithus rarus* NN	Rock Canyon Anticline, Colo. & Rebecca Bounds Core, Kansas
	L				
93.5		LO *Watinoceras devonense* AMM	Rock Canyon Anticline, W of Pueblo, Colo., base of Bed 86 in Bridge Ck. Mbr.	from: Watkins *et al.* (1993) HO *Microstaurus chiastius* NN or LO *Quadrum gartneri* NN	Rock Canyon Anticline, W. of Pueblo, Colo., base of Bed 86
Cenomanian	U				

Fig. 9. Turonian–Maastrichtian stage and substage boundary criteria: global v. North Sea (after Burnett 1996).

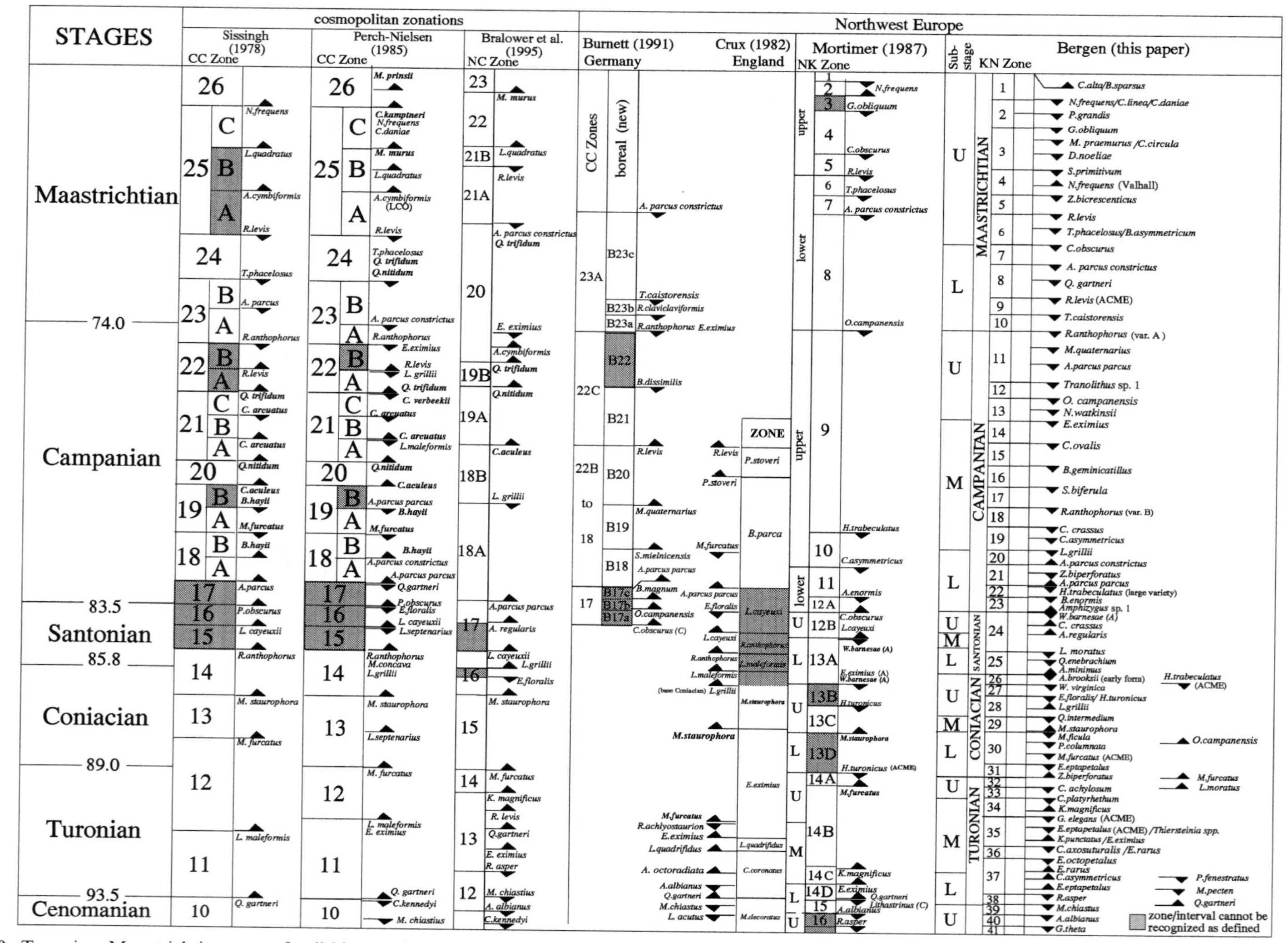

Fig. 10. Turonian–Maastrichtian nannofossil biozonations.

of all these biozonations (Fig. 10) to operational work in the North Sea is reduced by their reliance on first occurrence events and the use of several low-latitude events in the cosmopolitan biozonations, especially for the Campanian–Maastrichtian.

Mortimer (1987) published the only Upper Cretaceous nannofossil biozonation for the southern Norwegian and Danish sectors of the North Sea basin, which relied heavily on acme events on first occurrence data points in the Santonian–Turonian (Fig. 10). Mortimer (1987) included two summary range charts illustrating the occurrences on many useful secondary events. The Turonian–Maastrichtian biozonation presented herein (see below) is for the southern sector of the Norwegian North Sea and includes three-fold subdivisions of the Coniacian–Campanian stages (Fig. 10).

Of particular importance is the inability to reproduce any of the existing Santonian nannofossil biozonations schemes. When coupled with North Sea microfossil biozonations founded on benthic foraminifera successions in outcrop (King *et al.* 1989), this has lead to erroneous correlations within the North Sea basin. Also important is the definition of the Campanian–Maastrichtian boundary. An older definition of this boundary is maintained in this paper (highest occurrence of the calcareous nannofossil *Reinhardtites anthophorus*) consistent with historical usage within the North Sea basin (Mortimer 1987; Burnett 1991), as opposed to correlations to the potential boundary stratotype near Tercis, France (Odin 1996). This situation is further complicated by the condensed nature of the boundary interval within the study area and palaeogeographic endemism.

The biozonation described below will be revised to: (1) extend to the base of the Upper Cretaceous; (2) include subzone definitions; (3) emend taxonomic concepts and assignments; (4) illustrate all taxa; and (5) establish authorship. The nannofossil binomials used herein can be found in Perch-Nielsen (1985), Varol (1992), Varol & Girgis (1994) and Bralower & Bergen (1998).

Arkhangelskiella cymbiformis *zone (KN1)*

Age: late Maastrichtian.

Definition: interval from the LDO of *Cyclagelosphaera alta* or *Biantholithus sparsus* to the FDO of *Nephrolithus frequens*, *Cribrosphaerella linea* or *Cribrosphaerella daniae*.

Remarks: this zonal definition differs from Mortimer (1987), who used the FDO of common Cretaceous nannofossils to define the Cretaceous–Tertiary boundary.

Nephrolithus frequens *zone (KN2)*

Age: late Maastrichtian.

Definition: interval from the FDO of *Nephrolithus frequens*, *Cribrosphaerella linea* or *Cribrosphaerella daniae* to the FDO of *Gartnerago obliquum*.

Remarks: this zonal definition differs from Mortimer (1987), who used the total range of *Nephrolithus frequens*.

Gartnerago obliquum *zone (KN3)*

Age: late Maastrichtian.

Definition: interval from the FDO of *Gartnerago obliquum* to the FDO of *Seribiscutum primitivum*.

Remarks: Mortimer (1987) used the HO of *Calculites obscurus* to mark the base of this zone.

Seribiscutum primitivum *zone (KN4)*

Age: late Maastrichtian.

Definition: interval from the FDO of *Seribiscutum primitivum* to the FDO of *Zeugrhabdotus bicrescenticus*.

Zeugrhabdotus bicrescenticus *zone (KN5)*

Age: late Maastrichtian.

Definition: interval from the FDO of *Zeugrhabdotus bicrescenticus* to the FDO of *Reinhardtites levis*.

Remarks: *Zeugrhabdotus bicrescenticus* (Stover, 1966) is a senior synonym of the basionym *Zygolithus compactus* Bukry (1969).

Reinhardtites levis *zone (KN6)*

Age: late Maastrichtian.

Definition: interval from the FDO of *Reinhardtites levis* to the FDO of *Calculites obscurus*.

Remarks: Mortimer (1987) used the HO of *Tranolithus phacelosus* to mark the base of this zone.

Calculites obscurus *zone (KN7)*

Age: early Maastrichtian.

Definition: interval from the FDO of *Calculites obscurus* to the FDO of *Aspidolithus parcus constrictus*.

Remarks: Mortimer (1987) used different criteria for definition of this zone.

Aspidolithus parcus constrictus *zone (KN8)*

Age: early Maastrichtian.

Definition: interval from the FDO of *Aspidolithus parcus constrictus* to the FDO of the acme of *Reinhardtites levis*.

Markalius apertus *zone (KN9)*

Age: early Maastrichtian.

Definition: interval from the FDO of the acme of *Reinhardtites levis* to the FDO of *Tortolithus caistorensis*.

Tortolithus caistorensis *zone (KN10)*

Age: early Maastrichtian.
Definition: interval from the FDO of *Tortolithus caistorensis* to the FDO of *Reinhardtites anthophorus* (var A).
Remarks: smaller forms of *Reinhardtites anthophorus* having greater rim to central area width ratios are referred to variation A and mark the extinction of the species.

Reinhardtites anthophrous *zone (KN11)*

Age: late Campanian.
Definition: interval from the FDO of *Reinhardtites anthophorus* (var. A) to the FDO of *Tranolithus* sp. 1.
Remarks: *Tranolithus* sp. 1 is reserved for forms having four bright (first-order white birefringence) blocks in the central area.

Chiastozygus fessus *zone (KN12)*

Age: late Campanian.
Definition: interval from the FDO of *Tranolithus* sp. 1 to the FDO of *Orastrum campanensis.*

Orastrum campanensis *zone (KN13)*

Age: late Campanian.
Definition: interval from the FDO of *Orastrum campanensis* to the FDO of *Eiffellithus eximius.*
Remarks: Mortimer (1987) used the FDO of *Helicolithus trabeculatus* to mark the base of this zone, but this species was described from the Maastrichtian.

Eiffellithus eximius *zone (KN14)*

Age: middle Campanian.
Definition: interval from the FDO of *Eiffellithus eximius* to the FDO of *Calculites ovalis.*

Calculites ovalis *zone* (KN15)

Age: middle Campanian.
Definition: interval from the FDO of *Calculites ovalis* to the FDO of *Bifidalithus geminicatillus.*

Bifidalithus geminicatillus *zone (KN16)*

Age: middle Campanian.
Definition: interval from the FDO of *Bifidalithus geminicatillus* to the FDO of *Saepiovirgata biferula.*

Saepiovirgata biferula *zone (KN17)*

Age: middle Campanian.
Definition: interval from the FDO of *Saepiovirgata biferula* to the FDO of *Reinhardtites anthophorus* (var B).
Remarks: larger forms of *Reinhardtites anthophorus* having lower rim to central area width ratios are referred to as variation B.

Reinhardtites anthophorus *(var. B) zone (KN18)*

Age: middle Campanian.
Definition: interval from the FDO of *Reinhardtites anthophorus* (var. B) to the FDO of *Cylindralithus crassus.*

Cylindralithus crassus *zone (KN19)*

Age: middle Campanian.
Definition: interval from the FDO of *Cylindralithus crassus* to the FDO of *Lithastrinus grillii.*

Lithastrinus grillii *zone (KN20)*

Age: early Campanian.
Definition: interval from the FDO of *Lithastrinus grillii* to the LDO of *Aspidolithus parcus constrictus.*

Aspidolithus parcus parcus *zone (KN21)*

Age: early Campanian.
Definition: interval from the LDO of *Aspidolithus parcus constrictus* to the LDO of *Aspidolithus parcus parcus.*

Helicolithus trabeculatus *zone (KN22)*

Age: early Campanian.
Definition: interval from the LDO of *Aspidolithus parcus parcus* to the FDO of *Broinsonia enormis.*
Remarks: Mortimer (1987) used different criteria for the definition of this zone.

Broinsonia enormis *zone (KN23)*

Age: early Campanian.
Definition: interval from the FDO of *Broinsonia enormis* to the FDO of abundant *Watznaueria barnesae.*
Remarks: Mortimer (1987) defined this zone.

Watznaueria barnesae *zone (KN24)*

Age: late–middle Santonian.
Definition: interval from the FDO of abundant *Watznaueria barnesae* to the FDO of *Lithastrinus moratus*.

Lithastrinus moratus *zone (KN25)*

Age: early Santonian.
Definition: interval from the FDO of *Lithastrinus moratus* to the LDO of *Amphizygus minimus* or the FDO of *Amphizygus brooksii* (early form).
Remarks: early forms of *Amphizygus brooksii* have a broad, bright inner rim cycle and a relatively narrow, faint outer rim cycle in cross-polarized light (see Bralower & Bergen 1998, plate 2, Fig. 7).

Kamptnerius punctatus *zone (KN26)*

Age: late Coniacian.
Definition: interval from the LDO of *Amphizygus minimus* or the FDO of *Amphizygus brooksii* (early form) to the FDO of *Watznaueria virginica*.

Watznaueria virginica *zone (KN27)*

Age: late Coniacian.
Definition: interval from the FDO of *Watznaueria virginica* to the FDO of *Eprolithus floralis* or *Helicolithus turonicus*.

Helicolithus turonicus *zone (KN28)*

Age: late Coniacian.
Definition: interval from the FDO of *Eprolithus floralis* or *Helicolithus turonicus* to the FDO of *Quadrum intermedium*.

Quadrum intermedium *zone (KN29)*

Age: middle Coniacian.
Definition: interval from the FDO of *Quadrum intermedium* to the FDO of *Miravetesina ficula* or LDO *Micula staurophora*.

Miravetesina ficula *zone (KN30)*

Age: early Coniacian.
Definition: interval from the FDO of *Miravetesina ficula* or LDO *Micula staurophora* to the FDO of *Eprolithus eptapetalus*.

Eprolithus eptapetalus *zone (KN31)*

Age: early Coniacian.
Definition: interval from the FDO *Eprolithus eptapetalus* to the LDO of *Zeugrhabdotus biperforatus* or *Marthasterites furcatus*.

Cylindralithus coronatus *zone (KN32)*

Age: late Turonian.
Definition: interval from the LDO of *Zeugrhabdotus biperforatus* or *Marthasterites furcatus* to the FDO of *Stoverius achylosus*.

Stoverius achylosus *zone (KN33)*

Age: late Turonian.
Definition: interval from FDO of *Stoverius achylosus* to the FDO of *Chiastozygus platyrhethum*.

Chiastozygus platyrhethum *zone (KN34)*

Age: middle Turonian.
Definition: interval from FDO of *Chiastozygus platyrhethum* to the FDO of the acme of *Glaukolithus elegans* (*sensu* Roth & Thierstein, 1972).

Glaukolithus elegans *zone (KN35)*

Age: middle Turonian.
Definition: interval from the FDO of the acme of *Glaukolithus elegans* (*sensu* Roth & Thierstein 1972) to the FDO of *Calculites axosuturalis*.

Calculites axosuturalis *zone (KN36)*

Age: middle Turonian.
Definition: interval from the FDO of *Calculites axosuturalis* to the FDO of *Eprolithus octopetalus*.

Eprolithus octopetalus *zone (KN37)*

Age: middle–early Turonian.
Definition: interval from the FDO of *Eprolithus octopetalus* to the FDO of *Rhagodiscus asper*.

Rhagodiscus asper *zone (KN38)*

Age: early Turonian.
Definition: interval from the FDO of *Rhagodiscus asper* to the FDO of *Microstaurus chiastius*.
Remarks: Mortimer (1987) defined the base of this zone on the FDO of small to medium-sized *Seribiscutum primitivum*.

Microstaurus chiastius zone (KN39)

Age: late Cenomanian.

Definition: interval from the FDO of *Microstaurus chiastius* to the FDO of *Axopodorhabdus albianus*.

Axopodorhabdus albianus zone (KN40)

Age: late Cenomanian.

Definition: interval from the FDO of *Axopodorhabdus albianus* to the FDO of *Gartnerago theta*.

Gartnerago theta zone (KN41)

Age: middle Cenomanian.

Definition: interval from the FDO of *Gartnerago theta* to the FDO of *Gartnerago nanum*.

Appendix B: Microfossil biostratigraphy

The following microfossil biozonation for the Valhall Megaclosure (Fig. 11) employs a mixture of planktic and benthic foraminifera events, as well as a single radiolarian event. Abundance events have been avoided because of potential diachronism and redeposition. The stratigraphic ranges of the benthic foraminifera are not based solely upon their ranges in outcrop, but have been calibrated through other microfossil groups (nannofossils and palynomorphs) to the stage and substage terminology. Many zonal and subzonal taxa have rare, yet consistent, occurrences within the basin, such as the highest occurrences of *Globotruncana fornicata* in the middle late Maastrichtian and *Dicarinella imbricata* in the late Coniacian. Other marker taxa occur more frequently, but are obscured by more abundant occurrences of longer-ranging species within the same genus (e.g. *Whiteinella aprica* or *Marginotruncana coronata*). Thus, careful examination of washed residues and the use of thin sections in more indurated lithologies is required. Although this is more time-consuming, the use of more quickly determined and easily recognized data points (such as abundance events) can result in reduced stratigraphic resolution and erroneous correlations.

Pseudotextularia elegans zone (NCF1)

Age: late Maastrichtian.

Definition: interval from the FDO of the Tethyan heterohelicids *Pseudotextularia elegans* and *Racemiguembelina fructicosa* to the FDO of *Globotruncana fornicata*.

Remarks: the top of this zone is often marked by the FDO of several other Tethyan planktic species including *Pseudoguembelina excolata*, *Pseudoguembelina palpebra*, *Abathomphalus mayaroensis* and *Contusotruncana contusa*. Benthic species having the highest occurrences at the top of this zone include *Bolivinoides draco*, *Cibicidoides voltziana* and *Stensioina pommerana*. Recognition of zone NCF1 is often problematic in the North Sea chalk. In fine-grained, allochthonous chalks found well off-structure in the Valhall Megaclosure, the primary zonal markers have often been winnowed out. The appearances of the Tethyan planktic foraminifera taxa are diachronous regionally. Finally, the characteristic assemblage is often abundantly reworked into the basal Danian in the Valhall Megaclosure, so that careful comparisons must be made with other microfossil disciplines so that the top of the zone is not identified prematurely in well sections.

Subzone definition. *Pseudotextularia elegans* subzone (NCF1a): interval from the FDO of *P. elegans* and/or *R. fructicosa* to the FDO of the planktic foraminifer *Abathomphalus intermedia*.

Abathomphalus intermedia subzone (NCF1b): interval from the FDO of *A. intermedia* to the FDO of *G. fornicata*.

Rugoglobigerina pennyi zone (NCF2)

Age: late–early Maastrichtian.

Definition: interval from the FDO of *Globotruncana fornicata* to the FDO of the planktic species *Globotruncana lapparenti* and/or the morphologically similar taxon *Globotruncana linneiana*.

Remarks: the top of the zone is marked by the FDO of *Globotruncana fornicata*, but is also characterized for most of its duration by common occurrences of the another planktic species, *Rugoglobigerina pennyi*. Subsidiary components include often diverse *Globotruncanella* species, including *G. havanenesis* and *G. petaloidea*. The zone is easily recognized both on- and off-structure throughout the Valhall Megaclosure, although much of the lower portion of the zone is usually composed of poorly fossiliferous, fine-grained allochthonous chalks.

Subzone definition. *Globotruncana fornicata* subzone (NCF2a): interval from the FDO of the nominate species to the FDO of the planktic foraminifer *Rugoglobigerina pennyi* and/or the benthic foraminifer *Bolivinoides miliaris*.

Rugoglobigerina pennyi subzone (NCF2b): interval from the FDO of *R. pennyi* and/or *B. miliaris* to the FDO of *G. lapparenti* and/or *G. linneiana*.

Globotruncana lapparenti zone (NCF3)

Age: middle early Maastrichtian.

Definition: interval from the FDO of *G. lapparenti* and/or *G. linneiana* to the FDO of the large agglutinated benthic foraminifer *Tritaxia capitosa*.

Remarks: this zone is characterized by autochthonous chalks with diverse assemblages of planktic and benthic foraminifera. This facies marks a prominent change from the fine-grained allochthonous chalks characteristic of the lower portion of the overlying zone NCF2.

Tritaxia capitosa zone (NCF4)

Age: early Maastrichtian.

Definition: interval from the FDO of *T. capitosa* to the FDO of the benthic foraminifer *Gavelinella monterelensis*.

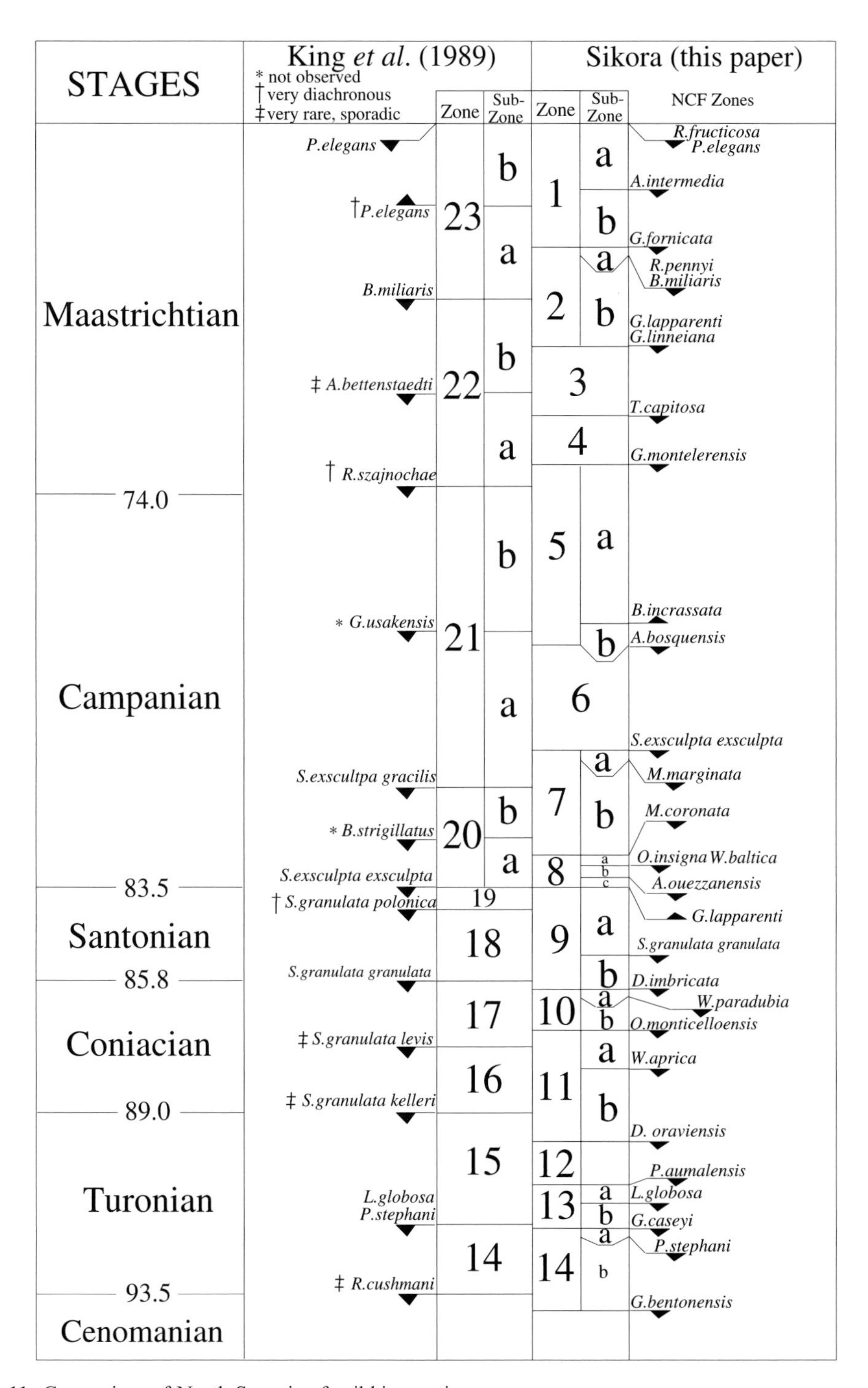

Fig. 11. Comparison of North Sea microfossil biozonations.

Remarks: the extinction of *Tritaxia capitosa* has previously been used to approximate the terminal Campanian in the North Sea basin (King *et al.* 1989), but correlation to the nannofossil succession shows it to range higher than *Reinhardtites anthophorus* (zone KN11) and roughly coincident with the FDO of common *Reinhardtites levis* (zone KN9). Zone NCF4 is similar in foraminiferal composition to the overlying *G. lapparenti* zone (NCF3), but is much more stratigraphically and geographically discontinuous across the Valhall Megaclosure.

Gavelinella monterelensis *zone (NCF5)*

Age: earliest Maastrichtian–late middle Campanian.

Definition: interval from the FDO of *G. monterelensis* to the FDO of the planktic foraminifer *Archaeoglobigerina bosquensis*.

Remarks: although the highest occurrence of *Gavelinella monterelensis* has been used to approximate the terminal Campanian in outcrop (Hart *et al.* 1989), this event consistently occurs immediately above the FDO of the calcareous nannofossil *Reinhardtites anthophorus* (var. A) in the wells studied. This interval includes autochthonous chalks that are often condensed, but are characterized by abundant and diverse assemblages of planktic and benthic foraminifera assemblages. Most of these foraminifer taxa range into the overlying zones NCF3 and NCF4. Zone NCF5 also includes discontinuous occurrences of pyritic chalks containing *Stensioina granulata polonica*, whose extinction previously has been used to mark the top of the middle Santonian.

Subzone definition. Gavelinella monterelensis subzone (NCF5a): interval from the FDO of *G. monterelensis* to the LDO of the benthic foraminifer *Brizalina incrassata*.

Brizalina incrassata subzone (NCF5b): interval from the LDO of *B. incrassata* to the FDO of *A. bosquensis*.

Archaeoglobigerina bosquensis *zone (NCF6)*

Age: middle Campanian.

Definition: interval from the FDO of the planktic foraminifer *Archaeoglobigerina bosquensis* to the FDO of the benthic foraminifer *Stensioina exsculpta exsculpta*.

Remarks: consistent, yet rare, occurrences of *A. bosquensis* mark the top of this zone. Otherwise, the zone is generally characterized by diverse assemblages of planktic and benthic foraminifera very similar to those noted for zone NCF5.

Stensioina exsculpta exsculpta *zone (NCF7)*

Age: early middle–late early Campanian.

Definition: interval from the FDO of the benthic foraminifera *Stensioina exsculpta exsculpta* to the FDO of the planktic foraminifer *Marginotruncana coronata*.

Remarks: *Stensioina exsculpta exsculpta* and the commonly associated taxon *S. exsculpta gracilis* extend into much younger section in the basin than in outcrop. The respective highest occurrences of *Stensioina exsculpta exsculpta* and *S. exsculpta gracilis* in outcrop occur at the top of the Santonian and within the lower Campanian (Hart *et al.* 1989), whereas both species occur well into the middle Campanian in the basin. Furthermore, the only consistent occurrences of *S. exsculpta gracilis* in the basinal chalk sections are within this zone.

Subzone definition. Stensioina exsculpta exsculpta subzone (NCF7a): interval from the FDO of *Stensioina exsculpta exsculpta* to the FDO of the planktic species *Marginotruncana marginata*.

Marginotruncana marginata subzone (NCF7b): interval from the FDO of *M. marginata* to the FDO of *M. coronata*. The top of subzone NCF7b is often marked by the first down-section occurrence of common spherical radiolarians.

Marginotruncana coronata *zone (NCF8)*

Age: earliest Campanian.

Definition: interval from the FDO of the planktic foraminifer *M. coronata* to the LDO of the planktic foraminifer *Globotruncana lapparenti*.

Remarks: although of short duration, this zone is characterized by condensed, autochthonous chalks containing extremely diverse planktic foraminifera and deep-water benthic foraminifera. It marks both a major flooding event and the culmination of a period of climatic warming.

Subzone definition. Marginotruncana coronata subzone (NCF8a): interval from the FDO of *M. coronata* to the FDO of the planktic foraminifer *W. baltica* and/or the cosmopolitan benthic foraminifer *Osangularia insigna*.

Whiteinella baltica subzone (NCF8b): interval from the FDO of *W. baltica* and/or *Osangularia insigna* to the FDO of the benthic foraminifer *Aragonia ouezzanensis*.

Aragonia ouezzanensis subzone (NCF8c): interval from the FDO the nominate species to the LDO of *Globotruncana lapparenti*. This subzone contains the total stratigraphic range within the basinal chalk section of the benthic foraminifer *A. ouezzanensis*. The basal Campanian subzone NCF8c was encountered in only two well sections in the Valhall Megaclosure.

Stensioina granulata granulata *zone (NCF9)*

Age: Santonian–latest Coniacian.

Definition: interval from the LDO the planktic foraminifer *Globotruncana lapparenti* to the FDO of the planktic foraminifer *Dicarinella imbricata*.

Remarks: the lowest occurrences of several other taxa also mark the top of this zone, including the benthic foraminifera *Aragonia ouezzanensis* and *Stensioina pommerana*.

Subzone definition. Gavelinella cristata subzone (NCF9a): interval from LDO of *G. lapparenti* to FDO of the benthic foraminifer *Stensioina granulata granulata*. This subzone is recognized in outcrop (Whitecliff, Isle of Wight, UK), but has not been encountered in the Valhall Megaclosure.

Stensioina granulata granulata subzone (NCF9b): interval from the FDO of *S. granulata granulata* to the FDO of *Dicarinella imbricata*. In the basin subzone NCF9b is only present in thin intervals located off the anticlinal crest.

Dicarinella imbricata *zone (NCF10)*

Age: late–middle Coniacian.

Definition: interval from the FDO of the planktic foraminifer *D. imbricata* to the FDO of the radiolarian *Orbiculoformis monticelloensis*.

Remarks: the top of zone NCF10 is immediately below the terminal Coniacian and marks the first down-section occurrence of the typical highly pelagic facies of the Hod Formation. The interval is usually dominated by stratigraphically long-ranging planktic foraminifera, such as *Marginotruncana marginata* and *Whiteinella baltica*. The zone is often absent from crestal Valhall due to erosional truncation.

Subzone definition. Dicarinella imbricata subzone (NCF10a): interval from the FDO of the nominate species to the FDO of the planktic foraminifer *Whiteinella paradubia*.

Whiteinella paradubia subzone (NCF10b): interval from the FDO of *W. paradubia* to the FDO of the radiolarian *Orbiculoformis monticelloensis*.

Whiteinella aprica *zone (NCF11)*

Age: early middle Coniacian–latest Turonian.

Definition: interval from the FDO of the radiolarian *Orbiculoformis monticelloensis* to the FDO of the planktic foraminifer *Dicarinella(?) oraviensis* (a provisional species designation by Robaszynski *et al.* 1990).

Remarks: the zone is characterized by monotonous, expanded sections of massive chalks dominated by planktic foraminifera and radiolarians.

Subzone definition. Orbiculoformis monticelloensis subzone (NCF11a): interval from the FDO of *O. monticelloensis* to the FDO of the planktic foraminifer *Whiteinella aprica*.

Whiteinella aprica subzone (NCF11b): interval from the FDO of the nominate species to the FDO of the planktic foraminifer *Dicarinella(?) oraviensis*.

Dicarinella(?) oraviensis *zone (NCF12)*

Age: late Turonian.

Definition: interval from the FDO of the planktic foraminifer *D.(?) oraviensis* to the FDO of the planktic foraminifer *Praeglobotruncana aumalensis*.

Remarks: a regional Turonian–Coniacian unconformity is present in the Valhall Megaclosure. Therefore, calibration of the boundary between zones NCF11 and NCF12 is provisional and will be refined with study of additional sections. The specific composition of zone NCF12 is similar to that of zone NCF11, although there is a general increase in the abundance of keeled planktic foraminifera indicating deeper water deposition and/or warmer water masses.

Praeglobotruncana aumalensis *zone (NCF13)*

Age: late middle Turonian.

Definition: interval from the FDO of the planktic foraminifer *P. aumalensis* to the FDO of the planktic foraminifer *Globigerinelloides caseyi*.

Remarks: this zone is characterized by an influx of *Praeglobotruncana* species, also including *P. kalaati* and *P. hilalensis*. Many of the constituent planktic foraminifera have Tethyan affinities and this zone marks the first major Tethyan influx down-section in the chalk succession below the upper Maastrichtian.

Subzone definitions. Praeglobotruncana kalaati subzone (NCF13a): interval from the FDO of *P. aumalensis* to the FDO of the deep-water benthic foraminifera *Lingulogavelinella globosa*. The LDO the planktic foraminifer *Marginotruncana coronata* is near the base of the subzone.

Lingulogavelinella globosa subzone (NCF13b): interval from the FDO of *L. globosa* to the FDO of *Globigerinelloides caseyi*. Rare and sporadic occurrences of the planktic foraminifer *Helvetoglobotruncana helvetica* also occur within this zone.

Praeglobotruncana stephani *zone (NCF14)*

Age: early middle Turonian–latest Cenomanian.

Definition: interval from the FDO of the planktic foraminifer *Globigerinelloides caseyi* to the FDO of the planktic foraminifer *Globigerinelloides bentonensis*.

Remarks: Planktic diversity remains high within this zone, which is also often marked by an increase in radiolarian abundance.

Subzone definition. Globigerinelloides caseyi subzone (NCF14a): interval from the FDO of the nominate species to the FDO of the planktic foraminifer *Praeglobotruncana stephani*.

Praeglobotruncana stephani subzone (NCF14b): interval from the FDO of the nominate species to the FDO of *G. bentonensis*. *P. stephani* is common throughout the lower portion of this subzone.

References

BERGGREN, W. A., KENT, D. V., SWISHER, C. C. & AUBRY, M.-P. 1995. A revised Cenozoic geochronology and chronostratigraphy. *In*: BERGGREN, W. A., KENT, D. V., AUBRY, M.-P. & HARDENBOL, J. (eds) *Geochronology, Time Scales and Global Stratigraphic Correlation*. Society of Economic Paleontologists and Mineralogists, Special Publication, **54**, 129–212.

BRALOWER, T. J. & BERGEN, J. A. 1998. Cenomanian–Santonian calcareous nannofossil biostratigraphy of a transect of cores drilled across the Western Interior Seaway. *In*: *Stratigraphy and Paleoenvironments of the Cretaceous Western Interior Seaway*. Society of Economic Paleontologists and Mineralogists Concepts in Sedimentology & Paleontology, **6**, 59–77.

——, LECKIE, R. M., SLITER, W. V., & THIERSTEIN, H. R. 1995. An integrated Cretaceous microfossil biostratigraphy. *In*: BERGGREN, W. A., KENT, D. V., AUBRY, M.-P. & HARDENBOL, J. (eds) *Geochronology, Time Scales and Global Stratigraphic Correlation*. Society of Economic Paleontologists and Mineralogists, Special Publication, **54**, 65–79.

BURNETT, J. A. 1991. New nannofossil zonation scheme for the Boreal Campanian. *International Nannoplankton Association Newsletter*, **12**(3), 67–70.

——1996. Nannofossils and Upper Cretaceous (sub-) stage boundaries – state of the art. *International Nannoplankton Association Newsletter*, **18**(1), 23–32.

CRUX, J. 1982. Upper Cretaceous (Cenomanian to Companion) calcareous nannofossils. *In*: LORD, A. R. (ed.) *A Stratigraphical Index of Calcareous Nannofossils*. British Micropalaeontological Society. Ellis Horwood, Chichester, 81–135.

DEEGAN, C. E. & SCULL, B. J. 1977. *A Standard Lithostratigraphic Nomenclature for the Central and Northern North Sea*. Institute of Geological Science Report 77/25. *Norwegian Petroleum Directorate Bulletin*, **1**.

GALE, A. S. 1996. Turonian correlation and sequence stratigraphy of the chalk in southern England. *In*: HESSELBO, S. P. & PARKINSON, D. N. (eds) *Sequence Stratigraphy in British Geology*. Geological Society, London, Special Publications, **103**, 177–195.

GRADSTEIN, F. M., AGTERBERG, F. P., OGG, J. G., HARDENBOL, J., VAN VEEN, P., THIERRY, J. & HUANG, Z. 1995. A Triassic, Jurassic and Cretaceous time scale. *In*: BERGGREN, W. A., KENT, D. V., AUBRY, M.-P. & HARDENBOL, J. (eds) *Geochronology, Times Scale and Global Stratigraphic Correlation*. Society of Economic Paleontologists and Mineralogists, Special Publication, **54**, 95–126.

HART, M. B., BAILEY, H. W., CRITTENDEN, S., FLETCHER, B. N., PRICE, R. J. & SWIECICKI, A. 1989. Cretaceous. *In*: JENKINS, D. G. & MURRAY, J. W. (eds) *Stratigraphical Atlas of Fossil Foraminifera* (2nd edn). Ellis Horwood, Chichester, 273–371.

HUBER, B. T. & WATKINS, D. K. 1992. Biogeography of Campanian–Maastrichtian calcareous plankton in the region of the Southern Ocean: palaeogeographic and palaeoclimatic implications. *American Geophysical Union, Antarctic Research Series*, **56**, 31–60.

ISAKSEN, D. & TONSTAD, K. 1989. *A Revised Cretaceous and Tertiary Lithostratigraphic Nomenclature for the Norwegian North Sea. Norweigan Petroleum Directorate Bulletin*, **5**.

JENKYNS, H. C., GALE, A. S. A. & CORFIELD, R. M. 1994. Carbon- and oxygen-isotope stratigraphy of the English Chalk and Italian Scaglia and its palaeoclimatic significance. *Geological Magazine*, **131**, 1–34.

KING, C., BAILEY, H. W., BURTON, C. A. & KING, A. D. 1989. Cretaceous of the North Sea. *In*: JENKINS, D. G. & MURRAY, J. W. (eds) *Stratigraphical Atlas of Fossil Foraminifera* (2nd edn). Ellis Horwood, Chichester, 372–417.

KOCH, W. 1977. Biostratigraphie in der Oberkreide und Taxonomie von Foraminiferen. *Geologisches Jahrbuch*, **A38**, 11–123.

MORTIMER, C. P. 1987. Upper Cretaceous nannofossil biostratigraphy of the southern Norwegian and Danish North Sea area. *Abhandlungen Geol. B.-A.*, **39**, 143–175.

OBRADOVICH, J. D. 1993. A Cretaceous time scale. *In*: CALDWELL, W. G. E. & KAUFMANN, E. G. (eds) *Evolution of the Western Interior Basin*. Geological Association of Canada, Special Paper, **39**, 379–396.

ODIN, G. S. 1996. Definition of a global boundary stratotype section and point for the Campanian/Maastrichtian boundary. *In*: RAWSON, P. F., DHONDT, A. V., HANCOCK, J. M. & KENNEDY, W. J. (eds) *Proceedings of the Second International Symposium on Cretaceous Stage Boundaries, Brussels, 8–16 September 1995. Sci. Terre Aardwetenschappen*, **66**, Supplement, 111–117.

MANN, K. & LANE, H. R. (eds) 1989. *Graphic Correlation*. Society of Economic Paleontologists and Mineralogists, Special Publication, **53**.

PERCH-NIELSEN, K. P. 1985. Mesozoic calcareous nannofossils. *In*: BOLLI, H. M., SAUNDERS, J. B. & PERCH-NIELSEN, K. (eds) *Plankton Stratigraphy*. Cambridge University Press, Cambridge, 329–426.

POSPICHAL, J. J. & WISE, S. W. 1990. Maastrichtian calcareous nannofossil biostratigraphy of the Maud Rise ODP Leg Sites 689 and 690, Weddell Sea. *Proceedings of the Ocean Drilling Program*, **113**, 465–487.

RAWSON, P. F., DHONDT, A. V., HANCOCK, J. M. & KENNEDY, W. J. (eds) 1996. *Proceedings of the Second International Symposium on Cretaceous Stage Boundaries, Brussels, 8–16 September 1995. Sci. Terre Aardwetenschappen*, **66**, Supplement.

ROBASZYNSKI, F., CARON, M., DUPUIS, C., AMEDRO, F., GONZALEZ DONOSO, J., LINARES, D., HARDENBOL, J., GARTNER, S., CALANDRA, F. & DELOFFRE, R. 1990. A tentative integrated stratigraphy in the Turonian of central Tunisia: formations, zones, and sequential stratigraphy in the Kalaat Senan area. *Bulletin des Centres de Récherches Exploration–Production Elf-Aquitaine*, **14**, 213–384.

ROTH, P. H. 1978. Cretaceous nannoplankton biostratigraphy and oceanography of the northwestern Atlantic Ocean. *In*: BENSON, W. E., SHERIDAN, R. E. *et al.* (eds) *Initial Reports of the Deep Sea Drilling Project*, Volume 44. US Government Printing Office, Washington DC, 731–759.

—— & THIERSTEIN, H. R. 1972. Calcareous nannoplankton: Leg 14 of the Deep Sea Drilling Project. *In*: HAYES, D. E., PIMM, A. C. *et al.* (eds) *Initial Reports of the Deep Sea Drilling Project*, Volume 14. US Government Printing Office, Washington DC, 421–485.

SCHONFELD, J. & BURNETT, J. 1991. Biostratigraphical correlation of the Campanian–Maastrichtian boundary: Lagerdorf–Hemmoor (northwestern Germany), DSDP Sites 548A, 549 and 551 (eastern North Atlantic) with palaeobiogeographical and palaeoceanographical implications. *Geological Magazine*, **128**, 479–503.

SHAW, A. B. 1964. *Time in Stratigraphy*. McGraw-Hill, London.

SIKORA, P. J., BERGEN, J. A., & FARMER, C. L. 1999. Chalk palaeoenvironments and depositional model, Valhall–Hod fields, southern Norwegian North Sea. *This volume*.

SISSINGH, W. 1977. Biostratigraphy of Cretaceous calcareous nannoplankton. *Geologie en Mijnbouw*, **56**, 37–65.

——1978. Microfossil biostratigraphy and stage-stratotypes of the Cretaceous. *Geologie en Mijnbouw*, **57**, 433–440.

VAROL, O. 1989. Paleocene calcareous nannofossil biostratigraphy. *In*: CRUX, J. & VAN HECK, S. E. (eds) *Nannofossils and Their Applications*. British Micropalaeontological Society. Ellis Horwood, Chichester, 267–310.

——1992. Taxonomic revision of the Polycyclolithaceae and its contribution to Cretaceous stratigraphy. *Newsletters in Stratigraphy*, **27**, 93–127.

—— & GIRGUS, M. 1994. New taxa and taxonomy of the Jurassic and Cretaceous calcareous nannofossils. *Neues für Jahrbuch Geologie Palaeontologie Abhandlungen*, **192**, 221–253.

WATKINS, D. K. 1992. Upper Cretaceous nannofossils from Leg 120, Kerguelen Plateau, Southern Ocean. *Proceedings of the Ocean Drilling Program*, **120**, 343–370.

——, BRALOWER, T. J., COVINGTON, J. M., & FISHER, C. G. 1993. Biostratigraphy and palaeoecology of the Upper Cretaceous calcareous nannofossils in the Western Interior Basin, North America. *In*: CALDWELL, W. G. E. & KAUFMANN, E. G. (eds) *Evolution of the Western Interior Basin*. Geological Association of Canada, Special Paper, **39**, 521–537.

——, WISE, S. W., POSPICHAL, J. J. & CRUX, J. 1995. Upper Cretaceous calcareous nannofossil biostratigraphy and palaeoceanography of the Southern Ocean. *In*: *Microfossils and Oceanic Environments*. British Micropalaeontological Society. University of Wales, Aberystwyth Press, Aberystwyth, 355–381.

WISE, S. W. 1988. Mesozoic and Cenozoic history of calcareous nannofossils in the region of the Southern Ocean. *Palaeogeography, Palaeoclimatology and Palaeoecology*, **67**, 157–179.

Chalk palaeoenvironments and depositional model, Valhall–Hod fields, southern Norwegian North Sea

P. J. SIKORA, J. A. BERGEN & C. L. FARMER

Amoco Exploration and Production Company, P.O. Box 3092, Houston, TX 77253, USA

Abstract: Palaeoenvironmental analysis in the North Sea chalk is hampered by the prevalence of widespread allochthonous units, intervals of massive reworking of microfossils and an often monotonous lithology of massive chalks of uncertain depositional history. Most difficult of all is devising precise regional biostratigraphic correlations in a complex section with large, but very subtly expressed, disconformities and condensed zones. Facies change quickly both laterally and vertically. Without very precise regional correlations, it is impossible to formulate any detailed palaeoenvironmental model.

Through a multidisciplinary palaeontological approach, employing graphic correlation methodology and a composite standard database, a highly accurate and detailed regional chronostratigraphy was generated. Biofacies recognized within the section were then scaled in absolute time, allowing for more reliable recognition of facies trends at any one time horizon. Biofacies were formulated based upon recovered benthic foraminifera, planktic foraminifera, calcispheres, siliceous microfossils and macrofossil debris.

Biofacies mapping for several higher-order sequences recognized in the Upper Cretaceous chalks has allowed recognition of four major periods of deposition: (1) early Turonian–early Santonian high pelagic productivity and rapid deposition of chalks over a relatively continuous ridge from the Valhall to Hod fields; (2) early Campanian–earliest Maastrichtian deep-water autochthonous chalk deposition off-structure with structural crests mantled by submarine hardgrounds; (3) very heterogeneous autochthonous and allochthonous middle Maastrichtian chalk deposition controlled by local grabens formed from crestal collapse; and (4) widespread sediment flows and redeposition of sediments with crestal shoaling and winnowing in the latest Maastrichtian culminating in crestal erosion of Cretaceous chalks in the early Danian.

The Valhall and Hod fields (Fig. 1) are located in the southern part of the Norwegian Central Graben just north of the Danish border in blocks 2/8 and 2/11. All production is from chalk. The Tor Formation (Maastrichtian–early Danian) is the primary reservoir with secondary production from the Hod Formation (Fig. 2). Overlying middle Danian–early Selandian chalks and calcareous shales (Ekofisk Formation and Våle Claystone) are very thin, condensed, and exhibit high water saturation and/or low porosity. The chalk section was deposited during a tectonically active time beginning in the middle Turonian and continuing into the Danian. Depositional patterns were strongly controlled by tectonics, both regional (e.g. highly variable palaeobathymetry due to the Valhall anticline) and local (e.g. syn-depositional faulting caused by extensional collapse of the anticlinal crest) (Farmer *et al.* 1996).

The importance of lithofacies and diagenesis on the quality of chalk reservoirs has long been recognized and has been the focus of much study (e.g. Scholle 1977; Nygaard *et al.* 1983; Kennedy 1987; Brasher & Vagle 1996). Nevertheless, past well studies have often underutilized, if not ignored, the fossil content, the key tool for palaeoenvironmental interpretation within chalks. A few investigations have integrated palaeontological data with lithofacies and depositional mechanisms, but have strongly concentrated on ichnofacies and have treated the fossil content purely as an extension of lithofacies; that is, as bioclasts (e.g. Siemers *et al.* 1994). Furthermore, only very recently has a detailed integration of chalk biostratigraphy and sequence stratigraphy been proposed (Green *et al.* 1996). Nevertheless, detailed palaeoenvironmental analyses present an opportunity for substantially improving our understanding of chalk depositional mechanisms and, when integrated with sedimentology and structural geology, the overall regional depositional model.

Sikora, P. J., Bergen, J. A. & Farmer, C. L. 1999. Chalk palaeoenvironments and depositional model, Valhall–Hod fields, southern Norwegian North Sea. *In:* Jones, R. W. & Simmons, M. D. (eds) *Biostratigraphy in Production and Development Geology*. Geological Society, London, Special Publications, **152**, 113–137.

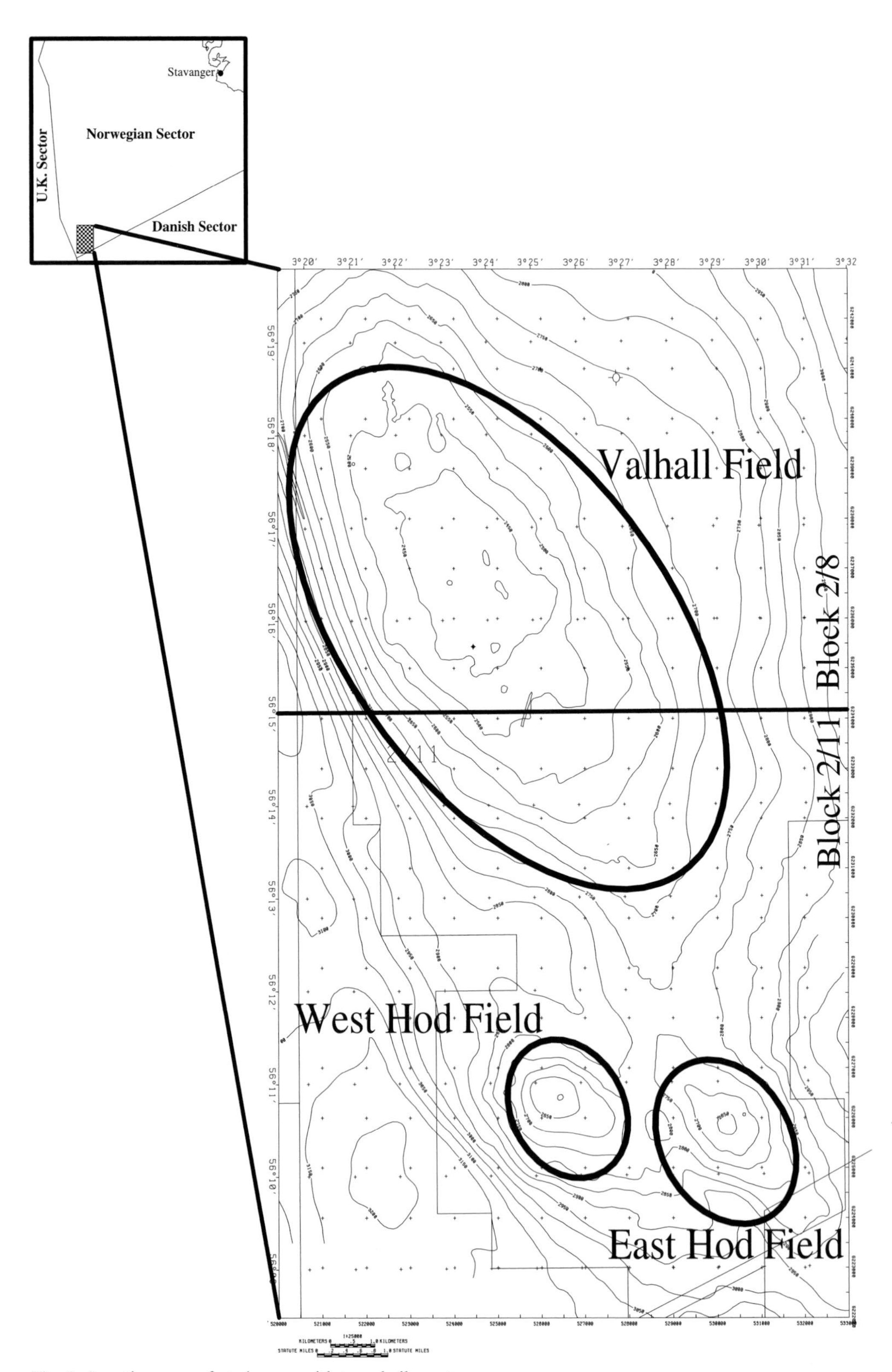

Fig. 1. Location map of study area with top chalk contours.

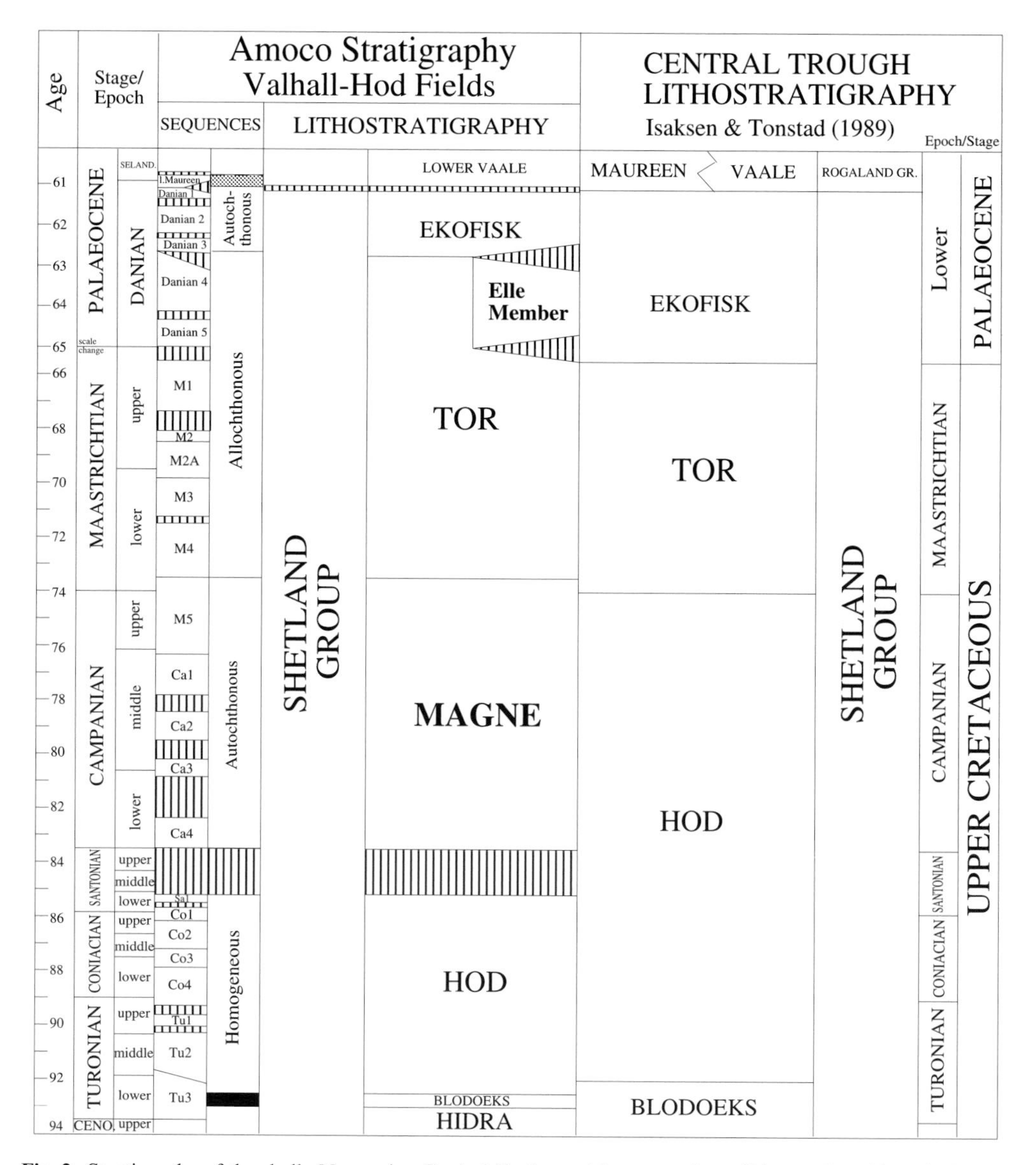

Fig. 2. Stratigraphy of the chalk, Norwegian Central Graben, with proposed new lithostratigraphic units in bold (time scale from Gradstein *et al.* 1995).

Stratigraphy

The Hod and Tor formations have traditionally been interpreted as representative of a relatively continuous period of pelagic deposition from the Turonian to the Maastrichtian (Fig. 2) (Isaksen & Tonstad 1989). However, improved chronostratigraphic control from graphic correlation analysis of several Valhall and Hod field wells has revealed a highly discontinuous stratigraphic section which can be divided into three major intervals (Fig. 2) (Bergen & Sikora 1999).

The lower interval is composed of the Turonian–lowermost Santonian Hod Formation and represents very rapid deposition of highly pelagic, mainly autochthonous chalks with a moderate amount of organic content. The near absence of benthic fossils seriously limits palaeobathymetric resolution as well as depositional modelling. The middle interval is characterized by Campanian–lowermost Maastrichtian, very condensed, organic-rich, argillaceous, autochthonous chalks previously assigned to either the Hod or Tor formations or to various informally named units (e.g. Supra-Hod Formation). This variable stratigraphic nomenclature makes it difficult to compare stratigraphic results from different sources, and is a point of confusion and potential cause for miscorrelation. It may therefore be desirable to designate this distinct stratigraphic interval as a new formation. The authors have held discussions with the Norwegian Petroleum Directorate on defining this new unit as the Magne Formation. However, the consensus was that more information is needed on the nature of this stratigraphic interval beyond the Valhall–Hod fields before formally proposing the formation. Nevertheless, for ease of reference, this interval will be informally referred to in the following discussion as the Magne Formation. Absent from crestal areas of Valhall, the Magne Formation progressively onlaps structure during Campanian time. Numerous condensed zones and disconformities representing periods of non-deposition occur throughout the unit. The contact between the Magne Formation and underlying Hod Formation is a major unconformity representing erosional truncation across the crest and upper flanks of Valhall.

The third unit is represented by largely Maastrichtian Tor Formation, composed of heterogeneous, partially allochthonous chalks of very low organic content and constituting the primary reservoir facies of the Valhall and Hod fields. Also included in this unit within the Valhall and Hod fields is a lowermost Danian interval previously incorrectly dated as Maastrichtian due to massive Cretaceous reworking of microfossils and nannofossils. This Danian section has reservoir and log parameters nearly identical to the Maastrichtian Tor Formation and is highly divergent from the overlying, condensed Ekofisk chalks. It is therefore included within the Tor Formation and may later warrant separation from the underlying Cretaceous as a member unit.

Relative palaeobathymetry

The late Cretaceous North Sea Chalk basin is a depositional system with no close past or modern analogue. Many benthic foraminifera common in the chalk have been assigned absolute palaeobathymetric limits based upon their occurrences in terrigenous passive margin sequences of open marine basins (e.g. Nyong & Olsson 1977; Van Morkhoven *et al.* 1986). Most of these are deep-water forms indicative of outer shelf to upper slope water depths of 200–500 m (e.g. *Stensioina beccariiformis*) or deeper (e.g. *Nuttallides truempyi*). However, it is not possible to conclude that these palaeoecological markers have the same palaeobathymetric ranges along the flanks of isolated highs in a largely enclosed basin under a radically different depositional setting as represented by the North Sea chalk.

Although absolute palaeobathymetric ranges cannot be made, it is essential to comprehend the relative palaeobathymetric succession of fossil assemblages within the chalk in order to fully understand chalk depositional mechanisms. However, even with relatively detailed transects from crestal areas into the basin, reconstructing the palaeobathymetric succession of the chalk microfossil assemblages is difficult. The Hod Formation contains few benthic facies markers. The Tor Formation is largely composed of allochthonous facies containing microfossils which are not indicative of palaeoenvironmental conditions at the site of deposition. A complete autochthonous palaeobathymetric succession across the region has not been recognized. Yet, given adequate age control, the succession can be reconstructed from incomplete successions occurring in different wells at different stratigraphic levels. Two major types of palaeobathymetric successions are recognized.

The first type is a low-resolution succession in the Turonian–Lower Santonian Hod Formation necessarily based upon variations in rich assemblages of planktic foraminifera and radiolarians. Shallower-water, non-keeled planktic species and low-diversity radiolarian assemblages dominate in crestal and upper flank regions, whereas deeper-water, keeled planktic foraminifera and high-diversity radiolarian assemblages predominate off-structure.

Greater palaeobathymetric resolution is possible in the Campanian–lower Danian (Magne and Tor formations) with four major palaeoecological zones: central crest; crest–flank transition; main flank; and lower flank–basin (Fig. 3). A distinct palaeobathymetric assemblage distinguishes each of the four zones. The central crestal area is dominated by abundant macrofossil debris, principally echinoid and bivalve, but with occasionally frequent bryozoan fragments. Benthic foraminifera are rare to common but always exhibit a low-diversity assemblage, strongly dominated by a few species such as *Stensioina beccariiformis* and *Cibicidoides hyphalus*. Calcispheres are rare to frequent and no planktic foraminifera are present. This ecological zone was absent during the time of deposition of the Magne Formation when the crestal area was marked by submarine hardground formation representing a geologically long period of non-deposition. The crest–flank transition is characterized by abundant, diverse benthic

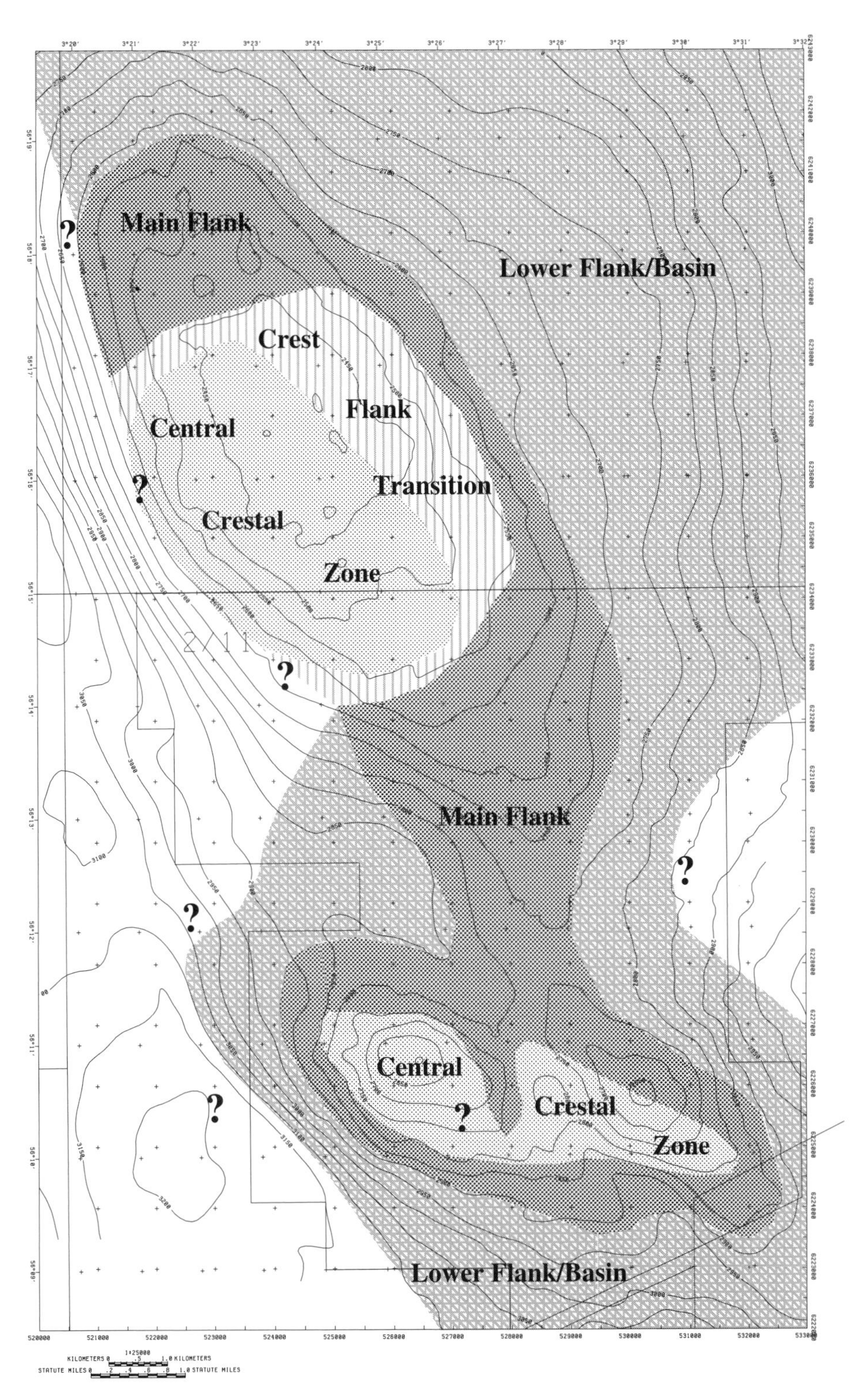

Fig. 3. Tor–Magne palaeobathymetric zones (see text for discussion).

foraminifera with consistently common to abundant calcispheres. Macrofossil debris is less common and dominated by *Inoceramus* prisms. Few to frequent planktic foraminifera appear. The crest–flank transition zone is poorly defined on the Hod Formation structures due to a lesser amount of data than are available for the Valhall Field. The main flank region is dominated by planktic foraminifera. Although *Rugoglobigerina* and *Archaeoglobigerina* specimens generally dominate, keeled species are occasionally common, especially in the Magne Formation. Benthic foraminifera are somewhat less common, but their diversity continues to rise. Calcispheres are uncommon and macrofossil debris is limited to rare *Inoceramus*. The lower flank–basin zone is marked by extremely diverse benthic foraminifera assemblages. Deep-water species appear which were previously absent, including *Nuttallides truempyi* and *Nuttallinella florealis*; and *Osangularia* spp. are common. Planktic foraminifera are also common, but no longer dominant and are outnumbered by benthic species. Macrofossil debris and calcispheres are absent.

Principal chalk biofacies

The fossil content of the chalk is the single most important component for reconstruction of chalk depositional palaeoenvironments. As bioclasts, fossils provide the basic lithological framework of the chalk, allow differentiation of allochthonous from autochthonous facies, and allow identification and differentiation of different types of sediment flows. As the remains of living organisms, fossils serve as palaeoecological markers, providing direct information on various parameters of the depositional environment including bathymetry, oxygen-content, organic flux and productivity. Thus, the distinction between biofacies and lithofacies in chalk deposits is blurred, and observed biofacies provide information well beyond palaeoecology alone. The biofacies of the present study were recognized primarily from cored section and are principally based upon benthic foraminifera, planktic foraminifera, calcispheres, siliceous microfossils and macrofossil debris (as observed in thin section or washed residue). The biofacies have been correlated to detailed sedimentological descriptions from chalk core. However, once established, the biofacies are recognizable from ditch cuttings samples provided that downhole contamination is not severe.

Chalk biofacies (Table 1) can be subdivided into two major types: allochthonous and autochthonous. Although sometimes indicative of substantial transportation from the original sediment source (e.g. the mudflow–turbidite biofacies, below), more frequently Valhall allochthonous sediments are indicative of only a small to moderate amount of transportation. Examples include extensively reworked and winnowed sediment, which is restricted to the central crestal zone, and debris flows restricted to the crest–flank transition zone. However, all allochthonous biofacies are noted for a mixing of two or more palaeoecological assemblages. They are thus distinguishable from autochthonous facies which contain *in situ* microfossils indicative of a single palaeoenvironment.

Allochthonous biofacies

Winnowed crestal biofacies. The biofacies is limited to restricted areas of the Valhall central crestal zone in the upper Maastrichtian and lowermost Danian. These chalks are marked by dominant macrofossil debris, mainly medium- to fine-grained fragments of *Inoceramus*, echinoid plates and spines, and indeterminate mollusc shell hash. Subsidiary components include bryozoan debris, holothurian sclerites and sponge spicules. Calcispheres are never common and exhibit highly variable abundance inversely proportional to the amount of winnowing which has occurred. Foraminifera are limited to rare benthic species with thick, abrasion-resistant walls such as *Stensioina beccariiformis* and *Eponides beisseli*. Planktic foraminifera are absent.

The biofacies is indicative of the shallowest water environments of the Valhall Megastructure, above storm wavebase in the late Maastrichtian and probably above mean wavebase in the earliest Danian when considerable erosion of older Cretaceous chalks occurred. Local slumping is common, although occasional incipient lithification surfaces are also noted. The most common associated lithofacies are intraclastic wackestones and, much more rarely, skeletal grainstones and packstones. The biofacies was probably more regionally extensive prior to erosional truncation of crestal Valhall during the early Danian.

Debris flow biofacies. Poorly sorted chalks exhibiting relatively coarse-grained bioclasts indicative of a mixture of palaeobathymetric assemblages are interpreted as debris flow deposits. Generally, elements of only two adjoining assemblages are present, usually central crest and crest–flank transition, or crest–flank transition and main flank. The limited mixing indicates only a low to moderate amount of transportation

Table 1. *Major characteristics of the principal chalk biofacies. See text for discussion*

Biofacies, sub-biofacies	Sorting	Palaeoecological mixing	Predominant bioclasts	Associated lithofacies
Winnowed crestal	Extensive	Minor	Inoceramus, echinoid debris	Intraclastic wackstone, grainstone
Proximal debris flow	Minor	Moderate	Calcispheres, benthic forams, macrofossils	Floatstone, intraclastic wackestone
Intermediate debris flow	Minor	Moderate	Calcispheres, benthic forams, macrofossils, planktic forams	Massive, mottled wackestone
Distal debris flow	Moderate	Major	Planktic forams, benthic forams, calcispheres, macrofossils	Intraclastic, massive and mottled wackestone
Mudflows/turbidites	Moderate–extensive	Minor	Calcispheres (proximal) or tiny planktic forams (distal)	Very indurated, massive wackestone/mudstone
Low-productivity crestal (Hod)	None	None	Small, non-keeled planktic forams	Very indurated, massive wackestone/mudstone
Low-productivity crestal (Tor)	None	None	Calcispheres	Very indurated, massive wackestone/mudstone
Open marine platform	None	None	Benthic forams, calcispheres, Inoceramus, planktic forams	Highly burrowed chalks, massive chalks
Upper slope high productivity	None	None	Planktic forams, benthic forams	Burrowed, massive chalks, periodites
Basinal	None	None	Benthic forams, planktic forams	Periodites, massive and intraclastic chalks
Carbonaceous – eutrophic	None	None	Infaunal benthic forams, planktic forams	Pyritic massive and bioturbated chalks
Carbonaceous – dysoxic	None	None	Benthic forams strongly dominated by *Stensioina*	Pyritic laminated chalks
Shallow-water pelagic	None to moderate(?)	None	Dominant non-keeled planktic forams; spherical rads	Massive and mottled chalks
Deep-water pelagic	None	None	Dominant keeled planktic forams; diverse rads	Massive and mottled chalks

of the sediment from its source. Mean bioclast size and bioclast composition varies with both the source area of the sediment and the amount of transportation and abrasion which has occurred. Three principal debris flow types are recognized: (1) proximal, sourced from the central crestal zone and involving little transportation; (2) intermediate, sourced from the outer crestal zone and crest–flank transition with variable amounts of transportation; and (3) distal, sourced from the upper flank and marked by a moderate amount of transportation. These deposits are very difficult to distinguish from lithofacies alone. As with the winnowed crestal biofacies, debris flow deposits are mainly restricted to the upper Maastrichtian and lowermost Danian Tor Formation.

Proximal debris flows are marked by abundant macrofossil debris, mainly echinoid and *Inoceramus*, mixed with common to abundant calcispheres, frequent, thick-walled benthic foraminifera and rare planktic foraminifera. Calcispheres and foraminifera exhibit a moderate amount of fragmentation. The most common associated lithofacies are floatstone, intraclastic wackestone and massive wackestone. Proximal debris flows mostly characterize the central crestal zone and crest–flank transition. However, along the north flank of Valhall, they also extend across the main flank zone.

Intermediate debris flow deposits are marked by a near equal mixture of macrofossil debris (again principally echinoid and *Inoceramus* fragments), calcispheres and a diverse, deep-water benthic foraminifera assemblage. Calcispheres are common to dominant. Fragmentation is common. The most common benthic foraminifera include *Stensioina beccariiformis*, *Cibicidoides hyphalus*, *Globorotalites multisepta* and *Eponides* spp. However, planktic foraminifera generally remain rare and it is the combination of low planktic abundance coupled with common benthic foraminifera which most clearly differentiates the intermediate type from other debris flow deposits. Intermediate debris flows are not lithologically distinct, most commonly characterized by massive and mottled wackestone. They predominate from the outer crestal to the lower flank zones.

Distal debris flows are characterized by abundant planktic foraminifera of wide size distribution, common benthic foraminifera exhibiting high diversity, common to dominant calcispheres, and rare, but often coarse-grained, macrofossil debris composed only of echinoid and inoceramid fragments. Fragmentation of foraminifera and calcispheres is often pervasive. Also frequent are very small chalk intraclasts (approximately 0.5–2.0 mm) usually composed of densely fossiliferous, calcisphere wackestone. As with the intermediate debris flow, distal debris flows are lithologically indistinct, composed primarily of massive and mottled wackestone. Distal debris flows are common along the main flank and lower flank–basin zones of Valhall. They probably originated near the crest–flank transition and involved a moderate amount of transportation down the structure flank and occasionally out into the basin.

Mudflow/turbidite biofacies. Chalks with well-sorted, very small bioclasts are interpreted as deposits of sediment flows involving extensive sorting and transportation from the original source. Such chalks in which bioclast distribution is homogeneous are interpreted as mudflow deposits, whereas rarer chalks exhibiting bioclast grading are interpreted as turbidites. Mudflows are very fine-grained debris flows. Although representing a different hydraulic regime from that of gravity flows, both types of deposits in the chalk exhibit a very similar bioclast composition and are indistinguishable if any subsequent bioturbation occurs. They are therefore included within a single biofacies. Mudflows likely represent fine-grained material winnowed from the central crestal and crest–flank transition zones and transported a considerable distance into the basin. Two major subdivisions based upon bioclast composition can be made, reflecting different sediment sources: calcisphere-dominated chalks and chalks dominated by very small planktic foraminifera.

Calcisphere-dominated mudflow/turbidite chalks contain very few bioclasts other than calcispheres. Tiny planktic foraminifera are the only significant subsidiary component. The chalks represent mudflows and turbidity currents originating near the crest–flank transition and having undergone extensive transportation and the sorting out of coarse-grained crestal bioclasts and larger planktic foraminifera of the main flank zone. Planktic foraminifera-dominated mudflow/turbidite chalks contain almost no bioclast other than tiny specimens of the planktic genera *Globigerinelloides*, *Heterohelix* and *Hedbergella*. The only significant subsidiary component is very small (silt-sized) *Inoceramus* fragments. The lack of calcispheres in these chalks indicates mudflow/turbidity currents originating in the main flank zone or more distal areas of the Valhall structure.

The mudflow/turbidite biofacies is the predominant chalk of the lower flank and basin in the upper lower Maastrichtian to lowermost Danian section of the Valhall–Hod region. Off-structure wells exhibit monotonous, thick intervals of stacked mudflow and turbidite chalks. The predominant associated lithologies are very indurated, massive and mottled wackestones with occasional laminated wackestones in turbidite intervals. In the uppermost Maastrichtian and lowermost Danian, interbedding with intermediate and distal debris flows is common.

Autochthonous biofacies

Low productivity crestal biofacies. Poorly fossiliferous mudstones and wackestones occasionally occur in the central crestal zone in Turonian and lower Coniacian Hod section, as well as initial crestal graben fill in the middle Maastrichtian Tor section. The Hod and Tor expression of the biofacies differ considerably in bioclast composition.

The Hod Formation biofacies is more poorly fossiliferous than the typical pelagic-rich chalks which characterize the bulk of the formation (see below). Planktic foraminifera are often numerically common but always exhibit very low diversity. The planktic genera *Globigerinelloides* and *Hedbergella* are uncharacteristically common, but keeled planktic species are absent and the normally abundant planktic foraminifer *Whiteinella* is rare. The only other numerically significant bioclasts are small, spherical radiolarians and rare, nodosariid benthic foraminifera.

The biofacies represents the shallowest water palaeoenvironment evident in the Hod Formation. However, deposition was still below storm wavebase and probably the photic zone as well. It is the rarest biofacies of the Hod Formation.

In the Tor Formation, the biofacies is characterized by poorly fossiliferous calcisphere-dominated mudstones and, less commonly, wackestone. It shows a highly discontinuous distribution in the central and outer crestal areas of Valhall, generally associated with the initial transgressive deposits of local graben fill. The absence of any other bioclast suggests possibly restricted, abnormal marine conditions.

In both the Tor and Hod formations, the most typical lithology for the biofacies is highly indurated, massive mudstone–wackestone. In the Tor Formation, it is also associated with intraclastic chalk and floatstone indicative of local slumping.

Open marine platform biofacies. The assemblage is characterized by very fossiliferous chalks containing moderately diverse, open shelf benthic foraminifera dominated by species of *Gyroidinoides*, *Globorotalites*, *Reussella*, *Bolivinoides* and nodosariids. *Stensioina* is often frequent but never common. Planktic foraminifera can also be frequent, dominated by *Globigerinelloides* and *Rugoglobigerina*, but never outnumbering the benthic species. Macrofossil debris is limited to rare to frequent *Inoceramus* fragments. Calcispheres are frequent to common. The biofacies is indicative of open marine, autochthonous deposition on the outer areas of crestal Valhall and Hod during times of relatively high local sea level.

The biofacies is associated with argillaceous wackestones indicative of a relatively slow rate of deposition; that is, highly burrowed and reburrowed, massive to mottled wackestones, and interbedded chalk and claystone. It is most commonly found in the crest–flank transition zone in the Magne Formation interval and as maximum flooding deposits in crestal grabens during Tor Formation deposition.

'Upper slope' high productivity biofacies. The assemblage dominates the main flank zone through most of the time that the Magne Formation was deposited, as well as rarely occurring in the same zone in Tor time during brief, sporadic periods of autochthonous deposition. Planktic foraminifera are dominant and diverse. The most distinguishing characteristic of the biofacies is a high abundance of keeled planktic species which are otherwise rare or absent in the Tor Formation and the upper Magne Formation. Benthic foraminifera are diverse and resemble assemblages characteristic of the upper half of the slope in normal terrigenous marine basins. Species of *Stensioina* and *Eponides* are most common. Calcispheres and macrofossil debris are absent. The upper slope biofacies is intermediate in palaeobathymetry between the open marine platform assemblage and the basinal biofacies (see below). It is commonly associated with lithologies indicative of slow depositional rates such as intensely burrowed mottled chalk, massive bioturbated chalk, and interbedded chalks and claystones (periodites).

Basinal biofacies. The biofacies is restricted to the lower flank–basin zone throughout much of the time that the Magne Formation was deposited, as well as rarely occurring in the same zone in Tor time during brief, sporadic periods of autochthonous deposition. It is easily recognized by an extremely diverse deep-water benthic foraminifera assemblage including species of *Stensioina*, *Cibicidoides*, *Aragonia*, *Alabamina*, *Gavelinella*, *Dorothia*, *Spiroplectammina*, *Pullenia* and *Osangularia*. The assemblage contains the highest abundances of *Osangularia*, as well as the only frequent occurrences of very deep-water ('lower slope') species, such as *Nuttallides truempyi* and *Nuttallinella florealis*. Planktic foraminifera, dominated by *Rugoglobigerina*, are common to abundant but, unlike the upper slope biofacies, do not outnumber benthic species. There are no calcispheres or macrofossil debris present. Overall, the presence of a lower slope palaeobathymetric indicators together with frequent calcareous agglutinated taxa and non-dominant abundance of planktic species serve to distinguish the basinal biofacies from other high-diversity assemblages occurring in the chalk section. It represents the deepest palaeo-water-depths observed in the chalks of the Valhall Megaclosure.

Interbedded chalks and claystones (i.e. periodites) are most frequently associated with this biofacies. It is also occasionally characterized by intraclastic chalks indicative of incipient break-up zones, probably associated with aborted, submarine hardground formation during times of sediment starvation.

Carbonaceous biofacies. Generally restricted to the Magne Formation, the biofacies occurs at various times in the crest–flank transition, main flank and lower flank–basin zones. The assemblage is dominated by benthic foraminifera of variable diversity and exhibits much lower planktic foraminifera abundance than either the upper slope or basinal biofacies with which it is commonly interbedded. Two sub-biofacies are distinguished by benthic diversity.

The eutrophic sub-biofacies is characterized by moderate diversity benthic assemblages and is easily recognized by the greatest abundance of infaunal benthic foraminifera occurring in the chalk succession. Most common of these infaunal taxa are *Bolivina incrassata*, *Coryphostomum plaitum*, *Neobulimina* spp. and *Praebulimina* spp., with *Stensioina* being the most common epifaunal form. Planktic foraminifera can be frequent, but are strongly dominated by thick-walled, encrusted forms likely reflecting differential preservation due to dissolution. The eutrophic sub-biofacies is generally associated with massive, pyritic bioturbated chalks. It is most common in the middle Magne Formation (middle Campanian) in the main flank zone, probably marking a time of high surface water productivity and upwelling associated with a major transgression of the Valhall structure.

The dysoxic sub-biofacies is marked by much lower diversity of benthic foraminifera with a great dominance of a few species of *Stensioina*, especially *S. granulata polonica*. Infaunal benthic species are rare to absent, and planktic foraminifera are few. The sub-biofacies is commonly associated with finely laminated, pyritic chalks as well as massive pyritic chalks, and probably represents dysoxic bottom water conditions with anoxic pore water. Framboidal pyrite is common. The few species of *Stensioina* which dominate in this assemblage are probably opportunistic species adapted to low-oxygen conditions. The dysoxic sub-biofacies is very discontinuous in both vertical and lateral distribution. In the lower Magne Formation (lower Campanian), the sub-biofacies generally occurs in the crest–flank transition zone, and represents an oxygen-minimum zone intermediate in palaeobathymetry between the shallower open marine platform biofacies and the deeper eutrophic sub-biofacies. In the upper Magne Formation (upper Campanian and lowermost Maastrichtian) the sub-biofacies is widespread in the lower flank–basin zone, indicative of partially restricted deep-water circulation.

'Shallow-water' pelagic biofacies. Restricted to the Hod Formation, the biofacies is dominated by non-keeled planktic foraminifera (especially *Whiteinella* but also *Hedbergella*, *Clavihedbergella* and *Globigerinelloides*). Keeled species are frequent to absent. Benthic foraminifera are very rare, diluted by the high pelagic content. Radiolarians are often an important subsidiary component, but exhibit low diversity, strongly dominated by *Cenosphaera*. The biofacies is generally centred on the crest–flank transition zone. Associated lithofacies are highly variable, but all indicative of low rates of sedimentation (i.e. laminated chalk, parallel banded chalk, intensely burrowed chalk, interbedded laminated chalk and claystone). The assemblage represents shallower-water, lower-productivity conditions during Hod time. It intergrades with the deep-water pelagic biofacies (see below) downslope and, occasionally, with the low-productivity crestal biofacies upslope. At times of higher local sea level, however, the shallow-water pelagic biofacies extends across the crestal zone of the Valhall and Hod structures. Also the assemblage occasionally occurs well off-structure in massive wackestones, possibly indicative of highly pelagic sediment flow deposits. However, poor palaeobathymetric control in the Hod Formation, due to lack of benthic taxa, makes it very difficult to distinguish allochthonous from autochthonous deposits.

'Deep-water' pelagic biofacies. Also restricted to the Hod Formation, the biofacies is marked by very pelagic-rich intervals with extremely abundant, highly diverse planktic foraminifera. Keeled species are dominant, especially *Marginotruncana marginata*, but non-keeled species are also abundant, including species of *Whiteinella*, *Archaeoglobigerina* and *Heterohelix*. Benthic foraminifera are rare to extremely rare and usually represented by deep-water species of the genera *Stensioina*, *Osangularia* or *Ammodiscus*. Radiolarians are usually common to abundant but, unlike in the shallow-water pelagic biofacies, exhibit high species diversity. The biofacies is generally associated with thick sections of massive chalk and massive bioturbated chalk of uncertain depositional mechanism. It probably represents both periods of autochthonous, rapid pelagic deposition and various sediment flows derived from such autochthonous sediments. Although most frequently found in the main flank and lower flank–basin zones, the biofacies also sporadically occurs across the crest–flank transition and central crestal zones, indicative of discrete flooding events.

Overall, the shallow-water pelagic and deep-water pelagic biofacies are probably indicative of different water masses within the Valhall–Hod area during the Turonian and Coniacian. The distribution of these water masses was tied to oceanographic conditions influenced by eustatic sea-level fluctuations and the configuration of the basin as controlled by tectonic movements. The precise nature of these oceanographic controls remain unclear, however, due to poor palaeoenvironmental control in these highly pelagic facies.

Chalk depositional model

Detailed multidisciplinary biostratigraphic data points generated for each of the study wells were graphically correlated against a Valhall Composite Standard (Bergen & Sikora 1999), calibrated to stages based on Gradstein *et al.* (1995). One of the strengths of the graphic correlation method is that however the well is scaled, the scale can be correlated to absolute time by correlation to the composite. Thus, biofacies recognized in each study well were scaled in

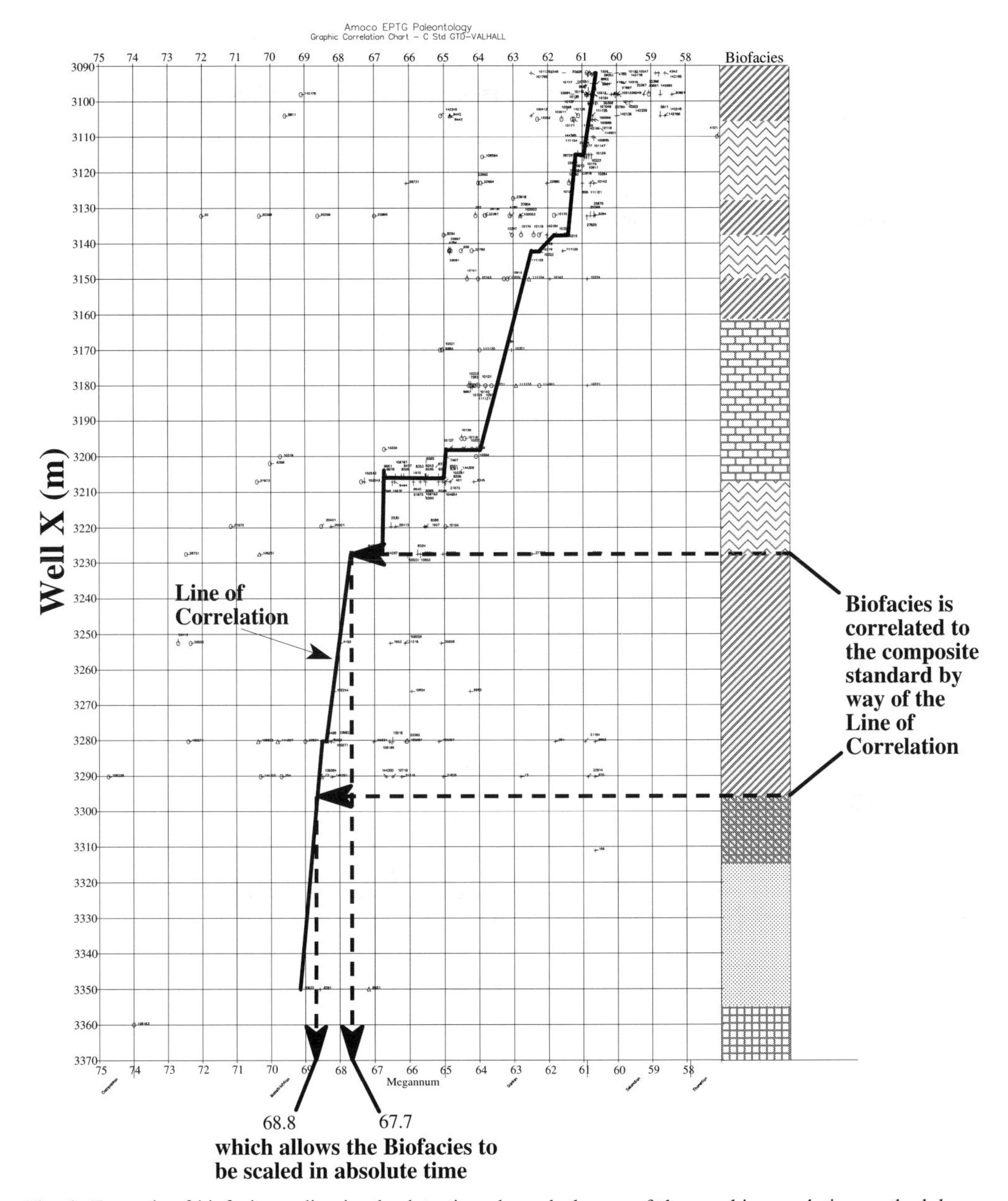

Fig. 4. Example of biofacies scaling in absolute time through the use of the graphic correlation methodology (time scale after Gradstein *et al.* 1995).

absolute time (Fig. 4). Then, for any one time line, the regional distribution of biofacies for the Valhall–Hod region could be easily mapped. By constructing a succession of time-slice biofacies distribution maps from Hod through Tor time, the depositional history of the basin becomes more readily apparent. Four major depositional regimes can be discerned in the Turonian–Danian chalk succession.

The early Turonian–early Santonian (Hod Formation) was characterized by deposition of chalks containing rich and well-preserved pelagic

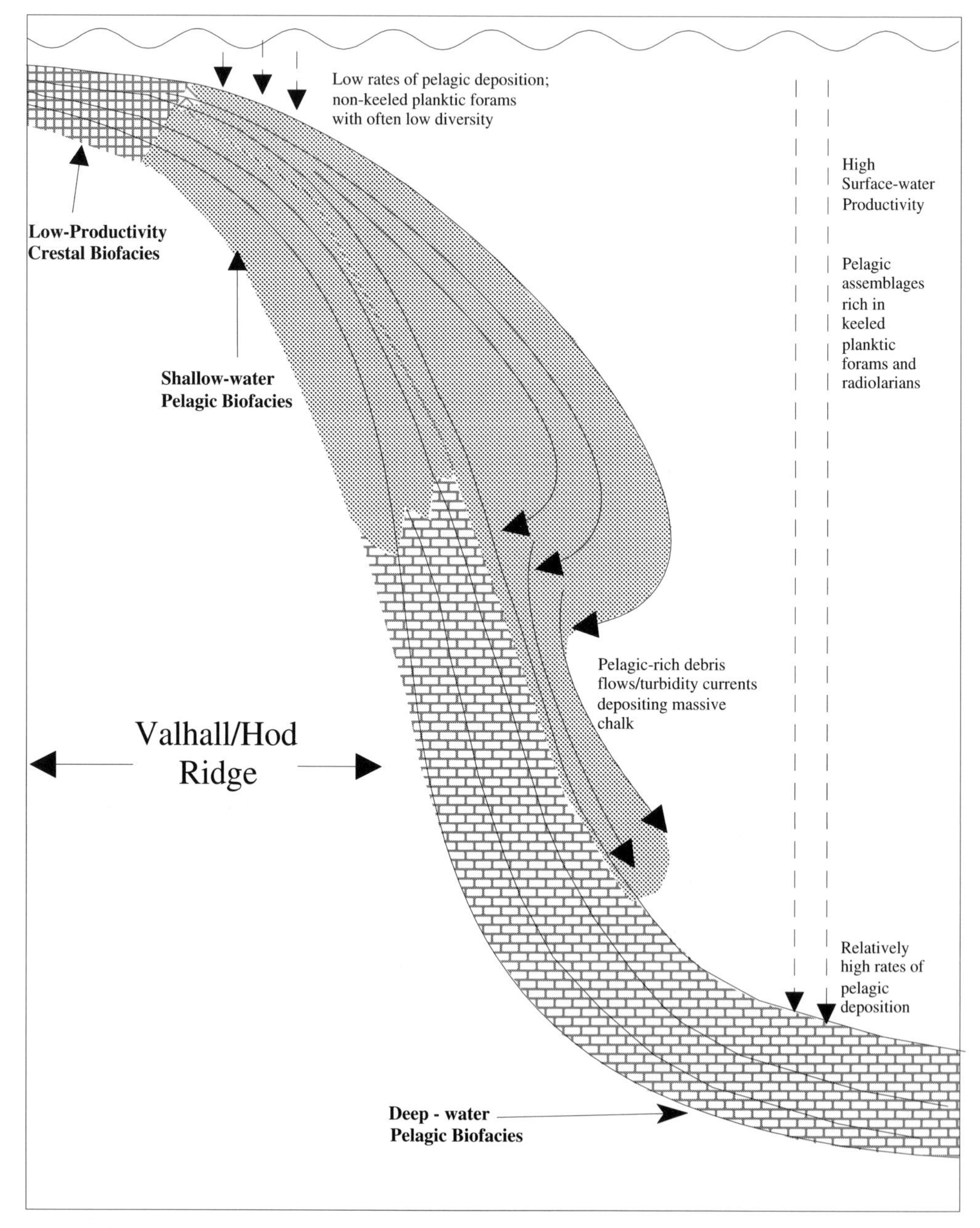

Fig. 5. Palaeoenvironmental summary for the Turonian–Coniacian Hod, a time of high pelagic productivity.

assemblages. Palaeobathymetric resolution is very limited due to the rarity of benthic microfossils. Nevertheless, based on the composition and diversity of the pelagic assemblages, some rudimentary palaeobathymetric resolution is possible (Fig. 5). The shallowest water depths are indicated by the low-productivity crestal biofacies succeeded downslope in progressively deeper-water deposits by the shallow-water pelagic and deep-water pelagic biofacies. This simple biofacies model holds for most of the upper Turonian–Coniacian section. However, for any one locality, the environmental succession can be more complex. The deep-water pelagic biofacies often occurs across the central crestal zone, indicative of discreet flooding events. The shallow-water pelagic biofacies is often observed off-structure, indicative of sea-level falls and/or sediment flows. Nevertheless, the regional biofacies distribution indicates a relatively continuous ridge from the Valhall to Hod structures for most of Hod time (e.g. the middle Coniacian; Fig. 6). The first regional differentiation of pelagic biofacies into this typical pattern occurs in the Middle Turonian (approximately 92 Ma BP; Fig. 2), marking the time of initial uplift of the Valhall Megastructure.

A major period of regional uplift follows, marked stratigraphically by a large regional unconformity (83.5–85 Ma BP; Fig. 2) and erosional truncation of the upper Hod Formation across the central crestal and crest–flank transition zones. Resumption of deposition in the early Campanian (approximately 83 Ma BP) marks the initiation of the second major period of chalk deposition, the Magne Formation. Initially, Magne Formation deposition was limited in the early Campanian to deep-water, autochthonous chalks; the basinal biofacies in the lower flank–basin zone and the upper slope biofacies along the main flank zone. The shallowest-water palaeoenvironment is the open marine platform biofacies at the central crestal zone of East Hod flanked by the dysoxic sub-biofacies indicative of a local oxygen minimum zone (Fig. 7). Later in the early Campanian (approximately 80 Ma BP), the carbonaceous biofacies becomes more widespread, with the dysoxic sub-biofacies indicating an extention of the oxygen-minimum zone across the central crestal zone of the Hod structure. In addition, the eutrophic sub-biofacies replaces the upper slope biofacies of the earlier Campanian throughout the main flank and lower flank–basin zones of the Hod and southern Valhall structures (Fig. 8). This increase in carbonaceous facies likely indicates higher surface water productivity and upwelling associated with the beginning of a transgression of the Valhall structure as evidenced by a progressive Campanian onlap by Magne Formation facies. There is no lower Magne Formation section from the central crestal and crest–flank transition zones of Valhall (Figs 7 and 8). However, the absence of allochthonous debris within the basin indicates that the structure crest was not aerially exposed or undergoing erosion. Rather, it is likely that the crestal area was below storm wavebase during this period, mantled by submarine hardgrounds and marked by geologically long periods of non-deposition.

Lower and middle Campanian biofacies are generally regional in extent (e.g. Fig. 7). However, beginning in the upper Campanian, some local differentiation of biofacies appeared along the crest–flank transition of Valhall and the central crestal zone of East Hod (Fig. 9). Structural analysis indicates that with continued Valhall uplift, an extentional regime developed causing crestal collapse and the formation of fault-bounded horsts and grabens (Farmer *et al.* 1996). The local differentiation of biofacies first noted in the upper Campanian probably indicates the earliest evidence of this crestal collapse (approximately 77 Ma BP). This differentiation includes flooding of local grabens along eastern Valhall (the open marine platform biofacies) and the sudden appearance of shallow-water deposits marking a horst on East Hod (i.e. the low-productivity crestal biofacies).

Off-structure during the late Campanian–earliest Maastrichtian, the main flank zone continued to be characterized by the upper slope high-productivity biofacies. This deep-water oxic facies graded downslope into widespread, low-oxygen deposits marked by the dysoxic sub-biofacies (Fig. 9). The Magne Formation is marked by major stratigraphic discontinuities, often marked by hardgrounds, indicative of lengthy periods of non-deposition (Bergen & Sikora 1999). Continued subsidence during these periods of non-deposition created a very deep basin. The prevalence of the dysoxic sub-biofacies in the lower flank–basin zone indicates partial restriction of the basin. Sluggish deep-water circulation and low turnover rates resulted in the development of low-oxygen conditions and the widespread deposition of organic-rich, pyritic chalks. Magne Formation depositional facies are summarized in Fig. 10.

The argillaceous, condensed chalks which constitute the great majority of the Magne formation are marked by low porosity and permeability. Reservoir facies are limited to restricted areas on the north slope of Valhall where local graben fill of the upper Magne Formation was

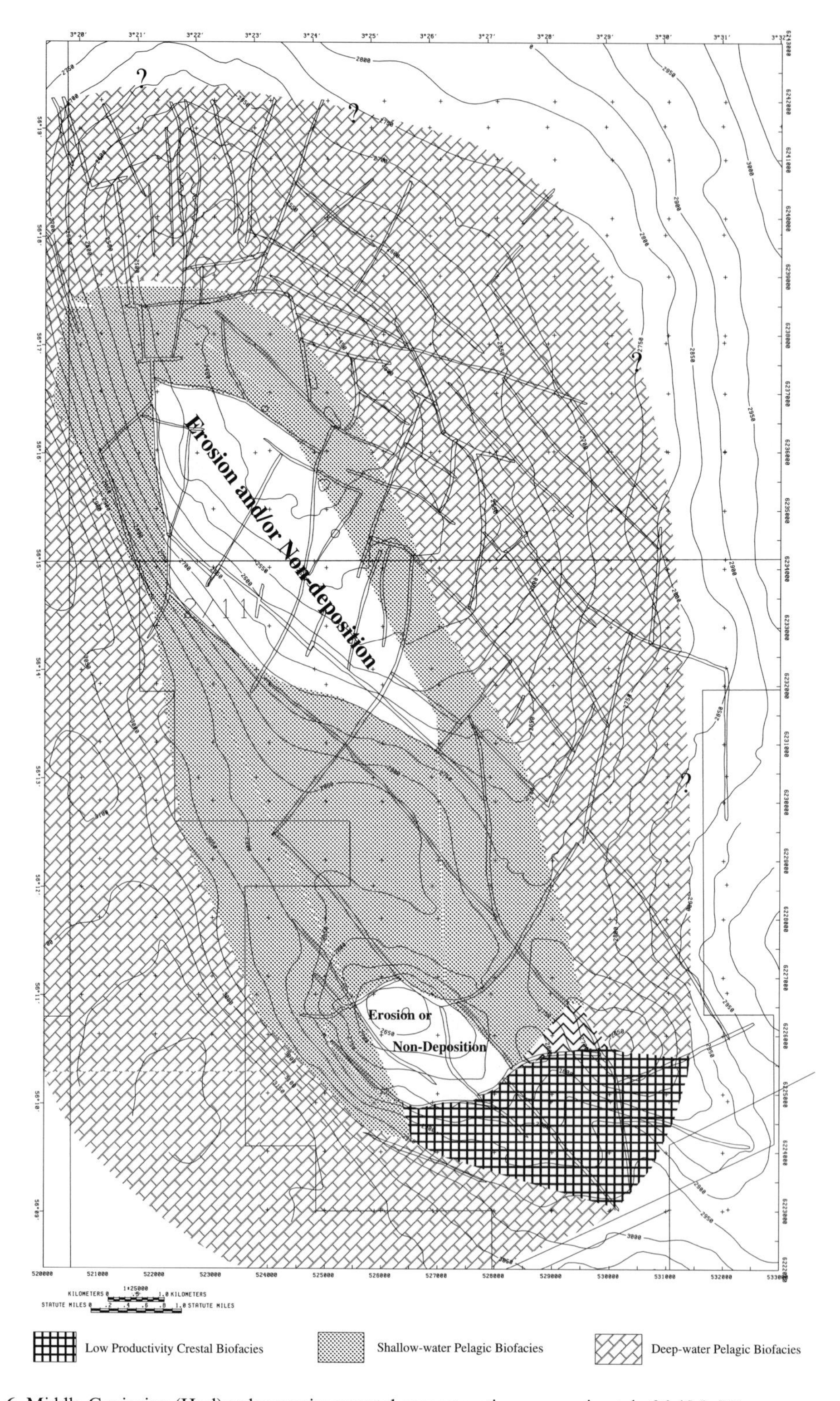

Fig. 6. Middle Coniacian (Hod) palaeoenvironmental reconstruction, approximately 86.4 Ma BP.

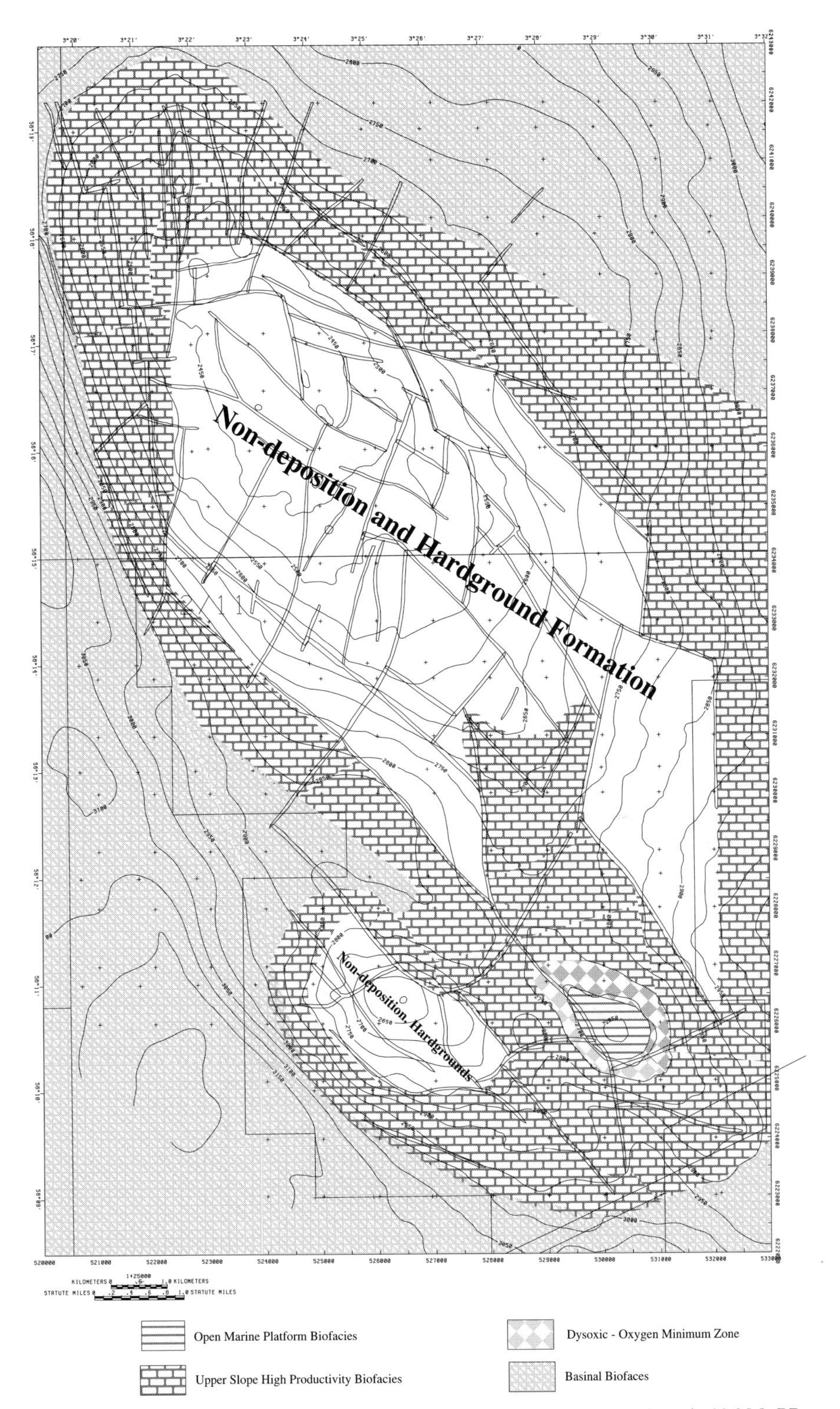

Fig. 7. Early Campanian (basal Magne) palaeoenvironmental reconstruction, approximately 83.2 Ma BP.

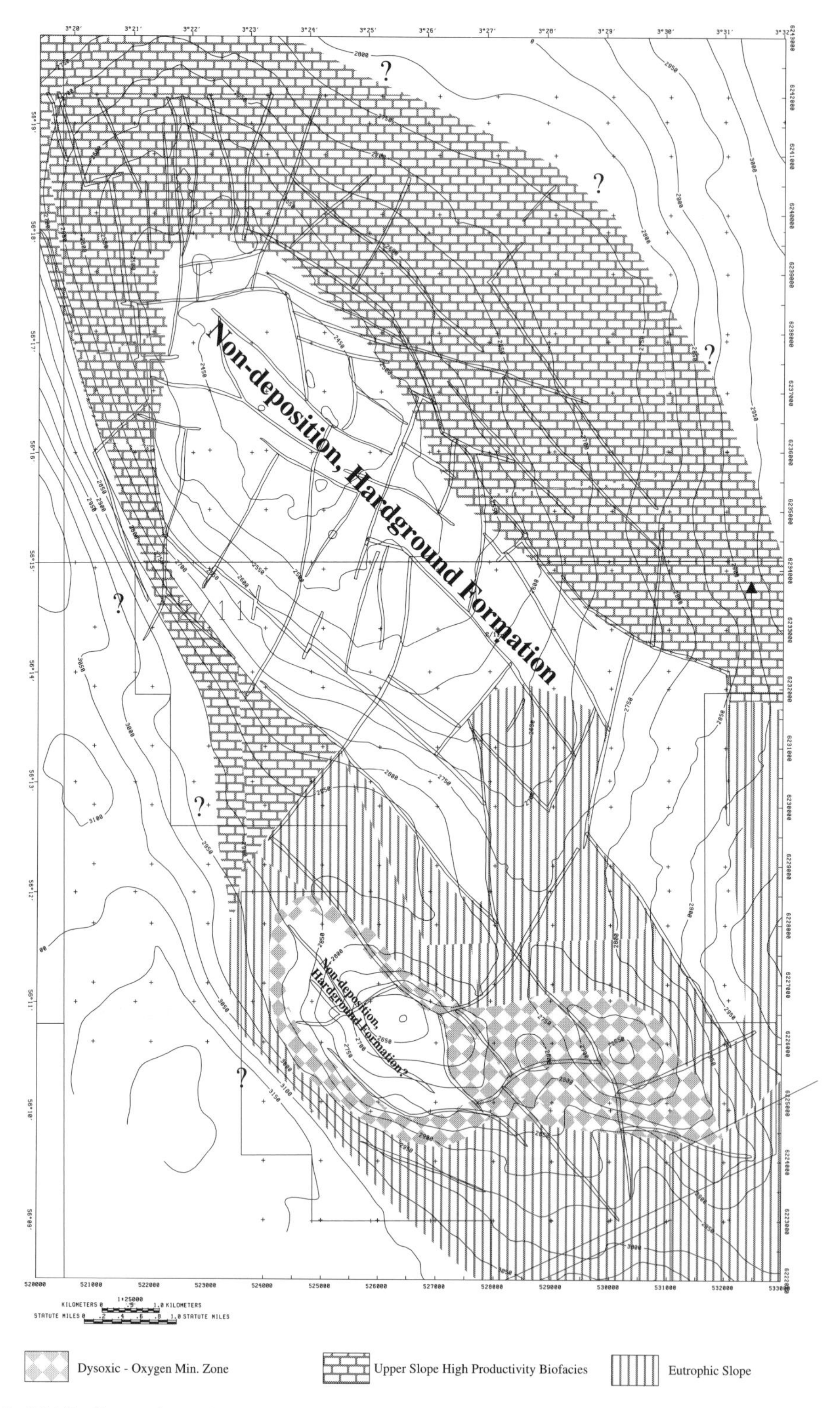

Fig. 8. Middle Campanian palaeoenvironmental reconstruction, time of the initial Magne transgression.

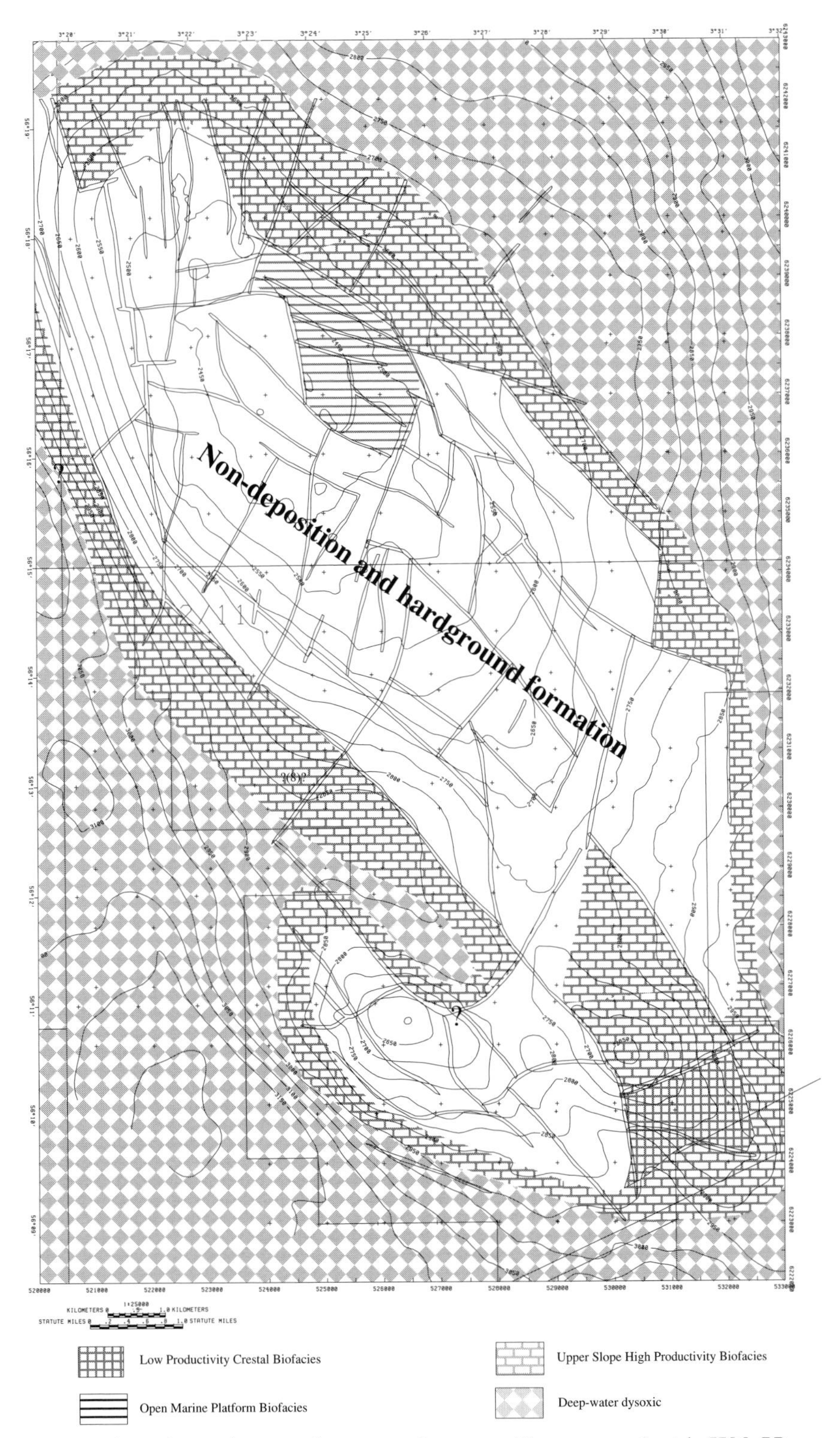

Fig. 9. Late Campanian palaeoenvironmental reconstruction; upper Magne, approximately 77 Ma BP.

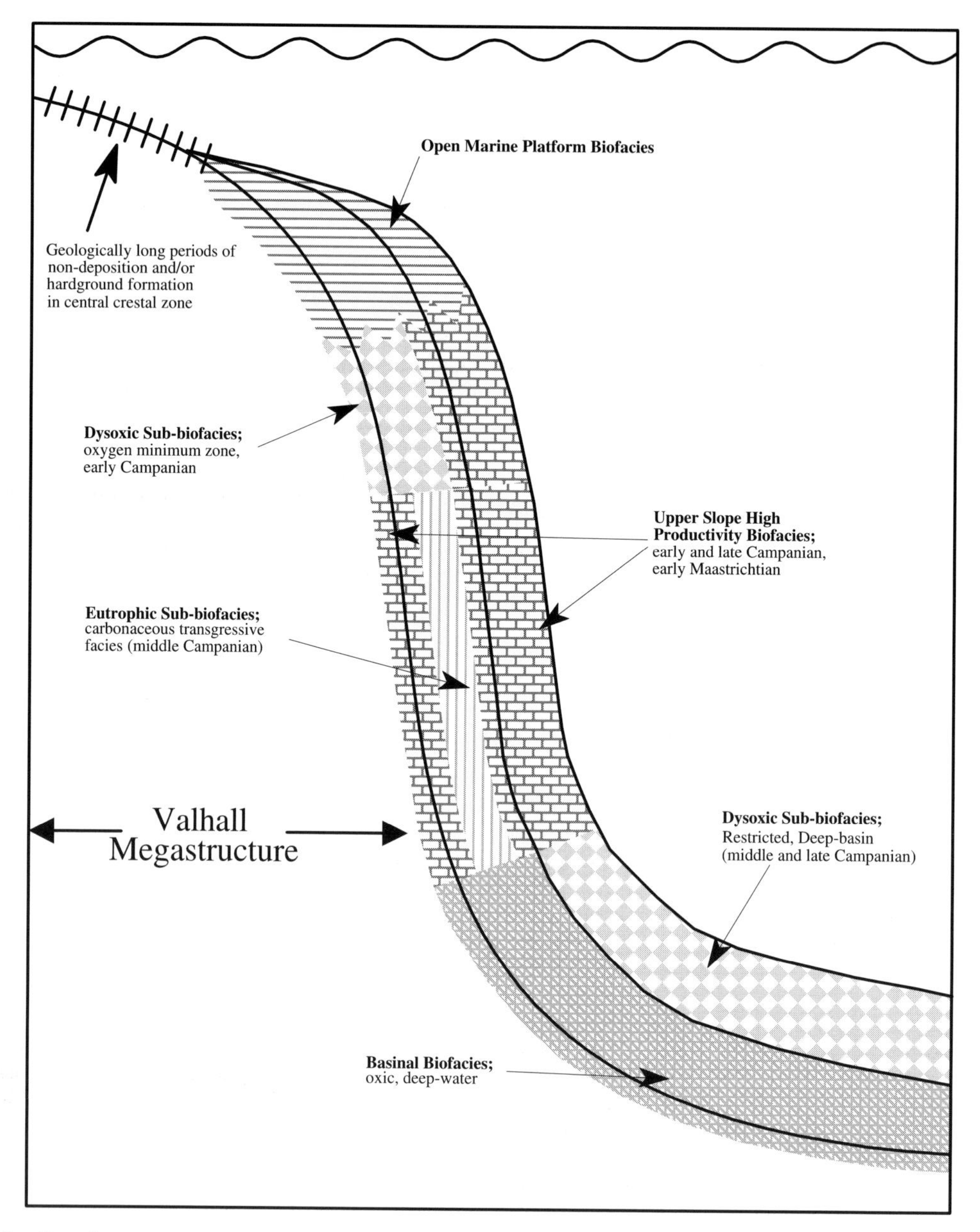

Fig. 10. Palaeoenvironmental summary of Magne time (early Campanian–earliest Maastrichtian).

sometimes characterized by relatively coarse-grained debris flow deposits (Ca1/M5 time). However, the Magne Formation more typically is a seal, especially on crestal Valhall where it is marked by a series of condensed zones and hardgrounds. These dense, thin intervals are easily identifiable by log response and are tempting datums to use in regional correlation. However, the condensed sedimentation during Magne deposition was ended by the process of graben formation and flooding, and the upper Magne Formation is indicative of much more rapid rates of sedimentation. Because grabens formed locally at different times from the late

Campanian to the middle Maastrichtian, from the flanks to the central crest of Valhall, the age represented by these condensed zones is highly variable and they do not constitute an isochronous correlation horizon.

With continued Valhall uplift, the process of crestal collapse became the dominant control on deposition during the middle Maastrichtian (68.5–73 Ma BP; Fig. 11). This lower Tor section marks the third major period of chalk deposition. Sedimentation became more widespread in the central crestal and crest–flank transition zones reflecting multiple graben flooding episodes. Allochthonous deposition predominated on the main flank and lower flank–basin zones as crestal deposition increased, creating a source for sediment flows of various types.

Unlike in the Campanian and earlier Maastrichtian (Magne Formation), lower Tor biofacies distribution was very heterogeneous both in time and area, controlled by local structure (Figs 11 and 12). With no surviving section from this time, the southwest area of Valhall was probably indicative of the highest crestal area and the site of sediment winnowing and erosion. In surviving graben deposits to the north and east, initial transgressive deposits are characterized by partially restricted, shallow-water autochthonous chalks of the low-productivity crestal biofacies overlain by maximum flooding deposits of the open marine platform biofacies. However, the main period of graben infilling began at the beginning of the late Maastrichtian (approximately 68.5 Ma BP) when autochthonous chalks were replaced by allochthonous units indicative of proximal and intermediate debris flows and local slumps. Debris flow deposition also characterized the main flank zone of northern Valhall throughout this period. However, the main flank along eastern Valhall as well as at East Hod is characterized by complex, bioturbated mixes of what originally were interbedded autochthonous upper slope high-productivity biofacies and well-sorted, very fine-grained bioclasts of the mudflow/turbidite biofacies. The latter biofacies comes to dominate further off-structure within the lower flank–basin zone. Sites from this region evidence thick, lower Tor section composed of stacked mudflow deposits and turbidites, rarely interbedded with deep-water autochthonous chalks of the basinal biofacies (Figs 11 and 12).

The debris flow deposits characteristic of the later stages of graben fill on the crest and upper flank of Valhall form the primary reservoir facies in the lower Tor Formation (sequences M2a, M3 and M4; Bergen & Sikora 1999). However, the facies are highly discontinuous due to the heterogeneity of depositional facies controlled by local structure. Laterally, facies porosity/permeability can decrease markedly into the graben as progessively finer-grained deposits are encountered farther from the bioclast source on the neighbouring horst(s) (e.g. proximal debris flow grading into intermediate debris flow grading into proximal mudflow). The same effect can occur vertically at any one location located away from the graben margin in fining-upward sequences.

The fourth major period of chalk deposition is represented by the latest Maastrichtian–early Danian upper Tor Formation (approximately 63–68 Ma BP), the primary reservoir facies in both Valhall and Hod fields. By the middle part of the late Maastrichtian, the crestal grabens had largely filled. The central crestal zone was marked by shallow-water environments near mean wavebase and characterized by extensive winnowing of skeletal debris (i.e. the winnowed crestal biofacies). These shallow-water crestal environments shed large amounts of sediment into the basin, fueling widespread sediment flows across the crest–flank transition, down the main flank zone and into the basin (Figs 13 and 14). These sediment flows are progradational, becoming progressively more proximal in derivation up-section. Crestal environments continued to shallow into the early Danian when extensive erosion and possible exposure occurred.

Initially, mudflows and relatively fine-grained intermediate and distal debris flows dominated the region from the crest–flank transition zone into the basin. However, by the latest Maastrichtian (approximately 66.5 Ma) and continuing into the early Danian, the winnowed crestal biofacies dominated across the crestal zone and coarser-grained, proximal debris flows characterized the main flank zone (Fig. 13). The shedding of coarser-grained bioclasts intensified in the early Danian with erosion of the central crestal zone, whereupon allochthonous sediments contained a very high percentage of reworked Cretaceous material. The amount of reworked material varies with the amount of sorting and transportation which has occurred (Fig. 14). Lower Danian chalks from the crestal area are marked by near 100% reworked Maastrichtian microfossils and nannofossils. The amount of nannofossil reworking remains nearly constant across the main flank and into the basin. However, the amount of microfossil reworking decreases down the structural gradient to reach a minimum value of about 10% in fine-grained mudflow deposits in the lower flank–basin zone. The larger reworked microfossils were thus substantially sorted out during

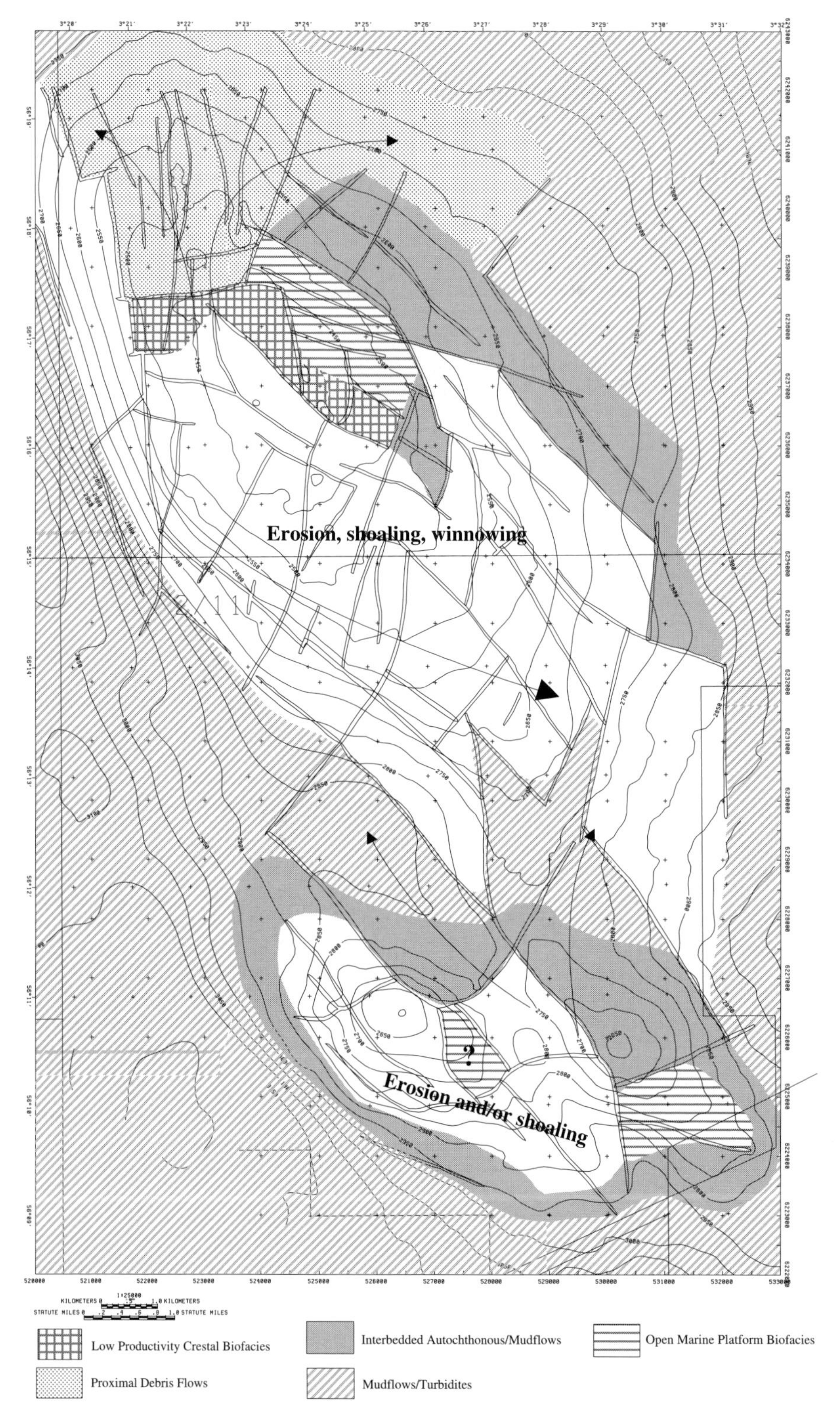

Fig. 11. Lower Tor (middle Maastrichtian) palaeoenvironmental reconstruction; early stage of graben fill, approximately 70 Ma BP.

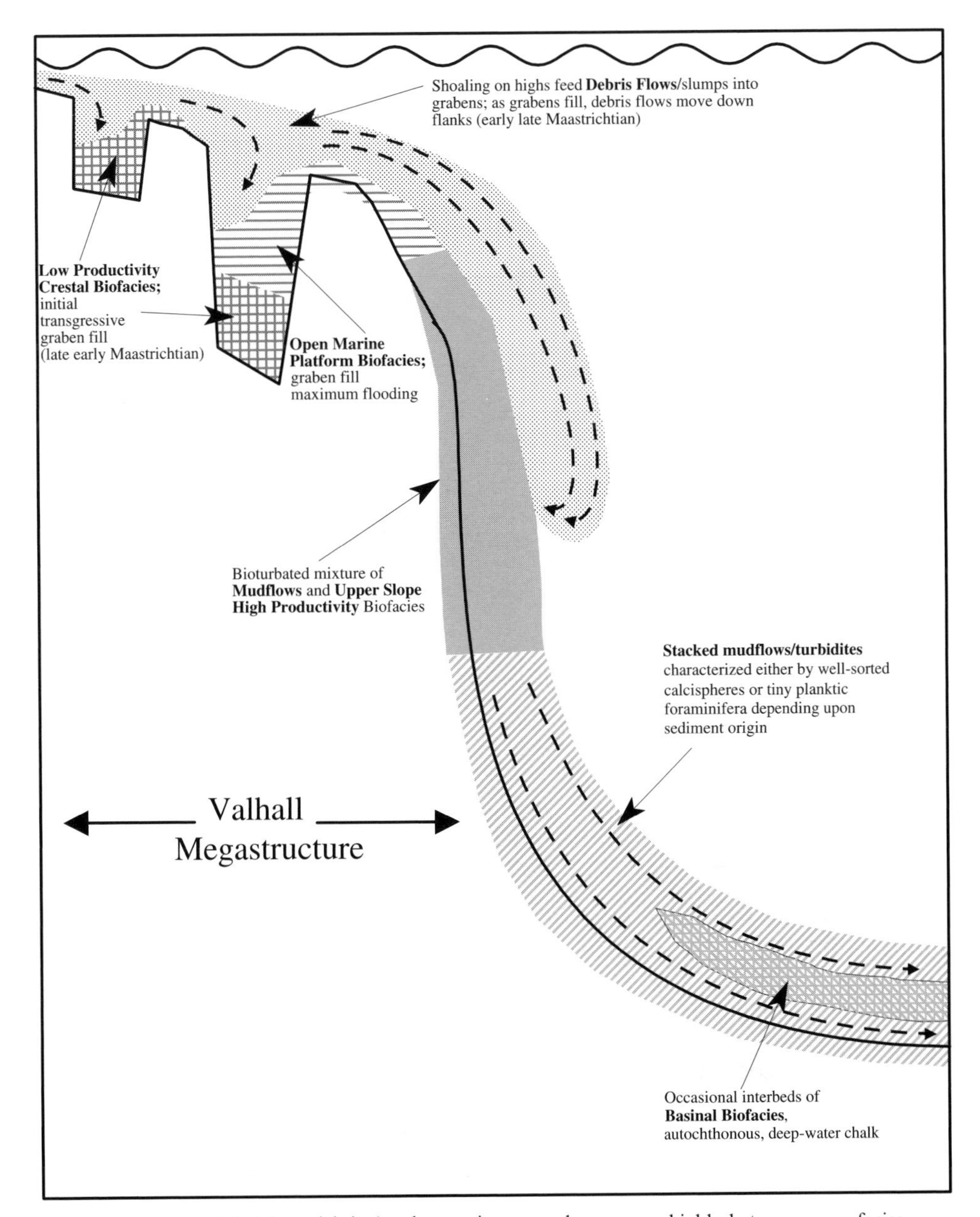

Fig. 12. Lower Tor (middle Maastrichtian) palaeoenvironmental summary; highly heterogeneous facies controlled by local structure associated with crestal collapse of the Valhall anticlinal crest.

transportation from the central crestal zone into the basin.

The sequences of the upper Tor Formation (Da5, M1 and M2; Bergen & Sikora 1999) represent widespread deposition of proximal and intermediate debris flows on the Valhall flank, and into the basin coupled with winnowing of sediment and formation of skeletal lag deposits across much of crestal Valhall. The reservoir facies of the upper Tor Formation are thus more continuously distributed than those of the lower Tor Formation. Nevertheless, continued local structural movements have led to a chronostratigraphically discontinuous section with major discontinuities (Bergen & Sikora 1999). Small faults beyond seismic resolution are

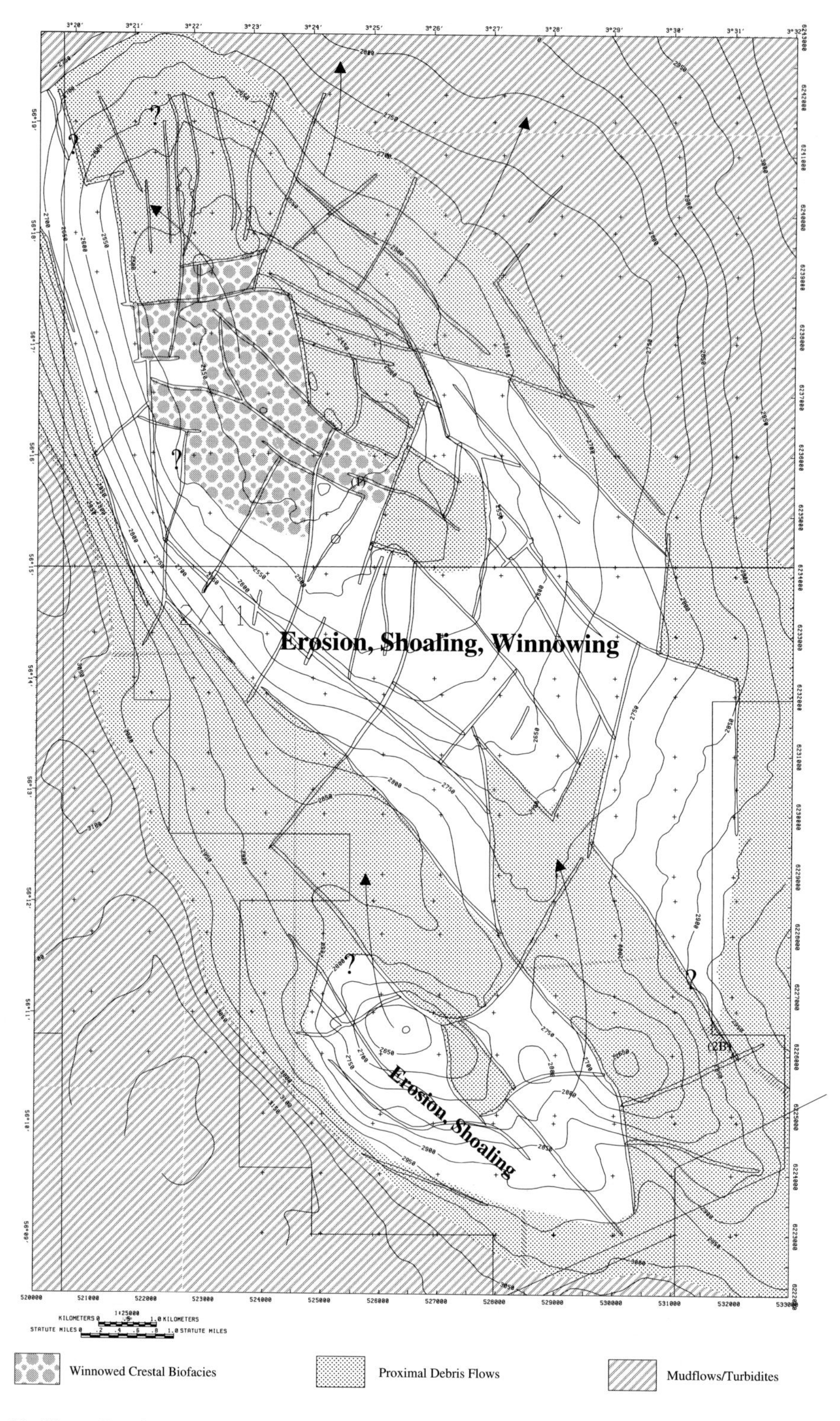

Fig. 13. Upper Tor (latest Maastrichtian) palaeoenvironmental reconstruction, approximately 66.5 Ma BP.

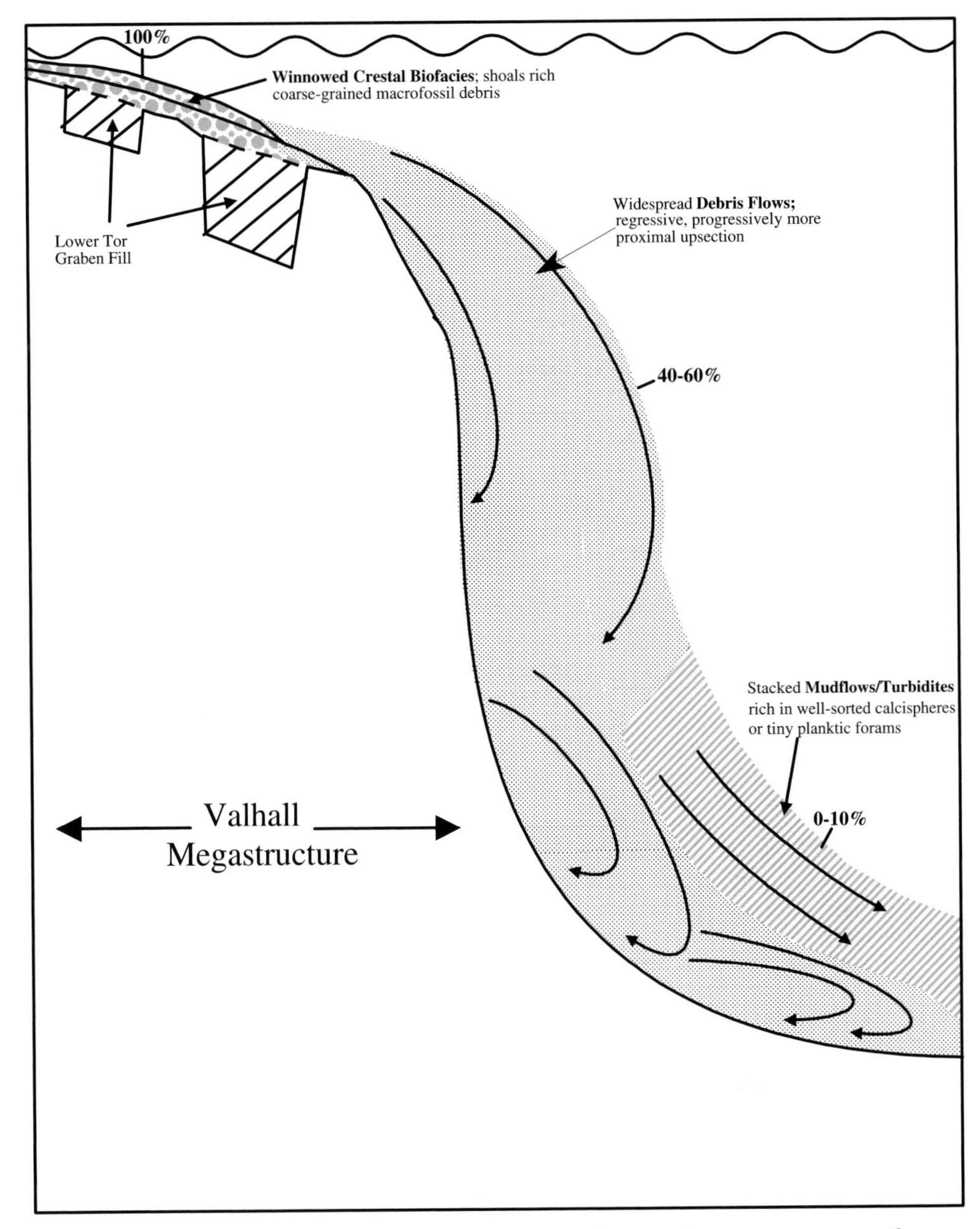

Fig. 14. Upper Tor (latest Maastrichtian–early Danian) palaoenvironmental summary; percentage figures indicate amount of reworked Cretaceous species as percentage of total recovered microfossils in Danian Tor.

common (throws of 30 m or less). These discontinuities are often associated with water flood zones and abrupt changes in biofacies and porosity. Thus, what first appears as a relatively homogeneous reservoir facies is in actuality a mosaic of smaller reservoirs and subtle structural and stratigraphic traps.

A major flooding event at about 63 Ma BP (Fig. 2) brought to an end this widespread allochthonous sedimentation. Younger Danian chalk (Ekofisk Formation) is a highly condensed, stratigraphically discontinuous, deep-water autochthonous deposit generally comparable to the upper slope, high-productivity biofacies and the open marine platform biofacies. At about 61 Ma BP, a major terrigenous influx occurred, ending chalk deposition in the basin and draping the entire region with the calcareous lower Våle

Claystone. Nevertheless, other than containing much more common agglutinated benthic foraminifera, the lower Våle Claystone is very similar in biofacies to the underlying Ekofisk chalks, indicating no major environmental change in this part of the North Sea basin. However, from 59 to 60 Ma BP, the depositional basin became restricted resulting in a major rise in the calcite compensation depth. Throughout the remainder of the Palaeocene calcareous sedimentation ceased and widespread anoxia occurred in the basin (Lista–Sele–Balder time).

This middle Danian–lower Selandian unit serves as an effective seal to the Tor Formation reservoirs. The thin Ekofisk Formation chalks are marked by prominately higher water saturation levels than the underlying Tor Formation and the Våle Claystone forms a terrigenous drape over the entire basin.

Conclusions

Detailed study of microfossil assemblages within the chalk allow recognition of numerous biofacies. When correlated to chalk lithofacies and sedimentology, such biofacies are observed to be indicative of fine-scaled depositional environments. Graphic correlation methodology allows for enhanced chronostratigraphic control, allowing the biofacies to be scaled in absolute time for each well. The regional distribution of palaeoenvironments for any one time line can then be reliably mapped. Construction of a series of palaeoenvironmental maps through time allows for construction of a more detailed depositional model than previously possible. Palaeoenvironmental reconstruction reveals a complex regional depositional history of the chalks of the Valhall–Hod area. A better understanding of the depositional history of the basin aids in defining the configuration of the basins and recognition of the timing of local and regional tectonic movements.

Turonian–Danian chalk deposition occurred in four major episodes:

- lower Turonian–lower Santonian (Hod Formation) pelagic chalks; the middle Turonian marks the first differentiation of facies reflecting Valhall–Hod uplift, culminating in thinning and truncation over crestal Valhall during the late Coniacian and Santonian. Reservoir parameters in the Hod Formation have not yet been fully integrated into the biofacies model;
- thin Campanian–lower Maastrichtian predominately autochthonous units (Magne Formation) onlapping structure and usually separated by hardgrounds and/or condensed zones. The Magne Formation often serves as an effective seal especially on crestal Valhall. It occurs as a series of condensed zones and hardgrounds. However, the timing of formation of these thin, dense intervals is in part governed by the timing of graben formation which varies over the crest and upper flank of Valhall. The condensed Magne Formation facies are thus local seals for the graben reservoirs, not a stratigraphic horizon isochronous across the Valhall Megastructure;
- allochthonous deposition in local sub-basins indicative of crestal collapse in the middle Maastrichtian (lower Tor Formation). Reservoir facies and quality are heterogeneous, both laterally and vertically within the crestal grabens. Facies porosity/permeability can decrease both laterally into the graben (as progressively finer-grained deposits are encountered farther from the bioclast source) and vertically at any one location within the graben in fining-upward sequences;
- widespread sediment flows and redeposition of sediments with crestal shoaling and winnowing in the latest Maastrichtian (upper Tor Formation) culminating in crestal erosion of Cretaceous chalks in the early Danian. The upper Tor Formation forms the main reservoir facies in both the Valhall and Hod fields. Although superficially a widespread, homogeneous coarse-grained chalk facies, the interval is actually a mosaic of smaller individual reservoirs in a chronostratigraphically discontinuous section with subtle stratigraphic traps.

A major flooding in the middle Danian resulted in highly condensed and discontinuous deep-water autochthonous deposition of Ekofisk chalk. The lower Våle Claystone calcareous shale rests nearly conformably on the Ekofisk Formation in the Valhall Megaclosure, draping the entire region and marking the cessation of chalk deposition.

Overall, reliable reservoir prediction and estimates of reservoir volume are very difficult to make in a section as structurally and as stratigraphically complex as that of the Valhall and Hod chalk fields. Successful reservoir management can only be achieved with very fine-scaled chronostratigraphic control allowing accurate biofacies definition and detailed depositional modelling through time.

The authors would like to thank Amoco Norway Oil Company (ANOC), Stavanger, and our partners Elf Aquitaine, Enterprise and Amerada Hess for their long-term support of this study. We would also like to thank ANOC and Amoco Exploration and Production Technology Group (EPTG), Houston, for permission to publish. Grateful acknowledgement is also due for the critical and thorough reviews by our colleagues, Ingrid Øxnevad (ANOC), and J. A. Stein and D. S. Van Nieuwenhuise (EPTG). Finally, we would like to thank Åse Moe of the Norwegian Petroleum Directorate for many fruitful discussions and constructive criticisms.

References

BERGEN, J. A. & SIKORA, P. J. 1999. Microfossil diachronism in southern Norwegian North Sea chalks: Valhall and Hod fields. *This volume*.

BRASHER, J. E. & VAGLE, K. R. 1996. Influence of lithofacies and diagenesis on Norwegian North Sea Chalk reservoirs. *AAPG Bull.*, **80**, 746–769.

GRADSTEIN, F., AGTERBERG, F. P., OGG, J. G., HARDENBOL, J., VAN VEEN, P., THIERRY, J. & HUANG, Z. 1995. A Triassic, Jurassic and Cretaceous time scale. *In*: BEGGREN, A. A., KENT, D. V., AUBREY, M. P. & HARDENBOL, J. (eds) *Geochronology, Time Scales and Global Stratigraphic Correlation*. Society of Economic Paleontologists and Mineralogists, Special Publication, **54**, 95–128.

GREEN, M., JUDGE, N. C., BRAMWELL, N. P., ADAM, P., MECIANI, L. & CAILLET, G. 1996. An allostratigraphic sequence scheme for the chalk. *In*: *Chalk Geoscience Workshop, Extended Abstracts, Norwegian Petroleum Directorate, Stavanger, Norway, 2–3 December 1996*.

FARMER, C. L., PEARSE, C. H. & BARKVED, O. I. 1996. Influence of syn-depositional faulting on thickness and facies variations in chalk reservoirs – Valhall and Hod Fields. *In*: *Fifth North Sea Chalk Symposium, Joint Chalk Research Program, Reims, France, 7–9 October 1996*.

ISAKSEN, D. & TONSTAD, K. 1989. A revised Cretaceous and Tertiary lithostratigraphic nomenclature for the Norwegian North Sea. *Norwegian Petroleum Directorate Bulletin*, **5**.

KENNEDY, W. J. 1987. Sedimentology of Late Cretaceous–Palaeocene chalk reservoirs, North Sea Central Graben. *In*: BROOKS, J. & GLENNIE, K. W. (eds) *Petroleum Geology of North West Europe*, Volume 1. Graham and Trotman, London, 469–481.

NYGAARD, E., LIEBERKIND, K. & FRYKMAN, P. 1983. Sedimentology and reservoir parameters of the Chalk Group in the Danish Central Graben. *Geologie en Mijnbouw*, **62**, 177–190.

NYONG, E. E. & OLSSON, R. K. 1984. A paleoslope model of campanian to lower Maestrichtian foraminifera in the North American Basin and adjacent continental margin. *Marine Micropalaeontology*, **8**, 437–477.

SCHOLLE, P. A. 1977. Chalk diagenesis and its relation to petroleum exploration: oil from chalks, a modern miracle? *American Association of Petroleum Geologists Bulletin*, **61**, 982–1009.

SIEMERS, W. T., CALDWELL, C. D., FARRELL, H. E., YOUNG, C. R., YANG-LOGAN, J. & HOWARD, J. J. 1994. Chalk lithofacies, fractures, petrophysics and paleontology: an interactive study of chalk reservoirs. *In*: *EAPG/AAPG Special Conference on Chalk, Copenhagen, Denmark, 7–9 September 1994*, Publisher??, 126–128.

VAN MORKHOVEN, P. C. M., BERGGREN, W. A. & EDWARDS, A. S. 1986. Cenozoic cosmopolitan deep-water benthic foraminifera. *Bulletin des Centres de Recherches Exploration–Production Elf-Aquitaine*, **11**.

Towards a stable and agreed nomenclature for North Sea Tertiary diatom floras – the '*Coscinodiscus*' problem

M. D. BIDGOOD,[1] A. G. MITLEHNER,[2] G. D. JONES[3] & D. J. JUTSON[4]

[1] *Grampian Stratigraphic Services, Unit 39 Howe Moss Avenue, Kirkhill Industrial Estate, Dyce, Aberdeen AB21 0GP, UK*

[2] *Micropalaeontology Unit, Department of Geological Sciences, University College London, Gower Street, London WC1 6BT, UK*

[3] *Pangaea Consulting, 14 Wallacebrae Crescent, Danestone, Aberdeen AB22 8YE, UK*

[4] *RWE-DEA AG, Laboratorium für Erdölgewinnung, Industrie Str. 2, D-29323, Weitze, Germany*

Abstract: Diatoms are one of the most useful microfossil groups to be found in Tertiary (particularly Palaeogene) sequences of the North Sea subsurface for their biostratigraphic utility, especially where other mineralized-walled microfossil groups (e.g. foraminifera, calcareous nannoplankton, etc.) are absent. An example from one offshore borehole (Shell UK; 29/25-1) is given. However, their further use in biostratigraphic and sequence correlation is hampered by the lack of any stable nomenclature applied to their taxonomy. The numbers of so-called 'in-house' taxa are legion with duplication of forms almost inevitable. Few of these many and varied taxa are directly comparable between these schemes, and this leads to almost inevitable confusion for exploration, production and development geoscientists trying to correlate between the various schemes.

One stratigraphically important and familiar diatom taxon (*Coscinodiscus* sp. 1) is formally described in this paper as *Fenestrella antiqua* (Grunow) Swatman. In the example shown here it is demonstrated that four forms, previously identified as four independent taxa in open nomenclature (with distinctive stratigraphic ranges), are in fact separate manifestations of the life habitat of this one single species. It is hoped that through an awareness of the biological complexity of these forms, and their relationships to morphology, a stable taxonomy will eventually arise. It is furthermore hoped that this will lead to a stimulus in the study of the biostratigraphic and palaeoenvironmental applications of this important microfossil group to exploration, production and development geoscience.

There can be few stratigraphers working on the commercial applications of micropalaeontology in the North Sea who are unfamiliar with the 'benchmark' diatom known as *Coscinodiscus* sp. 1 (in the sense of Bettenstaedt *et al.* 1962). It occurs widely in and around the North Sea basin, in both onshore and offshore sections (see Mitlehner 1996 for a diatom-based correlation of these strata). The highest stratigraphic occurrence (extinction) of a pyritized morphotype of this form (non-pyritized forms range higher) has been used as a reliable guide fossil for the downhole penetration of the Balder Formation (Moray Group) – a strong seismic reflector in close stratigraphic proximity to both exploration targets and hydrocarbon accumulations. Fluctuations in its abundance can be used to mark various levels within the Balder and Sele formations throughout most of the North Sea basin. In addition to this form, there have been perhaps several hundred additional diatom taxa recorded from various formations, which have potentially more or less stratigraphic utility. A detailed understanding of diatom stratigraphy has impacted subsurface issues on many producing reservoirs of the UK North Sea including Harding (Quadrant 9 Block 23), Gryphon (9/18), Sedgwick (16/6) and West Brae (16/7) fields (see also Payne *et al.* 1999).

Most of these forms have, for convenience, been placed within the genus *Coscinodiscus* Ehrenberg. However, it is readily apparent that '*Coscinodiscus*' has been used in this sense as a convenient 'bucket term' for almost any circular diatom-like form, and an entire pantheon of species numbering systems have been erected by the various oil companies, stratigraphic laboratories and consultants working on them. In fact,

BIDGOOD, M. D., MITLEHNER, A. G., JONES, G. D. & JUTSON, D. J. 1999. Towards a stable and agreed nomenclature for North Sea Tertiary diatom floras – the '*Coscinodiscus*' problem. *In:* JONES, R. W. & SIMMONS, M. D. (eds) *Biostratigraphy in Production and Development Geology*. Geological Society, London, Special Publications, **152**, 139–153.

there may be very few 'true' species of *Coscinodiscus* in the fossil record at all (Sims 1989). Yet it remains the fact that these important microfossils, which may number up to several hundred individual taxa, have *never been documented in one place* within the public domain. We consider this to be a serious omission, restricting the enormous potential value of this microfossil group in both large- and small-scale exploration, development and production biostratigraphy. This problem is also prevalent in the literature with, for just one example, *Coscinodiscus* sp. 3 (*sensu* Thomas & Gradstein 1981) synonymous with the aforementioned important morphotype *Coscinodiscus* sp. 1 (*sensu* Bettenstaedt *et al.* 1962).

Different aspects of this paper were undertaken by the various authors. The taxonomic foundation was constructed by A. Mitlehner, with M. Bidgood, G. Jones and D. Jutson providing information on the stratigraphic and palaeoenvironmental applications of the flora to North Sea stratigraphy. Overall coordination of the various contributions was undertaken by M. Bidgood. Stratigraphic queries should be addressed to M. Bidgood, G. Jones or D. Jutson. Taxonomic queries should be addressed to A. Mitlehner.

From a pragmatic to a more standardized approach

The task of the micropalaeontologist is first to place all of the many morphotypes of diatoms recorded into a stable, scientific taxonomic framework, rather than into an arbitrary system of open nomenclature (normally based around a simple numbering scheme for vaguely similar forms). It can be justifiably argued that this latter 'pragmatic' approach to taxonomy can have advantages in the day-to-day analysis of samples from commercial boreholes – speed of analysis and interpretation, together with local (i.e. in-house) consistency results in an efficient service to the line customer. The authors themselves have experienced this system at first hand and, indeed, some of the diatom taxa illustrated in this paper remain in open nomenclature. However, we believe that the 'pragmatic' approach to diatom taxonomy has probably reached the limits of its usefulness in that true taxonomic (i.e. biological) relationships are becoming unclear which is potentially damaging to biostratigraphic, palaeoenvironmental and sequence stratigraphic interpretation. In addition, 'pragmatic' schemes differ between in-house and external contractors, and changing a biostratigraphic subcontractor can result in painful and time-consuming reconstructions of the in-house database.

Preservation problems

The fact that most fossil forms recovered from North Sea offshore boreholes are preserved as pyritized infilled moulds ('*steinkerns*'), rather than the opal-A silica which forms the original diatom frustule, provides a major challenge to classical taxonomic interpretation. Consequently few authors (e.g. Mitlehner 1994, 1996) have made attempts to understand the relationship between original internal structures and the external appearance of the pyritized fossil; extensive use of the scanning electron microscope (SEM) is essential in this respect as many taxonomically defining features cannot be seen with the light microscope (LM). There is a pressing need to publish and illustrate these pyritized morphologies, as it is appreciated that industrial micropalaeontologists may not have ready access to SEMs or do not have time or resources to devote to extensive SEM studies on which some of the diagnostic taxonomic features are based. However, in this publication we have attempted to include features that can also be seen with the LM, for example the ridges on either side of valves of *Fenestrella antiqua* which mark the positions of the rows of labiate processes (see below) which are of taxonomic importance.

Diatom biostratigraphy: an example

Although the use of diatoms in North Sea Palaeogene biostratigraphy is well known amongst workers, there are few published references (e.g. Jacqué & Thouvenin 1975; King 1983; Mudge & Copestake 1992; Mitlehner 1996; Jones 1999), as the majority of datasets are based on offshore commercial boreholes generated by commercial companies; their presentation in the public domain is therefore not encouraged. Added to this problem is the lack of a stable nomenclature, which is detrimental to an effective appreciation of the group's biostratigraphic potential.

As a basic example of diatom stratigraphic utility, North Sea well 29/25-1 is located some 25 km southwest of the Auk and Fulmar (Shell) oil fields, and lies near the northeastern margin of the Western Platform and Mid North Sea High structural elements of the UK sector of the North Sea (Fig. 1). It was drilled in 1970 by the Shell UK–Esso partnership at position 56°20′N; 01°50′E. The well provides important data with respect to diatom biostratigraphy in that extensive parts of the Lower Palaeogene succession in the well were sampled by sidewall

Fig. 1. Offshore wells and localities mentioned in text.

cores as well as drilled ditch cuttings, and therefore caving effects (the contamination of *in situ* assemblages by stratigraphically younger material as a by-product of the drilling process) are minimized.

In broad terms, the micropalaeontological assemblages comprise an apparently normal sequence of Lower Palaeogene pyritized diatoms, interspersed with three, perhaps four, distinctive horizons of reworking as indicated by very small though abundant mid-Early Cretaceous planktonic foraminifera. As is typical for this interval across much of the North Sea basin, *in situ* microfossils (e.g. foraminifera, ostracods, radiolaria, etc.) were extremely rare, thus illustrating the usefulness of the diatoms. The data distribution chart for the Lower Palaeogene section of the well is shown in Fig. 2 (after Bidgood 1995).

The precise dating of the reworked horizons may have important implications to sequence stratigraphic analysis which can impact on the understanding of play fairways and the prediction of lowstand fan deposition. The highest two reworked horizons occur between the highest stratigraphic occurrences of the diatom species *Fenestrella antiqua* and *Coscinodiscus morsianus moelleri*. The third reworked horizon lies between the latter event and the lowest recorded occurrence (in sidewall cores) of *F. antiqua*. A possible fourth reworked horizon lies near the lowest recorded occurrences of common and consistent *F. antiqua* and *C. morsianus moelleri*.

The position of well 29/25-1 lies on the Western Platform–Mid North Sea High, rather than within the boundaries of the Central Graben itself. Therefore it may be that sequences resulting from sea-level fluctuations (which globally during Late Palaeocene–Early Eocene times were considerable; Haq *et al.* 1988) were better expressed in this well, rather than many of those from deeper parts of the basin (see also Neal *et al.* 1994; Bidgood 1995). A refined diatom biostratigraphy should therefore provide better calibration of such sequence events.

Throughout this paper the authors have used the chronostratigraphic and lithostratigraphic nomenclature of Knox & Holloway (1992) – the UKOOA standard – as the primary *convenient* framework on which to base their taxonomic conclusions. In this case, the Palaeocene–Eocene boundary is placed within the lower part of the Sele Formation. Biostratigraphically in that publication, this boundary is placed at the acme horizon of the palynomorph *Apectodinium* spp. The present authors, however, recognize that there is still no clear agreement as to the position of the Palaeocene–Eocene boundary within offshore subsurface Balder, Sele or adjacent sections (see Knox *et al.* 1996).

There is a good alternative case, for example, to include the whole of the Sele Formation within the uppermost part of the Late Palaeocene. It is thought that the influx of *Fenestrella antiqua* (the 'normal'/vegetative cells herein) at the top of the Sele Formation (a discrete influx which occurs stratigraphically below the first main downhole influx of this taxon normally associated with the upper Balder Formation) may represent the beginnings of a change in water-circulation pattern prior to the main Early Eocene transgressive phase in northwest Europe. In such a case, this event – the downhole disappearance of abundant *F. antiqua* – may be regarded as a useful proxy for the Palaeocene–Eocene chronostratigraphic boundary (Mitlehner 1996). This event is approximately coincident with the first downhole appearances

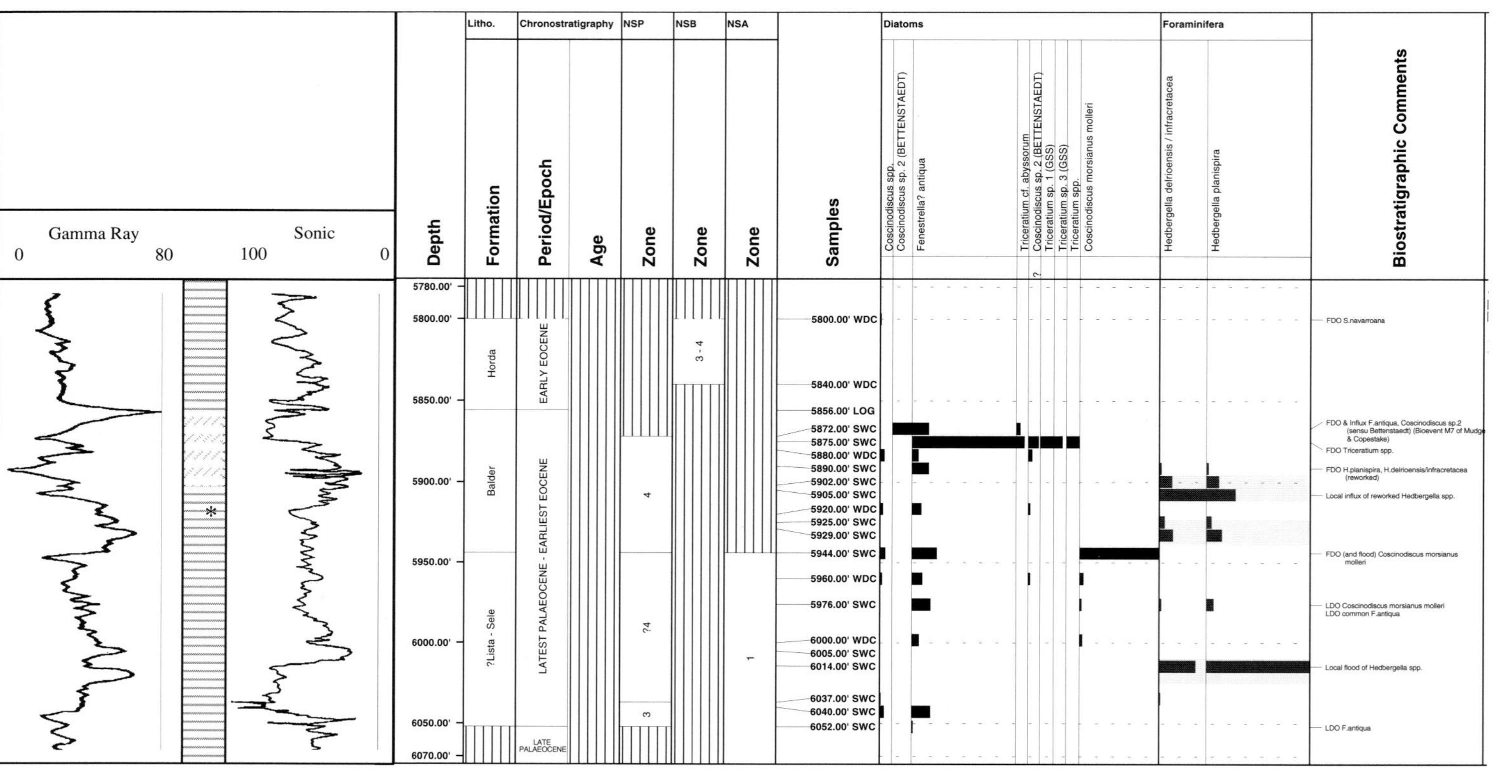

Fig. 2. Biostratigraphy of Shell UK well 29/25-1 (scale 1: 750).

Table 1. *Cartoons of some North Sea Palaeogene diatom morphotypes, with Linnean taxonomic identity where established*

	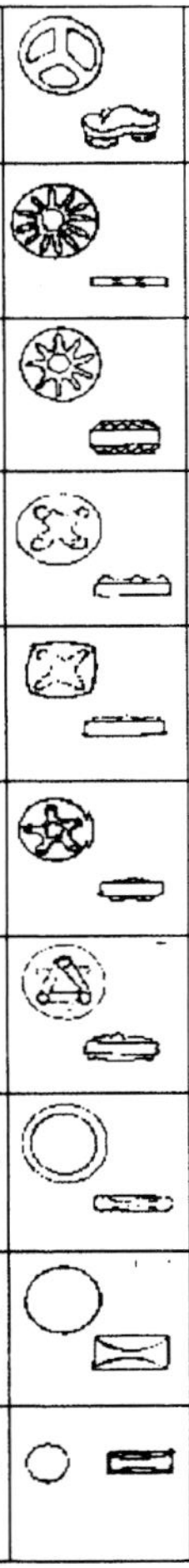		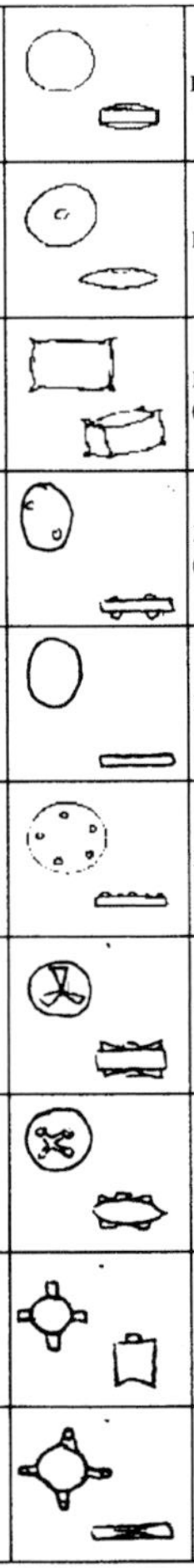		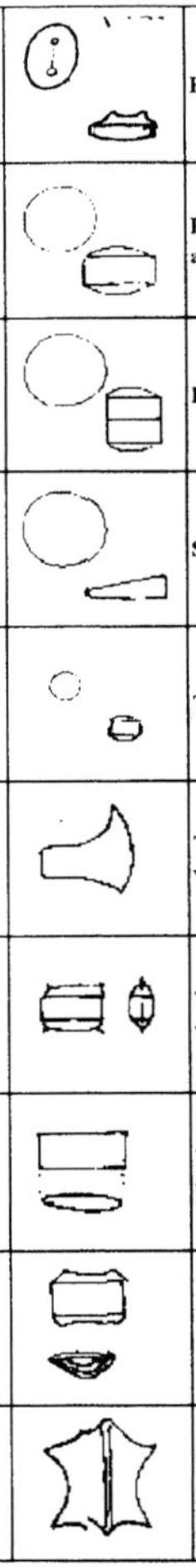		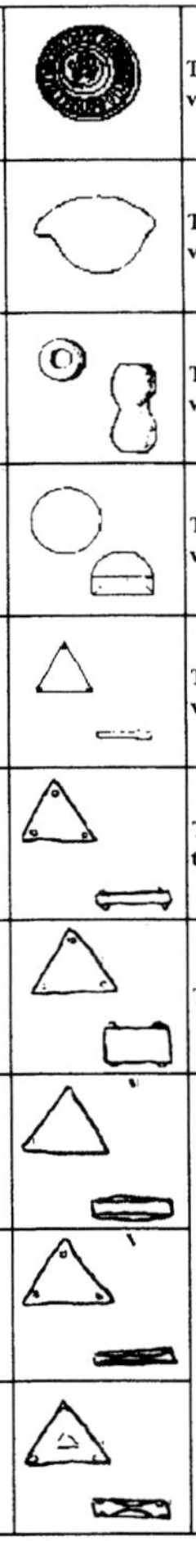		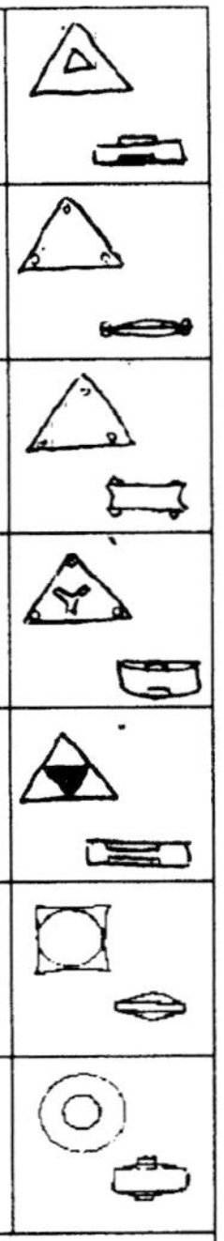
Actinoptychus senarius (resting spores)		Stellarima microtrias		Hemiaulus elegans		Paralia ornata		Trinacria regina (chain var.)	
Arachnodiscus indicus		Thalassiosiropsis wittiana		Fenestrella antiqua		Pseudostictodiscus angulatus		Trinacria regina (chain var.)	
Arachnodiscus indicus (variable no. of rays)		Cellataulus weisflogii		Fenestrella antiqua (autospore)		Pterotheca sp.		Trinacria regina (chain var.)	
Aulacodiscus allorgei		Diatom 'ovalis bituberculatum'		Fenestrella antiqua (resting spore)		Stephanogonia danica		Trinacria regina (chain var.)	
Aulacodiscus insignis var. quadrata		Diatom 'ovalis'		Fenestrella cf. antiqua		Triceratium sp.		Trinacria regina (chain var.)	
Aulacodiscus singiliewskyanus		Aulacodiscus aemulans		Hemiaulus ?elegans		Trinacria regina (chain var.)		Trinacria? regina tetragona	
Aulacodiscus subexcavata		Triceratium gibbosum		Hemiaulus elegans		Trinacria regina (chain var.)		Trochospira spinosa	
Coscinodiscus morsianus moelleri		Aulacodiscus hirtus		Odontella heidbergii		Trinacria regina (chain var.)			
Coscinodiscus morsianus var. morsianus		Solium exsculptum		Odontotropis cf. cristata		Trinacria regina (chain var.)			
Coscinodiscus morsianus var. morsianus (small)		Solium exsculptum		Odontotropis cristata		Trinacria regina (chain var.)			

(which may be better suited for industrial application) of *Coscinodiscus morsianus molleri* and *Odontotropis cristata* (see Table 1). Whatever the chronostratigraphic interpretation placed on these approximately coincident events, the fact remains that they are correlatable horizons which lie close to or at the Sele–Balder Formation boundary.

Formal diatom taxonomy

Background

Until recently, diatom classification has been based almost entirely on the siliceous cell wall, mainly because early workers who devised the classification system in the 19th century studied dead specimens and/or fossil material preserved without alteration (see Mitlehner 1996 for references to early studies). This situation has been extremely fortuitous for palaeontologists, stratigraphers and Quaternary palaeoecologists, but it must be emphasized that although the outline of the valve has been an important feature historically, this is not invariably reliable and the totality of the valve structure must be considered (Round *et al.* 1990, p. 29). In this study we illustrate an important example where the intricacies of pore structure, and the arrangement of wall organelles, are important criteria for identification. However, it is also demonstrated that SEM features need to be illustrated together with low-resolution photographs of the same specimens, in order to obviate the need for painstaking studies by industrial workers who may have limited or no access to the SEM. By this method workers need only examine the size and shape of a specimen, as well as looking for any other readily apparent features, in order to be confident of a species designation.

Diatom taxonomy: a state of flux

A number of authors have commented on the relatively chaotic present state of diatom taxonomy (Nikolaev 1990; Round *et al.* 1990; Edwards 1991; Cox 1993; Williams 1993). This situation is a function of the often conflicting criteria used to distinguish taxa, both at species level and variation within species. For example, the use of the concept of subspecies has not gained universal acceptance among diatom workers, who use the terms *variety* and *forma* to distinguish intraspecific morphological variation, which is displayed by many diatoms and thus may potentially further confuse commercial workers. Some taxa have been split into a bewildering number of varieties and formas. In the case of chain-forming diatoms (such as *Paralia* and *Trinacria* which occur in the North Sea Palaeogene – see Table 1) this is due to a lack of study of complete chains, as the valves at the ends of these chains (separation valves, see Crawford *et al.* 1990) have different features to those within the chain itself (linking valves).

Scope for further studies

A complete taxonomic treatment of northwest European Tertiary diatom microfloras is monographic in extent and therefore lies outside the scope of this paper; it will be subject to a separate approach by the authors elsewhere. However, one well-known North Sea diatom taxon (and its associated morphotypes) is formally described here; in addition, a table of 'cartoon' drawings of some recorded taxa, together with Linnean nomenclature where appropriate, is also presented (Table 1). Table 1 is not intended to be a definitive taxonomic treatment as such, rather as a statement of what the authors believe to represent the present status of European Palaeogene diatoms *where some degree of formal taxonomy has already been assigned.* The authors recognize that there are potentially many more diatom taxa which remain to be adequately taxonomically treated.

Diatom biology and ecology – importance for taxonomy

Lack of understanding of diatom biology and ecology in living forms can have undesirable consequences for the interpretation of fossil specimens. In this example four forms, previously regarded as separate (and stratigraphically important) open taxonomic entities, are shown to be the results of adaptations of a single species (i.e. *Fenestrella antiqua*) to life-cycle strategies and responses to environmental stress. It is therefore necessary to appreciate these biological principles when considering the revised taxonomic descriptions which follow.

The diatom life-cycle: 'Russian dolls'

Diatom reproduction involves alternating sexual (meiotic) and asexual (mitotic) stages (Fig. 3). The asexual stage results in progressive reduction in mean *normal* cell size, in the manner of 'Russian dolls', so that eventually a generation

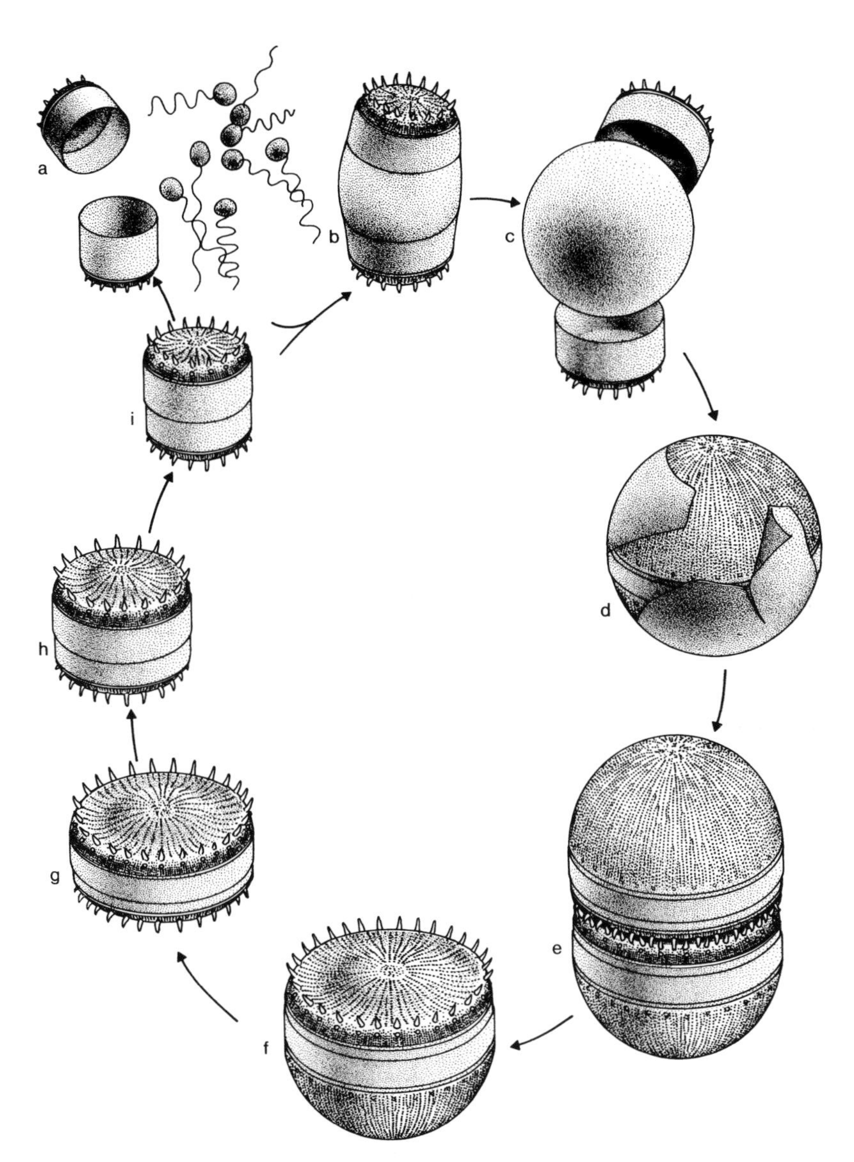

Fig. 3. The diatom life-cycle (reproduced from Round *et al.* 1990, with the permission of the authors and Cambridge University Press). (**a**) Formation of motile gametes, (**b**) and (**c**) formation of auxospore, (**d**) auxospore breaking open to reveal initial cell, (**e**) first division of initial cell, (**f**) one of the cells from (**e**) with a normal cell valve and an initial cell valve, (**g**) a cell formed following several divisions of (**f**), (**g**)–(**i**) vegetative size reduction.

results which must find some means of size restitution if it is to continue to sustain itself. This is achieved by the process of sporulation, resulting from the production and fusion of gametes from parent cells. An *auxospore* is formed inside which new large, or *initial*, cells develop. When mature, the auxospore divides and two new cells are produced. In this way the cell size increases and a new period of asexual reproduction can begin. Various stages in this reproductive cycle can be incorporated into the fossil record and must be borne in mind when a taxonomic identification is attempted. The following descriptions illustrate this phenomenon.

Normal (vegetative) cells. Usually equivalvar, with two equally sized valves separated by one or more *girdle bands*. Typical representatives of *F. antiqua* (= *Coscinodiscus* sp. 1) are recognized by all North Sea Palaeogene workers, although smaller examples (which probably represent successive stages in the normal growth period) may be referred to as *Coscinodiscus* cf. sp. 1 in some schemes.

Initial cells. The first stage in the diatom vegetative cell division cycle, formed after the fusion of gametes during sexual reproduction, inside an *auxospore* (see Fig. 3c–e). The shape of these cells is invariably cylindrical or 'barreliform' due to a very wide girdle, often with distinct cingula. The valves are typically highly domed, and do not exhibit the pronounced angle with the girdle seen in vegetative cells. Initial cells of *Fenestrella antiqua* ('*Coscinodiscus barreliformis*' of M. A. Charnock and others) often occur in conjunction with vegetative cells.

Initial cells, after first cell division. A stage in the vegetative life-cycle of diatoms, revealed after the auxospore breaks open. Two cells are thus formed, with an *initial cell valve* and a *normal cell valve*, the former being more markedly domed than the other (Fig. 3f). After several divisions, the cells take on the form of the normal, equally shaped, vegetative valves. Distinguishable by having one valve more markedly domed than the other. In addition, the valve–girdle junction varies on either valve, with that formed by the less highly domed valve showing a clear angle.

Resting spores: an environmental adaptation. A further mechanism which helps to increase the likelihood of a population surviving is via the production of *resting spores*. These are mainly formed in modern oceans by planktonic centric species, and take the form of thickly silicified cells which may, or may not, bear spines. Resting spores enable the diatom to sink through the water column and begin a phase of dormancy, which normally lasts until strong current action triggers the resuspension of the resting spore and the diatom cell can return to the photic zone. This phenomenon is a feature of populations which occur in upwelling zones along the western margins of the main ocean basins, and occurs in response to seasonal fluctuations in nutrient levels; but it is also characteristic of diatom populations in high latitudes which undergo strong seasonal variations in light intensity (Hargraves & French 1983). Resting spores are of widespread occurrence in the Upper Palaeocene and Lower Eocene diatom assemblages in and around the North Sea basin (Mitlehner 1994, 1996), and reflect the stressed environment present at the time, with widespread periods of anoxia, basin stratification, 'greenhouse warming' and volcanic ash falls.

Systematic descriptions/micropalaeontology

This section, by its very nature, is highly technical in content but is necessary to establish the taxonomic basis by which palaeontologists work. It is provided as a formal requirement of the procedure to establish new or emended taxa (i.e. species) in publications. If biostratigraphy is taken as one of the key underpinning agents of stratigraphy/sequence stratigraphy, then taxonomy is the key underpinning agent of biostratigraphy. Without a descriptive, illustrated taxonomy there can be no uniformity in species concepts between different workers and it is for this reason that this exercise is undertaken here.

Pyritized morphologies – rationale for description

Terminology. In recent years, a number of diatom taxa have been revised to take into account SEM observations (see Round *et al.* 1990). These are referred to under 'Revised diagnosis and description'. In addition, it has sometimes been necessary to further emend descriptions, as pyritized morphotypes usually preserve the original frustule shape with girdle bands intact, a feature not often observed in non-pyritized specimens which are usually preserved as isolated valves. Any further revision is referred to under 'Emended diagnosis (herein)'. Synonomy listings in bold type refer to pyritized forms.

Sample derivation. Well 15/28a-3, Outer Moray Firth, North Sea (see Fig. 1); comparative material from the Fur Formation diatomite, Island of Fur, Denmark, on a strewn slide housed in the collections of the Natural History Museum, London (see Mitlehner 1995, 1996 for further information on the Fur Formation diatomite).

Division **Bacillariophyta**
Class **Coscinodiscophyceae** Round & Crawford, in Round *et al.* (1990)
Subclass **Coscinodiscophycidae** Round & Crawford, in Round *et al.* (1990)
Order **Coscinodiscales** Round & Crawford, in Round *et al.* (1990)
Family **Stellarimaceae** Nikolaev ex Sims & Hasle 1990
Included genera: *Fenestrella*, *Stellarima*

Remarks. A recently defined family, comprising two genera formerly included within the Coscinodiscaceae. Nikolaev (1983, p. 1124) introduced the name 'Stellarimaceae Nikolaev, nom. nov' for a monotypic family but did not formally publish the name. Sims & Hasle (1990, p. 207) subsequently made a formal designation for the new family. Both of the included genera feature prominently in North Sea Palaeogene assemblages and have been described in detail only recently (but not in pyritized form), and so are given a comprehensive description. *Fenestrella* is described here; *Stellarima* is covered in the unpublished PhD thesis of Mitlehner (1994), but will be further considered in a future work. Specimens from both genera have been recovered which represent stages in the diatom life-cycle and are thus important from a palaeoecological viewpoint.

> A survey of *Stellarima* and *Fenestrella* show that the characters they have in common are overall shape, lack of a distinct central area, labiate processes that are identical in shape and structure and that are neither marginal, rarely central but usually positioned in a ring on the valve face. (Sims 1990, p. 287)

Stratigraphic range. Upper Cretaceous (Campanian)–Recent.

Genus *Fenestrella* Greville 1864
Type species *Fenestrella barbadensis* Greville 1864

Original diagnosis

> Frustules free, disciform: disc with a minute, radiant cellulation, interrupted in the middle by linear bands, composed of parallel lines of cellules, each band terminating in a flat ocellus. (Greville 1861–66, 1864, Vol. 9, 67 [80])

Remarks. Sims (1990, p. 278) considers that the main diagnostic features of *Fenestrella* are the lack of a distinct central area, unusual in a centric diatom, and the presence of two 'ocelli' (actually labiate processes) lying opposite and mid-way between the valve centre and margin. A hitherto rarely recorded fossil genus, ranging from Palaeocene to Miocene, described accurately only recently (see above). Two species are known, one of which, *F. antiqua*, has been found to form a major component of diatom assemblages in the Lower Eocene Balder Formation of the North Sea and its onshore equivalents. Its common occurrence as a pyritized steinkern may explain why this large, otherwise very fragile diatom is so frequently encountered therein, the pyrite infilling having preserved the shape of the frustule.

Fenestrella antiqua (Grunow) Swatman, emend
(Figs 4a–6e)

1882 *Janischia antiqua* (Grunow in Van Heurch 1896, pl. XCV bis, figs 10 and 11).
1889 *Coscinodiscus ludovicianus* Rattray (Rattray 1889, p. 596).
1896 *Janischia antiqua* Grun. 1882 (Van Heurck 1880–85, p. 536, fig. 282).
1940 ***Coscinodiscus* sp. (bikonvex) (Staesche & Hiltermann 1940, p. 15, pl. 6, fig. 3).**
1943 ***Coscinodiscus* sp. 1 (bikonvex) (Wick 1943–50, p. 5, pl. 1, figs 47–66.**
1948 *Fenestrella antiqua* (Grunow) Swatman comb. nov. (Swatman 1948, p. 53, pl. 2, figs. 10 and 11).
1962 ***Coscinodiscus* sp. 1 (Bettenstaedt *et al.* 1962, p. 357, pl. 2, figs 18 and 19).**
1972 *Janischia antiqua* Grun (Benda 1972, pl. 2, figs 10 and 11).
1975 ***Coscinodiscus* sp. 1 (Jacqué & Thouvenin 1975, p. 462, pl. 2, figs. A–E).**
1983 *Coscinodiscus* sp. 1 Bettenstaedt *et al.* (Bignot 1983, p. 17, pl. 1, figs 5, 6 and 8).
1983 ***Coscinodiscus* sp. 1 Bartenstein and others (King, 1983, p. 20, figs 1 and 2).**
1984 ***Coscinodiscus* sp. 1. (Malm *et al.* 1984, p. 158, fig. 8a).**
1990 *Fenestrella antiqua* (Grunow) Swatman (Sims 1990, p. 179, figs. 1–14 and 22).
1991 *Fenestrella antiqua* (Grunow) Swatman (Homann 1991, p. 48, pl. 18, figs 1, 2, 4 and 5).

Revised diagnosis & description (abridged)

> Valves circular, weakly domed, fragile and golden in colour. A ring of well-spaced labiate processes present at approximately 1/3 of the distance from the valve margin to the valve centre, these more densely aggregated in two opposite areas to form rows of processes (the ocelli of Greville 1864). These rows of closely packed processes vary in length and, on most valves, lie parallel to the valve margin, slightly within the ring of well-spaced or individual processes. The number of processes in each patch varies from 5 to over 50, and there is apparently no close relationship between valve diameter and number of processes. No obvious central area to the valve is present. Rows of areolae arranged in fascicles or plates, extend to the valve margin. (Sims 1990, p. 279)

Emended diagnosis (herein). Outline of frustule biconvex in girdle view, with prominent,

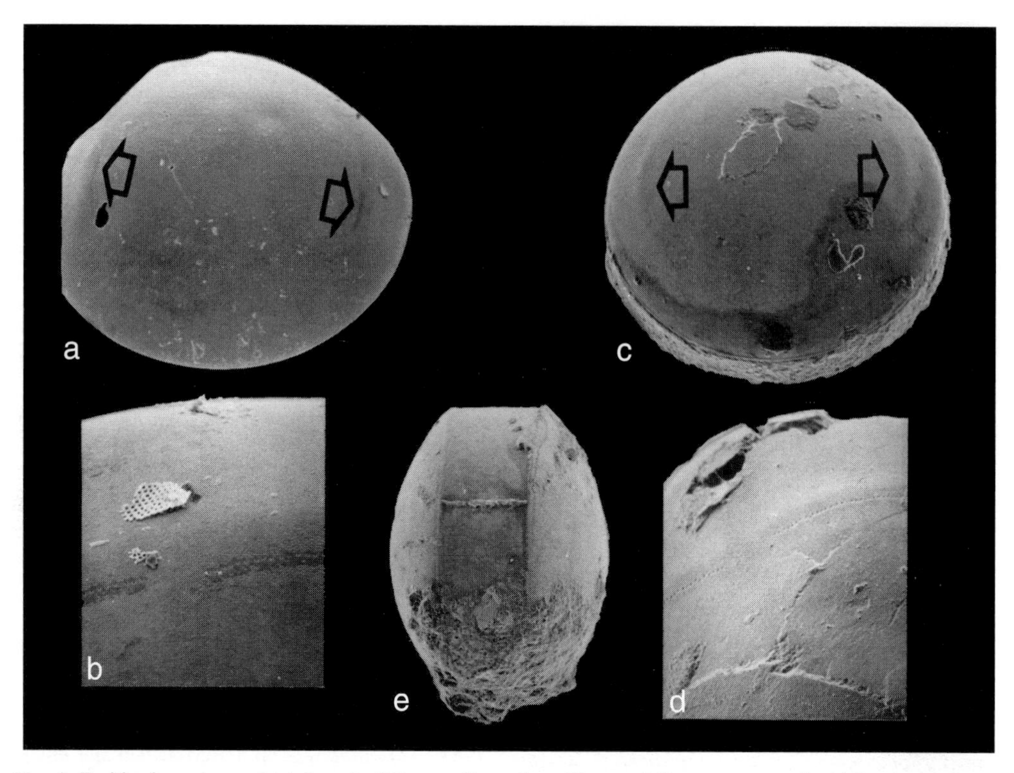

Fig. 4. Pyritized specimens (steinkerns) of *Fenestrella antiqua* (Grunow) Swatman, emend., with some non-pyritized specimens for comparison. (**a**)–(**e**) Vegetative cells. (**a**) and (**b**) Non-pyritized specimen. (**a**) Oblique valve view, showing lines of packed labiate processes on opposite sides of valve face (arrows). (**b**) Detail of (**a**). Note labiate processes, surrounded by fine areolae. Natural History Museum collections ('Fur Nykøbing'), ?lowermost Eocene. (**c**) and (**d**) Pyritized steinkern. (**c**) Oblique view of pyrite-infilled specimen, with a veneer of silica preserving areas of packed labiate processes (arrows). (**d**) Detail of (**c**). Note labiate processes (compare with b). BP well 15/28a-3, 6460 feet, ?uppermost Palaeocene. (**e**) Diameter 140 μm. Girdle view. Note prominent break in girdle band. BP well 15/28a-3, 6420 feet, ?lowermost Eocene.

heavily silicified girdle bands which form an angle with the valve margin (only observed in pyrite-infilled specimens). The girdle bands are of the 'open' or 'split' variety (Round *et al.* 1990, p. 48). Non-pyritized specimens are preserved as isolated valves, usually broken with girdle bands detached. The areas of packed labiate processes are normally only seen via SEM, under low magnification, in pyritized specimens and take the form of indentations.

Dimensions. Valve diameter 150–360 μm. Areolae 11–14 in 10 μm.

Remarks. Sims (1990, p. 279, figs 1–14 and 22) gave plates, both LM and SEM, of this taxon. Her descriptions were of well-preserved, non-pyritized specimens. The largest diatom in Palaeogene sediments from the North Sea basin, the pyritized biconvex frustules of *F. antiqua*, has hitherto been identified as *Coscinodiscus* sp. 1 (Bettenstaedt *et al.* 1962, see above). The distinctive labiate processes described above are preserved as indentations in pyritized specimens, thus forming an impression of the inside of the valve; these are normally only visible in the SEM in pyritized specimens, although they are clearly visible in the LM in non-pyritized valves. A veneer of original silica is preserved on the pyrite steinkerns of some specimens found in the Balder Formation, which gives the appearance of a prominent iridescent sheen to the specimen.

Occurrence (this work). Wells 15/28a-3 and 29/25-1, North Sea, Sele and Balder formations; Island of Fur, Denmark, Fur Formation (Fig. 1).

North Sea range. Upper Palaeocene–lowermost Eocene (Bettenstaedt *et al.* 1962; Jacqué & Thouvenin 1975; King 1983; Malm *et al.* 1984; Mudge & Copestake 1992).

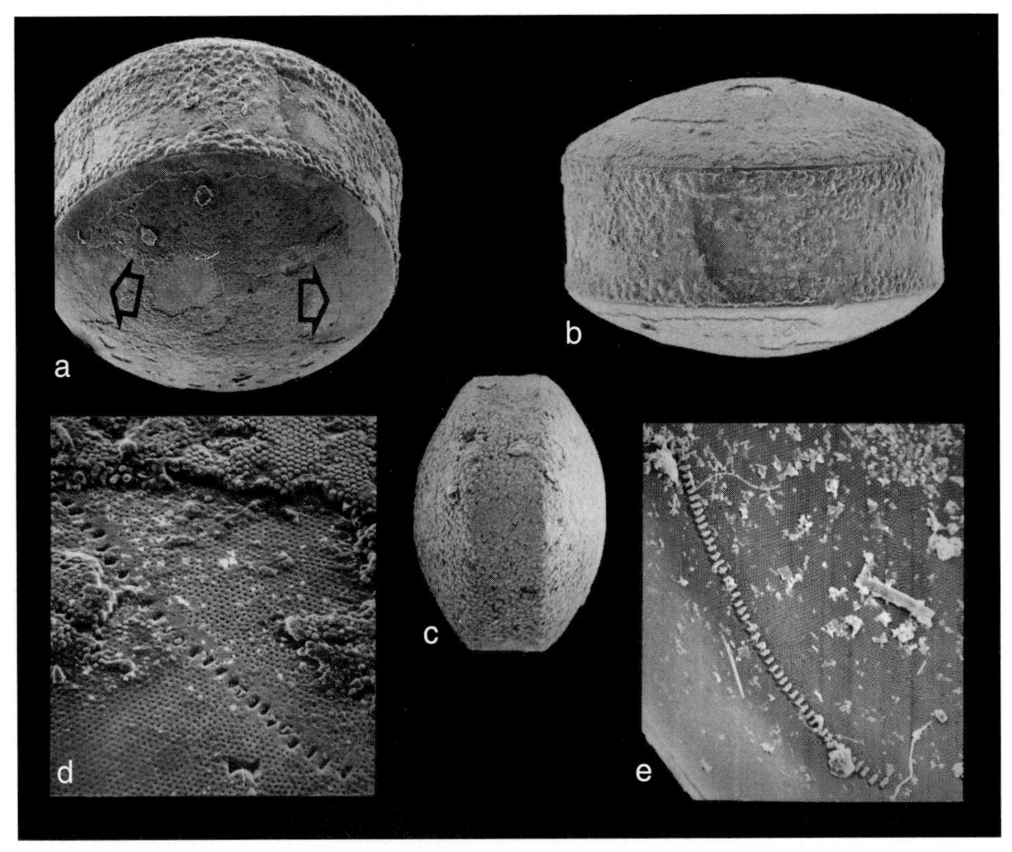

Fig. 5. Pyritized specimens (steinkerns) of *Fenestrella antiqua* (Grunow) Swatman emend., with some non-pyritized specimens for comparison. (**a**) and (**b**) Initial cell, after first cell division; (**c**)–(**e**) Vegetative cells. (**a**) Pyritized specimen. Oblique valve view. Note thickened girdle, and areas of packed labiate processes (arrows). BP well 15/28a-3, 6410 feet, ?lowermost Eocene. (**b**) Girdle view of (**a**). Note unequal convexity of valves, the uppermost (auxospore) valve being more highly domed than the lower (vegetative) valve (compare with Fig. 3f). (**c**) Pyritized specimen. Girdle view of normal cell. Ølst, Denmark ('Ølst D'), ?lowermost Eocene. (**d**) and (**e**) Detail of internal expression of packed labiate processes. (**d**) Specimen infilled and replaced by pyrite. Labiate proceses form impressions in non-crystalline pyrite, while framboidal pyrite has infilled areolae (top of photograph). BP well 15/28a-3, 6410 feet, ?lowermost Eocene. (**e**) Non-pyritized specimen. Note inwardly-projecting labiate processes. Natural History Museum collections ('Fur Nykøbing'), ?lowermost Eocene.

Range (literature). Upper Palaeocene–Lower Eocene: Fur Formation, Denmark (Rattray 1889; Schulz 1927; Benda 1972; Sims 1990; Homann 1991). Lower Eocene: North Germany (Schulz 1927); Russia (Glezer *et al.* 1974); Paris basin (Bignot 1983); Belgium (King 1990). ?Upper Eocene: Russia, 'Kamischev' (Chenevrière 1934; see also Ross & Sims 1985).

Fenestrella antiqua initial cells (Fig. 6a and b)

Description. Shape cylindrical due to a very wide girdle, often with distinct cingula. The valves are highly domed, and do not exhibit the pronounced angle with the girdle seen in vegetative cells. Rows of labiate processes not always clear due to poor preservation, but distinguishable in some specimens.

Dimensions. Valve diameter 120–300 μm width of cell 300–500 μm.

Remarks. A variant of *F. antiqua* commonly found in the volcaniclastic Sele Formation, central and northern North Sea. The unusually thickened girdle is strongly suggestive of an auxospore (see the section on 'Diatom biology and ecology' above), reflecting a period of vegetative reproduction with new, smaller diatom cells forming beneath the protective girdle band. Unpublished data from oil exploration and

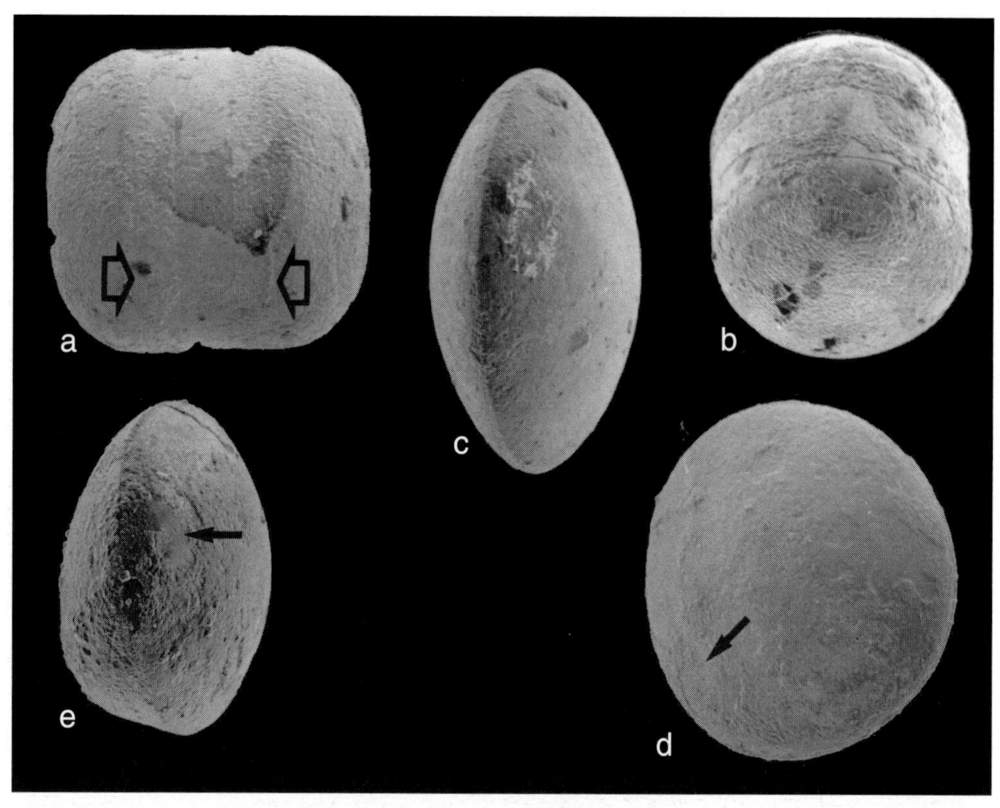

Fig. 6. Pyritized specimens (steinkerns) of *Fenestrella antiqua* (Grunow) Swatman emend., with some non-pyritized specimens for comparison. (**a**) and (**b**) Initial cells (compare with Fig. 3e); (**c**)–(**e**) Resting spores (all pyritized). (**a**) Girdle view. Note highly-rounded valves, merging with multiple girdle bands. Arrows mark valve–girdle junction. BP well 15/28a-3, 6360 feet, ?lowermost Eocene. (**b**) Oblique valve view. Specimen partially encrusted with overpyrite. BP well 15/28a-3, 6390 feet, ?lowermost Eocene. (**c**) Note lack of girdle bands, and smooth surface. BP well FC 22, 'Core 27', ?uppermost Palaeocene. (**d**) Valve view. Arrow marks position of packed labiate processes. BP well 15/28a-3, 6370 feet, ?uppermost Palaeocene. (**e**) Girdle view, with cell opening to reveal girdle of vegetative cell beneath (base of specimen). Arrow marks position of packed labiate processes. BP well 15/28a-3, 6470 feet, ?uppermost Palaeocene.

service companies allude to the widespread occurrence of this distinctive morphology in large enough numbers to form a marker for the upper part of the Sele Formation in proprietary zonation schemes. Known under various morphotype numbers in the informal diatom zonations of different companies, for example *Coscinodiscus* sp. 8 (RRI); *Coscinodiscus* N7 (British Petroleum). Informally named *Coscinodiscus barreliformis* by Charnock and others.

Occurrence (this work). Wells 15/28a-3 and 29/25-1, North Sea, Sele and Balder formations (Fig. 1).

North Sea range (unpublished). Upper Palaeocene (RRI), but Early Eocene based on presumed age of upper Sele Formation used herein (Knox & Holloway 1992).

Range (literature). The only published account of pyritized diatom auxospores is from the Middle Eocene of the Beaufort–Mackenzie basin, Canada (McNeil 1990). However, these clearly belong to the genus *Stellarima* as they exhibit labiate processes ('endocorona' of McNeil 1990) that are positioned at the centre of the valve face, and not nearer to the valve margins as in *Fenestrella*, or around the inside of the margin itself, as in *Coscinodiscus*.

Fenestrella antiqua initial cells, after first cell division
(Fig. 5a and b)

Description. A morphological variant of *F. antiqua* characterized by having one valve more markedly domed than the other. In addition, the valve–girdle junction varies on either valve, with that formed by the less highly domed valve showing a clear angle. The rows of packed labiate processes characteristic of *F. antiqua* are clearly visible on specimens unaffected by diagenetic recrystallization.

Dimensions. Valve diameter 120–280 μm.

Remarks. See the section on 'Diatom biology and ecology' above. This form often occurs in association with resting spores of *F. antiqua*.

Occurrence (this work). Wells 15/28a-3 & 21/9-1, central North Sea, Balder Formation.

Range (literature). No published account exists for this morphotype. It is often found in cuttings and core in the central and northern North Sea in the Balder Formation

Fenestrella antiqua resting spores
(Fig. 6c–e)

Description. Cells large (120–280 μm in diameter) with no girdle band, so that the frustule resembles a discus when seen in girdle view. Specimens are commonly very smooth, with a thickened frustule which carries fine areolae. The areas of packed labiate processes are much shortened by comparison to those of the normal cells, and do not curve with the valve face but are straighter, or curve slightly away from the valve margin. The rows of enlarged pores, which connect the two areas of packed labiate processes on the normal cells, are absent.

Remarks. A morphological variant of *F. antiqua* which is a resting spore, formed as a response to adverse environmental conditions (see the section on 'Diatom biology and ecology' above).

Occurrence (this work). BP wells 15/28a-3 and 21/9-1, Central North Sea, Sele and Balder formations, more abundantly in the Sele Formation. Isolated valves also observed in samples from the Fur Formation, Denmark, at the top of the Knudeklint Member.

Range (literature). There is no published description of this morphotype, but it is often encountered in ditch cuttings and cores from the central and northern North Sea in the Sele and Balder formations. It is normally referred to as *Coscinodiscus* N6 (British Petroleum), *Coscinodiscus* sp. 7 (RRI), *Coscinodiscus* sp. 15 (GEUS) or *Coscinodiscus* sp. 9 (Halliburton).

North Sea range. Mudge & Copestake (1992) refer to the abundance of this morphotype at the top of the Sele Formation in the northern North Sea.

Conclusions

Diatoms (in this case pyritized forms) are extremely useful biostratigraphic markers, particularly in northwest European Palaeogene sediments which lack calcareous microfossils, such as foraminifera and nannoplankton. However, their present degree of biostratigraphic utility has probably reached a limit due to the lack of understanding of diatom biology and the over-use of so-called 'pragmatic' or 'open' classification schemes. We have attempted to show that an understanding of the biological influences on diatom morphology requires the need for a comprehensive taxonomic revision of the group as it applies to northwest European Palaeogene biostratigraphy. For example, four diatom taxa (including the well-known *Coscinodiscus* sp. 1), previously thought to be distinct both taxonomically and biostratigraphically, are shown to be different life-cycle manifestations of one species, here identified as *Fenestrella antiqua*. The use of scanning electron microscopy is shown to be crucial in the discrimination of some important taxonomically significant features. Understanding of the organisms response to changes in environment, and the natural progression of its morphology through the 'normal' stages in its life-cycle, are shown to have an important bearing on palaeoenvironmental interpretations.

In addition to their respective institutions the authors would like to thank BP Exploration and Shell UK Exploration & Production for access to borehole samples. We would also like to thank the referees of this paper for their beneficial input. The taxonomic section was originally carried out by A. Mitlehner in the Department of Geological Sciences, University College, London, under the supervision of A. R. Lord, as part of a NERC/CASE funded PhD study with British Petroleum and the Natural History Museum. Both are thanked for access to material and facilities. Helpful comments were received from P. A. Sims (the Natural History Museum) and R. W. Battarbee (Environmental Change Research Centre, University College, London).

References

BENDA, L. 1972. The diatoms of the Moler Formation of Denmark (Lower Eocene). *In*: SIMONSEN, R. (ed.) *First Symposium on Recent and Fossil Marine Diatoms. Beiheft zur Nova Hedwigia*, **39**, 251–266.

BETTENSTAEDT, F., FAHRION, H., HILTERMANN, H. & WICK, W. 1962. Tertiär Norddeutschlands. *In*: SIMON, W. & BARTENSTEIN, H. (eds) *Arbeitskreis deutscher Mikropaläontologen: Leitfossilien der Micropaläontologie*. Gebruder Bontrager, Berlin.

BIDGOOD, M. D. 1995. *The Microbiostratigraphy of the Paleocene of the Northwest European Continental Shelf*. PhD thesis, University of Plymouth.

BIGNOT, G. 1983. Les Lagunes Sparnaciennes: Une étape dans la conquête des eaux douces par les diatomêes. *Revue de Micropaleontologie*, **26**, 15–21.

CHENEVRIÈRE, E. 1934. Sur un depôt fossile marin à diatomées situé à Kamischev (Russie centrale). *Bulletin de la Societé Francaise des Microscopie*, **3**, 103–107.

COX, E. J. 1993. Diatom systematics – a review of past and present practice and a personal vision for future development. *Beiheft zur Nova Hedwigia*, **106**, 1–20.

CRAWFORD, R. M., SIMS, P. A. & HAJOS, M. 1990. The morphology and taxonomy of the centric diatom genus *Paralia*. I. *Paralia siberica* comb. nov. *Diatom Research*, **5**, 241–252.

EDWARDS, A. R. (compiler) 1991. The Oamaru diatomite. *New Zealand Geological Survey Paleontological Bulletin*, **64**.

GLEZER, Z. I., JOUSE, A. P., MAKAROVA, I. P., PROSHKINA-LAVRENKO, A. I. & SHESHUKOVA-PORETSKAYA, V. S. 1974. *The Diatoms of the USSR Fossil and Recent, Volume 1*. Akademii Nauk SSSR, Botanical Institute, Leningrad (in Russian).

GREVILLE, R. K. 1861–66. Descriptions of new and rare diatoms. *Transactions of the Microscopical Society of London, New Series*, **I-XX**, 9–14.

HAQ, B. U., HARDENBOL, J. & VAIL, P. R. 1988. Mesozoic and Cenozoic chronostratigraphy and cycles of relative sea level change. *In*: WILGUS, C. K., HASTINGS, B. S., KENDALL, G. C. ST. C., POSAMENTIER, H. W., ROSS, C. A. & VAN WAGGONER, J. C. (eds) *Sea-level Changes: an Integrated Approach*. SEPM, Tulsa, Special Publication, **42**, 71–108.

HARGRAVES, P. E. & FRENCH, F. W. 1983. Diatom resting spores: significance and strategies. *In*: FRYXELL, G. A. (ed.) *Survival Strategies of the Algae*. Cambridge University Press, Cambridge, 49–68.

HOMANN, M. 1991. Die Diatomeen der Fur-Formation (Alt-tertiär) aus dem Limfjord-Gebeit, Nordjutland/ Dänemark. *Geologisches Jährbuch, Reihe A*, **123**.

JACQUÉ, M. & THOUVENIN, J. 1975. Lower Tertiary tuffs and volcanic activity in the North Sea. *In*: WOODLAND, A. W. (ed.) *Petroleum and the Continental Shelf of Northwest Europe, Volume 1: Geology*. Elsevier, Barking, 455–465.

JONES, R. W. 1999. Forties Field (North Sea) revisited: a demonstration of the value of historical micropalaeontological data. *This volume*.

JUTSON, D. J. 1995. *Paleogene Radiolaria and Diatoms From Three Exploration Wells From the Central Graben, North Sea (Danish Sector): Stratigraphic and Paleoenvironmental Significance*. EFP-92 Project Report 17, Århus University/Geological Survey of Denmark.

KING, C. 1983. *Cainozoic Micropalaeontological Biostratigraphy of the North Sea*. Institute of Geological Sciences Report 82/7. HMSO, London.

——1990. Eocene stratigraphy of the Knokke borehole (Belgium). *Toelichtende Verhandelinger Geologische en Mijjnkarten van Belgie*, **29**, 67–102.

KNOX, R. W. O'B. & HOLLOWAY, S. 1992. 1. Paleogene of the Central & Northern North Sea. *In*: KNOX, R. W. O'B. & CORDEY, W. G. (eds) *Lithostratigraphic Nomenclature of the UK North Sea*. British Geological Survey on behalf of, UKOOA, Nottingham.

——, CORFIELD, R. M. & DUNAY, R. E. (eds) 1996. *Correlation of the Early Paleogene in Northwest Europe*. Geological Society, London, Special Publications, **101**.

MALM, O. A., CHRISTENSEN, O. B., FURNES, H., LØVLIE, R., RUSELÅTTEN, H., & ØSTBY, K. L. 1984. The Lower Tertiary Balder Formation: an organogenic and tuffaceous deposit in the North Sea region. *In*: SPENCER, A. M. *et al.* (eds) *Petroleum Geology of the North European Margin*. Graham & Trotman, London, 149–170.

MCNEIL, D. H. 1990. Stratigraphy and paleoecology of the Eocene *Stellarima* assemblage zone (pyrite diatom steinkerns) in the Beaufort–Mackenzie Basin, Arctic Canada. *Bulletin of Canadian Petroleum Geology*, **38**, 17–27.

MITLEHNER, A. G. 1994. *The Occurrence and Preservation of Diatoms in the Paleogene of the North Sea Basin*. PhD thesis, University College, London.

——1995. *Cylindrospira*, a new diatom genus from the Paleogene of Denmark with Palaeoecological significance. *Diatom Research*, **10**, 321–331.

——1996. Palaeoenvironments in the North Sea Basin around the Paleocene–Eocene boundary: evidence from diatoms and other siliceous microfossils. *In*: KNOX, R. W. O'B., CORFIELD, R. M. & DUNAY, R. E. (eds) *Correlation of the Early Paleogene in Northwest Europe*. Geological Society, London, Special Publications, **101**, 255–273.

MUDGE, D. C. & COPESTAKE, P. 1992. Lower Paleogene stratigraphy of the Northern North Sea. *Marine and Petroleum Geology*, **9**, 287–301.

NEALE, J. E., STEIN, J. A. & GAMBER, J. H. 1994. Graphic correlation and sequence stratigraphy in the Palaeogene of NW Europe. *Journal of Micropalaeontology*, **13**(1), 55–80.

NIKOLAEV, V. A. 1983. On the genus *Symbolophora* (Bacillariophyta). *Botanicheskii Zhurnal*, **68**, 1123–1128 (in Russian).

——1990. The system of centric diatoms. *In*: *Proceedings of the 10th Diatom Symposium 1988*. *Beiheft zur Nova Hedwigia*, 17–22.

PAYNE, S. N. J., EWEN, D. F. & BOWMAN, M. J. 1999. The role and value of 'high-impact biostratigraphy' in reservoir appraisal and development. *This volume*.

RATTRAY, J. A. 1889. The revision of the genus *Coscinodiscus* Ehrb., and of some allied genera. *Proceedings of the Royal Society of Edinburgh*, **16**, 449–692.

ROSS, R. & SIMS, P. A. 1985. Some genera of the Biddulphiaceae (diatoms) with interlocking linking spines. *Bulletin of the British Museum, Natural History (Botany Series)*, **13**, 277–381.

ROUND, F. E., CRAWFORD, R. M. & MANN, D. G. 1990. *The Diatoms. Biology and Morphology of the Genera*. Cambridge University Press, Cambridge.

SCHULZ, P. 1927. Diatomeen aus norddeutschen Basalttuffen und Tuffgescheiben. *Zeitung Gescheibeforschung*, **3**, 66–78.

SIMS, P. A. 1989. Some Cretaceous and Paleocene species of *Coscinodiscus*: a micromorphological and systematic study. *Diatom Research*, **4**, 351–371.

——1990. The fossil diatom genus *Fenestrella*, its morphology, systematics and palaeogeography. *Beiheft zur Nova Hedwigia*, **100**, 277–288.

—— & HASLE, G. R. 1990. The formal erection of the Family Stellarimaceae Hasle & Sims, ex Nikolaev. *Diatom Research*, **5**, 173.

STAESCHE, K. & HILTERMANN, H. 1940. Mikrofaunen aus dem Tertiär Nordwest-deutschlands. *Abhandlungen Reichsanstalt Bodenforschung, Berlin*, **201**, 1–26.

SWATMAN, C. C. 1948. Note on the genus *Fenestrella* Greville. *Journal of the Royal Microscopical Society*, **68**, 51–54.

THOMAS, F. C. & GRADSTEIN, F. M. 1981. Tertiary subsurface correlations using pyritised diatoms, offshore eastern Canada. *Current Research, Part B. Geological Survey of Canada*, **81-1B**, 17–23.

VAN HEURCK, H. 1880–85. *Synopsis des Diatomées de Belgique*. Privately printed, Antwerp.

——1896. *A Treatise on the Diatomaceae*. Reprint 1962 by Wheldon & Wesley, & Verlag J. Cramer, Codicote and Weinheim.

WICK, W. 1943–50. Mikrofaunistiche Untersuchung des tieferen Tertiärs über ein Salzstock in der Nähe von Hamburg. *Abhandlungen Senckenberg Natürischen Geschichte*, **468**, 1–40.

WILLIAMS, D. M. 1993. Diatom nomenclature and the future of taxonomic database studies. *Beiheft zur Nova Hedwigia*, **106**, 21–31.

The Andrew Formation and 'biosteering' – different reservoirs, different approaches

N. A. HOLMES

Ichron Limited, 5 Dalby Court, Gadbrook Business Centre, Northwich CW9 7TN, UK

Abstract: Biosteering, the provision of real-time biostratigraphic data for the drilling of horizontal wells, has developed and evolved a wide range of approaches and applications in recent years. Its value to the success of a development project has grown beyond its core role as an effective and rapid method by which to define the relative stratigraphic position of the drill bit for steering purposes. High-resolution biostratigraphy and biosteering are now being used as part of an integrated service which is being incorprated in the planning of a horizontal drilling programme and individual well paths. This paper illustrates some of this diversity of applications by describing two very different styles of biosteering from age-equivalent turbidite sands of the Andrew Formation, on the United Kingdom continental shelf (UKCS).

The first example describes biosteering through thin (10–15 ft) turbidite sands from the Joanne Field. The reservoir is one of several thin allochthonous sand and limestone subunits, requiring a very detailed biostratigraphic framework to provide sufficient resolution to identify each of the subunits. The framework utilizes both *in situ* and derived/reworked assemblages and associations (microfacies) to identify up to 10 discrete and correlative lithostratigraphic subunits. Local dip variations, faults of subseismic resolution and thinness of sand hampered efforts to continuously remain within the reservoir during horizontal drilling operations. Well-site micropalaeontological interpretation correctly identified the position of the drill bit in relation to the reservoir sand, allowing the correct well-path deviation to be enacted. In the absence of this technique, hundreds of feet of additional non-reservoir section would have been drilled.

The second example describes the Andrew Field, again comprising deep-marine turbidite sandstones of the Andrew Formation. Development of the Andrew Field has incorporated both the biosteering function of well-site biostratigraphy as well as pre-drill, high-resolution facies analysis of the microfaunas to assist in depositional modelling of the field. Microfacies analysis provides data relating to the potential sealing qualities and lateral continuity of adjacent silt and claystones, and how these may act as barriers or baffles during production. This information is utilized in the planning of the well path to optimize production through well placement and stand-off from oil–water and gas–oil contacts, and at the well site during horizontal drilling to confirm, through micropalaeontological analysis, key horizons at which to geosteer.

This paper discusses the applications of micropalaeontology to development and production drilling, specifically horizontal drilling. These applications involve the recognition of subtle changes in microfossil assemblage and foraminiferal morphogroup composition (microfacies analysis) to help optimize well-path placement and 'biosteer' horizontal wells. This is illustrated by two examples from the North Sea (UK continental shelf); the Joanne Field and the Andrew Field (Fig. 1). Both fields are represented by Late Palaeocene turbidite systems, where the reservoir interval comprises the Andrew Sandstone Unit (or Andrew Formation) (Fig. 2). Although of equivalent age, both reservoirs have involved a very different biostratigraphic approach to field development and horizontal drilling. In order to describe the approach to microfacies analysis and its well-site applications, the discussion falls into two main areas:

- the utilization of foraminiferal morphogroups and microfacies analysis as a means to provide a very high-resolution lithofacies characterization; and
- to describe how these techniques are utilized in the drilling of horizontal wells from two fields, both at the pre-drill stage for well-path planning and at the well site to provide real-time data for steering discussions.

A distinction is made here between biosteering and geosteering. The term biosteering is applied

HOLMES, N. A. 1999. The Andrew Formation and 'biosteering' – different reservoirs, different approaches. *In:* JONES, R. W. & SIMMONS, M. D. (eds) *Biostratigraphy in Production and Development Geology*. Geological Society, London, Special Publications, **152**, 155–166.

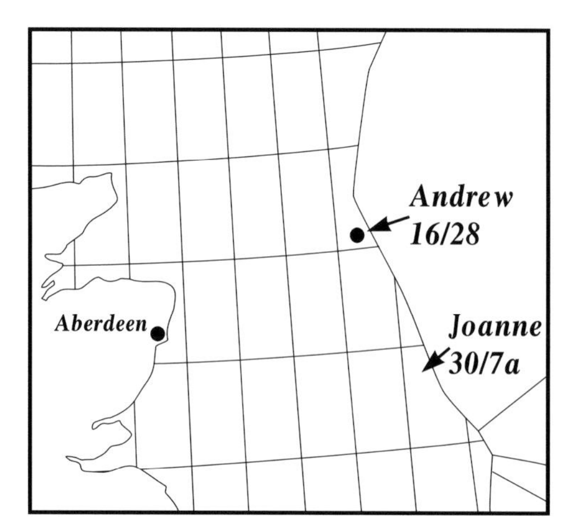

Fig. 1. Location map for the Andrew and Joanne fields, UKCS.

when well-site biostratigraphy provides constant real-time data to determine the relative stratigraphic position of the drill bit to reservoir, to determine and enable directional drilling changes to be made. This is typically the case when drilling thin sands or possibly very narrow targets (such as the Joanne Field). Geosteering is used here where the well path is predetermined, and well-site biostratigraphy is used only to monitor drill bit position and provide stratigraphic confidence by keying into specific marker horizons. This has applications to relatively thick reservoir sequences, such as the Andrew Field, where the drilling corridor is constrained by gas–oil and oil–water contacts.

The Joanne Field, operated by Phillips Petroleum (Block 30/7a), is a laterally extensive, thin (10–15 feet) gas condensate reservoir which was developed with eight horizontal wells. During the drilling of these wells there was a high risk of ‘losing’ the sands upon exiting out of either the upper or lower surface while drilling. In such cases, frequent (bio)steering decisions (up?, down?) were required to regain the reservoir, these were made largely on the basis of well-site micropalaeontology.

The Andrew Field, operated by BP Exploration (Block (16/28)), is, by contrast, represented by thick sands where the well path is geosteered along a ‘corridor’ with predetermined points of steering. These predetermined points were, in part, defined using microfacies techniques. In this example there is a high value attached to correctly identifying the points at which to geosteer and alter true vertical depth (TVD) within the corridor. Well-site micropalaeontology provides stratigraphic confidence by keying into specific marker horizons.

This paper discuses the micropalaeontological techniques developed to provide the above biostratigraphical control. The link to lithofacies is made where information on the relative amount pelagic or turbidite influence on deposition is

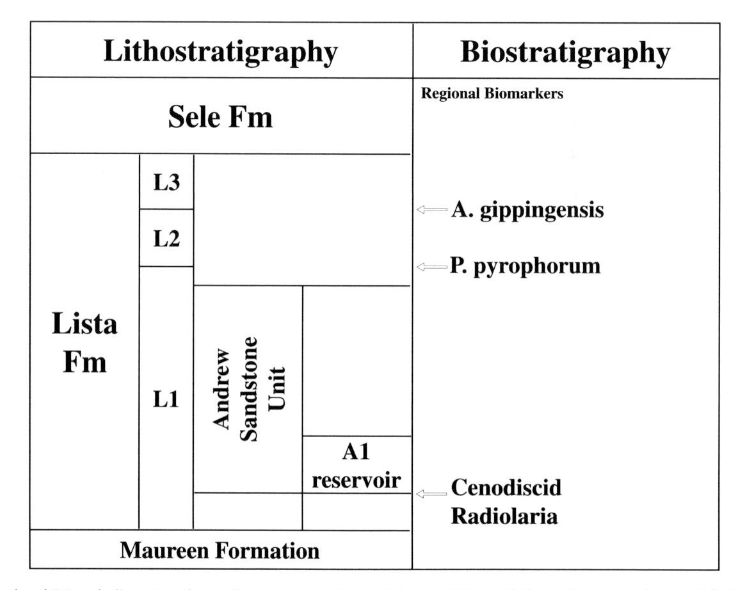

Fig. 2. Standard North Sea stratigraphy, and stratigraphic position of the A1 reservoir sand (of the Joanne Field) (after Knox & Holloway 1992).

given. This has a bearing on attempts to determine the relative extent of sediment distribution, the relative amount of geological time required for deposition and whether the sediment would be likely to act as a barrier or baffle to fluid migration.

Existing regional biostratigraphical control cannot provide the resolution required for the stratigraphic control and confidence demanded by horizontal drilling for the examples given here. Both reservoirs sit wholly within existing standard biomarker horizons and, importantly, no regional biozones occur within the Andrew Sandstone Unit, within which are located the reservoir sands of the Joanne and Andrew fields, (Fig. 2). In addition, for the Joanne Field, reworking of older Palaeocene regional biomarkers was a major consideration.

Joanne Field

Objectives

The primary objective was to improve the stratigraphical resolution of the reservoir interval. This was done through the establishment of a detailed lithostratigraphic framework, developed through microfacies analysis. The development of a microfacies zonation allowed a direct link to be made with lithofacies. At the pre-drill stage stratigraphic and facies modelling was able to determine potential sand connectivity, shale character and the distribution of other allochthonous units. The final objective was to apply this data to the well site in order to determine relative position of drill bit to reservoir (biosteering).

Existing biostratigraphical and lithostratigraphical framework

As the reservoir interval of the Joanne Field sits wholly within existing biomarkers (Fig. 2), and in the absence of chronostratigraphical events within the Andrew Sandstone Unit, resolution using existing biostratigraphy is clearly not sufficient for effective biosteering. It was therefore necessary to develop a reservoir zonation of sufficient resolution, specific to the Joanne Field area.

Reservoir zonation and microfacies analysis

The reservoir zonation developed for the Joanne Field is based on microfacies; in which microfaunal associations are considered indicative of definable palaeoenvironment and depositional settings. This involves the identification of *in situ* and derived fossil associations, and their relationship to particular periods of deposition. The important aspect to this is the direct link created between microfacies and lithofacies.

The approach to microfacies analysis of agglutinated foraminifera is illustrated in Fig. 3 and involves the use of foraminiferal morphogroups. A general relationship exists between agglutinated foraminiferal 'life position' (specifically feeding strategies) and test morphology (Jones & Charnock 1985). Consequently, taxonomic groupings, based on morphological similarities (morphogroups), are considered to be, at least in part, determined by the dynamic properties of the environment, with individual morphogroups having particular environmental preferences. See Jones & Charnock (1985) for a full discussion of morphogroups and a summary of foraminiferal palaeoecological studies.

A generalized 30/7a Block (Joanne Field) gamma and lithology log of the reservoir interval is shown in Fig. 3 with the A1 reservoir sand included. The overall section thickness approximates to about 100 feet, with the A1 sands averaging 10–15 feet thick. Sediments above and below consist of claystones, siltstones, reworked limestones and occasional sandstones. The middle column illustrates a range in foraminiferal morphogroup associations. We know that within interturbiditic claystones the simple, particularly the tubular, morphogroups are the initial colonizers of new surfaces, such as that following the deposition of turbidite sand (Kaminski & Schroder 1987). However, it is noted that in experimental work (Jones pers. comm.), under certain conditions, morphogroup 'C' is recorded as the pioneer colonizer. With time, as allochthonous influences wain, more complex and varied morphogroups colonize. The lowermost microfacies association (Fig. 3) can be considered as representing a pioneer (or colonizing) assemblage, reflecting allochthonous depositional influences. The uppermost association represents the typical basin plain assemblage at this time, reflecting a much greater pelagic influence upon deposition. This assemblage can be considered to represent the climax community. Several morphogroup associations are recognized between these end members – reflecting depositional conditions intermediate between that of a high allochthonous component and that of a high pelagic component. Within derived associations (right-hand column, Fig. 3) simple differentiation in specific ages and/or preservation also assist in identifying particular reworked lithofacies, as well as the recognition of contemporaneously derived associations.

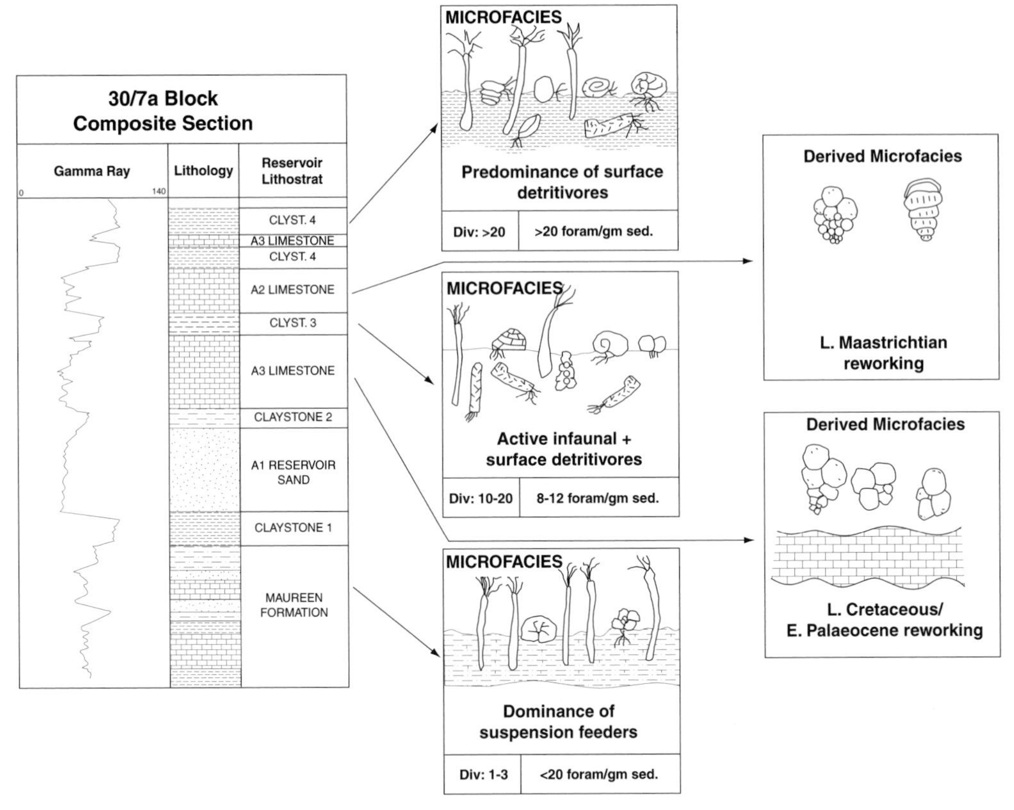

Fig. 3. Biofacies techniques in reservoir biostratigraphy.

Therefore, to summarize, a range from a high to a low hemipelagic component in the depositional setting results in a series of distinct microfacies (morphogroup) associations which allow the identification and categorization of a series of claystone types. Thereby establishing a critical relationship (as opposed to just association) between microfacies and lithofacies.

In defining the reservoir lithostratigraphic sequence for the Joanne Field, derived and *in situ* assemblages (identified as biozones, Fig. 4), defined through the analysis of microfacies associations and foraminiferal morphogroup associations, identify a series of allochthonous subunits from the Andrew Sandstone Unit (Fig. 4). In addition, several associated claystones of variable hemipelagic component are recognized.

Lithological correlation

From the framework outlined in Fig. 4 a lithological correlation showed rapid and very localized variations in lithofacies both above and below the A1 reservoir sand. Crucially, for horizontal drilling, this identified a high potential of difficulty in predicting sediment distribution and lithofacies above and below the reservoir. The requirement to biosteer was identified, whereby lithostratigraphical variations could be rapidly identified and interpreted.

Joanne Field and biosteering

The primary function of biosteering was to determine drill bit position relative to the reservoir using the biostratigraphic techniques described, so that on exiting the sands the appropriate directional changes could be enacted. To do this at the well site requires the constant monitoring of cuttings samples while drilling. As well as providing data on the 'amount' and the direction of re-orientation required, discontinuous sequences, such as faults including those of subseismic resolution, can be identified.

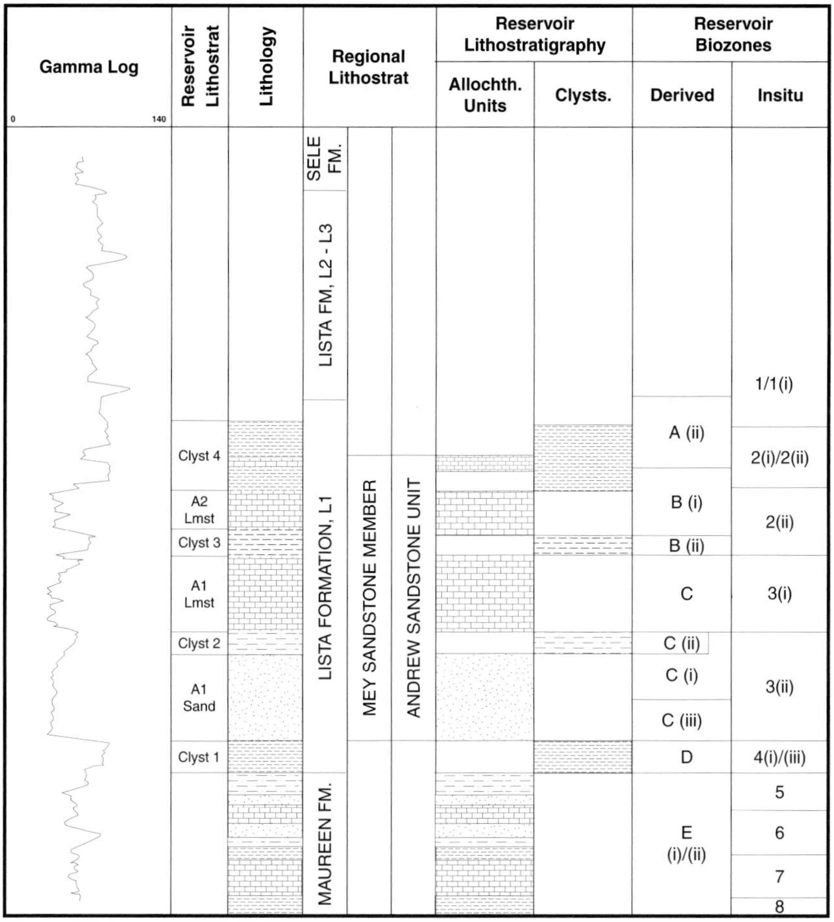

Fig. 4. Idealized reservoir biostratigraphy calibrated against wireline log data for the Joanne Field (Block 30/7a).

The pipe diagram (Fig. 5) illustrates a schematic well path for well A. Through constant monitoring of the cuttings samples, stratigraphic position relative to measured depth is determined. This diagram schematically represents (without any horizontal scale) the well path from casing point above the reservoir to terminal depth (TD). Numbers alongside the well path represent the measured depth of selected analyses. The initial dive section through the reservoir is shown to the left of the diagram adjacent to the lithostratigraphic interpretation. Subsequent samples are plotted against their particular lithofacies. A number of discontinuous sequences or breaks in the well path are identified between adjacent samples, for example between 12 580 and 12 590 feet and between 13 990 and 14 008 feet. These indicate a discontinuity in the vertical succession of lithofacies, and are interpreted as faults.

By plotting this well to scale (Fig. 6) the requirement for constant monitoring for biosteering is shown, particularly when having to react to an unpredicted fault 'throwing' the well out of the reservoir sand (fault 'b'). Of considerable note in this well section is the interpretation required at fault 'c'. Prior to this fault, the well path was running through a series of claystones with allochthonous sandstones and dominant limestones, clearly above the A1 reservoir sand and assigned to the A2 Limestone. Re-orientation, by dropping angle, was required to regain the reservoir. At a certain point, microfacies analyses (change in *in situ* morphogroup ratios) identified a position stratigraphically below the A1 Sand (within the

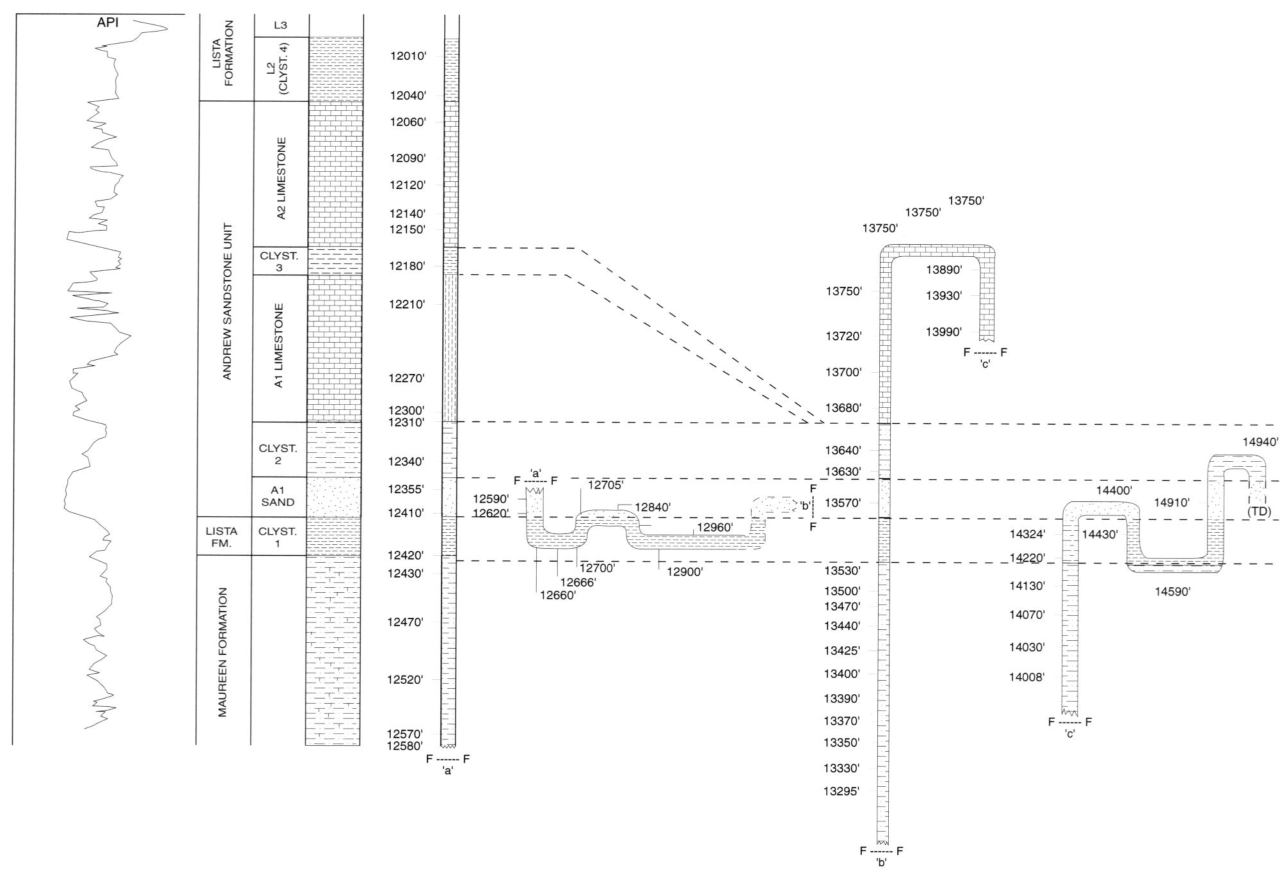

Fig. 5. Pipe diagram illustrating sample position relative to the A1 reservoir sand, well A horizontal section.

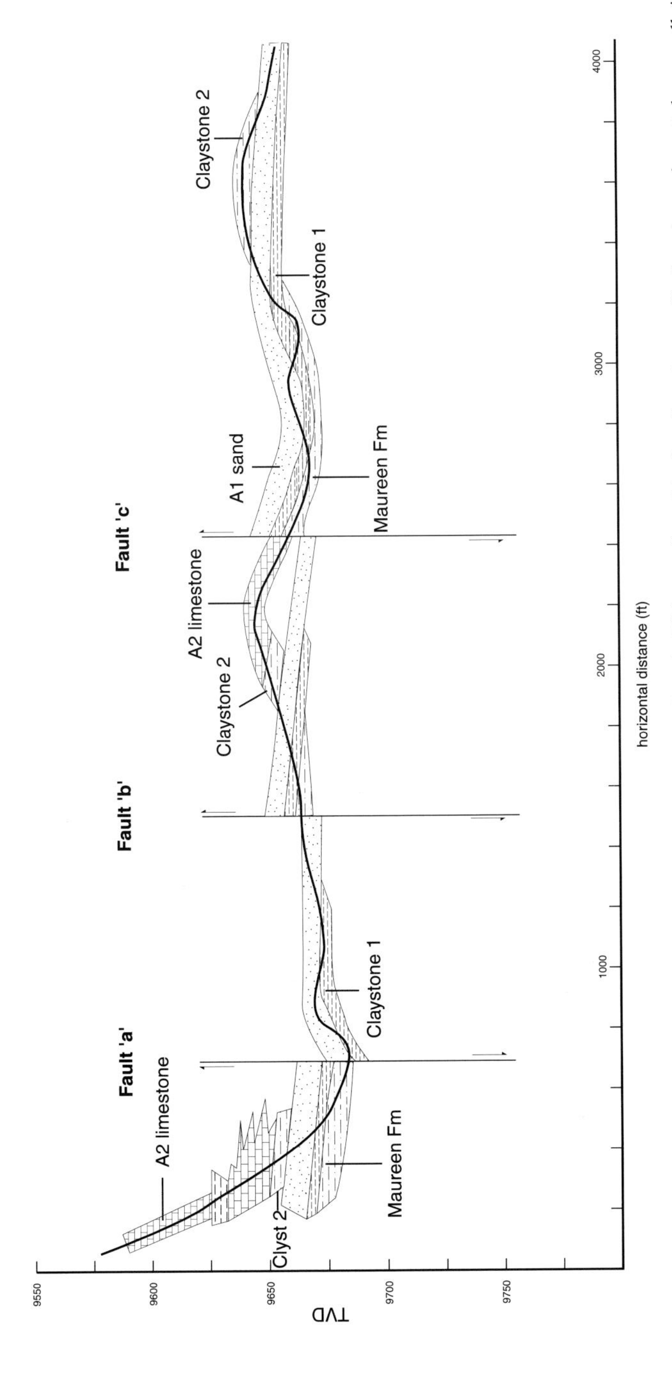

Fig. 6. Schematic cross-section and reservoir sand distribution relative to the well path. Lithostratigraphic data based on well-site micropalaeontology, well A.

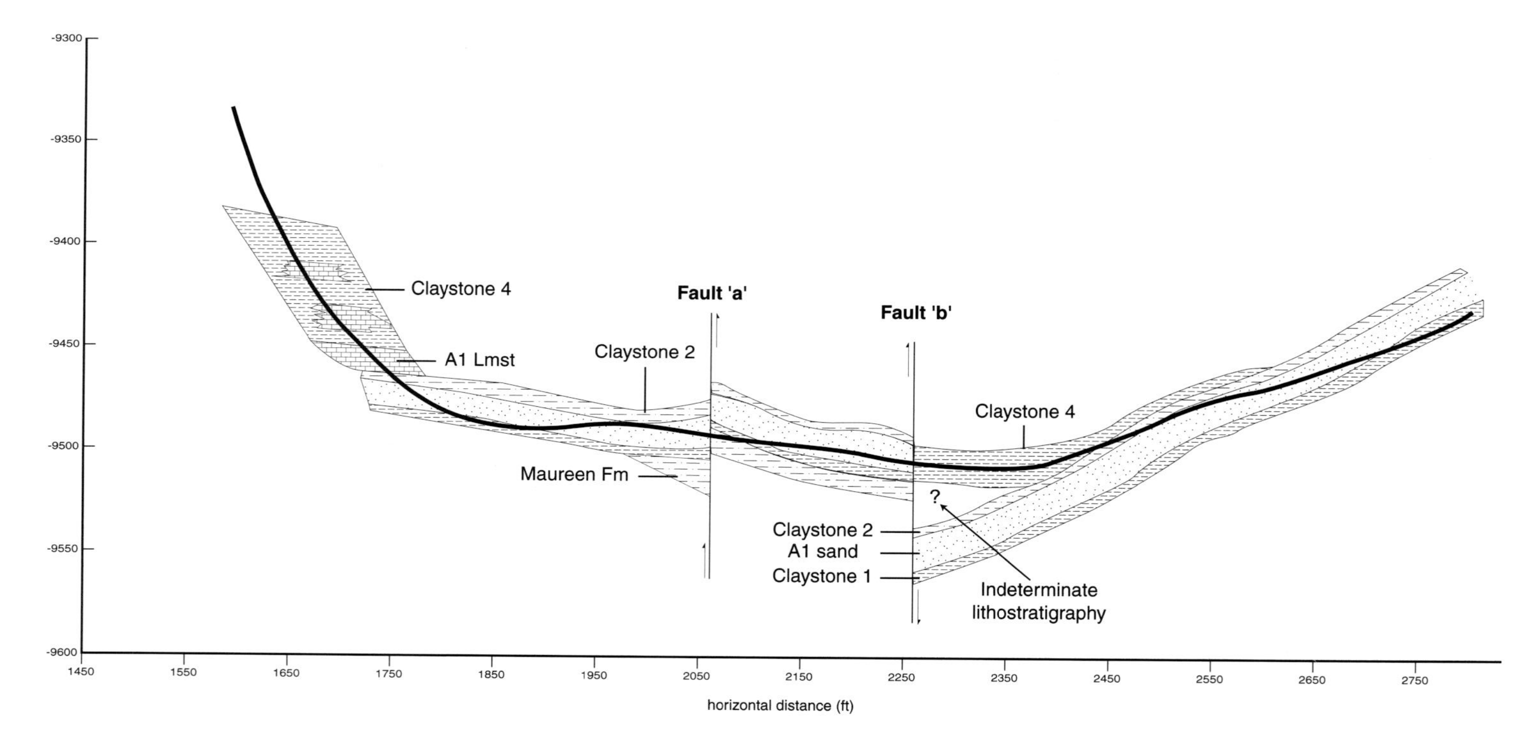

Fig. 7. Schematic cross-section and reservoir sand distribution relative to its well path. Lithostratigraphic data based on well-site micropalaeontology, Joanne Field horizontal well B.

Maureen Formation); even though the well path had not passed through the A1 Sand, nor was there any obvious break in lithology or 'logging while drilling' (LWD) data. A decision was then required as to whether to continue on the same well trajectory or re-orientate upwards. Through the full integration of biostratigraphy within the well-site team, the clients' decision to follow the microfacies interpretation resulted in the A1 Sand being re-entered and the remaining reservoir section was successfully drilled.

For operational reasons, abrupt changes in drilling direction (or dog-legs) are to be avoided. When a significant fault was passed in another horizontal well (well B) from the Joanne Field (Fig. 7 fault b), well-site micropalaeontological data indicated a stratigraphical position for the drill bit some way above the A1 Sand, requiring a significant increase in TVD and drop in angle to re-enter the sands. These data were incorporated into the decision not to drop the drilling angle and 'chase' the sands, but to drill on horizontally and allow the sands to subsequently cross the well path (as predicted seismically).

Andrew Field

The decision to develop the Andrew Field with horizontal wells required the availability of a detailed biostratigraphy of the reservoir interval (see also Payne *et al.* 1999). As with the Joanne Field, this involved the development of a field-specific microfacies zonation which allowed a link to be made with lithofacies, and subsequently enabled depositional modelling of the intra-reservoir claystones.

Initially, the approach to Andrew was very similar to that for the Joanne Field, but with very different geosteering applications. The main purpose of well-site micropalaeontology on the Andrew Field was to: (a) monitor the reservoir

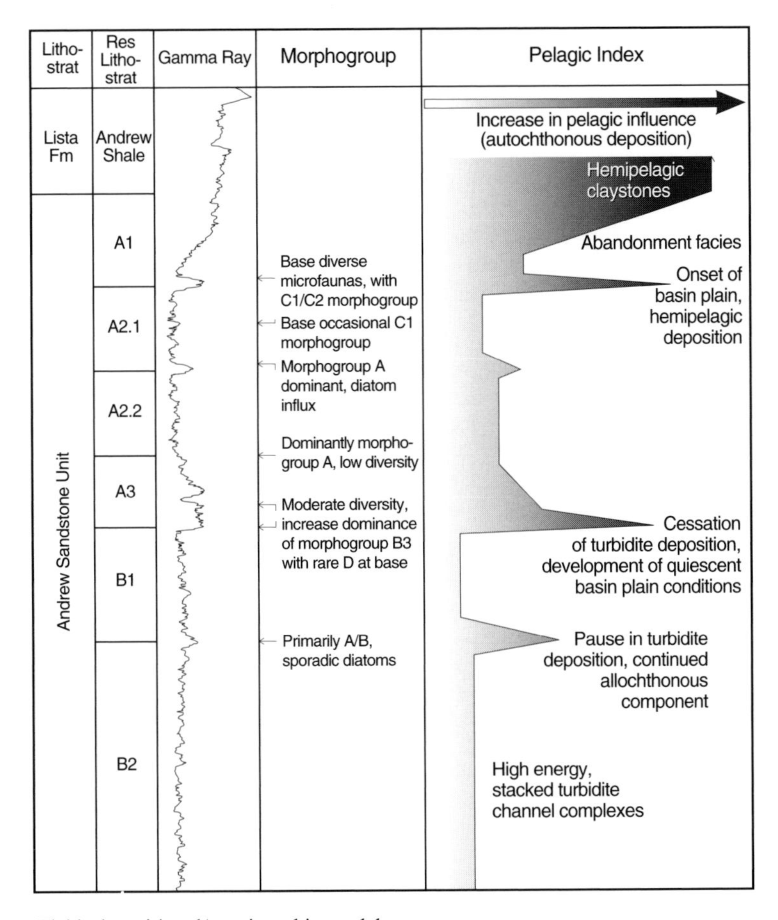

Fig. 8. Andrew Field: depositional/stratigraphic model.

stratigraphy; (b) to identify key horizons for geosteering (through the identification of specific intra-reservoir claystones; and (c) to identify the correct claystone in which to terminate drilling.

Reservoir zonation and microfacies interpretation

The Andrew reservoir interval is characterized by very high sand net: gross ratios capped by the informally termed Andrew Shale (or Lista Formation) (Fig. 8 left-hand column). From a microfacies interpretation of the intra-reservoir claystones, a palaeoenvironmental interpretation based on foraminiferal morphogroup analysis was carried out (Fig. 8) in a similar style to that for the Joanne Field. The underlying basis for interpretation of the Andrew Field microfacies is the perception that pioneering (or colonizing) and climax agglutinated foraminferal communities are end members that basically reflect a trend towards low energy or quiescence; in other words, reducing allochthonous influence (disturbance) and increasing pelagic influence (tranquillity, in the sense of Kaminski & Schroder 1987).

In the absence of any previously established lithological and foraminiferal specific zonation of the intra-reservoir claystones, this morphogroup connection to the degree of sea-floor quiescence (or hemipelagic influence) formed the basis of the intra-reservoir characterization. This characterization can be represented by a qualitative scale, referred to as the Pelagic Index (Fig. 8).

From a baseline of near zero Pelagic Index, small peaks in the index (Fig. 8 top of lithological unit A2.2), defined by foraminifera dominated by A and B (epifaunal) morphogroups (of Jones & Charnock, 1985) are interpreted as representing only temporary pauses in turbidite deposition of relatively short time span and where there is a continued allochthonous influence. This contrasts with a very high Pelagic Index, containing a spread of morphogroups, including C1 and C2 infaunal groups, recorded from the Andrew Shale and A1 subunit. This high index reflects the onset of basin plain, hemipelagic deposition within a quiescent setting. Within the Andrew reservoir one particular claystone (the A3 Claystone) consistently records a relatively high Pelagic Index. This high index indicates the temporary development towards quiescent basin plain conditions and the cessation of dominant turbidite influence. In comparison to other intra-reservoir claystones of a low index, the A3 Claystone is interpreted as having required a greater amount of geological time to develop and, by its depositional affinity to the overlying Andrew Shale, is likely to represent a laterally extensive deposit and a barrier to production (see also Jones 1999).

Andrew Field – geosteering and microfacies integration

The primary element in integration is that microfacies data were used to interpret claystone facies and depositional environment. This interpretation helped to assess the sealing qualities of certain claystones, some of which were determined to be potential barriers and baffles to hydrocarbon migration, and in turn, influenced well-path planning.

In the situation where a horizontal well is to be drilled parallel to the gas–oil and oil–water contacts, it is necessary to maintain a constant stand-off or position within the corridor. If shale is encountered, it is just a case of drilling through (Fig. 9). However, there may be production-related reasons to stay either above or below certain claystones within the corridor. This was the case for a number of the horizontal wells drilled on the Andrew Field.

The Andrew reservoir is a broadly dome-shaped structure with dipping flanks. The general well path of each horizontal well is to orientate horizontally within the lower lithostratigraphic subunit B (Fig. 8). As the flanks of the reservoir dip, the well path rises stratigraphically through the reservoir, to terminate in the overlying Andrew Shale. This situation is

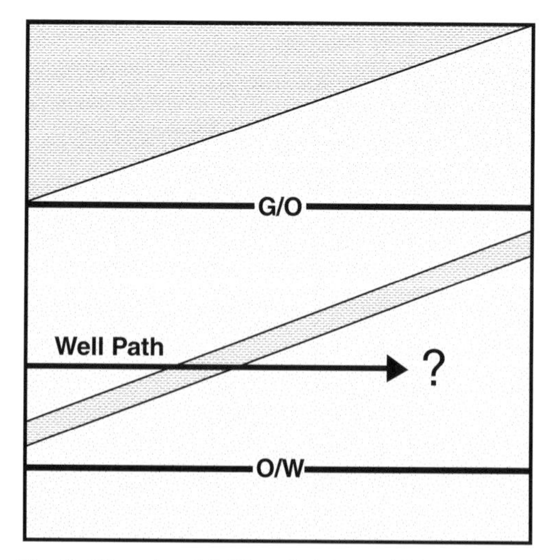

Fig. 9. Directional drilling decisions within a predetermined corridor.

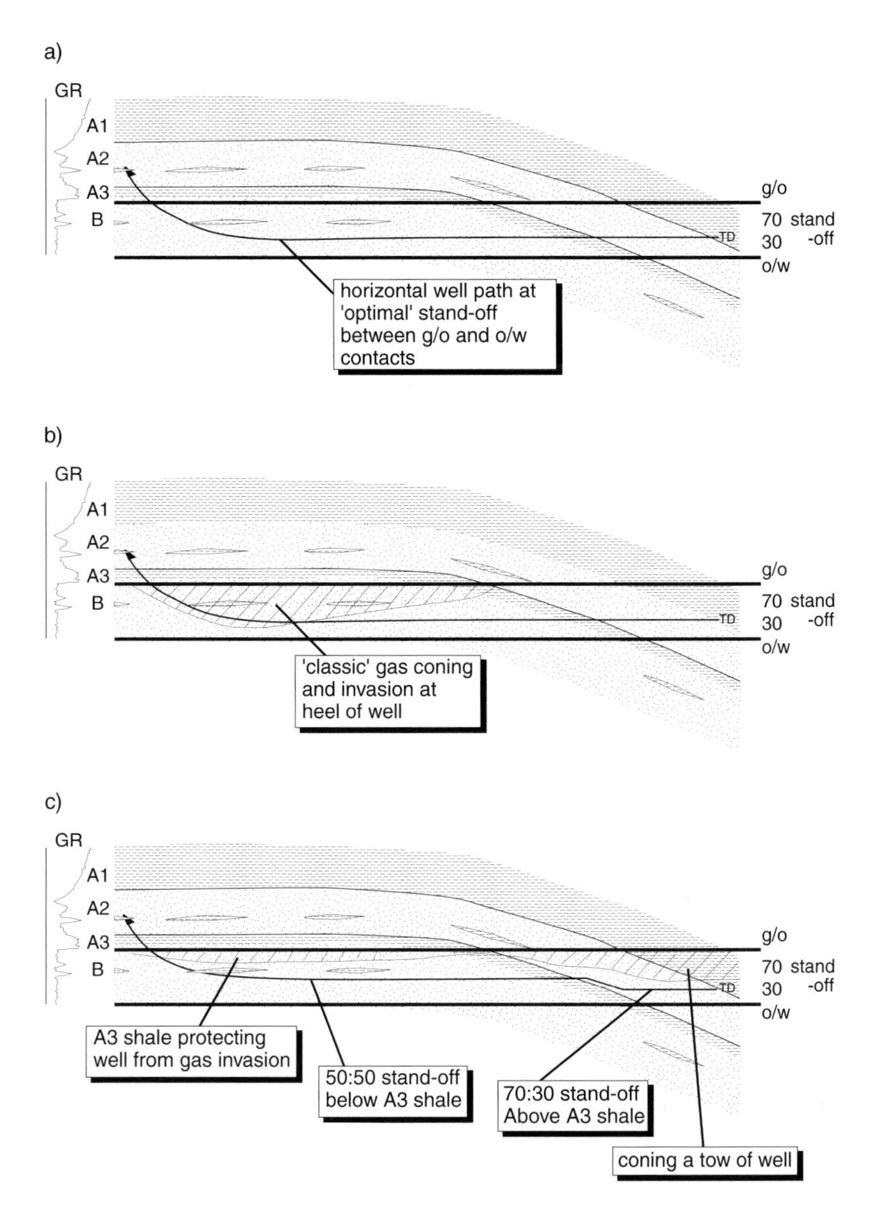

Fig.10. The role of microfacies analysis in optimization of well placement and stand-off. (**a**) 70 : 30 stand-off. (**b**) 70 : 30 stand-off. (**c**) 50 : 50 and 70 : 30 stand-off.

shown in Fig. 10a, where the well path is situated within a corridor of a 70 : 30 stand-off between the gas–oil and oil–water contacts, respectively. At this stand-off, gas coning and invasion at the heel of the well would be expected (Fig. 10b).

Utilizing the microfacies data (summarized in Fig. 8), the A3 Claystone is interpreted as a laterally extensive, basin plain deposit and also as a likely barrier to fluid migration. By incorporating the predictive microfacies techniques and interpreted lithofacies, the well stand-off can be optimized to enhance well productivity. This was achieved by using the A3 Shale as a protective 'umbrella' – to protect the well from gas drawdown and invasion (Fig. 10c), thereby allowing a well-path placement further away from the oil–water contact by taking up a 50 : 50 stand-off. As the flanks of the field dip and the well path eventually passes up through and above the A3 Shale, this protection from gas invasion is no longer available, so the well path is immediately lowered to a 70 : 30 stand-off. As well as being fundamental to this model, microfacies analysis at the well site allows the rapid identification of the A3 Claystone, to enable the drop in true vertical depth to be made.

Biostratigraphy has therefore provided a pre-drill depositional model for the Andrew Field, and at the well site provides the identification of where to adjust stand-off and identify the correct claystone to terminate drilling.

Concluding remarks

The aim of this paper is to demonstrate two contrasting applications of biostratigraphy and microfacies analysis to development and production drilling, even on age-equivalent reservoirs of similar depositional facies. It is clear that a gulf in resolution can exist between 'conventional' micropalaeontology (that is more reliant on potentially rare marker species) and microfacies analysis (which utilizes the entire assemblage and biofacies variations). The wider variation that exists in microfacies and morphogroup analysis can be used to an advantage by allowing a direct link to lithofacies characterization.

This paper has dealt with:

- the utilization of foraminiferal morphogroups and microfacies analysis;
- the provision of high-resolution lithofacies characterization;
- its use at the pre-drill stage for horizontal well-path planning and at the well site for biosteering;
- the uniqueness of every reservoir, even of age equivalence and general depositional setting, where each case requires different biostratigraphic solutions

Finally, it has been the intention to demonstrate the value of the operator being involved and having familiarity with reservoir biostratigraphic applications and its potential.

The author is indebted to Phillips Petroleum and their co-ventureres BG and Agip and to BP Exploration for their support and permission to publish. The author would like to acknowledge the assistance and thank the following personnel within these companies, Richard Bayes and Chris Holien of Phillips, and Simon Todd and Simon Payne of BP. Their early involvement and faith in these techniques has resulted in successful applications in these and other field areas.

The author would also like to acknowledge the work of Bob Jones (BP) and Mike Charnock (Norsk Hydro) for their contribution to microfacies analysis in demonstrating the potential of foraminiferal morphogroups to such studies, which has led to its key involvement in development drilling.

References

JONES, R. W. 1999. Forties Field (North Sea) revisited: a demonstration of the value of historical micropalaeontological data. *This volume*.

—— & CHARNOCK, M. A. 1985. 'Morphogroups' of agglutinating foraminifera. Their life positions and feeding habits and potential applicability in (paleo)ecological studies. *Revue de Paleobiologie*, **4**, 311–320.

KAMINSKI, M. A. & SCHRODER, C. J. 1987. Environmental analysis of deep-sea agglutinated foraminifera: Can we distinguish tranquil from disturbed environments? *In*: *Innovative Biostratigraphic Approaches to Sequence Analysis: New Exploration Opportunities*. GCSSEPM Foundation Eighth Annual Research Conference, Texas, 90–93.

KNOX, R. W. O'B. & HOLLOWAY, R. 1992. Palaeogene of the Central and Northern North Sea. *In*: KNOX, R. W. O'B. & CORDEY, W. G. (eds) *Lithostratigraphic Nomenclature of the UK North Sea*. British Geological Survey on behalf of UKOOA, Nottingham.

PAYNE, S. N. J., EWEN, D. F. & BOWMAN, M. J. 1999. The role and value of 'high-impact biostratigraphy' in reservoir appraisal and development. *This volume*.

High-resolution biostratigraphy and sequence development of the Palaeocene succession, Grane Field, Norway

G. MANGERUD, T. DREYER, L. SØYSETH, O. MARTINSEN & A. RYSETH

Norsk Hydro Research Centre, N-5020 Bergen, Norway

Abstract: A detailed biostratigraphic zonation and sequence stratigraphic framework has been developed for the Grane Field (PL 169), situated in Block 25/11 in the North Sea. The study is focused on the Upper Palaeocene interval, where a high level of chronostratigraphic resolution has been achieved and where the depositional history can be understood in terms of a finite set of depositional cycles and subcycles.

This work documents that the Grane Field reservoir succession, which consists of alternating turbidite sandstones and marine basin plain mudstones, fall within two main biozones, corresponding to a general Late Palaeocene age. The reservoir succession is developed between two regional datums: last stratigraphical common occurrence of *A. gippingensis* and last stratigraphical occurrence of *Isabellidinium*? *viborgense*. This corresponds to our major cycles T40 and T50, in which the reservoir sands comprise lowstand units filling in local accommodation during stepwise encroachment of the submarine fan system into this area.

Within the reservoir interval the zonation is based on the appearance and extinction of species with a regional distribution that also have marked frequency variations. These include *A. gippingensis*, *P. bulliforme*, *C. striatum*, *P. pyrophorum* and *I? viborgense*. In addition, events representing quantitatively marked peaks in the frequency of long-ranging taxa are used for correlation purposes. These events need to be constrained by other events, but are good for local correlations. Examples from the Grane Field include influxes of *Deflandrea denticulata*, *Cometodinium comatum* and *Subtilisphaera* spp.

The high-resolution biostratigraphy serves as a template for the sequence stratigraphic framework in the study area. Sixteen correlatable zones have been established by linking each of the fossil events to a specific sequence stratigraphic surface. As shown by previous workers, surfaces associated with condensation (regional and subregional maximum flooding surfaces) prove to be the most useful ones for correlating the Tertiary of the North Sea. In addition, some of the biostratigraphic events seem to coincide with transgressive surfaces and sequence bounding unconformities. Application of the sequence stratigraphic framework indicates that several reservoir sand units, which appear correlatable on well logs, actually are of different age. They may therefore contain potential flow barriers, formed in periods of condensation between the lowstand events.

The Grane Field (formly called Hermod) is situated in Block 25/11 (PL 169) in the Norwegian sector of the North Sea (Fig. 1), and was awarded after the 13th concession round with Norsk Hydro as the operator. Possible satellites of the field may exist in southwestern parts of 25/11 and may also extend north into Block 25/8. The Grane Field is separated from the nearby Balder Field by a N–S-oriented zone a few kilometres across in which reservoir sandstones are inferred to be absent.

The field is located on the western flank of the Utsira High, a structural element defining the eastern margin of the present South Viking Graben. The reservoir consists of turbidite sandstones of Palaeocene age, sourced from the East Shetland Platform to the west. Between this shallow-marine platform and the deep-marine basin in which the Grane Field is located, the areas in the vicinity of the Bruce–Beryl Embayment represented a slope-like transition zone.

The Grane Field was discovered by exploration well 25/11-15 in 1991. This well held an oil column of biodegraded 19 API undersaturated oil and was placed centrally on a mounded structure. Later, 25/11-16 and 25/8-4 were drilled back to back, testing two other structures also proving oil. Appraisal well 25/11-18 was drilled in 1993 to confirm the southward extension of the 25/11-15 discovery, while well 25/11-20 was drilled in 1995 to test a possible eastern satellite to the Palaeocene sandstones of the Grane Field.

MANGERUD, G., DREYER, T., SØYSETH, L., MARTINSEN, O. & RYSETH, A. 1999. High-resolution biostratigraphy and sequence development of the Palaeocene succession, Grane Field, Norway. *In*: UNDERHILL, J. R. (ed.) *Development and Evolution of the Wessex Basin*, Geological Society, London, Special Publications, **133**, 167–184.

(a)

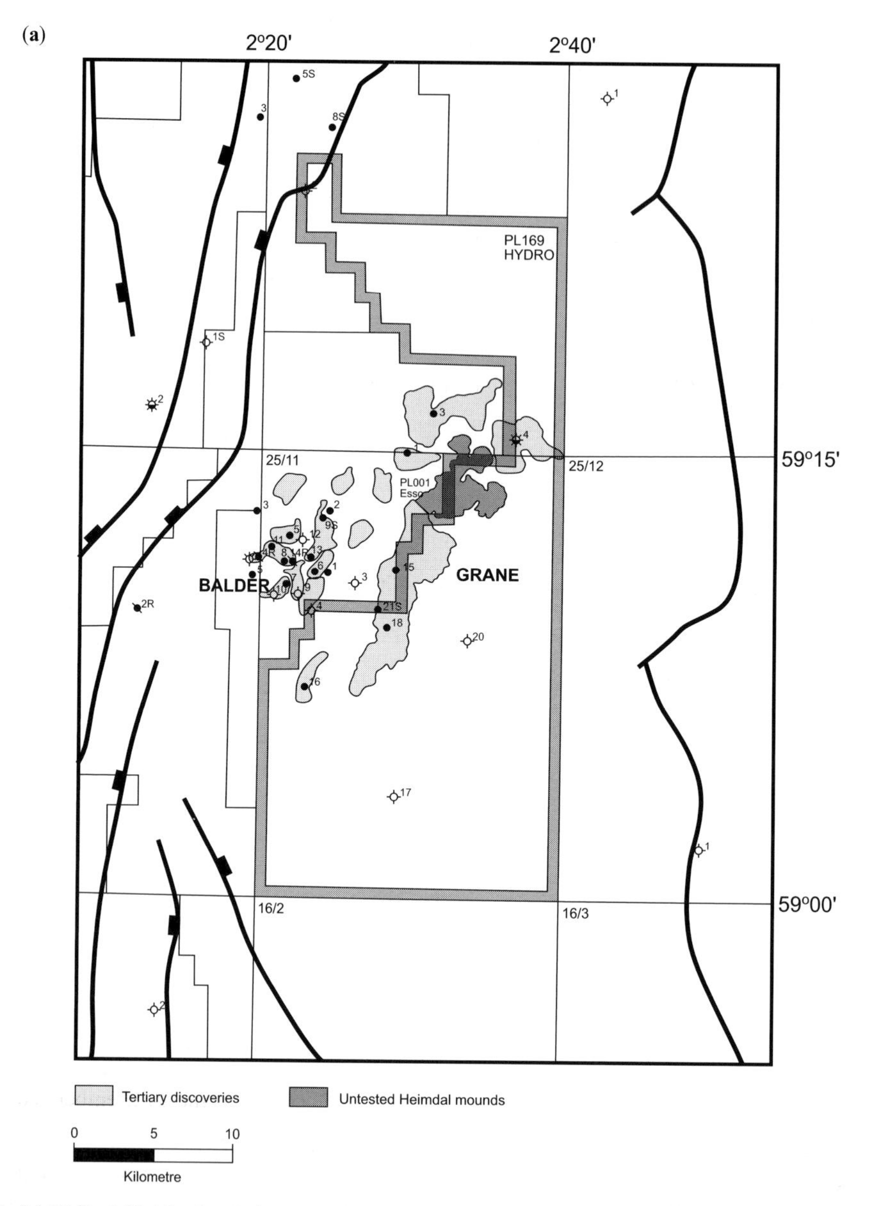

Fig. 1. (**a**) Wells drilled in the study area and their occurrence in relation to Tertiary discoveries and untested Tertiary mounds. (**b**) Main structural elements of the study area.

No sands were, however, found in this well. In 1995, well 25/11-21S was drilled as a pilot for an horizontal well (25/11-21A) in which test production was performed.

To understand the Late Palaeocene sand distribution of the Grane Field in time and space, a high-resolution biostratigraphic framework was developed in conjunction with detailed sedimentological and seismic studies. This framework is mostly based on palynomorphs, a fossil group recognized as very effective for dating and correlating the Palaeocene succession in the North Sea (Hansen 1977; Heilmann-Clausen 1985, 1994; Powell 1988, 1992; Gradstein *et al.* 1992, 1994; Stein *et al.* 1995; Mudge & Bujak 1996; Powell *et al.* 1996). The main focus of this paper is to document this stratigraphic framework. In particular, we would like to bring the attention to the high-resolution biostratigraphical zonation and how this has

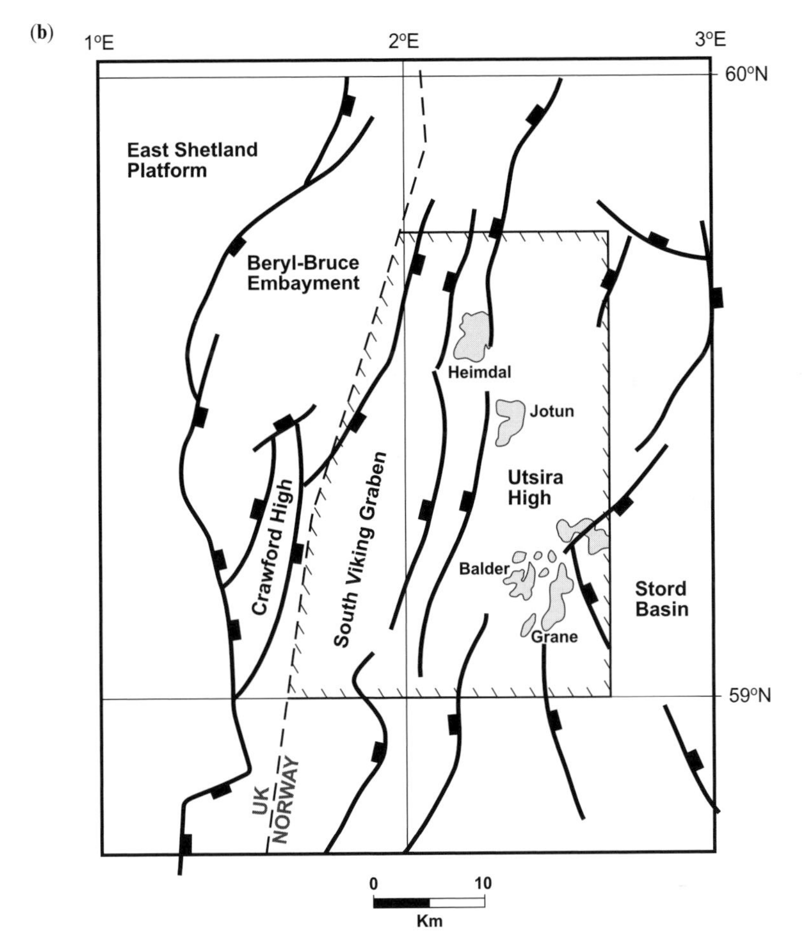

Fig. 1. (*continued*)

served as the basis for creating a sequence stratigraphic framework for the Grane Field.

Palaeogene stratigraphy

The lithostratigraphic scheme applied in the present study combines the two nomenclatures existing for the studied succession, namely those of Isaksen & Tonstad (1989) and Mudge & Copestake (1992). As can be seen in Fig. 2 our attention is on the Rogaland Group, which is divided into the Våle, Lista, Sele and Balder formations (Isaksen & Tonstad 1989). Major sandstones are seen as members within the respective formations (Fig. 2) (Mudge & Copestake 1992) highlighting the fact that these are locally appearing units.

The Våle Formation is relatively thin and argillaceous in the Grane area, due to the pinchout of the deep-sea fan sandstones further west. Sandstones appearing in the Våle Formation are referred to as the Ty Member. The Lista Formation is typically a varicoloured, greenish–reddish grey, massive to bioturbated mudstone. It is widespread across the North Sea with sandstones present throughout the South Viking Graben. These are referred to as the Heimdal Member, and comprises the main reservoir unit of the Grane Field. It is succeeded by the Sele Formation, which is a dark grey laminated mudstone interbedded with thin tuff beds. Intra-Sele sandstones are referred to as the Hermod Member in this study. The overlying Balder Formation is a multicoloured mudstone with interbedded sandy and pyritic volcanic tuffs. Intra-Balder sandstones are known mainly from the UK sector, where they are referred to as the Odin Member (Mudge & Copestake 1992). The boundaries between the formations tend to

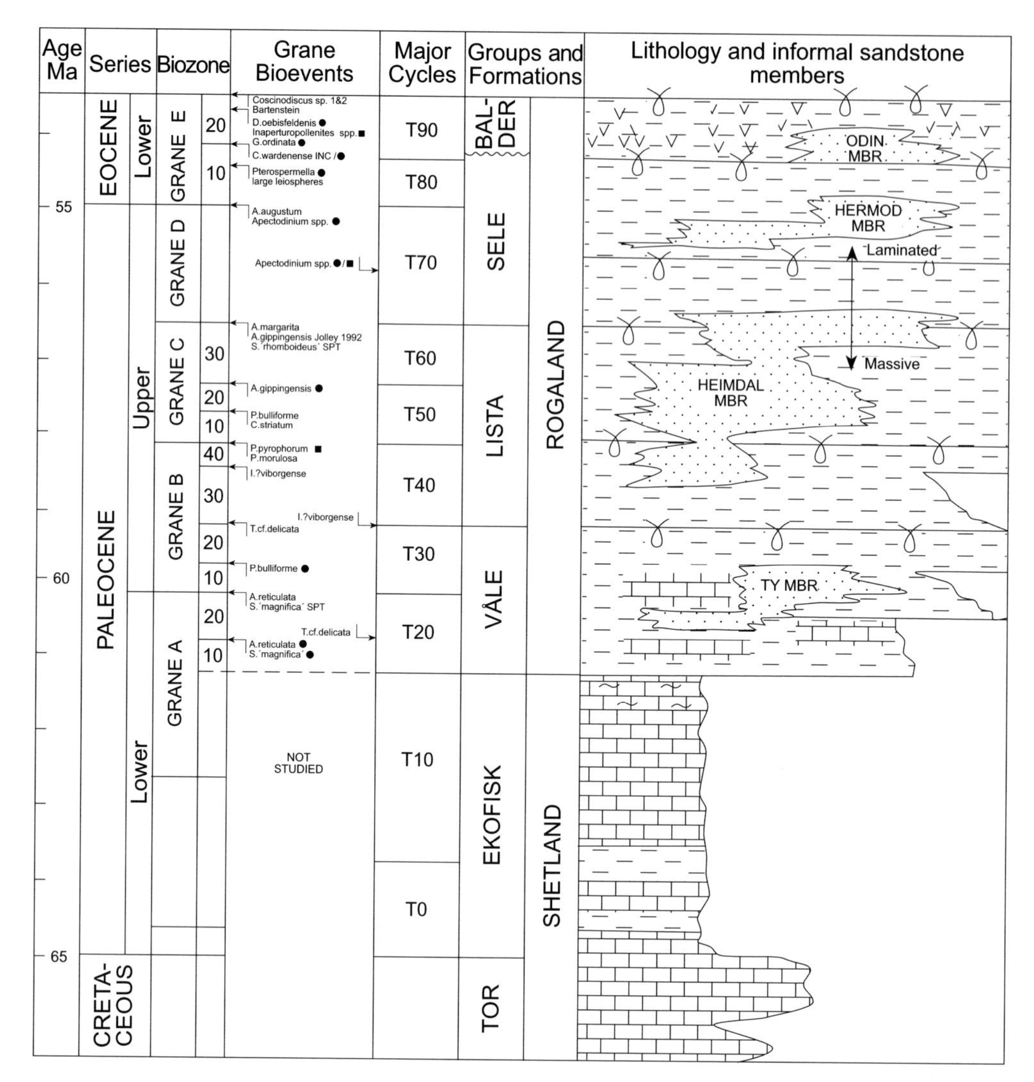

Fig. 2. The Palaeocene–early Eocene stratigraphy of the Grane–Balder area, with the lithostratigraphic nomenclature used in the present study, the relation to the biozones, the bioevents and the biostratigraphically constrained main sequence stratigraphic cycles.

be marked by a condensed deposit, recognizable by high gamma-ray log spikes. Such spikes may also be found within the formations (see below).

Database and methods

Biostratigraphic data obtained from 14 wells within blocks 25/8 and 25/11 form the basis for the high-resolution stratigraphic framework. Special emphasis is placed on wells drilled by the PL 169 licence. These were subjected to a densely spaced sampling programme followed by quantitative analysis and a more extensive analysis programme than in normal routine studies. Increased resolution was achieved by incorporating more taxa for correlation purposes, including, for example, frequency variation of taxa with long stratigraphic ranges, which proved to be successful in correlating wells on a local scale. The core coverage is very good, including two wells (25/11-17 and 25/11-20) with a cored shale succession time-equivalent to the reservoir interval in the Lista Formation. The former well was selected as the reference well for the Grane Field biozonation. The full results from this well is documented in a master thesis by one of the co-authors (Søyseth 1998).

A quality control system where the various biostratigraphic events were classified into three main categories was established. Category 1 events refer to tops and bases corresponding to the main zones and subzones shown in Fig. 2. Most of these correspond to the events published by Mudge & Bujak (1996). Category 2 and 3 events refer to frequency variations in short- and long-ranging taxa, respectively, and are always constrained by the regional category 1 events. Category 2 include the frequency variation of, for example, *A. gippingensis* using the full range of events related to this species life-cycle, including first occurrence (FO), acme occurrence, first abundant occurrence (FAO) and last occurrence (LO = FDO). Category 3 includes the frequency variation of long-ranging taxa, for example, *D. denticulata* and *C. comatum*. It is clear that particularly the category 3 events are very local, and may therefore be field-specific for Grane. A categorization like this therefore guides the biostratigraphers, but serves, in addition, as a 'warning' to other geo-personnel that many biostratigraphic events may be of local character.

Subsequent to the definition of biozones, the chronostratigraphic framework was used to establish a reservoir-scale sequence stratigraphic model for the studied succession. In this work we applied the sequence stratigraphic model introduced by Van Wagoner *et al.* (1988) and later elaborated upon by, for example, Posamentier *et al.* (1992) and Helland-Hansen & Martinsen (1996). Many of the category 1 bioevents were found to correspond well with surfaces of major condensation in the basin, and thus served as time lines defining events of maximum transgression and/or minimum basinal sediment supply. These events were therefore used to constrain the boundaries of our major cycles (mts to mts, see below). Some of the category 2 events also appear to be condensed surfaces of at least subregional importance, and have therefore been used in the definition of major cycles. A similar usage of maximum transgressive surfaces as main correlative surfaces in the North Sea Tertiary has been described by Galloway *et al.* (1993) and Dixon & Pearce (1995). Category 2 and 3 bioevents were then used to subdivide the sequence stratigraphic cycles into subcycles and systems tracts (see below). Moreover, a few category 1 events have also been used for this purpose (those which define condensation events of lesser magnitude than the main cycle boundaries).

Biostratigraphic results

This section deals with the biostratigraphic results obtained in the present study. Main emphasis is on the Lista Formation, but the biostratigraphic results from the Våle, Sele and Balder formations are included. The established Grane Field biozones are treated in stratigraphic order, starting with the oldest. The base of each biozone is defined by the top of the underlying biozone.

Biozone Grane A (Fig. 2)

Age: Early Palaeocene (Danian)

The top is defined by the last occurrence (LO) of *A. reticulata* and/or LO of *Spiniferites magnificus*. *A. reticulata* was also used by Mudge & Bujak (1996), and the reader is referred to their discussion of the stratigraphic and geographic importance of this species. Mudge & Bujak (1996) do, however, record *S. magnificus* slightly higher in the section in their study, but this cannot be shown in the studied wells, although the sampling density of this section in our reference well is about every 20 cm. Our division into two subzones is based on LO of common *A. reticulata* and/or *S. magnificus*.

The Grane A bio-zone is identified in all the studied wells except for 25/11-18 and 25/11-21A which did not penetrate the Våle Formation. The base of this zone is not known, as the underlying chalk section is not studied. The assemblages recorded within Biozone Grane A are, however, regarded to be late Danian in age, since there are no records of, for example, *D. californica*. A characteristic assemblage containing a.o. *D. californica* was reported by, for example, Heilmann-Clausen (1985) from Danian limestones in the Viborg 1 borehole in Denmark. This assemblage was overlain by a succession of marl containing an assemblage similar to the one recorded within Biozone Grane A in the Grane Field.

Biozone Grane B (Fig. 2)

Age: Early Late Palaeocene

The top is defined by LO of abundant *P. pyrophorum*, a widely used correlation event (e.g. Mudge & Bujak 1996). Apparently, different workers use various aspects of the *P. pyrophorum* frequency variation to define this correlation event . Mudge & Bujak (1996) define their top P4 zone based on the the *P. pyrophorum* acme, whereas Stein *et al.* (1995) report the LO of consistent *P. pyrophorum.* Laursen *et al.* (1995) apply FDC occurrence of *P. pyrophorum* as their correlative event. In this study, however, the top of biozone Grane B is taken at the first downhole occurrence of abundant *P. pyrophorum* counting 16–30% of the total assemblage (Fig. 3). This requires dense sampling to disinguish from the intra-Grane B-40 superabundant event occurring below (Fig. 3), with records of *P. pyrophorum* between 30 and 75% of the total assemblage.

A subdivision of the Grane B biozone into four subzones (Fig. 3) is based on (from oldest) the LO of common *P. bulliforme*, LO *T.* cf. *delicata*, LO, *I.*? *viborgense*.

The LO of common *P. bulliforme* is not recorded in all wells, but this event has proved useful as a regional, North Sea event according to our in-house experience. The event is easily observable in reference well 25/11-17, and defines the top of the Grane B-10 biozone.

The LO of *T.* cf *delicata* defines the top of the Grane B-20 biozone. *T.* cf. *delicata* was used by Mudge & Bujak (1996) to define the top of their P3a zone. The base of *I.*? *viborgense* is also used to define the top of this subzone and has also been applied in cases were we had core samples available. In several of the wells an influx of *P. morulosa* is observed just below this event (Fig. 3) representing an intra-Grane B-20 event.

The LO of *I.*? *viborgense* is also a regional marker used by many North Sea workers (e.g Laursen *et al.* 1995; Mudge & Bujak 1996), as well as in the western Barents Sea (Nagy *et al.* 1997). Mudge & Bujak (1996) use this species to define the top of their P3b zone.

Biozone Grane C (Fig. 2)

Age: Late Palaeocene

The top is defined by LO of *A. margarita* and/or *S. rhomboideus*. The last occurrence of *A. margarita* was used by Mudge & Bujak (1996) to define their zone P5b. *A. margarita* was also reported by Heilmann-Clausen (1985) having its last occurrence in the upper part of the Holmehus Formation slightly above the abundance event of his *Areoligera* cf. *coronata*, now assigned to *A. gippingensis* by Jolley (1992).

A further subdivision of Grane C biozone into three subzones is based on (from oldest) the LO of *P. bulliforme* and/or LO *C. striatum* and LO common *A. gippingensis.*

LO of *P. bulliforme* and *C. striatum* marks the boundary between the Grane C-10 and C-20 biozones. They are recognized as regional events in our in-house studies but has not been reported by many authors as good index species. *C. striatum* seems to be the most consistently occurring of the two, and as shown in Fig. 3 other local correlation events occur between this event and top of the underlying Grane B biozone. The subzone Grane C-10 therefore comprises a set of category 2 and 3 events suitable for local correlation.

Areoligera gippingensis has a short range within the Upper Palaeocene and we recognized four events in the frequency variation of this species, all occurring within subzone Grane C-20 (Fig. 3). Mudge & Bujak (1996) based their top P5A on LO of abundant *A. gippingensis* which is a category 2 event in our zonation scheme. We prefer instead to use the LO common *A. gippingensis*, and assigned this event to the boundary between biozone Grane C-20 and C-30. According to Mudge & Bujak (1996) this event definitely occurs above NP6 in the type Thanetian in England. As shown in Fig. 3, other local events including an influx of *D. denticulata* occur between the LCO and LAO of this species.

Biozone Grane D (Fig. 2)

Age: Late Palaeocene

The top is defined by the LO of *A. augustum* (Fig. 2). We have observed an intrazone event of the reappearance of *Rhizammina*/*Bathysiphon*, but this biozone is not further subdivided in this study. Dixon & Pearce (1995) demonstrated, however, the use of, for example, various *A. augustum* events dividing a time-equivalent zone into four subzones in the UK quadrant 9.

Biozone Grane E (Fig. 2)

Age: Early Eocene

The top is defined by the LO of *Coscinodiscus* sp. 1 and 2 occurring slightly above the LO of common *D. oebisfeldensis* and abundant *Inaperturopollenites* spp. The last common occurrence of *Deflandrea oebisfeldensis* is a regional event used as a subzone marker species by, for example, Mudge & Bujak (1994). This event is often associated with a distinct downhole increase of diverse miospores, including *Inaperturopollenites* spp. In condensed sections it appears close to the LCO *Cerodinium wardenense*, but as pointed out by Mudge & Bujak (1996) the two events appear in ascending downhole stratigraphic order in

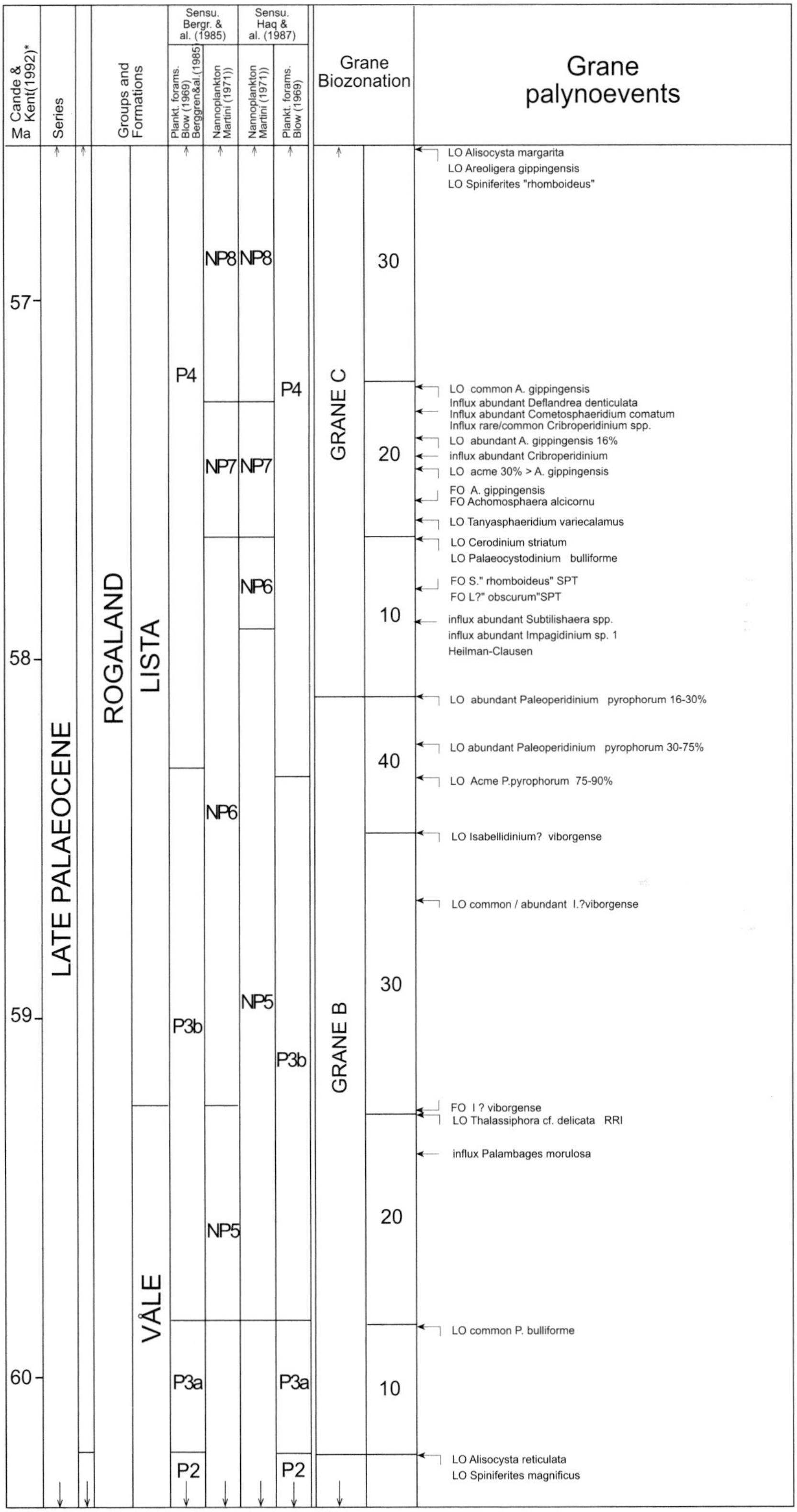

Cande & Kent(1992)* Ma	Series		Groups and Formations		Sensu. Bergr. & al. (1985): Plankt. forams. Blow (1969) Berggren&al.(1985)	Sensu. Bergr. & al. (1985): Nannoplankton Martini (1971))	Sensu. Haq & al. (1987): Nannoplankton Martini (1971))	Sensu. Haq & al. (1987): Plankt. forams. Blow (1969)	Grane Biozonation		Grane palynoevents
57	LATE PALAEOCENE		ROGALAND	LISTA	P4	NP8	NP8	P4	GRANE C	30	LO Alisocysta margarita LO Areoligera gippingensis LO Spiniferites "rhomboideus"
						NP7	NP7			20	LO common A. gippingensis Influx abundant Deflandrea denticulata Influx abundant Cometosphaeridium comatum Influx rare/common Cribroperidinium spp. LO abundant A. gippingensis 16% influx abundant Cribroperidinium LO acme 30% > A. gippingensis FO A. gippingensis FO Achomosphaera alcicornu LO Tanyasphaeridium variecalamus
58							NP6			10	LO Cerodinium striatum LO Palaeocystodinium bulliforme FO S." rhomboideus" SPT FO L?" obscurum"SPT influx abundant Subtilishaera spp. influx abundant Impagidinium sp. 1 Heilman-Clausen
						NP6			GRANE B	40	LO abundant Paleoperidinium pyrophorum 16-30% LO abundant Paleoperidinium pyrophorum 30-75% LO Acme P.pyrophorum 75-90%
59					P3b		NP5	P3b		30	LO Isabellidinium? viborgense LO common / abundant I.?viborgense
				VÅLE		NP5				20	FO I ? viborgense LO Thalassiphora cf. delicata RRI influx Palambages morulosa
60					P3a			P3a		10	LO common P. bulliforme
					P2			P2			LO Alisocysta reticulata LO Spiniferites magnificus

Fig. 3. Palynological event chart for the Grane Field biozones Grane B and Grane C.

complete sections. The assemblages in this part of the succession consist of low-diversity types, caused by basin-wide anoxia, but species like, for example, *G. ordinata* is relatively common.

A division into two subzones is done by applying the LO of common *C. wardenense*. Also for this part of the succession Dixon & Pearce (1995) gave a detailed description of assemblages corresponding to our biozone Grane E-10 .

The acme of the prasinophyceaean alga *Pterospemella* spp. and large leiospheres associated with the diatom *Coscinodiscus* sp. 7 represents a widely distributed regional event in the North Sea. Mudge & Bujak (1994) describe this as a biofacies assemblage that usually is associated with an earliest Eocene age, but which can also occur in other parts of the Eocene provided a stressed marine environment. In our study this event appears between the LCO of *Deflandrea oebisfeldensis* and the LCO of *A. augustum*. Although a tie to standard chronostratigraphy is difficult due to lack of planktonic foraminifera and calcareous nannofossils in this part of the succession, its stratigraphic position is quite well constrained.

Correlation of events

A correlation of biostratigraphic events between wells 25/11-15, 25/11-18, 25/11-16 and 25/11-17 is presented in Fig. 4. Well 25/11-17, containing no reservoir sand, holds the most complete stratigraphic record.

An important point shown in this correlation is that the reservoir sand bodies in wells 25/11-15 and 25/11-18 belong to different biozones. The sand body in well 25/11-15 lies within biozone Grane B-40 which has a top defined by the LAO of *P. pyrophorum*. In well 25/11-18 the main sand body lies within biozone Grane C-10/20, and the acme of *P. pyrophorum* clearly occurs below the sand. This demonstrates diachroneity with regard to distribution of reservoir sandstone within the main submarine fan.

Well 25/11-16 demonstrates clearly the importance of high-resolution biostratigraphy. In this well three different sandstones occur within subzone Grane C-20, but can be distinguished and correlated to other wells due to category 2 and 3 events. The main sand body lies within the middle part of biozone Grane C-20. In this case category 2 and 3 events, including the influxes of *Cribroperidinium* spp., *C. comatum* and the *A. gippingensis* abundant event, distinguishes this sand from the younger intra-Heimdal sand belonging to upper part of biozone Grane C-20. Although these three events show cross-overs (see Fig. 4) they always occur close to each other and they always occur above the *A. gippingensis* acme, which in well 25/11-16 occurs below the main sand body. In well 25/11-18 the *A. gippingensis* acme occurred above the sand (see Fig. 4). Category 2 and 3 events, including the *A. gippingensis* acme, the FO of *A. gippingensis* and the base of *A. alcicornu* distinguished the main sand body from an older, thinner Heimdal sand belonging to the the lower part of Grane C-20 (Fig. 4).

The age of the sands in well 25/11-16 and the diachroneity within the main fan thereby demonstrate that the sand deposition took place with a pronounced younging towards the south.

Sequence stratigraphic framework

Owing to the high resolution of the biostratigraphy on the Grane Field, the definition of sequences and their subunits relies heavily on the chronostratigraphic framework. However, sedimentological interpretations of cores and logs, together with observations of stratigraphic relationships on seismic data, were also used to construct a sequence stratigraphic framework for the Grane Field. In summary, the seismic and sedimentological data show that the studied succession formed in a deep-marine environment. Pelagic mudstones formed by suspension fallout alternate with high-density turbidite sandstones affected by both syn- and post-sedimentary soft-sediment deformation (Martinsen *et al.* 1998). The individual turbidite sandstones of the Grane Field were strongly influenced by sea-bottom topography, and were deflected in a N–S direction between the Utsira High in the east and the base of the East Shetland Platform to the west. Accumulation took place within the sea-bottom depressions created by slumping and channelization, and the massive and homogenous nature of the sands can be attributed to rapid dumping of a well-sorted sediment load upon ‘collision’ with the flank of the Utsira High. It is believed that the well-sorted nature of these sands was caused by repeated mass-flow surges along the extensive slope between the East Shetland Platform and the narrow basin plain.

The sequence stratigraphic framework contains three different types of genetic units as described below and illustrated in Fig. 5.

- Major cycles serving as regionally correlatable cyclothems (Fig. 2). In the studied

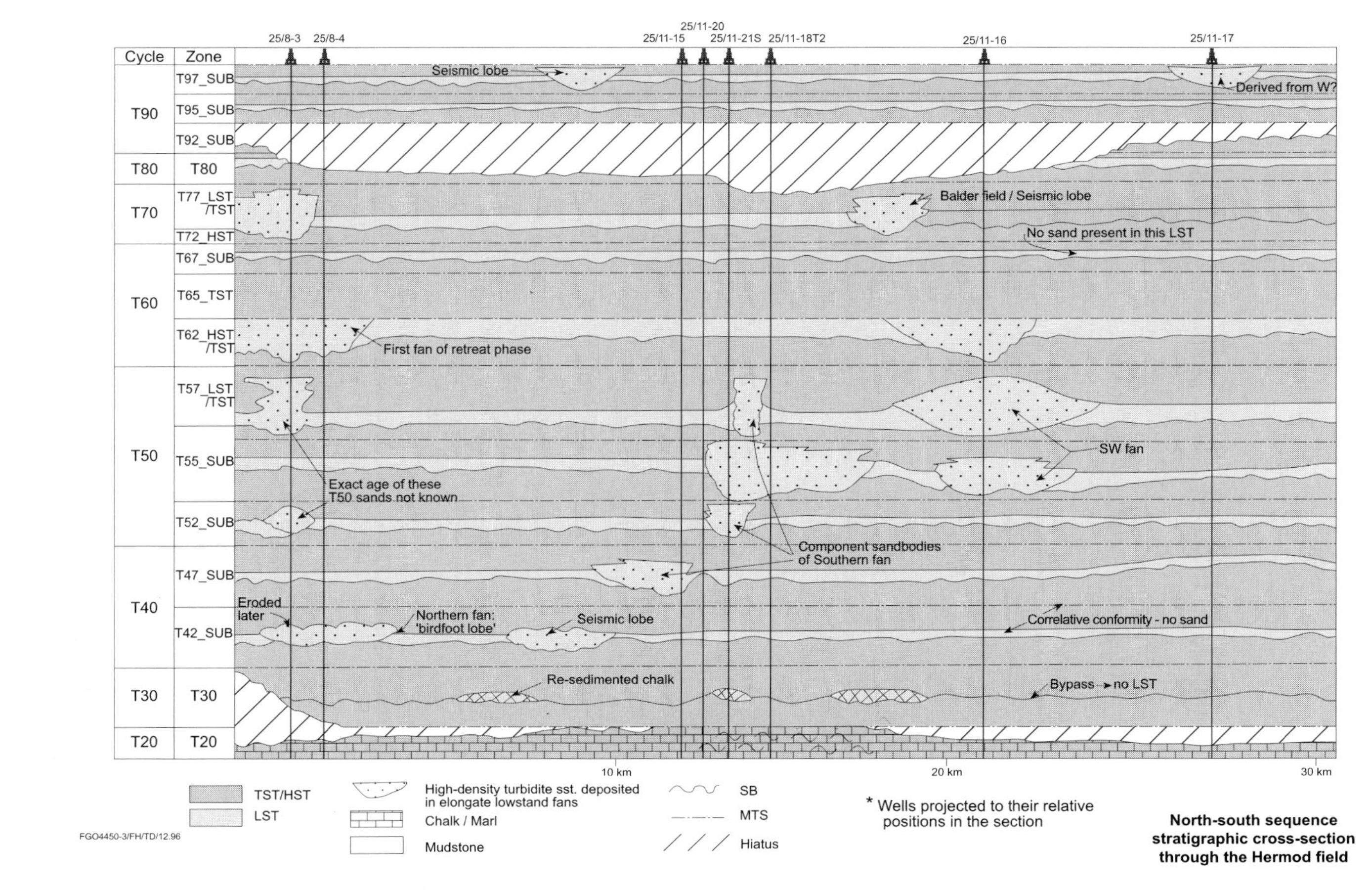

Fig. 5. N–S sequence stratigraphic cross-section through the Grane Field.

succession eight major cycles, named T20–T90, have been identified. The major cycle boundaries correspond to widespread condensed surfaces, probably formed under conditions of maximum transgression, and are located in 2–30 m-thick mudstone-rich zones. They are named by using the term mts and the number cycle, so that for instance the condensed surface at the base of cycle T30 is referred to as mts T30. In a number of cases, the condensed surfaces of the major cycles are associated with a high gamma-ray log spike (Mudge & Bujak 1996). In lithostratigraphic terms, T20–T30 correspond to the Våle Formation, T40–T60 belong to the Lista Formation, T70–T80 and the lowermost part of T90 correspond to the Sele Formation, while the remainder of T90 fall within the Balder Formation. The relationship between the major cycle boundaries and the biozones were discussed above, and can be seen on Fig. 2.

By comparing our major cycles with published biostratigraphically based sequence subdivisions of the studied succession, a great deal of similarity can be seen. Den Hartog Jager *et al.* (1993), Stein *et al.* (1995), Neal (1996) and Mudge & Bujak (1996) all operate with eight major cycles (or genetic sequences in their terminology), of which six are approximately equal to the ones presented here. The same six main cycles have also been recognized by BP in a number of recent papers (e.g. Dixon & Pearce 1995). The main difference between this study and the other four schemes is found in the subdivision of the Våle Formation (T20 and T30 cycles). Neal (1996) does not operate with mts T30 (top Grane B-20 biozone) as a major cycle boundary, whereas the three other studies do not recognize mts T30 as a regional correlation line. Instead, they define an additional major cycle *within* our T40 (Den Hartog Jager *et al.* 1993; Stein *et al.* 1995), T70 (Mudge & Bujak 1996), or T90 (Neal 1996).

- High-frequency cycles (Fig. 5) occurring where less pronounced condensed surfaces have been recognized *between* the condensed surfaces of the major cycles (e.g. the boundary between units T41 and T42 in the T40 major cycle). These are indicated by the abbreviation SUB (meaning subcycle) in the zonation column. Regional cycles T40, T50, T60 and T90 contain two or more high-frequency cycles. Their bounding surfaces are placed at subtle to well-defined high gamma-ray peaks within 0.5–40-m (average 5-m) thick shaly units. Biostratigraphically, they are associated with category 1 or 2 events.
- Systems tracts (Fig. 5) which can be correlated locally on the basis of category 2 or 3 biostratigraphic time lines and sedimentological development. The introduction of systems tracts into the sequence stratigraphic scheme is intended to make the correlation framework more useful in predicting the distribution of reservoir sand. Cycle correlations tend to highlight the stratigraphic evolution of the shale-prone parts of the succession by focusing on the maximum transgressive surfaces. Systems tracts which lack biostratigraphic control, are not included in the zonation. Systems tract boundaries correspond to surfaces separating the lowstand (LST), transgressive (TST) and highstand (HST) systems tracts. Hence, in addition to maximum transgressive surfaces, two additional surfaces are recognized: the *transgressive surface* and the *sequence bounding unconformity*. In cases where pronounced condensation has taken place (i.e. where cycles are thin), no system tract subdivision has been attempted. Transgressive surfaces constitute another variant of the condensed surface, but in this case associated with the transition from sandy–silty to more muddy sections (i.e. the boundary between zones T62HST–LST and T65TST).

Sequence boundaries (SBs) are picked at the base of prominent deep-water sandstone units, and represent erosional surfaces across which a significant increase in sediment supply occur. The thick Grane sandstones thus correspond to the 'classic' lowstand units incorporating both the forced regressive wedge and lowstand wedge systems tracts (Helland-Hansen & Martinsen 1996). Owing to resolution and reworking problems, biostratigraphy has played a more limited role in the definition of the sequence boundaries. The exceptions are the sequence boundaries at the base of zones T57_LST–TST and T77_LST–TST, which are biostratigraphically well-defined surfaces associated with monospecific acme occurrences. The pinchout of LST sands and the limited biostratigraphic control makes the SB less suitable than the MTS for regional correlation (hence the preferred use of cycles rather than sequences). Only two of the SBs are associated with widespread erosion and development of a biostratigraphically mappable hiatus. These are the SB below mts T20 (top Ekofisk unconformity) and the SB above mts T90 (base Balder unconformity). Local hiati may be developed in association with all the other SBs due to incision at the base of the submarine fan sand bodies, but the magnitude of erosion is usually limited and regionally discontinuous.

Cycle description and sand distribution

The major cycles and the biostratigraphic links to the sequence stratigraphic zonation are discussed in detail below. The main points will be summarized by considering each cycle and its subzones separately. Gross depositional environment maps which highlight the subregional sandstone distribution for some of the cycles are given in Fig. 6.

T20

This cycle comprises an upwards-fining succession of marls, sandstones and re-sedimented chalks of the Våle Formation, including basal parts of the sandy Ty Member. Its basal boundary, mts T20, is placed at a thin marly section just above the base Våle–top Ekofisk unconformity. Locally (i.e. well 25/11-18T2), chalk reworking appear to have dominated to such an extent that mts T20 is located within a re-sedimented chalk lithology. On palaeostructural highs T20 is missing. Biostratigraphically, mts T20 is located immediately above the upper occurrence of *E. aff. trivialis* Blow. For instance, a combination of erosion (slumping) and condensation along the western flank of the Utsira High has usually removed the T20 cycle in the Grane Field. Elsewhere in the study area T20 is thin and marly due to its distal position in the basin, although some turbidite sands are present to the west of the Utsira High.

T30

Located in the upper part of the Våle Formation T30 is dominated by condensed mudstones in the vicinity of the Grane Field, and alternations between turbidite sandstones and mudstones proximally (northwest). The base of this cycle is defined by biostratigraphic markers such as the upper occurrences of *A. reticulata* and *S. magnifica*, and is rarely associated with high gamma-ray readings. T30 varies in thickness from a few metres in the Grane area to 150 m to the west. Major turbiditic T30 sandstones occur in wells 25/7-2, 25/10-2 and 25/11-3. In the cases where the T30 sands are thick and blocky they are interpreted as channelized submarine fan deposits formed during an intra-T30 relative sea-level fall. These lowstand sandstones dominate the entire T30 cycle in the fan depocentre areas. Sands formed on the distal parts of the fan and the fan fringe tend to be interbedded with basin plain mudstone and show upwards fining tendencies both on bed and fanlobe scale. According to Neal (1996), a lowstand prograding deltaic wedge in UK quadrants 13, 14, and 16 represents time-equivalent up-dip deposits to the Ty Member submarine fan sand bodies.

T40

In this cycle, palaeogeographic reconstructions (Fig. 6a) indicate a change towards more NW–SE progradation of the slope to basin plain system. An increase in sediment supply and the long-term trend towards relative sea-level fall during deposition of the Våle and Lista formations (Jones & Milton 1994) led to the construction of a sand-rich lower slope in the northwestern parts of the study area. This prograding slope consists of numerous partly overlapping submarine fan sand bodies, and may have acted as a temporary sand repository from which the Grane area mass-flow sands were re-sedimented during subsequent relative sea-level falls. Both seismic images and correlation of closely spaced wells show that these elongated fans extend from the base of slope onto the basin plain, and represent the first Palaeogene mass-flow sand bodies to reach the Grane Field (Fig. 6a). Biostratigraphically, the base of cycle T40 is associated with the last stratigraphic occurrence of *T.* cf. *delicata* and an acme of *Cenodiscus* sp. 1. Its thickness varies from a few metres (well 25/8-3, top-truncated?) to more than 200 m in the west, the eastward thinning being associated with a pronounced decrease in sand content. Where sand is absent, cycle T40 has a log response characterized by an upwards increase in gamma-ray readings.

In most wells, T40 can be subdivided into two subcycles; T42_SUB and T47_SUB. These represent distinct reservoir zones, and the higher-order mts separating them is marked by the last stratigraphic occurrence of *I. ?viborgense*. Reservoir-quality sands belonging to the basal part of the Heimdal Member may be present in both subcycles, although they are most common in T47_SUB. The northern elongate fan on Fig. 6a was formed during the lowstand systems tract of T42_SUB, and is composed of overlapping depositional lobes which pinch out in a 'birdfoot' pattern. In contrast, the 'cigar-shaped' southern elongated fan is a lowstand unit formed in T47_SUB. The proximity to the axis of the

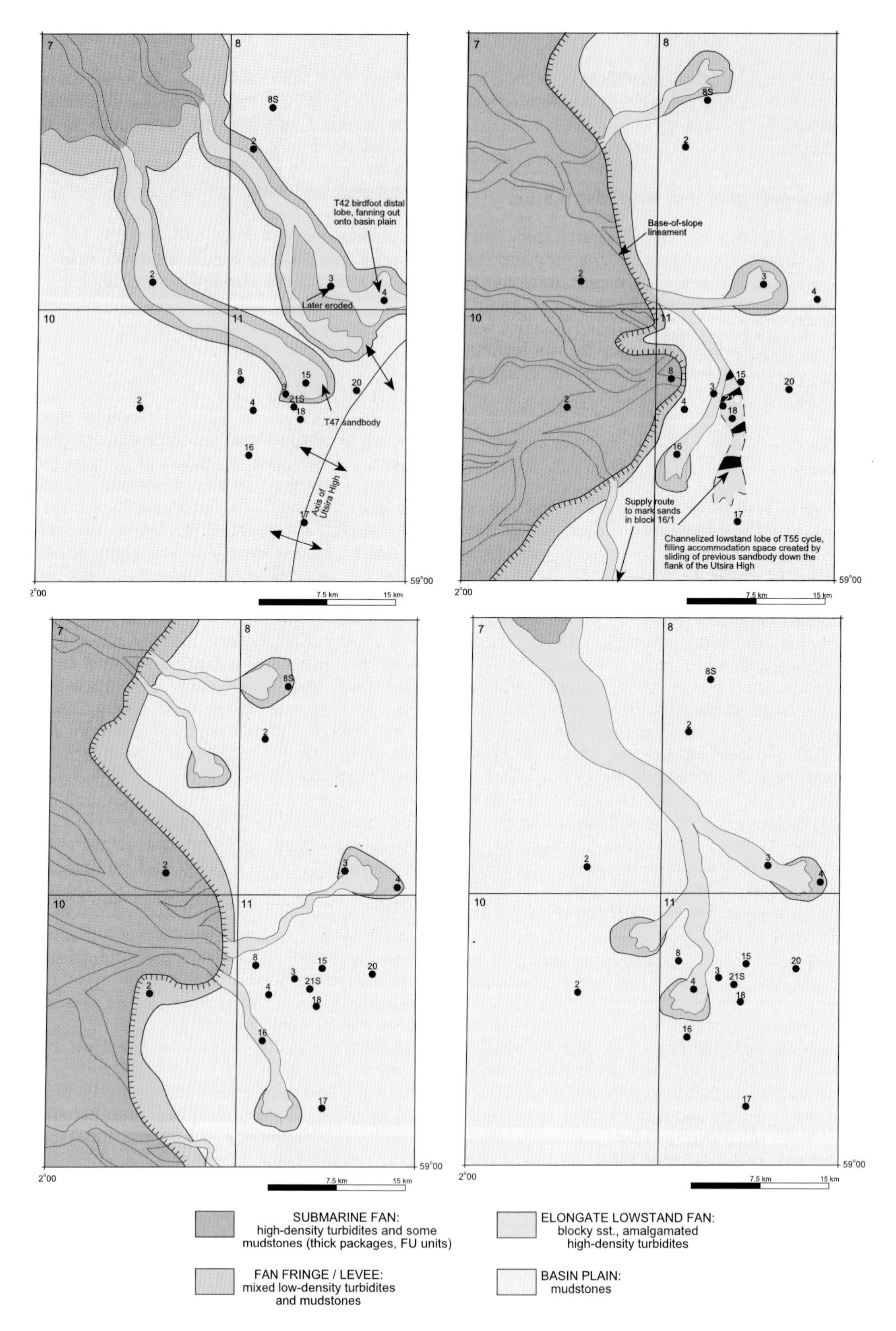

Fig. 6. Palaeogeographical maps showing the gross depositional environments of cycles T40, T57, T62 and T77. Note the advance and retreat of the submarine fan system.

Utsira High probably caused rapid 'snout-like' termination of this southern elongate fan, preventing it from fanning out 'freely' on an unconfined basin floor.

T50

This unit has been subdivided into three subunits, and is the most sand-rich cycle in the studied succession. Regional studies show that the forced regressive tendencies in the study area culminated towards the end of T50, implying that the sand-rich slope reached its maximum basinward extent in this period. Mts T50 at the base of the cycle is marked by a high gamma-ray peak, and is distinguished by the prominent bioevent corresponding to the last abundant stratigraphic occurrence of the *P. pyrophorum* dinocyst. The T50 cycle comprises the bulk of the reservoir sandstones in the Grane Field, but it should be noted that some Grane wells contain only low gamma-ray mudstone even in this cycle. In the cases where T50 is shaly, the cyclothem may be as thin as 10–20 m (e.g. wells 25/11-20 and 25/8-4), whereas it attains thicknesses in excess of 300 m in extremely sand-rich wells.

The three subunits of cycle T50 are referred to as T52_SUB, T55_SUB and T57_LST–TST. They are recognizable over most of the study area, and are biostratigraphically constrained by the last stratigraphic occurrence of *C. striatum* (base T55) and an acme of *A. gippingensis* (base T57). Whereas the former is associated with a condensed surface, this appears not to be the case for the base T57 event. There is usually a significant sand influx immediately above this event. Hence, we believe that the base T57 event is associated with a major regression, and it is therefore referred to as a sequence boundary. All three subunits are sand-rich in proximal wells, consisting of up to several hundred metres thick submarine fan sand bodies. These contain a mixture of upwards-fining, upwards-coarsening and massive intervals, often separated by mudstone-dominated bedsets 0.5–11 m in thickness. The more prominent of these are found in association with the condensed surfaces within the T50 cycle. Basinwards, the sand-body thickness decreases and the facies types diversifies. Some wells consist entirely of low gamma-ray mudstones with perhaps a few low-density turbidites (e.g. 25/11-4), whereas others contain a mixture of mudstone and thin turbidite sandstones (e.g. 25/8-8S). There is also a number of wells that contain 20–70-m thick massive sand bodies (e.g. 25/11-18T2). These are invariably located in the lowstand part of the cyclothems, and are believed to represent elongate fans derived from the submarine slope to the west. They contain stacked high-density turbidites, and their positions appear to be controlled by structural features such as minor top Chalk depressions and slump-scars along the western flank of the Utsira High. In the Grane area subunits T55 and T57 must be regarded as the most important reservoir zones because of the presence of sand-rich elongate lowstand fans even in this distal location. The palaeogeography in Fig. 6b is a composite reconstruction valid for both of these subunits. It is believed that the sand-rich slope facies association was located in approximately the same position for both T55_SUB and T57_LST–TST. From this major slope, 'finger-like' subsidiary fans were redeposited during the T55 and T57 lowstand periods. Combined well control and seismic indicators in the Hermod area allow us to predict the location of these elongate lowstand fans with some degree of accuracy. In the T55 subcycle, the main elongate fan extended N–S through wells 25/11-21S and 25/11-18 towards well 25/11-17, probably filling a newly created slump-scar along the flank of the Utsira High (stippled sand body in Fig. 6b). Sand is also present in 25/11–16 in this subunit, but it is not clear whether this sand is a side-arm of the main elongate lobe described above, or if it represents a separate lowstand fan derived from the nearby base-of-slope 'escarpment'. In T57, sand deposition was shifted more towards the west, with the main N–S-oriented lowstand lobe probably being located west of 25/11-21S and extending some distance to the south of 25/11-16.

T60

As is the case of T50, this cyclothem has been split into three subunits: T62_HST–LST, T65_TST and T67_SUB. In the cases where T60 is shaly, the cyclothem may be as thin as 15 m (25/8-2), whereas it attains a thickness of almost 300 m in extremely sand-rich wells to the west. The mts at the base of the cycle is marked by the last *abundant* stratigraphic occurrence of *A. gippingensis*, whereas the proposed transgressive surface at base T65 is located slightly above an influx of *C. comatum* and at or slightly above an influx of *D. denticulata* (see Fig. 3). The higher-frequency mts separating T65 and T67 is associated with the last *common* stratigraphic occurrence of *A. gippingensis*. In our study area, the T60 cycle is characterized by aggradation and initial landward-stepping of the submarine slope system. The depositional setting is otherwise unchanged, with submarine fans coalescing on the slope during lowstand periods, creating local instabilities from which slumping and turbidite re-sedimentation may occur (Fig. 6c). Studies by Jones & Milton (1994) and Dixon & Pearce (1995), conducted in the Palaeocene on the British side of the North Sea, did not reveal any landward retreat of the fan systems until the T70 cycle. These authors worked in a region that corresponds to the proximal parts of our study area, and one would therefore expect the relative sea-level curves to be similar. Apart from misinterpretations, this discrepancy may be due to local effects such as strike variations in sediment supply even over short distances. It is also theoretically possible that a regressive trend may persist in the vicinity of the platform margin whilst submarine fan retreat takes places distally on the basin plain. This can be accomplished by rotation around a hinge line located near the base of the slope (Mutti 1992). Up-dip of the hinge line, uplift causes accommodation decrease and sediment instability leading to regression and re-sedimentation, whereas accommodation increase and fan retreat may occur down-dip of the hinge line. As uplift and tilting has been documented for the Upper Palaeocene of the North Sea (e.g. Nadin & Kusznir 1995), the hinge-line rotation theory may well explain the regional differences in relative sea-level trends in the T60 cycle.

The effects of the initial slope retreat is felt first in the most distal part of the basin (e.g. Grane–Balder–Jotun area). Here, the amount of major elongated fan sand bodies decreases significantly from T50 to T60 and also upwards in T60. Only 25/11-16, 25/11-8 and 25/8-8 contain appreciable amounts of sandstone, all in the T62_HST–LST zone (Fig. 6c), but these sandstones are thinner than those in the cycles below. To the west, good sandstone development is seen in western parts of Block 25/7, again restricted to the T62 zone.

While the T67 zone is regarded as a 'traditional' subcycle due to the high-frequency mts at its base, the other two zones are interpreted as systems tracts of at least subregional extent. The reasons for this are partly pragmatic: the T62 unit is substantially more sandy than the succeeding units, and its upper bounding surface is associated with a bioevent corresponding to re-establishment of palynomorphs which had been suppressed during the period of sand influx. It thus seems likely that it represents deposits primarily formed under lowstand conditions. Moreover, the sand-poor zone between the base T65 bioevent and the high-frequency mts at the base of T67 usually shows an upwards increase in gamma-ray reading and fossil diversity, both characteristic features of the transgressive systems tract. Hence, the presence of a useful biostratigraphic time line at what seems to be a transgressive surface enables us to include these intra-T60 systems tracts in our sequence stratigraphic zonation (Fig. 5).

T70

The tendency towards northwestward-stepping of the submarine fan system continued in the T70 cycle. This resulted in sand starvation in the distal areas, with only a few very elongated lowstand fans being developed here (Fig. 6d). The deep-water mudstones of T70 belong to the lower part of the Sele Formation, and are characterized by darker colours, better developed lamination, and concentration of dinocysts and microfossils in distinct condensed horizons. In the Balder Field Jenssen *et al.* (1993) documented the presence of a few moderately thick elongate fan sand bodies with prominent deformational features. Seismic data suggested that these were derived from the north, and it is thought (Fig. 6d) that they belong to an (over)extended submarine fan which also encompasses the T77 sand body in the 25/8-3 well north of Hermod. The T70 cycle is normally quite thin (0–70 m), especially on the Utsira High where it occasionally appears to be absent due to erosion (25/11-18T2).

In this study T70 has been divided into two subunits: T72_HST and T77_LST–TST. The few major turbidite sandstones of this cycle are all present in the upper subunit. The base of the T70 cycle is defined by the last stratigraphic occurrence of *A. margarita* and *S. rhomboideus*. As we correlate the intra-T70 event *Rhizaminna*/*Bathysiphon* to a sequence-bounding unconformity, the two subunits within the T70 cycle correspond to systems tracts. The T72 unit, which is mudstone-dominated with a slight coarsening-upwards trend in many wells (e.g. 25/11-20, 25/8-4 and 25/8-3), then becomes a highstand systems tract, whereas the lower gamma-ray shales and blocky sands of T77 fall within a lowstand–transgressive systems tract.

T80

This cycle is generally sand-poor in the study area, and for this reason is relatively thin (0–70 m often truncated or removed by the base Balder sequence boundary on the Utsira High, Fig. 5). No major sand bodies are known to occur within this cycle in the greater Grane area, possibly due to sand-trapping in a base-of-slope environment in the Beryl trough. The base of T80 is often associated with an uranium-rich condensed section that is seen as a high gamma-ray peak on the logs. Biostratigraphically, this condensed surface can be recognized by the *A. augustum* event. The Sele Formation mudstones of the T80 cycle are well laminated and usually grey, with a few tuffaceous beds in the upper part.

T90

The base of this cycle (mts T90) is associated with perhaps the most prominent high gamma-ray peak in the North Sea Palaeogene section, and represents a major condensation event. Biostratigraphically, it is associated with the acme of the *Pterospermella* acritarch and the upper abundant occurrence of large *Leiospheres*. As these species have a fairly long time-range and are not consistently reported in biostratigraphic studies, one should also use the *C. wardenense* dinocyst acme to constrain the position of mts T90. The *C. wardenense* event is present shortly above the base Balder unconformity, and is a widespread and well-defined time line (top Grane E-10) defining the boundary between subcycles T95 and T97. The upper boundary of T90 (mts T100) is associated with the top of the Balder Formation, and is biostratigraphically defined by the *Coscinodiscus* sp. 1 and 2 event (top biozone Grane E). Towards the distal parts of the basin it is often associated with a prominent gamma-ray spike, but proximally (to the west) this peak becomes blurred. Cycle T90 varies in thickness from 170 m in the west to about 15 m (Utsira High area, e.g. Grane Field) due to lateral changes in accommodation space and sediment supply. It represent the later stages of onlap onto the Utsira High, smoothening out the basin-floor topography in the process (Laursen *et al.* 1995). In most wells, it is possible to subdivide T90 into three high-frequency cycles: T92_SUB, T95_SUB and T97_SUB (Fig 5). The higher-order condensed surfaces at the base of subcycles T95 and T97 are usually marked by a prominent high gamma-ray log spike attributable to a high content of uranium. Biostratigraphically, these surface are picked at the upper common occurrences of *C. wardenense* (base T95) and the *D. oebisfeldensis* abundance event (base T97).

The part of T90 located between the basal gamma-ray spike and the Balder tuffs belong to the Sele Formation, and comprises grey laminated mudstones with some tuff beds. Otherwise, T90 is dominated by tuffaceous mudstones, with subordinate amounts of sandstone (Odin Member) and high gamma-ray claystones. Both this and previous studies indicate that the sand bodies in T90 are laterally discontinuous, representing high-density turbidites formed within narrow deep-sea channels or confined submarine fans (den Hartog Jager *et al.* 1993; Jenssen *et al.* 1993; Timbrell 1993). Of the wells used in this study, only 25/11-17 contains a major T90 sand body. The thick and blocky sand body in 25/11-17 is present in the T97 subcycle, and seismically appears to turn southwards west of the well.

Within T90, shortly above the basal mts, a regional unconformity is present (e.g. Neal 1996). This unconformity is located at the base of the tuffaceous Balder Formation, and is locally observed to erode as far down as the T60 cycle (Fig. 5). Tuffaceous breccias are seen in cores at this unconformity (25/8-4). Although not documented in the literature to any extent, the base Balder unconformity features prominently in several biostratigraphic zonation schemes (M. Charnock, pers. comm. 1996). The seismic data indicate that the base Balder surface was topographically complex, with prominent highs (mounded features) separated by shallow to deep lows. The most striking of these lows is a N–S-striking feature located between the Balder and Hermod fields. Seismic studies show evidence of truncation of reflectors at the margins of this low, implying that it represents an erosional feature, possibly a major turbidite channel incised during formation of the base Balder unconformity. This is supported by biostratigraphic data from the only well within the shale-prone 'low'; 25/11-3. The events used as time-line markers for both mts T90 and T80 appear to be missing in 25/11-3, implying truncation of the upper part of the Sele Formation during formation of the base Balder unconformity. Jenssen *et al.* (1993), in their study of the Balder Field, suggest that an episode of fault rejuvenation and westward tilting took place at the Sele–Balder transition, creating an unconformity of tectonic origin. On the flanks of the Utsira High, mts T90 is usually removed by this unconformity and the basal subcycle of T90 is also missing, possibly due to bypassing during the lowstand period.

Grane area sequence stratigraphic summary

The eight major cycles and their subunits allow us to subdivide the studied succession into 16 sequence stratigraphical zones (Fig. 5) with a caculated average duration of 200 000 years. The lower two cycles, T20 and T30, are characterized by the total absence of sand and a combination of condensed deposition (chalk, marl, mudstone) and bypass/erosion down the flank of the Utsira High. The condensed nature of these cyclothems can be related to their distal position with respect to the sediment source. The first sands entered the Grane area in the T42 subcycle. Geometrically, this radial lowstand sand is unlike the other submarine fan sands of the Grane area, and the reason for this is probably that it: (a) represents the terminal lobe of an elongate lowstand fan that died out as it reached the Grane area; and (b) it is located in an area where the basin plain was less confined than is the case for the wells penetrating the main Grane sand in the trough between the Balder Field and the Utsira High.

The N–S elongated main Grane sand body was formed incrementally by amalgamation of lowstand sands during subcycle T47 and all of cycle T50. The first elongate lowstand fan entered the area from the northwest in subcycle T47, and the massive, stacked high-density turbidites in 25/11-15 belong to this unit (Fig. 5). Our model suggests that this sand body was draped by mudstone during the TST of subcycle T47 and in the entire T52 subcycle. It does *not* correlate to the similar-looking neighbouring sand bodies in wells 25/11-21S and 25/11-18T2. The sand in the latter wells belongs to the lowstand systems tract of the T55 subcycle.

In the lowstand systems tract of unit T57, the southern fan complex spreads further south, probably reaching its maximum extent at the pinchouts south of wells 25/11-18T2 and 25/11-16. The widely spread mudstone within the T57 TST is rich in palynomorphs, and forms the basal part of the thick muddy section that seals the main Grane sand body. It is thought to reflect a major transgressive event which brought the basinward progradation of the whole submarine fan system to a halt.

Continuous backstepping of cyclothems is inferred for cycles T60–T80. The basal unit of this mudstone-dominated succession is the T62 HST–LST. In the lowstand part of this unit, thin turbidite sands are seen in wells 25/8-3 and 25/8-4, and a more prominent elongate fan sand body is present in well 25/11-16. It is unlikely that these sandstones are in primary contact with the main Hermod sands below as there is an appreciable amount of HST–TST mudstone in the intervening interval. The next level containing high-density turbidite sands is believed to be the T77 LST–TST unit, which is separated from T62 by a prominent mudstone-prone interval encompassing T65 TST, T67 subcycle and T72 HST (Fig. 5). In the T72 unit, sand appear to be concentrated to the western (Balder Field; Jenssen *et al.* 1993) and northern part of the area, again reflecting the tendency towards sourceward retreat of the submarine fan system relative to the main sand-bearing T50 cycle. The widespread and strongly condensed mudstones present in the upper part of T77 and presumably in the whole T80 cycle represent the culmination of the westward retreat of these fans.

The base of the T90 cycle seems to be a subregional unconformity which could be related to uplift of the Utsira High. It is associated both with bypassing (deposits from the T92 subcycle are mostly absent from the area) and erosion (the T80 cyclothem is thought to be fully to partially truncated). After uplift had ceased, accelerated subsidence may have taken place, as indicated by the muddy deposits within the widespread T95 subcycle.

Conclusions

- Our study demonstrates the utility of a detailed biostratigraphic zonation for the Grane area and the establishment of a sequence stratigraphic framework from which the depositional history of the area can be understood in terms of a finite set of depositional cycles and subcycles. The biostratigraphic data calibrated to the numerical time-scale suggest average durations of the depositional period for each zone to be around 200 000 years.
- The improved stratigraphic resolution presented for biozones Grane A–Grane E with the recognition of a series of locally correlatable events, represents in itself one of the main results of the present study. This highlights the potential for high-resolution correlation resulting from quantitative biostratigraphic methods.
- The defined cycles can be summarized as follows: cycles T20–T30 show condensation without well-developed lowstand units; cycles T40–T50 show stepwise encroachment of the submarine fan system towards the Grane area, with well-developed but areally restricted lowstand units filling local accommodation; cycles T60–T80 show stepwise (?) retreat of the submarine fan system towards the west-northwest. Lowstand sand units become less prominent as the area once again is subjected to condensation; and cycle T90 shows that an unconformity is developed at the base of the Balder Formation, probably due to uplift/tilting of the Utsira High. After a brief phase of non-depositional forced regression, the rate of

relative sea-level rise accelerates. Lowstand sands associated with this uplift phase appear to be scarce.

- The sequence stratigraphic model presented contains 16 correlatable zones, and is well suited for reservoir zonation due to its ability to depict in detail how the various sand accumulations in the area are genetically related. It is concluded that all the main sandstone bodies (submarine fans) are emplaced during relative sea-level lowstands.
- Potential flow barriers may have formed in the TST–HST periods between lowstand sand deposition. One candidate is the southern Grane fan complex where biostratigraphical data demonstrate that the sand in the two wells are diachronous. This will, however, depend on the degree to which this mudstone unit was affected by erosion associated with emplacement of the next sand body.

The authors would like to thank Norsk Hydro and their partners in licence PL 169 for permission to publish this manuscript. Many colleagues have contributed to this work, and we would in particular like to thank Geir Indrevær who did the seismic interpretation. We also want to thank Lucy Costa at Robertson, Wales, for carrying out many of the palynological analyses and for fruitful discussions regarding many of the fossil events. M. A Kaminski and R. W. Jones are acknowledged for their reviews of the manuscript.

References

DIXON, R. J & PEARCE, J. 1995. Tertiary sequence stratigraphy and play fairway definition, Bruce-Beryl Embayment, Quadrant 9, UKCS. *In*: STEEL, R. J., FELT, V. L., JOHANNESSEN, E. P. & MATHIEU, C. (eds) *Sequence Stratigraphy on the Northwest European Margin*. Norwegian Petroleum Society, Special Publications, **5**, 443–470.

DEN HARTOG JAGER, D., GILES, M. R. & GRIFFITHS, G. R. 1993. Evolution of Paleogene submarine fans of the North Sea in space and time. *In*: PARKER, J. R. (ed.) *Petroleum Geology of Northwest Europe: Proceedings of the 4th Conference*. Geological Society, London, 59–71.

GALLOWAY, W. E., GARBER, J. L., LIU, X. & SLOAN, B. J. 1993. Sequence stratigraphic and depositional framework of the Cenozoic fill, Central and Northern North Sea basin. *In*: PARKER, J. R. (ed.) *Petroleum Geology of Northwest Europe: Proceedings of the 4th Conference*. Geological Society, London, 33–43.

GRADSTEIN, F. M., KAMINSKI, M. A., BERGGREN, W. A., Kristiansen, I. L. & D'IORO, M. A. 1994. Cenozoic biostratigraphy of the North Sea and Labroador Shelf. *Micropaleontology*, **40**, Supplement.

——, KRISTIANSEN, I. L., LØMO, L. & KAMINSKI, M. A. 1992. Cenozoic foraminiferal and dinoflagelate biostratigraphy of the Central North Sea. *Micropaleontology*, **38**, 101–137.

HANSEN, H. J. 1977. Dinoflagellate stratigraphy and echinoid distribution in Upper Maastrichtian and Danian deposits from Denmark. *Bulletin of the Geological Society of Denmark*, **26**, 1–26

HEILMANN-CLAUSEN, C. 1985. Dinoflagellate stratigraphy of the uppermost Danian to Ypresian in the Viborg 1 borehole, central Jylland, Denmark. *Danmarks Geologiske Undersogelse, Series* **A7**.

——1994. Review of Paleocene dinoflagellates from the North Sea region. Meeting proceedings "Stratigraphy of the Paleocene", GFF, **116**, 51–53

HELLAND-HANSEN, W. & MARTINSEN, O. J. 1996. Shoreline trajectories and sequences: description of variable depositional-dip scenarios. *Journal of Sedimentary Research*, **66**, 670–688.

ISAKSEN, D. & TONSTAD, K. 1989. A revised Cretaceous and Tertiary lithostratigraphic nomenclature for the Norwegian North Sea. *Norwegian Petroleum Directorate Bulletin*, **5**.

JENSSEN, A. I., BERGSLIEN, D., RYE-LARSEN, M. & LINDHOLM, R. M. 1993. Origin of complex mound geometry of Paleocene submarine-fan sandstone reservoirs, Balder Field, Norway. *In*: PARKER, J. R. (ed.) *Petroleum Geology of Northwest Europe: Proceedings of the 4th Conference*. Geological Society, London, 135–143.

JOLLEY, D. W. 1992. A new species of the dinoflagellate genus Areoligera Lejuene–Carpentier from the Late Paleocene of the eastern British Isles. *Tertiary Research*, **14**, 25–32.

JONES, R. W. & MILTON, N. J. 1994. Sequence development during uplift: Palaeogene stratigraphy and relative sea-level history of the Outer Moray Firth, UK North Sea. *Marine and Petroleum Geology*, **11**, 157–165.

LAURSEN, I., FUGELLI, E. & LERVIK, K. S. 1995. Sequence stratigraphic framework of the Paleocene and Eocene successions, Block 16/1, Norwegian North Sea. *In*: STEEL, R. J., FELT, V. L., JOHANNESSEN, E. P. & MATHIEU, C. (eds) *Sequence Stratigraphy of the Nortwest European Margin*. Norwegian Petroleum Society, Special Publications, **5**, 471–481.

MARTINSEN, O. J., INDREVÆR, G., DREYER, T., MANGERUD, G., RYSETH, A. & SØYSETH, L. 1998. Slumping, Sliding, and Basin Floor Physiography: Controls on Turbidite Deposition and Fan Geometries in the Paleocene Grane Field Area, Block, 25/11, Norwegian North Sea. *In*: *AAPG 1998, Extended Abstracts*, Volume 2, 435.

MUDGE, D. C. & BUJAK, J. P. 1994. Eocene stratigraphy of the North Sea Basin. *Marine and Petroleum Geology*, **11**, 166–181.

—— & ——1996. Palaeocene biostratigraphy and sequence stratigraphy of the UK central North Sea. *Marine and Petroleum Geology*, **13**, 295–312.

—— & COPESTAKE, P. 1992. Lower Paleogene stratigraphy of the northern North Sea. *Marine and Petroleum Geology*, **9**, 287–301.

MUTTI, E. 1992. *Turbidite Sandstones*. AGIP/University of Parma Special Publication.

NADIN, P. A. & KUSZNIR, N. J. 1995. Paleocene uplift and Eocene subsidence in the northern North Sea basin from 2D forward and reverse stratigraphic modelling. *Journal of the Geological Society, London*, **152**, 833–848.

NAGY, J., KAMINSKI, M. A., JOHNSEN, K. & MITLEHNER, A. G. 1997. Foraminiferal, palynomorph and diatom biostratigraphy and paleoenvironments of the Torsk Formation: A reference section for the Paleocene–Eocene transition in the western Barents Sea. *In*: HASS, H. C & KAMINSKI, M. A. (eds) *Contributions to the Micropaleontology and Paleoceanography of the Northern North Atlantic*. Grzybowski Foundation, Special Publication, **5**, 15–38.

NEAL, J. E. 1996. A summary of Paleogene sequence stratigraphy in northwest Europe and the North Sea. *In*: KNOX, R. W., CORFIELD, R. M. & DUNAY, R. E. (eds) *Correlation of the Early Paleogene in Northwest Europe*. Geological Society, London, Special Publications, **101**, 15–42.

POSAMENTIER, H. W., ALLEN, G. P., JAMES, D. P. & TESSON, M. 1992. Forced regressions in a sequence stratigraphic framework: concepts, examples, and exploration significance. *American Association of Petroleum Geologists Bulletin*, **76**, 1687–1709.

POWELL, A. J. 1988. A modified dinoflagellate cyst biozonation for latest Paleocene and earliest Eocene sediments from the Central North Sea. *Review of Paleobotany and Palynology*, **56**, 322–344.

——1992. Dinoflagellate cysts of the Tertiary System. *In*: POWELL, A. J. (ed.) *A Stratigraphic Index of Dinoflaggelate Cysts*. British Micropaleontological Society Publication Series. Chapman & Hall, London, 155–251.

——, BRINKHUIS, H. & BUJAK, J. P. 1996. Upper Paleocene–Lower Eocene dinoflagellate cyst sequence biostratigraphy of southeast England. *In*: KNOX, R. W., CORFIELD, R. M. & DUNAY, R. E. (eds) *Correlation of the Early Paleogene in Northwest Europe*. Geological Society, London, Special Publications, **101**, 145–183.

STEIN, J. A., GAMBER, J. H., KREBS, W. N. & LA COE, M. K. 1995. A composite standard approach to biostratigraphic evaluation of the North Sea Paleogene. *In*: STEEL, R. J., FELT, V. L., JOHANNESSEN, E. P. & MATHIEU, C. (eds) *Sequence Stratigraphy of the Northwest European Margin*. Norwegian Petroleum Society, Special Publications, **5**, 401–414.

SØYSETH, L. 1998. *High resolution biostratigraphy in well 25/11-17, Granc Field, North Sea*. Thesis, University of Bergen.

TIMBRELL, T. 1993. Sandstone architecture of the Balder Formation depositional system, UK Quadrant 9 and adjacent areas. *In*: PARKER, J. R. (ed.) *Petroleum Geology of Northwest Europe: Proceedings of the 4th Conference*. Geological Society, London, 107–121.

VAN WAGONER, J. C., POSAMENTIER, H. W., MITCHUM, R. M., VAIL, P. R., SARG, J. F., LOUTIT, T. S. & HARDENBOL, J. 1988. An overview of the fundamentals of sequence stratigraphy and key definitions. *In*: WILGUS, C. K. *et al.* (eds) *Sea-level Changes – An Integrated Approach*. Society of Economic Paleontologists and Mineralogists, Special Publications, **42**, 40–45.

Forties Field (North Sea) revisited: a demonstration of the value of historical micropalaeontological data

R. W. JONES

Reservoir Description Team, BP Exploration Operating Company Ltd, Chertsey Road, Sunbury-on-Thames, Middlesex TW16 7LN, UK

Abstract: The history of the discovery, appraisal and development of the Forties Field and the evolution of the reservoir model are briefly reviewed.

The process of integrated reservoir description (IRD) is still ongoing, iterating with the acquisition of new analytical or (re)interpreted biostratigraphic or sedimentological or other static (seismic, well) or dynamic data. IRD has already made a significant impact on the ongoing infill production drilling programme.

Even in the absence of newly acquired high-resolution palynostratigraphic and three-dimensional (3D) seismic data, historical high-resolution micropalaeontological data can be used to identify eight main sands in the reservoir system, separated vertically and/or laterally by shales (production barriers).

Historical micropalaeontological and sedimentological data can also be used to refine the interpretation of the depositional environment of the reservoir section. It indicates the development of both high-density channel-axis turbidite and low-density channel and overbank turbidite facies in pay zones, and interturbidite facies (with low potential as production barriers) and hemipelagite facies (with high potential as production barriers) in non-pay zones. The progressive elimination of infaunal and epifaunal benthonic foraminifera during the time the reservoir section was deposited indicates a progressive deterioration of the physics or chemistry of the benthonic environment (ultimately resulting in the development of a non-bioturbated laminite facies).

The Forties Field is located in 104–128 m of water, some 180 km east–northeast of Aberdeen in BP Licence Block 21/10 (main part of field) and Shell–Esso Block 22/6a (southeastern extension) in the United Kingdom continental shelf (UKCS) sector of the North Sea (Fig. 1).

It was discovered in October 1970 by well 21/10-1, which was drilled on the crest of an elongate anticline (mapped on regional 2D seismic) with approximately 180 m of closure at 'Base Tertiary' level, and which encountered oil in Forties Formation sands (see below) of Late Palaeocene age at a depth of approximately 2130 mbrt, and an oil–water contact at 2250 mbrt (2217 mss). The structure resulted from a combination of sedimentary drape, syn-sedimentary faulting and compaction over the Forties–Montrose Ridge.

Following the acquisition of 2D seismic data over a denser grid, four appraisal wells were drilled in 1971–1972. The appraisal wells demonstrated the existence of a giant oil field with an area of approximately 90 km^2, and with STOIIP of approximately 4.34 billion barrels. Estimates of oil in place have decreased to approximately 4.00 billion barrels with the acquistion and interpretation of new reservoir geological and engineering data since the initial appraisal. However, the projected recovery factor has increased from 40 to 57% (with bottom-water drive, artificial lift and the application of various enhanced oil recovery techniques such as surfactant flooding), resulting in an actual growth in ultimate recoverable reserves to approximately 2.47 billion barrels. Carmalt & St John (1986) quoted the recoverable reserves as 2.00 billion barrels, making Forties the 121st largest oil field in the world and the fifth largest (behind Troll, Statfjord, Brent and Ekofisk) in the North Sea (in terms of recoverable reserves of oil equivalent): a figure of 2.47 billion barrels would have made it the 95th largest in the world. Weimer & Link (1991) also quoted the recoverable reserves as 2.00 billion barrels, making Forties the equal sixth largest oil field in the world and the largest in the North Sea containing a turbiditic reservoir: a figure of 2.47 billion barrels would have made it the fourth largest in the world containing a turbiditic reservoir.

The development plan was put into operation between 1972 and 1975, during which time four fixed platforms (Alpha, Bravo, Charlie and

JONES, R. W. 1999. Forties Field (North Sea) revisited: a demonstration of the value of historical micropalaeontological data. *In:* JONES, R. W. & SIMMONS, M. D. (eds) *Biostratigraphy in Production and Development Geology*. Geological Society, London, Special Publications, **152**, 185–200.

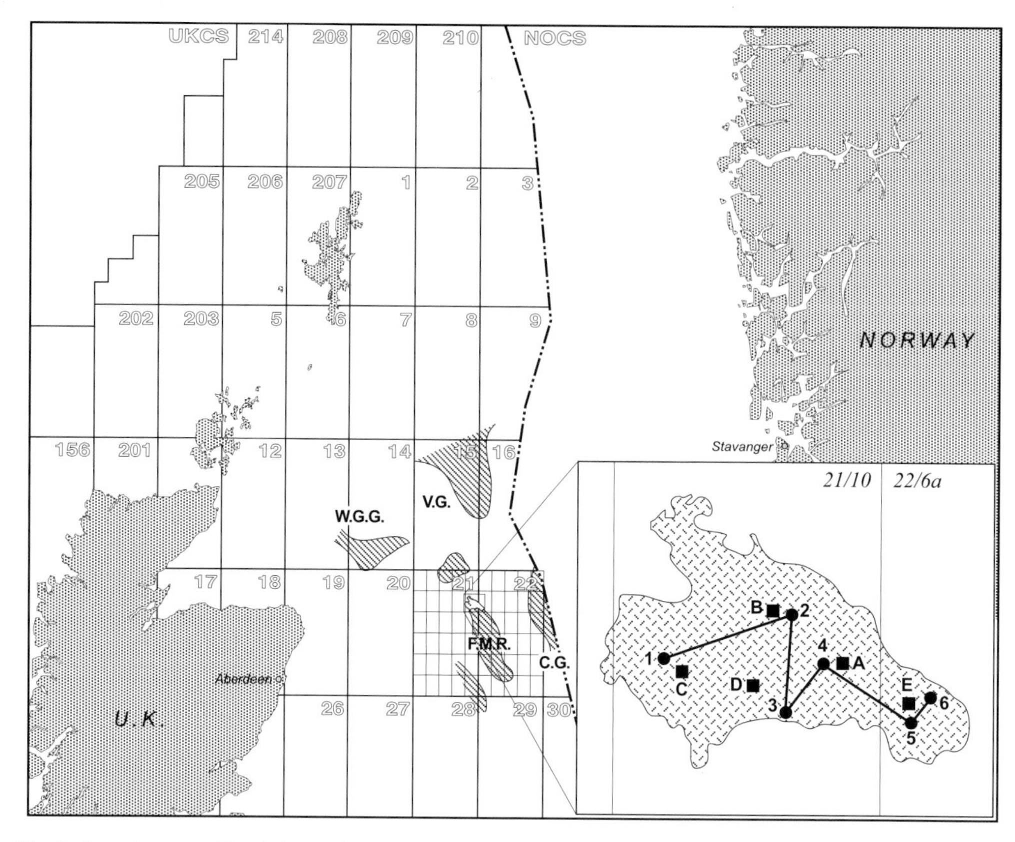

Fig. 1. Location map, North Sea. Principal structural elements also indicated (highs are shaded, lows are unshaded): C.G., Central Graben; F.M.R., Forties–Montrose Ridge; V.G., Viking Graben; W.G.G., Witch-Ground Graben. UKCS, United Kingdom continental shelf; NOCS, Norwegian continental shelf. Median line dashed. Numbers (1–30) UKCS quadrant numbers. Each quadrant divided into 30 blocks (denoted 21/10, 22/6a, etc.), each measuring 15 × 10 km. *Inset*: scale bar, 1 km; A–E, production platforms (A, Alpha; B, Bravo; C, Charlie; D, Delta; E, Echo); 1–6, cored wells (1, 21/10-FC22, 2, 21/10-FB55, 3, 21/10-5, 4, 21/10-FA12, 5, 22/6a-1, 6, 22/6a-2); solid line, line of correlation (Figs 3 and 4).

Delta) with facilities for handling 108 wells, producing 500 000 barrels of oil per day, and injecting 600 000 barrels of water per day, were installed in the main part of the field. A fifth (Echo) was added on Southeast Forties in 1986. Development drilling commenced in June 1975, production drilling in September of that year (and in Southeast Forties in March 1987). A water-injection programme was initiated in 1976. Oil production reached a plateau of 500 000 barrels per day in 1981. As of the beginning of 1996, cumulative production had totalled approximately 2.3 billion barrels of oil (92% of the recoverable reserves).

The life of the field is now at a late (decline management) stage, although production of the remaining 0.2 billion barrels of oil (8% of the recoverable reserves) is projected to extend well into the new millennium. Even at this late stage, the processes of integrated reservoir description (IRD) and reservoir simulation are still ongoing, iterating with the acquisition of new analytical or (re)interpreted biostratigraphic or sedimentological or other static (3D seismic, well) or dynamic data. IRD has already made a significant impact on the ongoing infill production drilling programme (in assisting with the siting of injector and producer wells such as to maximize sweep efficiency).

The value of the newly acquired high-resolution palynological biostratigraphic data (including its potential 'biosteering' applications) will be discussed elsewhere (Payne *et al.* 1999). This paper attempts to demonstrate the value of the historical high-resolution micropalaeontological biostratigraphic data. It also briefly reviews the evolution of the reservoir model. Readers interested in further details are referred to publications by Thomas *et al.* (1974), Walmsley (1975), Hillier *et al.* (1978), Hill & Wood (1980), Carman & Young (1981), Johnson & Stewart (1985), Kulpecz & van Geuns (1990) and Wills & Peattie (1990).

Historical reservoir model

Stratigraphy

Historical work showed that most of the oil in the Forties Field occurs in sands of the 'Forties Formation' of Late Palaeocene age, with a subordinate amount (especially in the southeastern extension (Southeast Forties)) in the Middle Eocene part of the Hordaland Group (Fig. 2).

Historical subdivision of the 'Forties Formation' reservoir was based on integration of static data (high-resolution (semi)quantitative biostratigraphic (micropalaeontological and palynological) data (principally from closely spaced samples from perceived key cored wells (e.g. 21/10-FC22), sedimentological data (also from cored wells), 2D seismic data (including 1980–1981 and 1986 vintage) and well data (calibrated by synthetic seismograms)) and dynamic (reservoir engineering (production barrier)) data. The earliest subdivision was four fold (into the 'Forties Shale' (strictly part of the Andrew Formation), 'Forties Sand', 'Charlie Shale' and 'Charlie Sand' (strictly part of the Sele Formation)). This was later succeeded by a nine-fold subdivision (into Units C–D (approximately equivalent to the 'Forties Shale'), E–F ('Forties Sand'), H ('Charlie Shale') and J–M ('Charlie Sand' in the main field; unconnected sand in Southeast Forties) (see also Payne *et al.* 1998, Figs 8 and 9)).

Historical correlation of reservoir units in wells lacking biostratigraphic control was essentially based on lithostratigraphy (i.e. correlation of sand with sand and shale with shale (assuming comparatively little lateral heterogeneity and complexity)), and is demonstrably wrong in places.

Depositional environment

Despite some contrary views, the consensus of historical opinion (based on regional geology and geophysics, and micropalaeontological and sedimentological work undertaken on cored wells) envisaged the Forties reservoir(s) as having been deposited in structurally controlled stacked NW–SE-oriented channels within a submarine fan sourced from uplifted areas in the hinterland.

Current reservoir model

The process of integrated reservoir description (IRD) is still ongoing, iterating with the acquisition of new analytical or (re)interpreted biostratigraphic or sedimentological or other static (3D seismic, well) or dynamic data. IRD iterating all currently available data (including historical high-resolution micropalaeontological biostratigraphic data and newly acquired high-resolution palynological biostratigraphic data (focusing on pollen and spores as against dinoflagellates) and 3D seismic data (1988 vintage)) has led to the recognition of a number of flaws in historical layering schemes (mis-ties in historical correlations, etc.) and of a higher degree of lateral heterogeneity and complexity (i.e. a lower degree of sand-body connectivity) than previously envisaged. IRD has already made a significant impact on the ongoing infill production drilling programme, in assisting with the siting of injector and producer wells such as to maximize sweep efficiency. It could also lead to the recognition, quantification and recovery of additional reserves of incremental oil.

The value of the newly acquired high-resolution palynological biostratigraphic data will be discussed elsewhere (Payne *et al.* 1999, fig. 10). The value of the historical high-resolution micropalaeontological biostratigraphic data is discussed below.

Value of historical micropalaeontological data

Stratigraphic interpretation

The micropalaeontological and palynological biostratigraphic zonation schemes used in the subdivision of the Palaeocene–Eocene of the North Sea, and (with modifications necessitated by the recognition of intervening barren intervals) in the subdivision of the reservoir interval in Forties Field, are shown in Fig. 2. The calibration of the zones against the absolute chronostratigraphic time scale is discussed in detail in Appendix A. The definitions of those recognized over the principal (Palaeocene (Forties Formation/sequence)) reservoir interval are given in Appendix B.

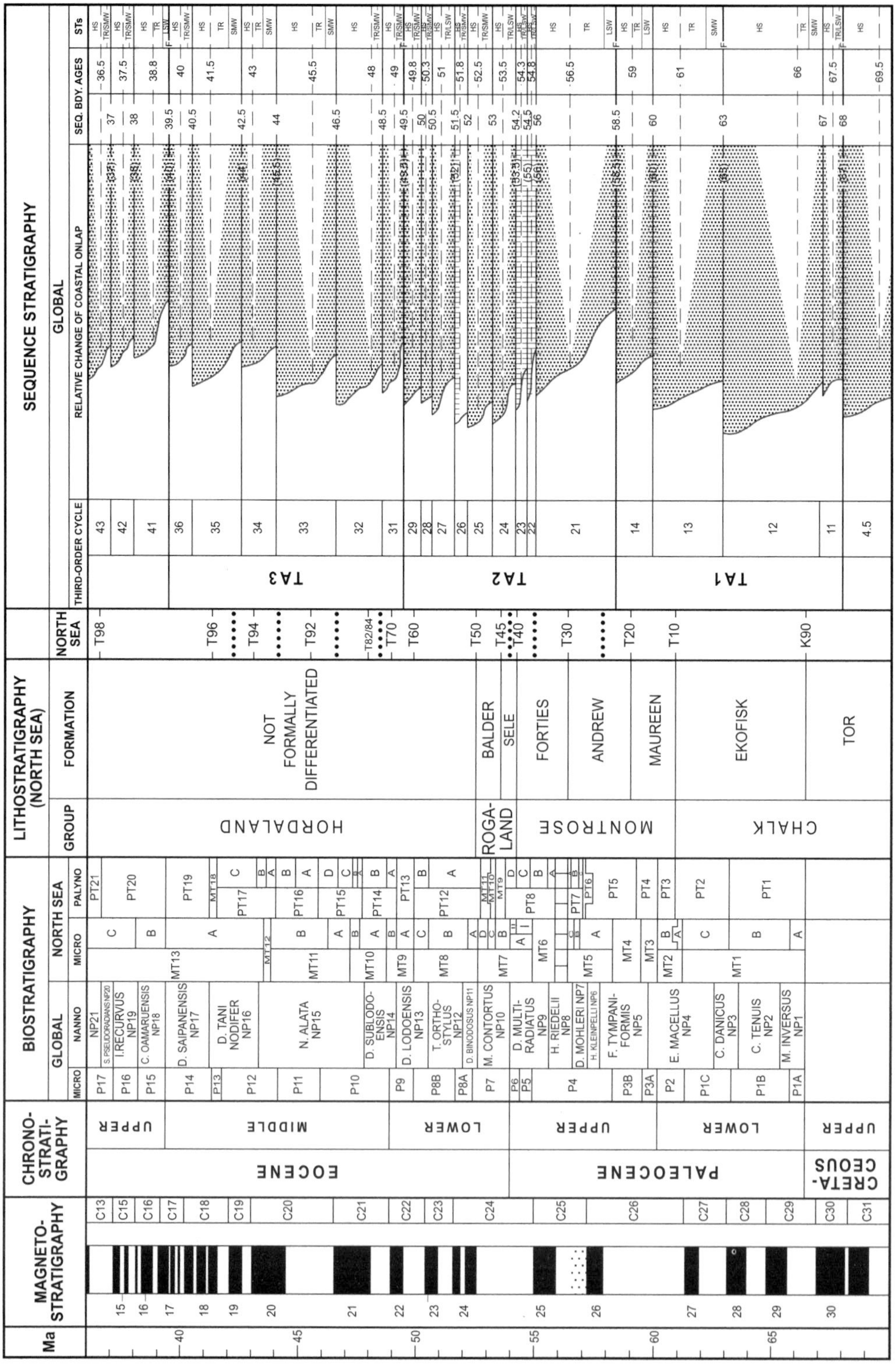

Ma: 40, 45, 50, 55, 60, 65

MAGNETO-STRATIGRAPHY

Anomalies: 15, 16, 17, 18, 19, 20, 21, 22, 23, 24, 25, 26, 27, 28, 29, 30

Chrons: C13, C15, C16, C17, C18, C19, C20, C21, C22, C23, C24, C25, C26, C27, C28, C29, C30, C31

CHRONO-STRATIGRAPHY

EOCENE: UPPER, MIDDLE, LOWER

PALEOCENE: UPPER, LOWER

CRETACEOUS: UPPER

BIOSTRATIGRAPHY

GLOBAL – MICRO: P17, P16, P15, P14, P13, P12, P11, P10, P9, P8B, P8A, P7, P6, P5, P4, P3B, P3A, P2, P1C, P1B, P1A

GLOBAL – NANNO: NP21; S. PSEUDORADIANS NP20; I. RECURVUS NP19; C. OAMARUENSIS NP18; D. SAIPANENSIS NP17; D. TANI NODIFER NP16; N. ALATA NP15; D. SUBLODOENSIS NP14; D. LODOENSIS NP13; T. ORTHOSTYLUS NP12; D. BINODOSUS NP11; M. CONTORTUS NP10; D. MULTIRADIATUS NP9; H. RIEDELII NP8; D. MOHLERI NP7; H. KLEINPELLI NP6; F. TYMPANIFORMIS NP5; E. MACELLUS NP4; C. DANICUS NP3; C. TENUIS NP2; M. INVERSUS NP1

NORTH SEA – MICRO: MT13 (C, B, A); MT12; MT11 (B, A); MT10 (B, A); MT9 (B, A); MT8 (C, B, A); MT7 (D, C, B, A, I, II); MT6; MT5 (C, B, A); MT4; MT3; MT2 (B, A); MT1 (C, B, A)

NORTH SEA – PALYNO: PT21; PT20; PT19; MT18; PT17 (C, B, A); PT16 (B, A); PT15 (D, C, B, A); PT14 (B, A); PT13; PT12 (B, A); MT11; MT10; MT9; PT8 (D, C, B, A); PT7 (B); PT6; PT5; PT4; PT3; PT2; PT1

LITHOSTRATIGRAPHY (NORTH SEA)

GROUP	FORMATION
HORDALAND	NOT FORMALLY DIFFERENTIATED
ROGALAND	BALDER
ROGALAND	SELE
MONTROSE	FORTIES
MONTROSE	ANDREW
MONTROSE	MAUREEN
CHALK	EKOFISK
CHALK	TOR

NORTH SEA: T98, T96, T94, T92, T82/84, T70, T60, T50, T45, T40, T30, T20, T10, K90

SEQUENCE STRATIGRAPHY – GLOBAL

THIRD-ORDER CYCLE: 43, 42, 41; TA3: 36, 35, 34, 33, 32, 31; TA2: 29, 28, 27, 26, 25, 24, 23, 22, 21; TA1: 14, 13, 12, 11; 4.5

RELATIVE CHANGE OF COASTAL ONLAP: (37), (38), (40), (44), (46.5), (49.5), (52), (53.5), (55), (56), (58.5), (60), (63), (67)

SEQ. BDY. AGES: 37, 38, 39.5, 40.5, 42.5, 44, 46.5, 48.5, 49.5, 50, 50.5, 51.5, 52, 53, 54.2, 54.5, 56, 58.5, 60, 63, 67, 68

Ages: 36.5, 37.5, 38.8, 40, 41.5, 43, 45.5, 48, 49, 49.8, 50.3, 51, 51.8, 52.5, 53.5, 54.3, 54.8, 56.5, 59, 61, 66, 67.5, 69.5

STs: HS, TR/SMW, HS, TR/SMW, HS, TR, LSW, HS, TR/SMW, HS, TR, SMW, HS, TR, SMW, HS, TR, SMW, HS, TR/SMW, HS, TR/SMW, HS, TR/SMW, HS, TR/LSW, HS, TR/SMW, HS, TR/SMW, HS, TR/LSW, HS, TR/LSW, TR/LSW, HS, TR, LSW, HS, TR, LSW, HS, TR, SMW, HS, TR, SMW, HS, TR/LSW, HS

Even in the absence of newly acquired high-resolution palynostratigraphic and 3D seismic data, historical high-resolution micropalaeontological data can be used to identify eight main sands in the 'Forties Formation' reservoir system, separated vertically and/or laterally by shales (production barriers) (Figs 3 and 4). In approximate ascending stratigraphic order, they are: the 'Unassigned Sands' ('2' and '1'); the 'Main Sand'; the 'Alpha–Bravo Sands' ('2' and '1'); the 'Charlie Sand'; and the 'Echo Sands' ('2' and '1') of Southeast Forties.

'Unassigned Sands'. The 'Unassigned Sands' ('2' and '1') are best developed in crestal parts of Forties and in Southeast Forties. They fall within microzone MT5 and the barren section between MT5 and MT6, respectively, and are thus approximately age-equivalent to Unit D (as identified in key well 21/10-FC22).

'Main Sand'. The 'Main Sand' is more or less ubiqitously developed, but shale-prone in 22/6a-1 in Southeast Forties. It falls within microzone MT6 and the barren section between MT6 and MT7, and is thus approximately age-equivalent to Units E and F and the lower part of Unit H (formerly Unit G) (in 21/10-FC22).

'Alpha-Bravo Sands'. The 'Alpha–Bravo Sands' ('2' and '1') are best developed around the Alpha and Bravo platforms of Forties. They fall within microzone MT7a(i) and the barren section between MT7a(i) and MT7a(ii), respectively, and are thus approximately age-equivalent to the upper part of Unit H and the lower part of Unit J (in 21/10-FC22).

'Charlie Sand'. The 'Charlie Sand' is best developed around the Charlie Platform of Forties. It falls within microzone MT7a(ii), and is thus approximately age-equivalent to the upper part of Unit J (in 21/10-FC22).

'Echo Sands'. The 'Echo Sands' ('2' and '1') are best developed around the Echo Platform of Southeast Forties. They fall within the barren section between microzones MT7a(i) and MT7a(ii) and within MT7a(ii), respectively, and are thus age-equivalent to the 'Alpha–Bravo (1)' and 'Charlie' sands of the main, respectively, but separated laterally from them by shales.

After many years of production and bottom-up water flooding most of the recoverable reserves of the comparatively homogeneous 'Unassigned' and 'Main' sands are likely to have been recovered, and most of the remaining recoverable reserves are likely to reside within the more heterogeneous (stratigraphically compartmentalized) 'Alpha–Bravo', 'Charlie' and 'Echo' sands.

Environmental interpretation

Refined, essentially micropalaeontologically based, interpretations of the stratigraphy and depositional environments of one well from the main part of Forties Field (21/10-FC22) and one from the southeastern extension (Southeast Forties) (22/6a-E9) are summarized in Figs 5 and 6.

Integrated micropalaeontological and sedimentological evidence indicates that Biofacies/Lithofacies A (Lithofacies A–B of Wills & Peattie 1990), characterized by an absence of benthonic foraminifera, represents a high-density (high net : gross) channel-axis turbidite (?debris flow (Shanmugam 1996)) facies. Biofacies/Lithofacies B (Lithofacies C–E of Wills & Peattie 1990), named after the most characteristic foraminiferal 'morphogroup' in the sense of Jones & Charnock (1985) (B (epifaunal)), represents a low-density (low net : gross) channel and overbank turbidite and interturbidite facies. Biofacies/Lithofacies C (Lithofacies G–N of Wills & Peattie 1990), also named after the

Fig. 2. Stratigraphic summary chart, North Sea (Palaeocene–Eocene). Global magneto-, chrono-, bio- and sequence stratigraphy modified after Haq *et al.* (1987); North Sea bio-, litho- and sequence stratigraphy follows BP usage. Stipple in North Sea sequence stratigraphy column indicates reservoir sand development. *Biostratigraphy* – North Sea biostratigraphic zones based on micropalaeontology (agglutinated, calc-agglutinated, calcareous benthonic and planktonic foraminifera, diatoms and radiolarians) and palynology (essentially dinoflagellates (high-resolution pollen and spore event palynostratigraphy over Forties Formation/sequence reservoir interval in the Forties Field not shown)). Notes on calibration of zones against absolute chronostratigraphic time scale are given in Appendix 1. Notes on definition of zones over Forties Formation/sequence reservoir interval are given in Appendix 2. *Sequence stratigraphy* – sequence boundaries (flooding surfaces) identified on seismic and/or wireline log evidence. All appear to correlate with global (?glacio-eustatically mediated) third-order flooding surfaces (Haq *et al.* 1987). Confident calibrations of flooding surfaces within the Forties Formation/sequence are impossible due either to insufficient biostratigraphic resolution or to tectonic rather than glacio-eustatic control on sequence stratigraphy and sedimentation.

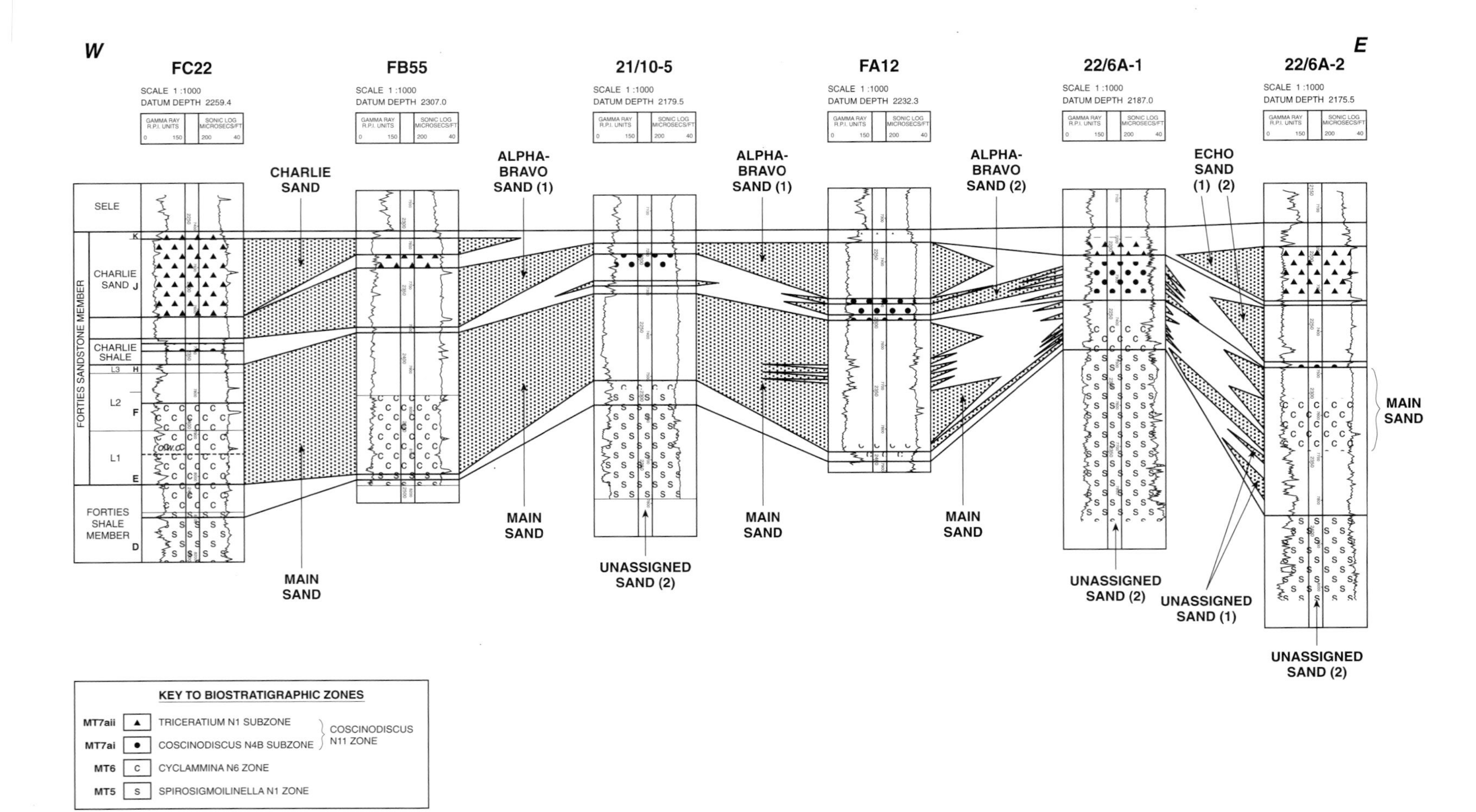

Fig. 3. W–E stratigraphic correlation across the Forties Field. Line of correlation shown in Fig. 1. Scale 1:1000. Datum 'Base Forties' (as defined on composite log). Stipple indicates sand development.

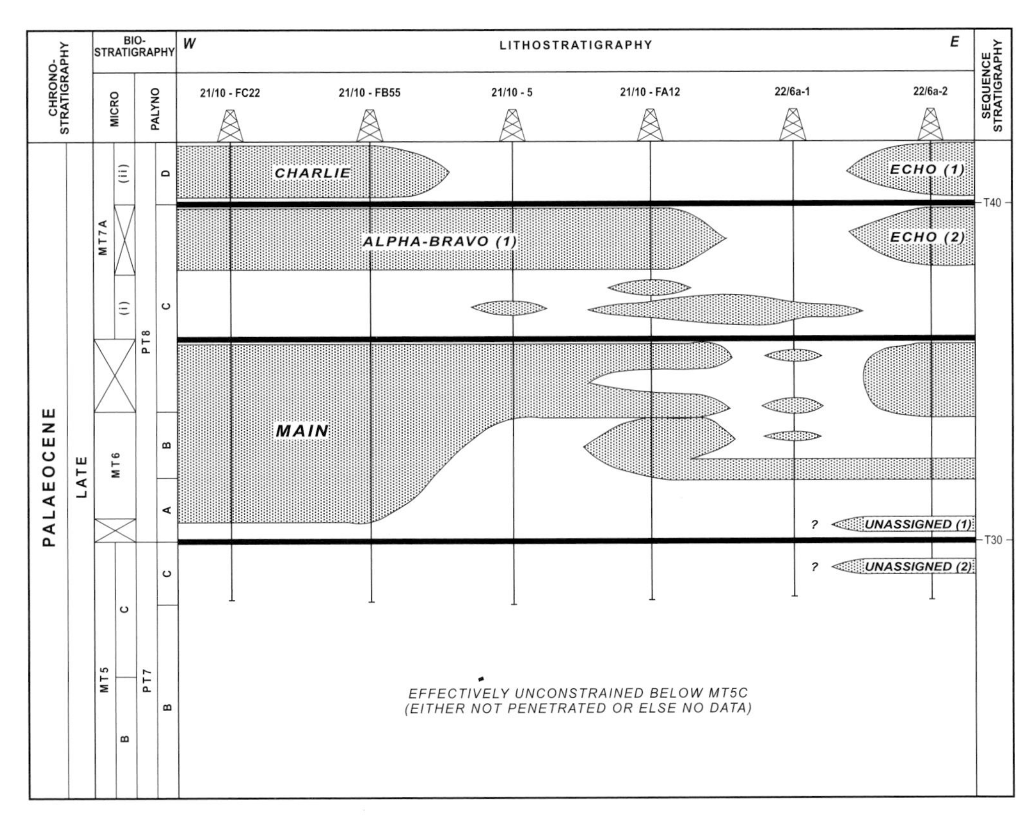

Fig. 4. Chronostratigraphic cross-section of the Forties Field. Line follows Fig. 3. Stipple indicates sand development.

most characteristic 'morphogroup' in the sense of Jones & Charnock (1985) (C (infaunal)), represents a basin-plain hemipelagite facies characterized by pervasive bioturbation (questionably foraminiferal in origin (compare the 'vermicelli' bioturbation of Kaminski *et al.* 1988)). Biofacies/Lithofacies D (Lithofacies G of Wills & Peattie 1990), characterized by an absence of benthonic foraminifera but an abundance of planktonic diatoms (and dinoflagellates), represents a basin-plain laminite facies. In regard to reservoir quality, field-wide, the high-density turbidites of Biofacies/Lithofacies A have mean porosities of the order of 27% and permeabilities of 300 mD, the low-density turbidites of Biofacies/Lithofacies B mean porosities of 23–25% and permeabilities of 100–200 mD. The interturbidites of Biofacies/Lithofacies B the hemipelagites of Biofacies/Lithofacies C and the laminites of Biofacies/Lithofacies D have very low mean porosities and permeabilities. Depositional facies appears to be the principal control on reservoir quality, with diagenetic effects exerting little influence. *In regard to production, it is worth emphasizing that essentially micropalaeontological evidence enables ready recognition of hemipelagic shales (Biofacies/Lithofacies C), with high potential as production barriers, as against interturbiditic shales (Biofacies/Lithofacies B), with low potential as production barriers* (see below; see also Holmes 1999).

MT5. At this time, the benthonic environment favoured the sustenance and/or preservation of non-calcareous agglutinating benthonic foraminifera at the expense of their calcareous counterparts, leading to the development of agglutinate-dominated assemblages ('*Rhabdammina*' or 'flysch-type' faunas) (while, interestingly, the pelagic environment favoured the proliferation of dinoflagellates at the expense of diatoms). Conditions that are thought to favour the development of agglutinate-dominated benthonic microfaunas (see, for instance, Jones 1996, and references therein) include turbiditic and associated sedimentation in bathyal to abyssal environments, and reduced oxygen concentration at the sediment surface (the latter

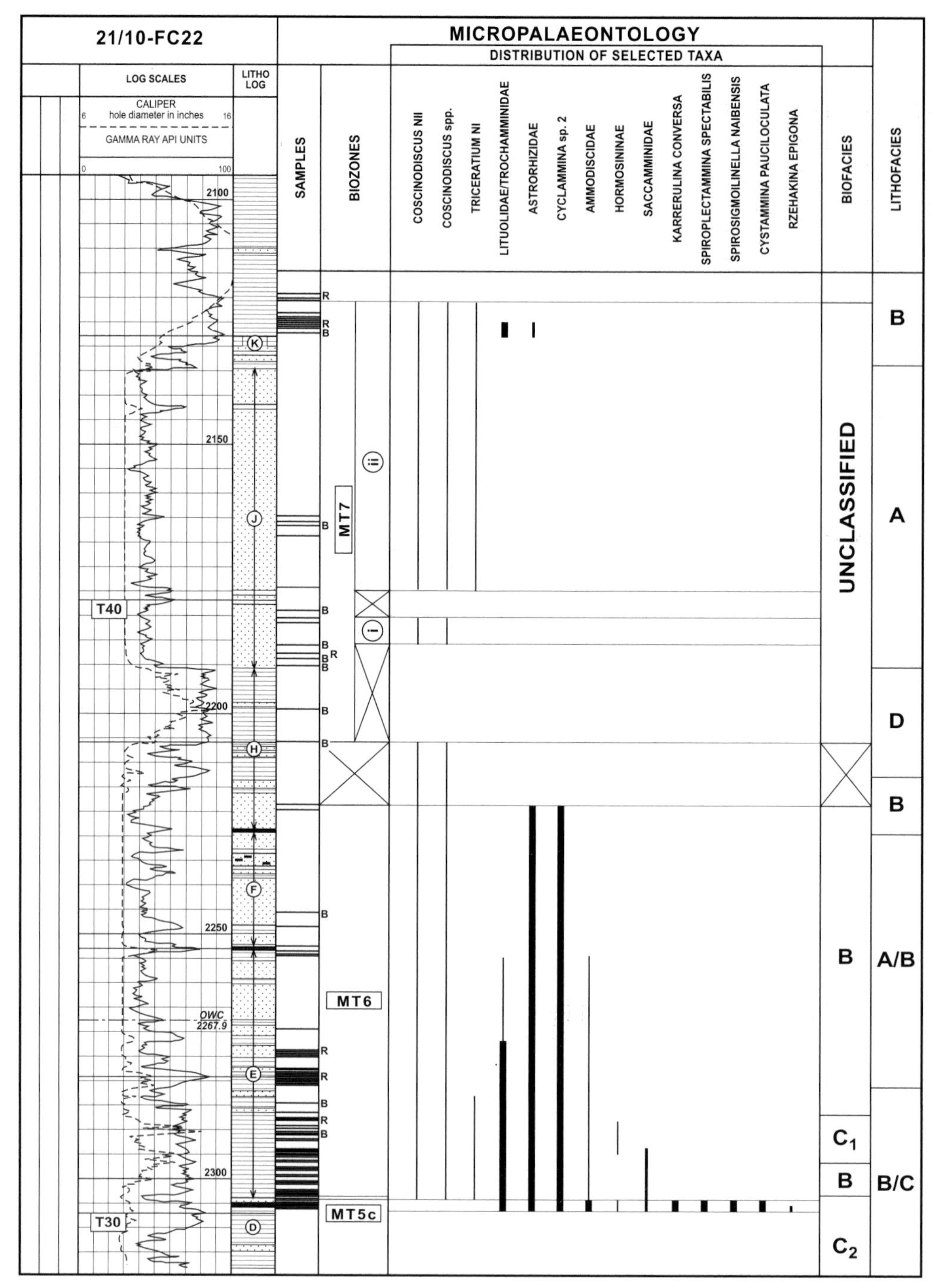

Fig. 5. Stratigraphic and palaeoenvironmental interpretation, 21/10-FC22. *Samples (core)* – B indicates barren sample; R indicates reworking. *Biozones* – see Fig. 2. *Distributions of selected taxa* – *Coscinodiscus* N11, *C.* spp. and *Triceratium* N1 diatoms, remainder agglutinating foraminifera. Taxonomy of agglutinating foraminifera follows Charnock & Jones (1990). *Cyclammina* sp. 1 equivalent to *C.* N6, *Karrerulina conversa* to *Karreriella subeocaena* and *Spirosigmoilinella naibensis* to *S.* N1 of former BP usage. Narrow bar indicates present (one–four specimens per sample), wide bar common to abundant (more than five specimens per sample). *Biofacies* – letter denotes most advanced 'morphogroup'. Classification follows Jones & Charnock (1985). *Lithofacies* – A, high-density turbidites; B, low-density turbidites and interturbidites (some debris flows and slumps); C, hemipelagites; D, laminites.

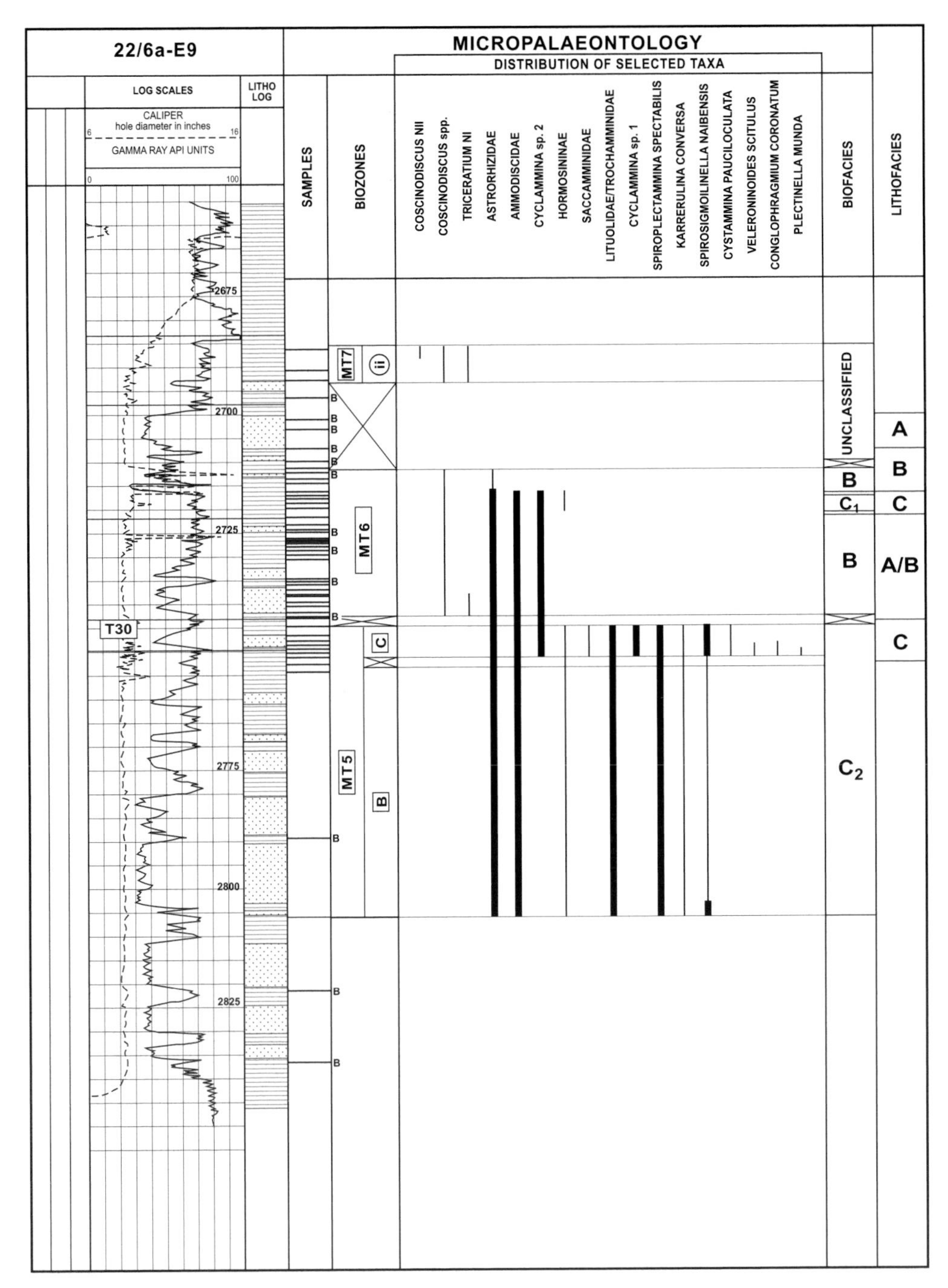

Fig. 6. Stratigraphic and palaeoenvironmental interpretation, 22/6a-E9. *Samples (core)* – B indicates barren sample. *Biozones* – see Fig. 2. *Distributions of selected taxa* – *Coscinodiscus* N11, *C.* spp. and *Triceratium* N1 diatoms, remainder agglutinating foraminifera. Taxonomy of agglutinating foraminifera essentially follows Charnock & Jones (1990), except that their *Labrospira scitula* is referred to *Veleroninoides scitulus*. *Conglophragmium coronatum* equivalent to *Glomospirella* N2, *Cyclammina* sp. 1 to *C.* N6, *C.* sp. 2 to *C.* N5, *Karrerulina conversa* to *Karreriella subeocaena* and *Spirosigmoilinella naibensis* to *S.* N1 of former BP usage. Narrow bar indicates taxon present (one–four specimens per sample), wide bar common to abundant (more than five specimens per sample). *Biofacies* – letter denotes most advanced 'morphogroup'. Classification follows Jones & Charnock (1985). *Lithofacies* – A, high-density turbidites; B, low-density turbidites and interturbidites (some debris flows and slumps); C, hemipelagites; D, laminites.

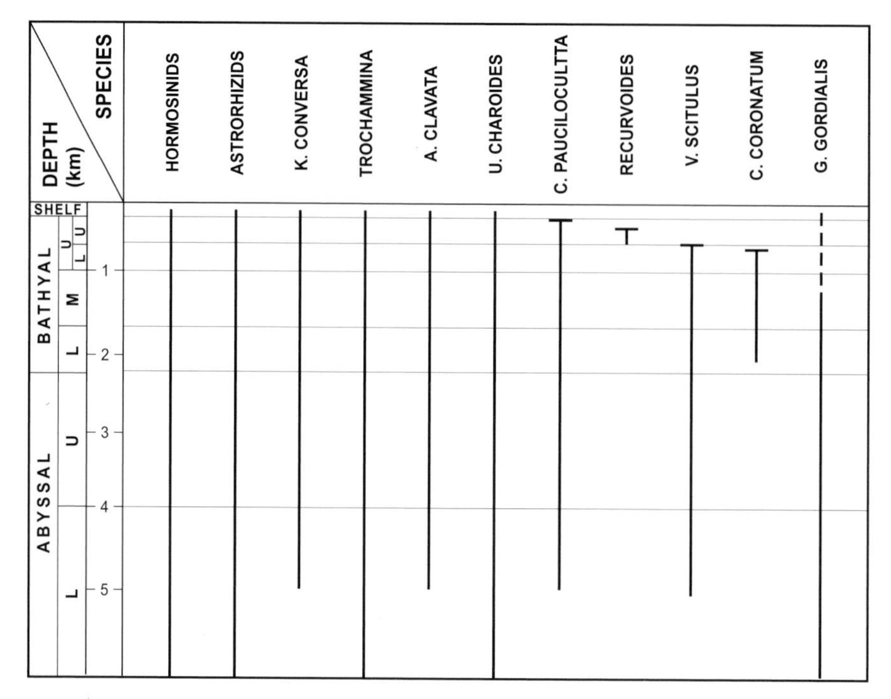

Fig. 7. Bathymetric ranges of selected agglutinating foraminifera (modified after Charnock & Jones 1990). *K. conversa, Karrerulina; A. clavata, Ammolagena; U. charoides, Usbekistania; C. pauciloculata, Cystammina; V. scitulus, Veleroninoides; C. coronatum, Conglophragmium; G. gordialis, Glomospira.* Taxonomy essentially follows Charnock & Jones (1990), except that their *Labrospira scitula* is referred to as *Veleroninoides scitulus.*

probably brought about by oxidation of rapidly produced and/or sedimented organic matter (which incidentally also liberates acids that could militate against the sustenance and/or preservation of calcareous forms), by freshwater runoff and salinity stratification, or by warming (warm water dissolving less oxygen than cool water)).

MT5b appears to have been characterized by turbiditic ('Unassigned Sand') sedimentation in parts of Southeast Forties (see below). MT5c also appears to have been characterized by turbiditic sedimentation in parts of Southeast Forties, in particular in 22/6a-7.

MT5 assemblages contains a number of agglutinating foraminifera that outside the North Sea range up to the Recent, and which therefore have known bathymetric distributions (summarized in Fig. 7). Many of these taxa have upper depth limits in the upper bathyal zone (between 200 and 1000 m) which confirm a deep-water environment.

MT5c assemblages from 21/10-FC22 fall in the 'middle–lower bathyal to abyssal' (1000–>2250 m) field on a plot of agglutinating foraminiferal 'morphogroups' A–B/C–D (as devised by Jones & Charnock 1985) (Fig. 8). In fact, on account of the particular (Northeast Atlantic) bias in the database used in the construction of the original field distributions, the 'middle-lower bathyal to abyssal' field probably represents little more than an area influenced by submarine (Barra and Donegal) fan sedimentation.

MT5c assemblages from 21/10-FC22, and MT5b and MT5c assemblages from 22/6a-E9, fall in separate fields from those from MT6 on plots of 'morphogroups' A–B–C (Figs 9 and 10). Most fall within the field interpreted as essentially hemipelagic, although the MT5b assemblages from 22/6a-E9 fall within the field interpreted as essentially interturbiditic (on the basis of plots of sedimentologically constrained agglutinating foraminiferal distribution data presented by Grun *et al.* 1964 and Hiltermann 1968). Significantly, they are characterized by comparatively high proportions of infaunal 'morphogroup' C.

MT5c assemblages from 21/10-FC22, and MT5b and MT5c assemblages from 22/6a-E9, also fall in separate fields on cross-plots of agglutinating foraminiferal abundance v. diversity (Figs 11 and 12, respectively).

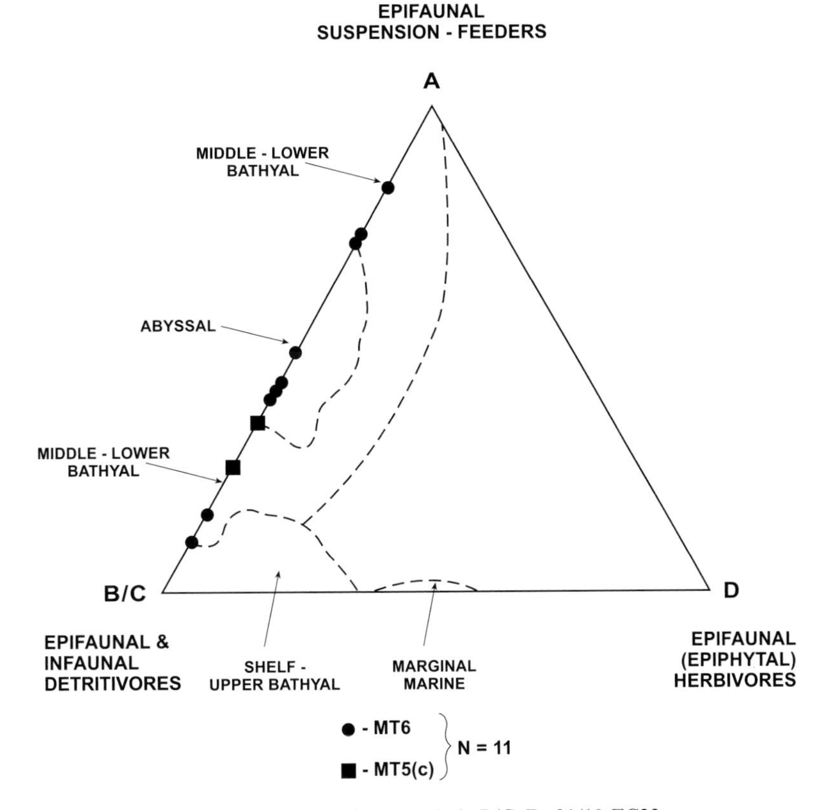

Fig. 8. Plot of agglutinating foraminiferal 'morphogroups', A–B/C–D, 21/10-FC22.

MT6. At this time, the physics or chemistry of the benthonic environment deteriorated to the point where it was only habitable by opportunistic epifaunal foraminifera. The pelagic environment favoured the proliferation of organic-walled and siliceous plankton (dinoflagellates and diatoms, respectively) at the expense of their calcareous counterparts (planktonic foraminifera and calcareous nannoplankton). The (admittedly, slightly later) proliferation of diatoms at the base of the Eocene (Balder and equivalent formations) was considered by Mitlehner (1996) to reflect eutrophication associated with seasonal (monsoon-driven) upwelling to the east (and downwelling or restricted circulation to the west) of the Ringkobing–Fyn High on the eastern margin of the Central Graben.

MT6 appears to have been characterized by a high incidence of turbiditic ('Main Sand') sedimentation in both Forties and Southeast Forties.

MT6 assemblages from 21/10-FC22 fall in the 'middle–lower bathyal to abyssal' (1000–>2250 m) field on a plot of agglutinating foraminiferal 'morphogroups' A–B/C–D (Fig. 8), which probably represents little more than an area influenced by submarine fan sedimentation.

MT6 assemblages from 21/10-FC22 and 22/6a-E9 fall in separate fields from those from MT5 on plots of 'morphogroups' A–B–C (Figs 9 and 10), and within the field interpreted as essentially interturbiditic as against hemipelagic (based on plots of sedimentologically constrained agglutinating foraminiferal distribution data presented by Grun *et al.* 1964 and Hiltermann 1968). Significantly, they are characterized by comparatively high proportions of epifaunal 'morphogroup' B.

MT6 assemblages from 21/10-FC22 and 22/6a-E9 also fall in separate fields from MT5 assemblages on cross-plots of agglutinating foraminiferal abundance v. diversity (Figs 11 and 12). They are characterized by locally higher abundance but lower diversity (i.e. higher dominance – a function of environmental stress).

MT7. At this time, the physics or chemistry of the benthonic environment deteriorated to the extent that it was no longer habitable by foraminifera. The pelagic environment generally continued to favour the proliferation of both

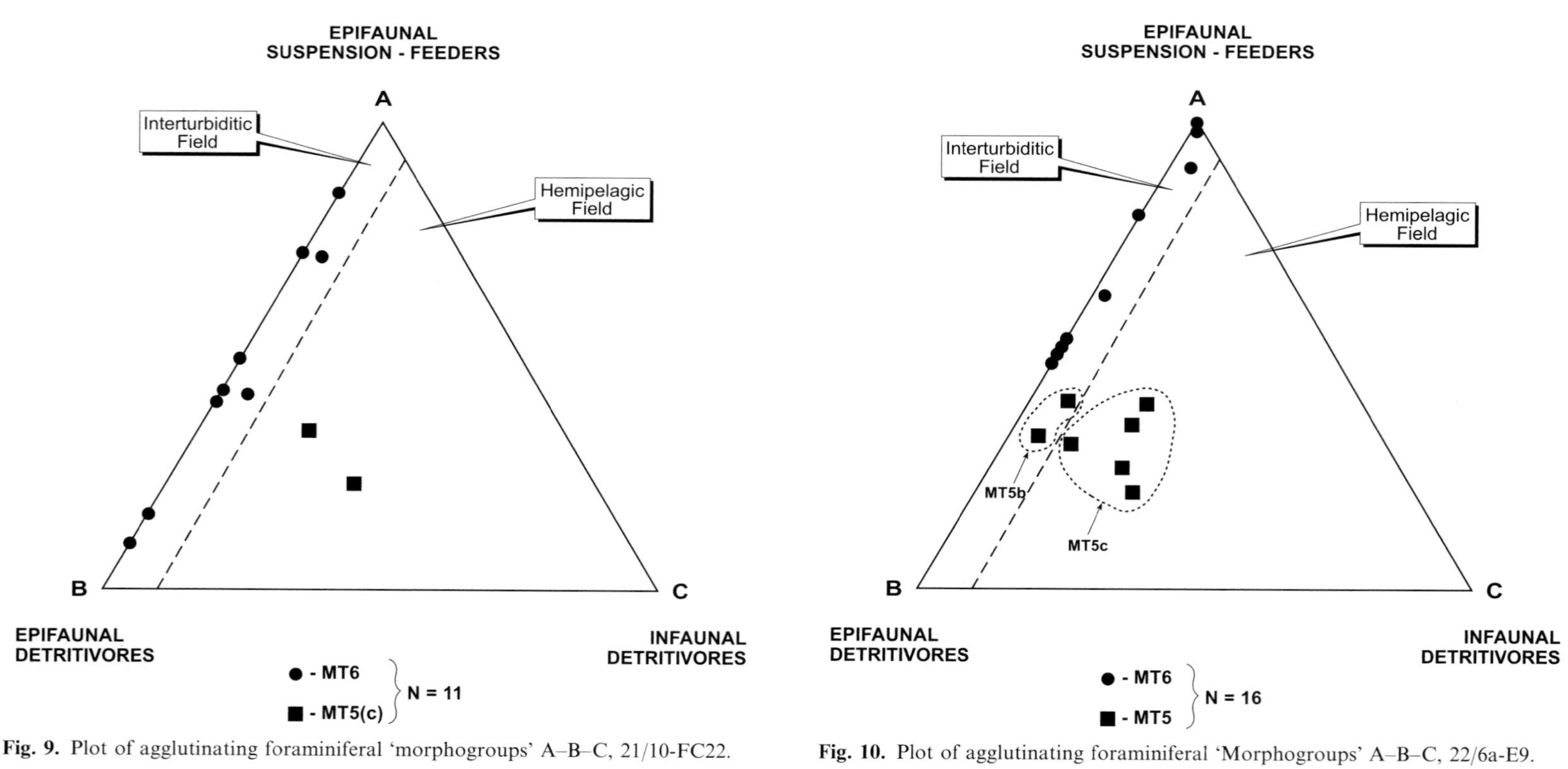

Fig. 9. Plot of agglutinating foraminiferal 'morphogroups' A–B–C, 21/10-FC22.

Fig. 10. Plot of agglutinating foraminiferal 'Morphogroups' A–B–C, 22/6a-E9.

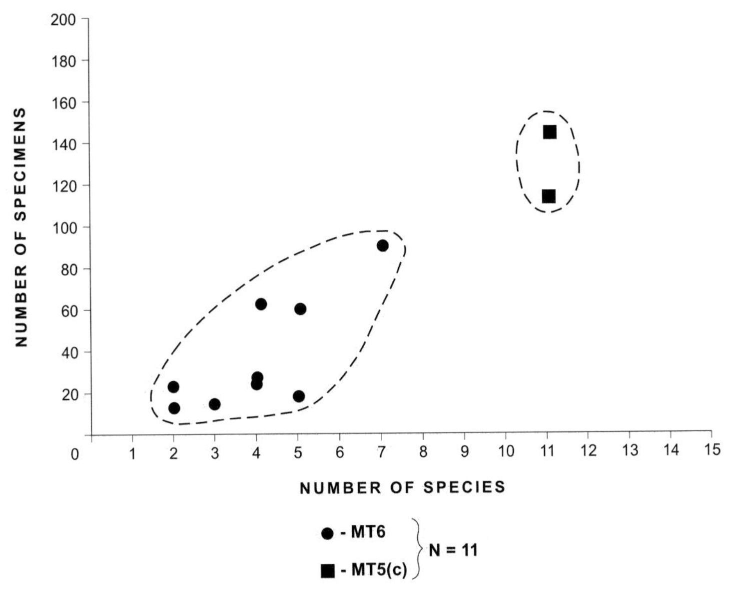

Fig. 11. Cross-plot of agglutinating foraminiferal abundance v. diversity, 21/10-FC22.

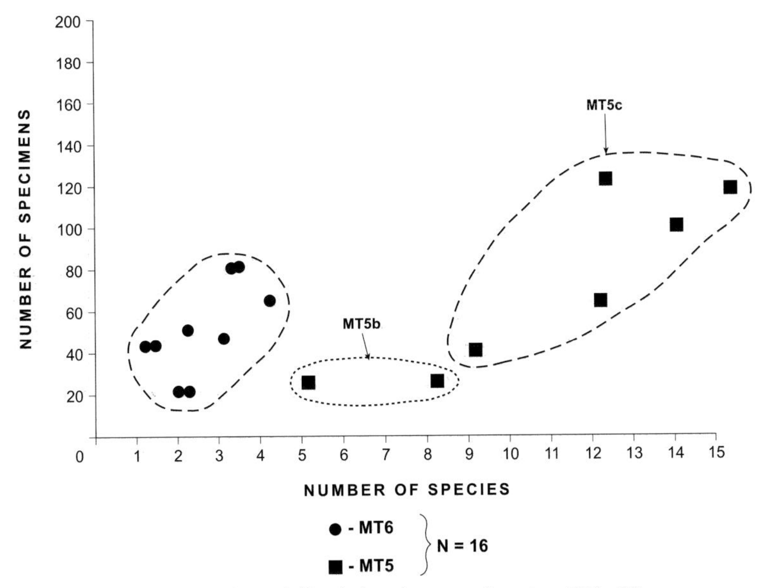

Fig. 12. Cross-plot of agglutinating foraminiferal abundance v. diversity, 22/6a-E9.

dinoflagellates and diatoms, but locally deteriorated to the extent that it was no longer habitable by diatoms.

MT7 appears to have been characterized by a high incidence of turbiditic sedimentation in both Forties ('Alpha–Bravo' and 'Charlie' sands) and Southeast Forties (unconnected 'Echo' Sands). Laminite sedimentation, reflecting the lack of a burrowing infauna, was also locally important.

Conclusions

The process of integrated reservoir description (IRD) of the 'Forties Formation' reservoir of the Forties Field (IRD) is still ongoing, iterating with the acquisition of new analytical or (re)interpreted biostratigraphic or sedimentological or other static (3D seismic, well) or dynamic data. IRD iterating all currently available data has led to the recognition of a higher degree of lateral heterogeneity and complexity (i.e. a lower degree of sand-body connectivity) than previously envisaged. IRD has already made a significant impact on the ongoing infill production drilling programme (in assisting with the siting of injector and producer wells such as to maximize sweep efficiency).

Even in the absence of newly acquired high-resolution palynostratigraphic and 3D seismic data, historical high-resolution micropalaeontological data can be used to identify eight main sands in the reservoir system, separated vertically and/or laterally by shales (production barriers). In approximate ascending stratigraphic order, they are: the 'Unassigned Sands' ('2' and '1'); the 'Main Sand'; the 'Alpha–Bravo Sands' ('2' and '1'); the 'Charlie Sand'; and the 'Echo Sands' ('2' and '1'). After many years of production, most of the remaining recoverable reserves are likely to reside within the stratigraphically compartmentalized 'Alpha–Bravo', 'Charlie' and 'Echo' sands.

Historical micropalaeontological and sedimentological data can also be used to refine the interpretation of the depositional environment of the reservoir section. It indicates the development of both high-density channel-axis turbidite and low-density channel and overbank turbidite facies in pay zones, and interturbidite facies (with low potential as production barriers) and hemipelagite facies (with high potential as production barriers) in non-pay zones. The progressive elimination of infaunal and epifaunal benthonic foraminifera during the time the reservoir section was deposited (in MT6 and MT7, respectively) indicates a progressive deterioration of the physics or chemistry of the benthonic environment (ultimately resulting in the development of a non-bioturbated laminite facies).

BP are thanked for permission to publish, and for providing word-processing and drafting facilities.

Mike Charnock of Norsk Hydro and Nick Holmes of Ichron are thanked for their constructive reviews.

Steve Davies (micropalaeontology) and Jamie Powell (palynology) generated some of the historical biostratigraphic data.

Appendix A: notes on the calibration of the North Sea biostratigraphic zones against the absolute chronostratigraphic time scale

Microzone MT13a can be calibrated against planktonic foraminiferal zone P14 of Blow (1969, 1979) (Middle Eocene) on the basis of the occurrence of *Pseudohastigerina micra*, which ranges no younger.

MT11 can be calibrated against P11 (early Middle Eocene) on the basis of the occurrence (within the zone) of *Globigerina (Subbotina) frontosa*, which ranges no younger.

MT9b can be calibrated against P9 (Early Eocene) on the basis of the ocurrence of *Globorotalia (Acarinina) pentacamerata*, which ranges no younger.

MT8b can be indirectly calibrated against calcareous nannoplankton zones NP11–NP12 of Martini (1971) (mid-Early Eocene) on the basis of a biostratigraphic correlation (by means of an abundance of *Globigerina (Subbotina)* ex gr. *linaperta)* with those parts of the London Clay Formation of the London basin independently dated as such (King's Divisions B–C) (King 1981; Aubry 1985, 1986; Knox 1990).

MT7 can be indirectly calibrated against calcareous nannoplankton zones NP9?–NP10/11 (latest Palaeocene–earliest Eocene) on the basis of a biostratigraphic correlation (by means of *Coscinodiscus* sp. 1) with the Oldhaven Sand Member of the Harwich Formation (formerly the Harwich Member (Division A1) of the London Clay Formation) and those parts of the London Clay Formation (Divisions A2–B1) of the London basin independently dated as such (King 1981; Knox 1990). It can also be indirectly calibrated against magnetostratigraphic polarity zone 24R (latest Palaeocene–earliest Eocene) by means of a biostratigraphic correlation (again by means of *Coscinodiscus* dsp. 1 (author's unpublished observations)) with the Harwich Formation in the British Geological Survey (BGS) borehole at Ormesby in Norfolk (Ali & Hailwood 1995).

MT5 can be indirectly calibrated against magnetostratigraphic polarity zones 26N–25R (Late Palaeocene) on the basis of a biostratigraphic correlation (by means of *Spirosigmoilinella naibensis* (author's unpublished observations)) with the Ormesby Clay of the Ormesby borehole (Ali & Hailwood 1995). Here, the upper part of polarity zone 25R (equivalent to the middle part of NP8) is absent.

MT2 can be calibrated against P2 (Early Palaeocene) on the basis of the occurrence of *Globigerina (Eoglobigerina) trivialis* and *Globorotalia (Planorotalites)* cf. *compressa*, which range no younger.

MT1 can be calibrated against P1 (earliest Palaeocene) on the basis of the occurrence of *Globigerina (Eoglobigerina) simplicissima* and *G.(E.) aff. trivialis*, which range no younger.

Palynozone PT20 can be calibrated against calcareous nannoplankton zone NP20 of Martini (1971) (Late Eocene) on the basis of the occurrence of *Areosphaeridium dictyoplokus*, which ranges no younger (Vinken 1988).

PT19 can be calibrated against NP17 (Middle Eocene) on the basis of the occurrence of *Rhombodinium porosum*, which ranges no younger (Powell 1992).

PT15 can be calibrated against the early part of NP15 (early Middle Eocene) on the basis of the occurrence of *Diphyes ficusoides*, which ranges no younger (Powell 1992).

PT14b can be calibrated against NP14 (earliest Middle Eocene) on the basis of the occurrence of *Eatonicysta ursulae*, which ranges no younger (Vinken 1988).

PT8d can be calibrated against the early part of NP10 (earliest Eocene) on the basis of the occurrence of *Apectodinium augustum*, which ranges no younger (Powell 1992).

PT8b can be calibrated against calcareous nannoplankton zone NP9 (latest Palaeocene) on the basis of the occurrence of *Apectodinium* spp., which range no older (Powell 1992).

PT7c can be calibrated against the early part of NP8 (Late Palaeocene) on the basis of the occurrence of *Alisocysta margarita*, which ranges no younger (Powell 1992). It can also be indirectly calibrated against magnetostratigraphic polarity zones 26N–25R (Late Palaeocene) on the basis of a biostratigraphic correlation (again by means of *Alisocysta margarita* (Jolley 1992; A. J. Powell pers. comm.)) with the Ormesby Clay of the BGS borehole at Ormesby in Norfolk and the stratotypical Thanet Formation of East Kent (Ali & Hailwood 1995). In the Ormesby borehole, the upper part of polarity zone 25R (equivalent to the middle part of NP8) is absent. At Thanet, both the upper part of zone 25R and the whole of 25N (together equivalent to middle NP8–lower NP9) are absent at the so-called 'Chron 25N Hiatus' (Ali 1993; Ali & Jolley 1996).

PT5 can be calibrated against NP6 (early Late Palaeocene) on the basis of the occurrence of *Isabelidinium viborgensis*, which ranges no younger (Powell 1992).

PT2 can be calibrated against the early part of NP4 (Early Palaeocene) on the basis of the occurrence of *Alisocysta reticulata*, which ranges no younger (Powell 1992).

PT1 can be calibrated against NP3 (earliest Palaeocene) on the basis of the occurrence of *Senoniasphaera inornata*, which ranges no younger (Powell 1992).

Appendix B: notes on the definition of zones over the Forties Formation/sequence reservoir interval

For notes on calibration refer to Appendix A.

Microzone MT7a(ii): top defined by top consistent *Coscinodiscus* N11 (diatom); base by local base *Triceratium* N1 (diatom).

MT7a(i): top defined by base overlying zone; base by top underlying zone.

MT6: top defined by top impoverished agglutinating foraminiferal fauna characterized by *Cyclammina* sp. 2 Charnock & Jones 1990 (*C.* N6 of former BP usage); base by top underlying zone.

MT5c: top defined by top diverse agglutinating fauna characterized by *Spirosigmoilinella naibensis* (*S.* N1 of former BP usage) (demonstrably facies-independent – occurs in prodeltaic, turbiditic and hemipelagic facies); base by base *Cyclammina* sp. 1 Charnock & Jones 1990 (*C.* N5 of former BP usage) or base common *Spirosigmoilinella naibensis*.

MT5b: top defined by base overlying zone.

Palynozone PT8d: top defined by top *Apectodinium augustum* (dinoflagellate); base by top underlying zone.

PT8c: defined by top acme, base by base acme *Apectodinium augustum*.

PT8b: top defined by base overlying zone; base by base *Apectodinium* spp.

PT8a: top defined by base overlying zone; base by top underlying zone.

PT7c: top defined by top *Alisocysta margarita* (dinoflagellate); base by top underlying zone.

PT7b: top defined by top acme, base by base acme *Areoligera regalis* (dinoflagellate).

References

ALI, J. R. 1993. Magnetostratigraphy of the Type Thanetian, SE England. *In*: *Abstracts, Symposium on the Correlation of the Early Palaeogene in Northwest Europe*. Geological Society, London.

—— & HAILWOOD, E. A. 1995. Magnetostratigraphy of Upper Paleocene through Lower Middle Eocene Strata of Northwest Europe. *In*: BERGGREN, W. A. (ed.) *Geochronology, Time Scales and Stratigraphic Correlation*. Society of Economic Paleontologists and Mineralogists Special Publications, **54**, 275–280.

—— & JOLLEY, D. W. 1996. Chronostratigraphic framework for the Thanetian and Lower Ypresian deposits of Southern England. *In*: KNOX, R. W. O'B., CORFIELD, R. M. & DUNAY, R. E. (eds) *Correlation of the Early Palaeogene in Northwest Europe*. Geological Society, London, Special Publications, **101**, 129–144.

AUBRY, M.-P. 1985. Northwestern European magnetostratigraphy, biostratigraphy and paleogeography: calcareous nannofossil evidence. *Geology*, **13**, 198–202.

——1986. Paleogene calcareous nannoplankton biostratigraphy of Northwestern Europe. *Palaeogeography, Palaeoclimatology, Palaeoecology*, **55**, 267–334.

BLOW, W. H. 1969. Late Middle Eocene to recent planktonic foraminiferal biostratigraphy. *In*: BRONNIMANN, P. & RENZ, H. H. (eds) *Proceedings of the First Planktonic Conference, Geneva, 1967*. E. J. Brill, Leiden 199–402.

——1979. *The Cainozoic Globigerinida*. E. J. Brill, Leiden.

CARMALT, S. W. & ST JOHN, B. 1986. Giant oil and gas fields. *In:* HALBOUTY, M. T. (ed.) *Future Petroleum Provinces of the World*. American Association of Petroleum Geologists, Memoir, **40**, 11–54.

CARMAN, G. J. & YOUNG, R. 1981. Reservoir geology of the Forties oilfield. *In:* ILLING, L. V. & HOBSON, D. G. (eds) *Petroleum Geology of the Continental Shelf of North-West Europe*. Heyden, London, 371–379.

CHARNOCK, M. A. & JONES, R. W. 1990. Agglutinated foraminifera from the Palaeogene of the North Sea. *In*: HEMLEBEN, CH., KAMINSKI, M. A., KUHNT, W. & SCOTT, D. B. (eds) *Paleoecology, Biostratigraphy, Paleoceanography and Taxonomy of Agglutinated Foraminifera*. Kluwer, Dordrecht, 139–244.

GRUN, W., LAUER, G., NIDERMAYR, G. & SCHNABEL, W. 1964. Die Kreide-Tertiar Grenze im Wienerwaldflysch bei Hochstrass (Niederosterreich). *Verhandlungen der Geologischen Bundesanstalt*, **1964**(2), 226–283.

HAQ, B. U., HARDENBOL, J. & VAIL, P. R. 1987. The new chronostratigraphic basis of Cenozoic and Mesozoic sea level cycles. *In*: ROSS, C. A. & HAMAN, D. (eds) *Timing and Depositional History of Eustatic Sequences: Constraints on Sequence Stratigraphy*. Cushman Foundation for Foraminiferal Research, Ithaca, New York, Special Publication, **24**, 7–13.

HILL, P. J. & WOOD, G. V. 1980. Geology of the Forties Field, UK Continental Shelf (North Sea). *American Association of Petroleum Geologists Bulletin*, **64**(8), 81–93.

HILLIER, G. R. K., COBB, R. M. & DIMMOCK, P. A. 1978. Reservoir development planning for the Forties Field. *In*: *Proceedings of the European Offshore Petroleum Conference*, Volume II. Society of Petroleum Engineers, Dallas, Texas, 325–335.

HILTERMANN, H. 1968. Neuere Palaontologische Daten zum Flysch-Problem. *Erdoel-Erdgas-Zeitschrift*, **84**, 151–157.

HOLMES, N. A. 1999. The Andrew Formation and 'biosteering' – different reservoirs, different approaches. *This volume*.

JOHNSON, M. D. & STEWART, D. J. 1985. Role of clastic sedimentology in the exploration and production of oil and gas in the North Sea. *In*: BRENCHLEY, P. J. & WILLIAMS, B. P. J. (eds) *Sedimentology, Recent Developments and Applied Aspects*. Geological Society, London, Special Publications, **9**, 249–310.

JOLLEY, D. W. 1992. Palynofloral association sequence stratigraphy of the Palaeocene Thanet Beds and equivalent sediments in Eastern England. *Review of Palaeobotany and Palynology*, **74**, 207–237.

JONES, R. W. 1996. *Micropalaeontology in Petroleum Exploration*. Oxford University Press, Oxford.

—— & CHARNOCK, M. A. 1985. 'Morphogroups' of agglutinating foraminifera: their life positions, feeding habits and potential applicability in (paleo)ecological studies. *Revue de Paleobiologie*, **4**, 311–320.

KAMINSKI, M. A., GRASSLE, J. F. & WHITLATCH, R. B. 1988. Life history and recolonization among agglutinated foraminifera in the Panama Basin. *Abhandlungen der Geologischen Bundesanstalt, Wien*, **41**, 229–244.

KING, C. 1981. The stratigraphy of the London clay and associated deposits. *Tertiary Research Special Publication*, **6**.

KNOX, R. W. O'B. 1990. Thanetian and Early Ypresian chronostratigraphy in south-east England. *Tertiary Research*, **11**(2–4), 57–64.

KULPECZ, A. A. & VAN GEUNS, L. C. 1990. Geological modeling of a turbidite reservoir, Forties Field, North Sea. *In*: BARWIS, J. H., MCPHERSON, J. G. & STUDLICK, J. R. J. (eds) *Sandstone Petroleum Reservoirs*. Springer, New York, 489–507.

MARTINI, E. 1971. Standard Tertiary and Quaternary calcareous nannoplankton zonation. *In*: FARINACCI, A. (ed.) *Proceedings of the Second Planktonic Conference, Rome 1969*. Edizioni Tecnoscienza, Rome, 739–785.

MITLEHNER, A. G. 1996. Palaeoenvironments in the North Sea Basin around the Palaeocene–Eocene boundary: evidence from diatoms and other siliceous microfossils. *In*: KNOX, R. W. O'B., CORFIELD, R. M. & DUNAY, R. E. (eds) *Correlation of the Early Palaeogene in Northwest Europe*. Geological Society, London, Special Publications, **101**, 255–274.

PAYNE, S. N. J., EWEN, D. F. & BOWMAN, M. J. 1999. The role and value of 'high-impact biostratigraphy' in reservoir appraisal and development. *This volume*.

POWELL, A. J. (ed.) 1992. *A Stratigraphic Index of Dinoflagellate Cysts*. Chapman & Hall, London.

SHANMUGAM, G. 1996. High-density turbidity currents: are they sandy debris flows? *Journal of Sedimentary Research*, **66**, 2–10.

THOMAS, A. N., WALMSLEY, P. J. & JENKINS, D. A. L. 1974. Forties Field, North Sea. *American Association of Petroleum Geologists Bulletin*, **58**, 396–405.

VINKEN, R. (ed.) 1988. The northwest European Tertiary basins: results of the International Geological Correlations Programme Project No. 124. *Geologisches Jahrbuch, Reihe A*, **100**.

WALMSLEY, P. J. 1975. The Forties Field. *In*: WOODLAND, A. W. (ed.) *Petroleum and the Continental Shelf of North-west Europe*. Wiley, New York, 477–486.

WEIMER, P. & LINK, M. H. 1991. Global petroleum occurrences in submarine fans and turbidite systems. *In*: WEIMER, P. & LINK, M. H. (eds) *Seismic Facies and Sedimentary Processes of Submarine Fans and Turbidite Systems*. Springer, New York, 9–70.

WILLS, J. M. & PEATTIE, D. K. 1990. The Forties Field and the evolution of a reservoir management strategy. *In*: *Oil and Gas Reservoirs – II*. Graham & Trotman, London.

Constraints on the application of palynology to the correlation of Euramerican Late Carboniferous clastic hydrocarbon reservoirs

D. McLEAN[1] & S. J. DAVIES[2]

[1] *Centre for Palynology, University of Sheffield, Dainton Building, Brook Hill, Sheffield S3 7HF, UK*

[2] *Department of Geology and Geophysics, Grant Institute, University of Edinburgh, West Mains Road, Edinburgh EH9 3JW, UK*

Abstract: Palynology represents the standard tool for subsurface biostratigraphical correlation of Euramerican Late Carboniferous sequences. While palynomorphs are usually absent from sandstone lithologies, they form diverse and abundant assemblages in fine-grained sediments and coals. The occurrence of inter- and intrasandstone horizons that contain distinctive palynological assemblages forms the basis for the correlation and discrimination of sandstone reservoirs, and the application of palynology to development drilling programmes. The successful application of development palynology is contingent upon many factors related to geology (sandstone-body type, occurrence and lateral consistency of distinctive horizons, nature of intrareservoir units, degree of reservoir incision, etc.) and to practical constraints (such as the nature of databases, sampling strategies with ditch cuttings, standardization of preparation and data acquisition techniques, and time and cost limitations). The development palynologist needs to be critically aware of the existence of such constraints, and should limit the confidence attached to interpretations and predictions made accordingly.

The term 'Euramerican' as used here corresponds to the geographical extent of the 'Euramerican Coal Province' as circumscribed by Calder & Gibling (1994). The province consists of a series of sedimentary basins (Fig. 1) which accumulated significant thicknesses of fluvio-deltaic and marginal-marine coal-bearing and deep-marine strata during the Late Carboniferous. In Europe this corresponds to the mid-Namurian–Stephanian series of the Silesian subsystem (Wagner 1974; Ramsbottom *et al.* 1978), and in the United States and eastern Canada to the Lower, Middle and Upper Pennsylvanian series of the Pennsylvanian system (Bradley 1956; Peppers 1996) (Fig. 2).

Carboniferous marine mudstones, lacustrine shales and coals provide sources for hydrocarbons (Cornford 1990; Cameron & Ziegler 1997; Maynard *et al.* 1997). These hydrocarbons may occur in traditionally sought clastic reservoirs (sandstones, conglomerates: Ritchie & Pratsides 1993; Maynard *et al.* 1997) and coal-bed reservoirs (MacCarthy *et al.* 1996; Murray 1996), also of Carboniferous age. As in all hydrocarbon exploitation situations, the effective technical development of these reservoirs requires an understanding of reservoir connectivity and compartmentalization (e.g. Mijnssen 1997). Such an understanding requires detailed stratal correlation between wells. Rapid lateral and vertical lithological variations in the Late Carboniferous fluvio-deltaic sequences means that accurate lithological and electric-log correlations can be extremely difficult. The same may be true even within relatively short stratigraphical intervals constrained by other methodologies (usually biostratigraphy and/or chemostratigraphy). Examples from the subsurface and from outcrop analogues to Late Carboniferous reservoirs indicate that biostratigraphy can provide detailed stratal correlations, and can be a valuable tool for reservoir correlation and discrimination.

Biostratigraphical correlation of Late Carboniferous strata

Ammonoids (Ramsbottom *et al.* 1978; Ramsbottom & Saunders 1985), non-marine bivalves (Calver 1969) and plant macrofossils (Wagner 1984; Cleal 1991) represent the standard for

McLEAN, D. & DAVIES, S. J. 1999. Constraints on the application of palynology to the correlation of Euramerican Late Carboniferous clastic hydrocarbon reservoirs. *In:* JONES, R. W. & SIMMONS, M. D. (eds) *Biostratigraphy in Production and Development Geology*. Geological Society, London, Special Publications, **152**, 201–218.

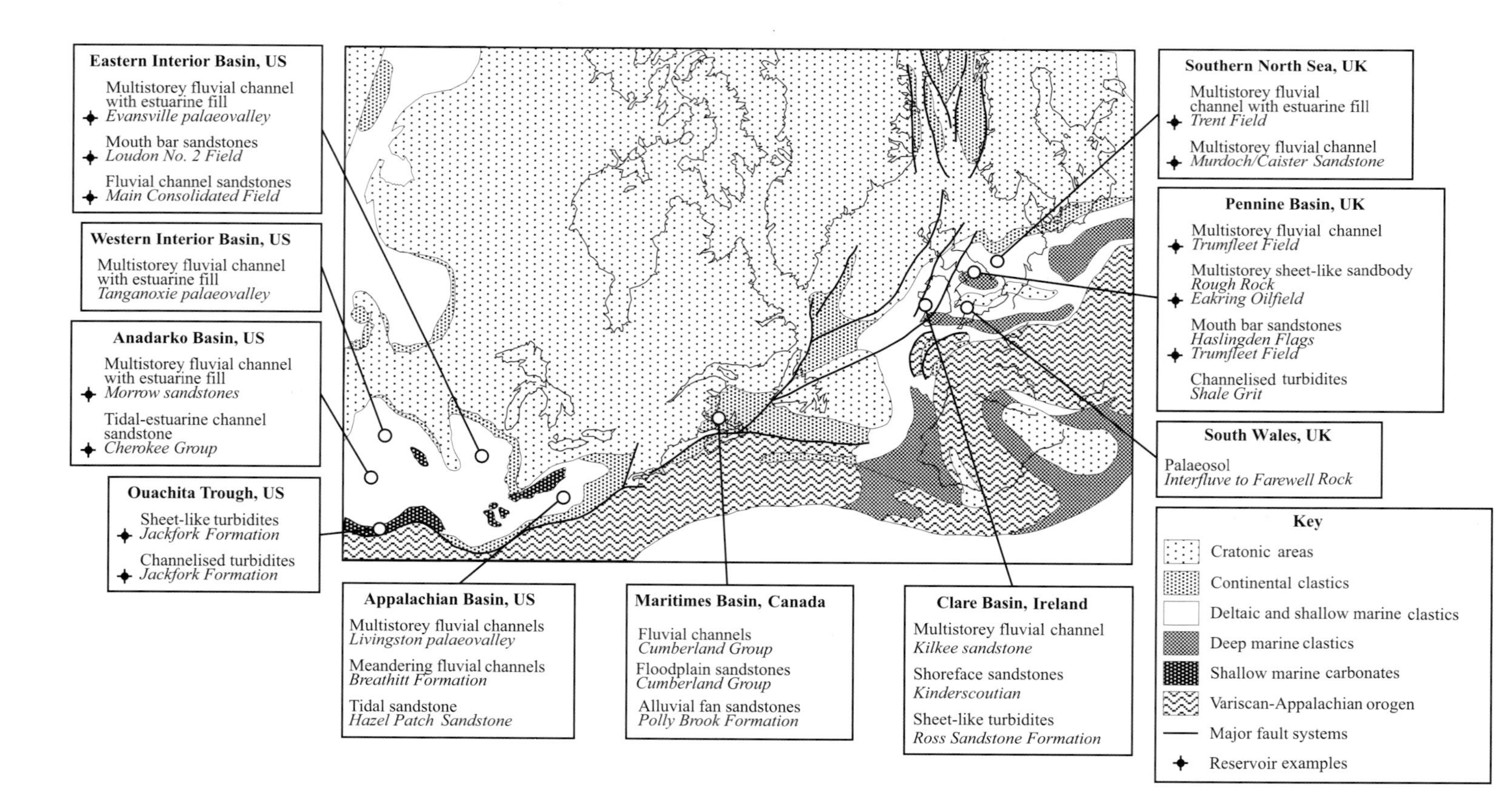

Fig. 1. Principal sedimentary basins within the Euramerican province showing locations of outcrop and subsurface examples. Palaeogeography after Ziegler (1989).

North America				Europe		
U.S.G.S.		Midcontinent U.S.				
System	Series	Series	Series	Subsystem	Series	Stages
PENNSYLVANIAN	Upper	Upper	Virgilian	SILESIAN (pars.)	Stephanian	D
						C
			Missourian			B
						Baruellian
						Cantabrian
	Middle	Middle	Desmoinesian		Westphalian	D
			Atokan			Bolsovian
						Duckmantian
		Lower	Morrowan			Langsettian
	Lower				Namurian (pars.)	Yeadonian
						Marsdenian
						Kinderscoutian
						Alportian
						Chokierian

Fig. 2. Late Carboniferous chronostratigraphical units (after Peppers 1996).

correlation of the Late Carboniferous at outcrop. However, their utility in borehole biostratigraphy is limited. While they are occasionally recovered from conventional core (McLean & Murray 1996) and rotary sidewall cores (Bowler *et al.* 1995), they are rare in conventional sidewall cores and never preserved in fragmentary ditch cuttings. The latter provide the standard type of subsurface rock sample and, in development drilling, usually the only type. Consequently, the onus of subsurface biostratigraphy falls upon microfossils which are recoverable from ditch cuttings. In the Late Carboniferous these include conodonts, ostracodes, foraminifera and palynomorphs. Other than palynomorphs these are subject to strong facies controls, being dominantly marine. Late Carboniferous palynomorphs are mostly the products of terrestrial vegetation which are widely and abundantly distributed through most medium- or fine-grained terrestrial and marine rocks, other than red beds. Palynology (the study of palynomorphs) consequently represents the standard tool for biostratigraphic correlation of Euramerican Late Carboniferous strata in the subsurface.

Biostratigraphy in exploration v. biostratigraphy in development

Micropalaeontology (including palynology) is routinely applied in hydrocarbon exploration. Data are related to chronostratigraphic units via relatively coarse-resolution, regional, biostratigraphical zonations (e.g. Clayton *et al.* 1977) and to finer-resolution, basinal-scale zonations (e.g. McLean 1995). These zonations use standard types of biozone as defined by national and international stratigraphic guides (e.g. Hedberg 1976; Whittaker *et al.* 1991). With the recognition of a hydrocarbon reservoir, and the shift to a development phase, attention is focused upon a particular stratigraphical interval. Development requires fine-resolution correlations of inter- and intrareservoir and adjacent stratal units which are often beyond the resolving power of the widely applicable zonations used during exploration. The stratigraphical resolution of a biostratigraphical zonation is generally in inverse relation to its geographical applicability (Fig. 3). Regional biostratigraphies consist of stratigraphically extensive biozones recognizable over tens or hundreds of kilometres. At the field scale (units of kilometres) these biozones remain recognizable and are associated with other biostratigraphical datums which are consistently recognizable only over a limited geographical area. For example: stratigraphically localized apparent range tops and bases; assemblage and acme events associated with individual beds or facies; and assemblage diversity or dominance criteria are commonly used in reservoir-scale biostratigraphy. Such criteria are rarely defined in detail and often lack accurate definition, but their application will continue if only because of the pragmatic assertion that, in many cases, it appears to provide the required correlations.

The conception that changes in geographical scale influence the degree of stratigraphical resolution is central to appreciating the difference between biostratigraphy in exploration and in development. The (often implicit) aim of exploration biostratigraphy is to relate rock sequences to chronostratigraphy and so effect broad correlations. For example, Quirk (1997) recognizes palynostratigraphical resolution (using data from different sources) across the northern part of the southern North Sea of the order of chronostratigraphic stages within the Westphalian. At this geographical scale he is unable to positively identify specific coals, marine bands or other marker horizons. This exploration-related resolution should be compared to that of the development situation provided in the case study below.

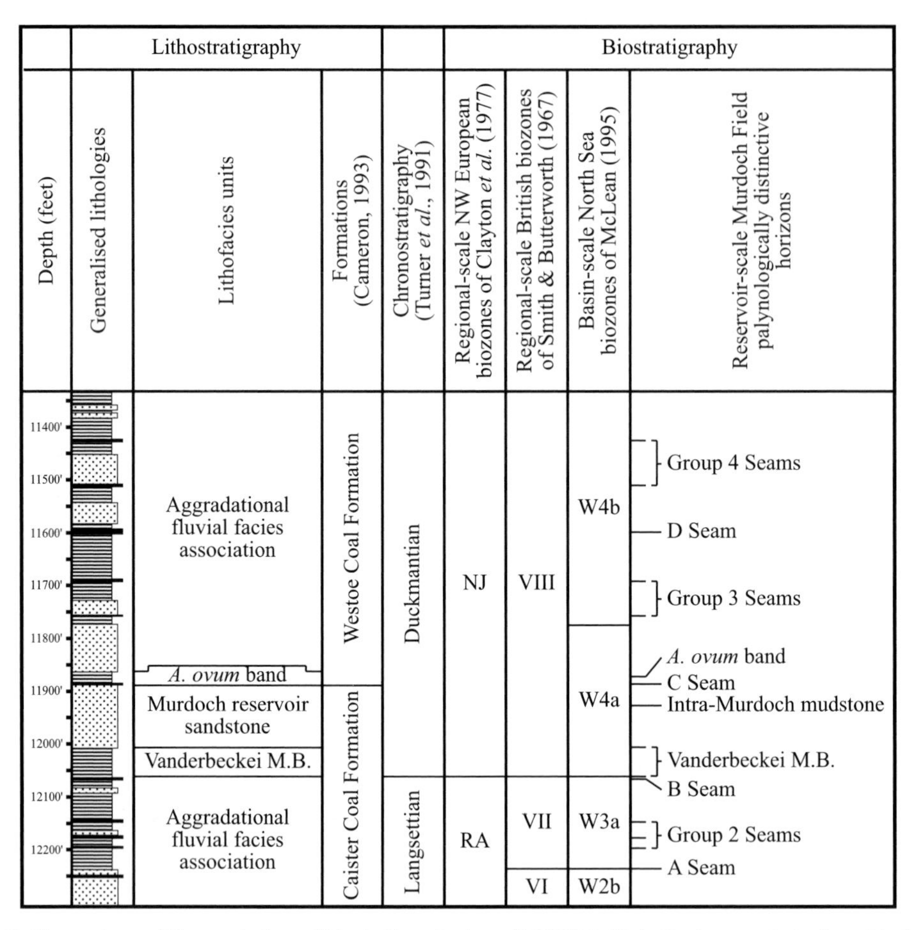

Fig. 3. Comparison of the resolution of biostratigraphy in well 44/22-1. Note the increased stratigraphical resolution from left to right with decreasing geographical scale.

Biostratigraphy in development involves stratigraphical *discrimination* of strata and geo graphical *correlation*. The chronostratigraphic validity of such correlations is generally irrelevant provided that they are lithostratigraphically accurate. In fact the ability to correlate certain diachronous lithological units can be advantageous. For example, correlation of mud drapes over migrating point bar sandstones can identify baffles to reservoir connectivity.

Palynostratigraphy in development situations

The application of palynology in development geology requires a reservoir-specific biostratigraphic model which identifies the distribution of palynomorphs in and around a reservoir unit. This is intended to be applied consistently only over limited geographical scales to predict the distribution of stratal units in new wells in relation to the modelled distribution of palynomorphs in old wells. Accuracy of the model is contingent upon the removal of as many biases to the palynological data as possible. The model is based on all available data from relevant well sections in the field area. Data are ideally derived from resampling, repreparation and restudy of material, but practical or economic constraints may preclude this. As variable preparation of original sample material may introduce artificial differences in palynological assemblages (Alpern 1963; Hughes *et al.* 1964) standardized preparation and recording techniques should be applied. Degrees of confidence should be made very clear when standardization does not occur.

Development palynostratigraphy essentially involves correlation of lithological facies as represented by palynological assemblages (palynofacies *sensu lato*). This ideally takes place within a framework constrained by evolutionary and/or palaeoecological events which have time significance. Limitations to correlations will be determined by the lateral consistency of palynological assemblages, and the ease with which correlative assemblages can be identified and discriminated from others. Raistrick (1935, p. 146) identified two requirements for the use of palynology in the correlation of coals:

(a) The microspore-content of the coal of different seams must be sufficiently different in its proportions for the seams to be distinguished from one another, and to be recognizable by definite characteristics of the microspore assemblage; and (b) the characteristics of each seam must be recognizably constant over a wide area of the coalfield.

These premises apply equally to correlation of mudstone and siltstone horizons. Coal seams can be correlated over large (coalfield-scale) areas using palynology (Raistrick 1935) largely due to regional palaeoclimatic controls producing a geographically homogeneous coal mire vegetation. As assemblage components in clastic rocks potentially originate from many disparate palaeoecological habitats, it may be unrealistic to expect them to be correlatable over large areas. However, certain assemblage characteristics may remain constant over several kilometres (Davies & McLean 1996; McLean & Chisholm 1996). In development situations there is an assumption that subbiozone-scale correlations are realistic. They are difficult or impossible to test because of a lack of independent (fossil, geochemical) evidence and because they are much finer than can be resolved using seismic data. Consequently, the palynostratigraphical development

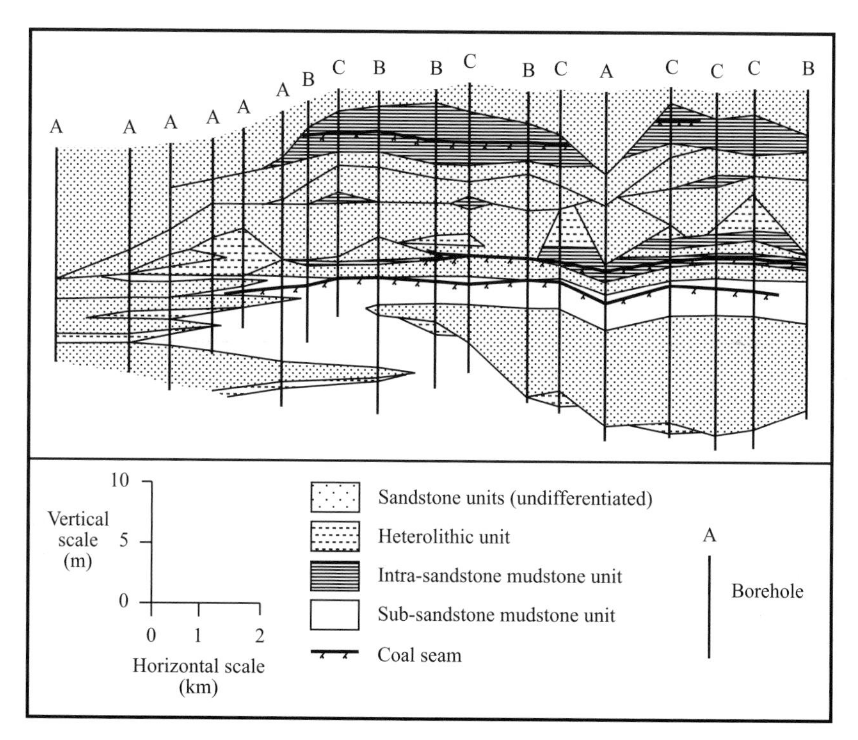

Fig. 4. Rapid lateral intra- and subsandstone facies variations in the Morrowan New River Formation of West Virginia (modified after Staub 1994 with permission from Elsevier Science). Boreholes in different geographical locations encounter a variety of facies associations which would influence interpretation of the complexity of the sandbodies. Boreholes marked 'A' encounter relatively simple, sand-prone sections. Those marked 'B' encounter slightly greater intra- and subsandstone complexity. Boreholes marked 'C' encounter highly complex facies variations. Any one set of boreholes, taken in isolation, would provide an unrealistic database upon which to produce a development model. Intra- and subsandstone mudstone and coal units will provide different and distinctive palynological assemblages. The two subsandstone coals rest upon different (sandstone or mudstone) substrates, and so differences in palaeoecological mire drainage will have produced different palynological assemblage characteristics for each coal.

model must be produced and assessed critically. The model only represents the true distribution of palynomorphs around a reservoir in so far as limitations of the database allow, and this is limited by many factors such as: the placement of wells (Fig. 4); the subsequent representation of inter- and intrareservoir facies within these sections; and the nature and quality of the original palynological database. The only way to test the model is to compare data from new development wells to it. This occurs in situations which have economic, as well as scientific, imperatives. Again, it is important that data are assimilated critically and that degrees of certitude are attached to interpretations made.

Late Carboniferous coal-bed reservoirs

Hydrocarbons occur in Late Carboniferous coal beds. Palynological correlation of these is achieved using standard techniques of coal-seam assemblage correlation (Smith & Butterworth 1967; Peppers 1996). The principal geological constraints on these techniques are related to autotrophic development of the coal mire and allogenic factors, such as flooding, delta switching, local basin development and regional palaeoclimate. While fine-resolution correlations of coal seams can be achieved (with examples dating back to the work of Raistrick 1935) the utility of many of the techniques are limited by practical considerations. Several techniques, such as the correlation of seam splits (Smith & Butterworth 1967), depend upon the existence of complete sections through individual seams. The absence of core material in development drilling effectively precludes such applications. Considerations for and examples of coal-bed reservoir correlation are reviewed by McLean & Murray (1996) and are not considered further here.

Late Carboniferous clastic reservoirs

A large number of types of sandstone bodies are represented in the Euramerican Late Carboniferous. Certain characteristics of these are given in Figs 5 and 6 with outcrop and subsurface examples. Five of these types are considered to be significant as proven hydrocarbon reservoirs.

Incised valley fill sandstones

Ramsbottom (1977) interpreted the stratigraphical distribution of British Carboniferous marine facies and sandstone bodies in terms of fluctuations in sea level. These sea-level changes are generally accepted as glacio-eustatic in origin. Periods of sea-level rise as represented by marine bands can be traced across the Euramerican area (Ross & Ross 1985). Sequence stratigraphy predicts the occurrence of valley systems which were incised at times of low sea level, and are filled, at least in part, by sandstone bodies deposited during subsequent sea-level rise. Multistorey fluvial systems interpreted as lying within incised valleys represent the most widely exploited Euramerican Carboniferous hydrocarbon reservoirs. Numerous such incised valley fills are described from the UK (Hampson *et al.* 1997) and North America (Gibling & Wightman 1994; Aitken & Flint 1995; Archer & Greb 1995). However, not all multistorey fluvial sand bodies are interpreted as incised valley fills (Guion *et al.* 1995). Euramerican incised valley fills are most commonly 10–30 km wide. Their down-dip continuity is poorly understood. Estimates from the lower Kinderscout Grit suggest minimum lengths of 70 km (Hampson 1997). In the Appalachian and Eastern Interior basins (Fig. 1) palaeovalleys can be traced for hundreds of kilometres (Chestnut 1988; Archer & Greb 1995). Preserved stratal thicknesses within valleys are generally between 25 and 40 m, although discrete palaeovalley fills up to 80 m (Jones & Chisholm 1997) and 120 m (Howard & Whittaker 1988) are known. Vertical amalgamation of a number of palaeovalleys can produce anomalous sandstone successions in excess of 150 m thick.

Palaeovalleys may be grouped into two types: those with an entirely fluvial fill; and those with a lower, fluvial part and an upper, heterolithic, estuarine- or marine-influenced part to the fill.

Multistorey fluvial channels with an entirely fluvial fill. Most Late Carboniferous palaeovalleys in the UK contain strata deposited entirely by low-sinuosity fluvial systems which were braided at low flow stages. These fluvial deposits consist of vertically and laterally stacked, 6–10 m thick channel units with erosive bases. Entirely fluvial fills are sand-rich with in excess of 90–95% sandstone. The distal extremities in some of the deepest incised valleys contain coarse pebble conglomerates with 'giant' (up to 35 m high) foresets (Hampson 1997; Jones & Chisholm 1997) indicative of deposition by braid deltas prograding into deep water at the valley mouths. These sandstones have very high reservoir potential. Smaller foresets present near the base of the palaeovalley fills are interpreted as the products of migrating in-channel bars (McCabe 1977; Hampson 1997). The occurrence

Sandbody type	Lateral continuity	Connectivity	Reservoir potential	Sequence stratigraphic significance	Outcrop analogues	Reservoir examples
Alluvial fans	10s of km	Good to very good	Moderate to high	Present along basin margins not restricted to any systems tract	Polly Brook Formation, Namurian/Westphalian, Maritimes Basin, Nova Scotia, Canada (Calder 1994)	
Multistorey fluvial channels without estuarine fill	1 km to 60 km	Very good	Very high	Late lowstand and early transgression (following a significant fall in sea-level)	Kilkee Sandstone, Kinderscoutian, Clare, Ireland (Hampson *et al.* 1997) Livingston Palaeovalley, Appalachian Basin, US (Archer & Greb 1995)	Murdoch/Caister Fields, early Langsettian, North Sea (Bailey *et al.* 1993, Ritchie & Pratsides 1993) Eastern Interior Basin, US (Howard & Whitaker 1988) Trumfleet Field, Marsdian, Yorshire (Cowan & Shaw 1991)
Mulitstorey fluvial channels with estuarine fill	1 km to 60 km	Good in lower fluvial part. Very poor in upper estuarine part (may form a seal)	High in lower fluvial part. Very low in upper estuarine part	Late lowstand and early transgression (following a significant fall in sea-level) Presence of upper estuarine fill may result from a low sediment supply and/or a relatively down-dip location	Tonganoxie palaeovalley, early Virgillian (Stephanian) NE Kansas,US (Feldman *et al.* 1995)	Morrow sandstones, Namurian, Colorado, US (Bowen *et al.* 1995) NW Eva Field, Morrow sandstones, Namurian, Oklahoma, US (Puckette *et al.* 1996) Trent Field, Marsdenian, Southern North Sea, UK (O'Mara *et al.* 1997)
Multistorey fluvial sheet-like channel sandstones	70 to 90 km	Good	High	Late lowstand systems tract to early rise in sea-level	Rough Rock, Yeadonian, Pennine Basin, UK (Bristow 1988, 1993)	Relatively poor reservoir, Rough Rock, Yeadonian, Eakring Oilfield, Nottinghamshire. UK (Storey & Nash 1993)
Meandering fluvial channel sandstones (with lateral accretion)	<1 km	Poor	Low	Highstand systems tract	Highstand sandstones in the Breathitt Formation, Westphalian B/C, Appalachian Basin, US (Aitken & Flint 1995)	
Fluvial channel sandstones	<200 m	Variable	Low	High subsidence and high sediment supply regimes in all systems tracts	Cumberland Group, Namurian/Westphalian, Maritimes Basin, Nova Scotia, Canada (Rust *et al.* 1984)	Main Consolidated Field, Illinois (Gogarty & Sorkado 1971)
Tidal-estuarine channel sandstones	100s m to 10s km	Low	Low	Early transgressive systems tract	Morrowan succession, Appalachian Basin, Kentucky, US (Greb & Chesnut 1995)	Cherokee Group, Pennsylvaninan (Westphalian D), Tucemseh Field, Oklahoma (Brown *et al.* 1996)
Palaeosols	Variable up to kms	Poor	Very low	Mature paleosol may represent an interfluve to an incised valley fill	Interfluve to the Farewell Rock, Yeadonian, South Wales (Hampson *et al.* 1996)	
Floodplain sandstones	Variable 10s m to 10s km	Poor	Low	High supply, high subsidence regime in lowstand, transgressive and highstand systems tracts	Cumberland Group, Namurian/Westphalian, Maritimes Basin, Nova Scotia, Canada (Rust *et al.* 1984)	
Mouthbar sandstones	Variable, 4-5 km May amalgamate up to 25 km	Variable	Moderate	a) Top of the prograding lowstand wedge (Elliott & Davies 1994) b) Transgressive systems tracts (Aitken & Flint 1995)	Haslingden Flags, Yeadonian, Pennine Basin, UK (Collinson & Banks 1975)	Loudon 2 Field, Illinois, US (Bragg *et al.* 1982) Trumfleet Field, Beacon Hill Flags, Marsdenian, South Yorkshire (Cowan & Shaw 1991)
Shoreface sandstones	10s km	Moderate	Low	Late transgressive systems tract, where sediment supply is high (Pulham 1987) & late highstand systems tract	Kinderscoutian, Clare Ireland (Pulham 1987)	
Tidally influenced sandstones	100s m to 10s km	Moderate to good	Moderate	Lowstand systems tract/ early transgressive systems tract	Hazel Patch Sandstone, Appalachian Basin, Kentucky, US (Greb & Archer 1995)	
Channelised sand-rich turbidites	10s m to 100s m	Good	Good	Lowstand systems tract	Shale Grit, Kinderscoutian, Derbyshire, UK (Clark & Pickering 1996)	Jackfork Formation, Morrowan - Pennsylvanian, Ouachita Trough, Oklahoma/Arkansas (Shirley 1997)
Sheet-like sand-rich turbidites	100s m	Good lateral, fair to poor vertical connectivity	Moderate	Lowstand systems tract	Ross Sandstone Formation, Alportian - earliest Kinderscoutian, Clare, Ireland (Chapin *et al.* 1994)	Jackfork Formation, Morrowan - Pennsylvanian, Ouachita Trough, Oklahoma/Arkansas (Shirley 1997)
Sand-poor turbidites	10s m to 100s m	Fair to poor	Low	Lowstand systems tract	Brejeira Formation, Namurian, Southwest Portugal (Oliveira 1990)	

Proximal ↑ — Distal ↓

Fig. 5. Summary of Late Carboniferous sandstone types, with their relevant sedimentological and hydrocarbon potential characteristics.

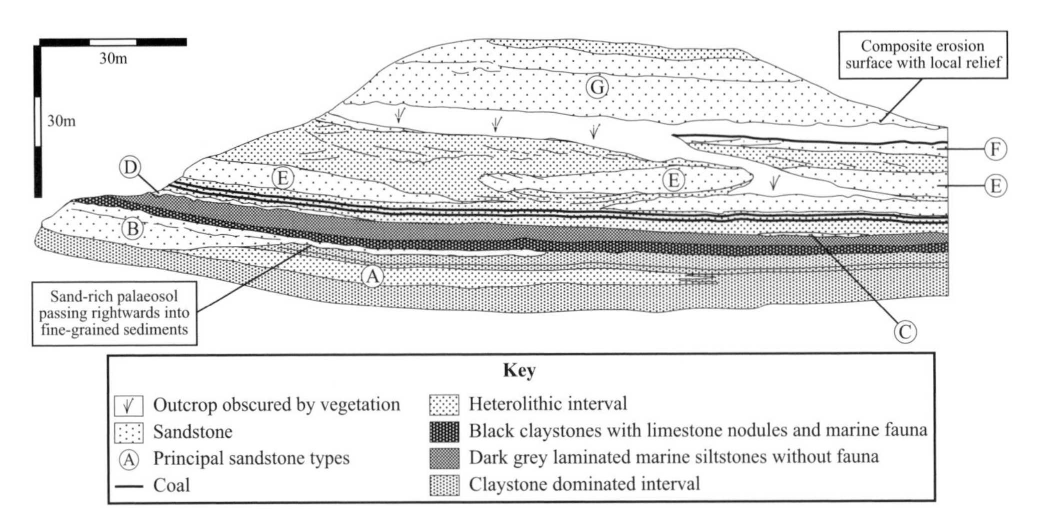

Fig. 6. Field sketch of an outcrop of the Pennsylvanian Breathitt Formation, Central Appalachian Basin, highway US23, 6 km east of Prestonburg, Kentucky. The sketch illustrates the scale of lateral variability in the extent, continuity and internal characteristics of several sandstone bodies. The main sandstone bodies are: (A) mouth-bar sandstones which become increasingly heterolithic to the right; (B) a minor distributary channel less than 200 m in lateral extent. The base of the channel erodes into the most sand-rich part of the mouth bar and shows *c.* 5 m of local relief; (C) a discontinuous, intensely bioturbated ?marine sandstone overlying fossiliferous and non-fossiliferous marine mudstones; (D)–(F) stacked meandering sandstones characterized by common heterolithics between poorly connected sandstones; (G) 12 m thick, slightly fining-upwards sandstone with a high net to gross ratio. Regional data indicate that this represents a major braided fluvial system above a sequence boundary (Aitken & Flint 1995). The marine mudstones will contain distinctive palynological assemblages. While the fine-grained material in the complex heterolithic meandering unit will not be easy to differentiate internally, they will, as a whole, be distinguishable from those of the delta-front/mouth-bar siltstone and marine mudstone units.

of fine-grained drapes associated with some of these foresets (Hampson 1997) serves to reduce the reservoir potential of this part of the palaeovalley fill.

Intra-channel, fine-grained facies occur rarely within these multistorey successions. Channel-margin, channel-plug or bank-collapse facies are of limited continuity where they are present, and so would not be expected to represent significant baffles to hydrocarbon migration. Localized fine-grained material, in the form of mud-rich intraclast conglomerates may be abundant in the basal part of the fill and in succeeding erosive-based channel members. Overbank or flood-plain deposits are not recognized in these multistorey channel complexes. The most significant intrachannel mudstones occur in the upper part of the valley fill where they drape dune bedforms and serve to greatly reduce reservoir potential. The fluvial complexes are either capped by palaeosols (of local or regional extent) or have an intensely bioturbated upper surface. In some cases thin (<50 cm) carbonaceous shales or cannel coals occur above the palaeosol cap. These are interpreted as the deposits of shallow, stagnant lakes (Hampson *et al.* 1997).

Multistorey fluvial channels with a bipartite fill. Palaeovalleys documented from North America generally have a bipartite fill consisting of a lower, fluvial part and upper, heterolithic, marine/tidally influenced part. Good examples include the network of narrow (individual valleys are less than 2 km wide) palaeovalleys in the Morrowan of Colorado, Kansas and Oklahoma (Bowen *et al.* 1993). The basal parts of these provide exploited hydrocarbon reservoirs. Major trunk palaeovalleys, such as the Evansville palaeovalley in the Eastern Interior basin (Howard & Whittaker 1988), are wider (up to 40 km) and contain only a minor fluvial component which is nevertheless also a significant producing reservoir. The Tonganoxie palaeovalley of Kansas has a basal, coarse-grained fluvial fill which provides the best reservoir potential (Feldman *et al.* 1995). The upper, estuarine component makes up the greater proportion of this palaeovalley fill. Discontinuous and isolated sand bodies encased in mudstones in this part of the fill have sedimentary structures demonstrating palaeoflow reversals and diverse trace ichnofabrics indicating deposition within tidal channels. The presence of foraminiferal tests in shale drapes within the sandstone bodies is further evidence for marine influence. These mudstone horizons act to restrict fluid flow. This in combination with the isolated occurrence of the sandstones results in a very low reservoir quality for the upper part of the palaeovalley fill. The Marsdenian reservoir sandstone in the North Sea Trent Field is interpreted as a bipartite incised valley fill by O'Mara *et al.* (1997). In contrast to the North American examples, the upper, tidally influenced part of this fill has reservoir qualities which have been improved (relative to the underlying fluvial fill) by tidal reworking and winnowing of fines

Multistorey sheet-like sandstones

Multistorey, multilateral sandsheets are documented from the Pennine basin, UK (Bristow 1988, 1993; Hampson *et al.* 1997) and the Central Appalachian basin, (Archer & Greb 1995). They are restricted to the late Yeadonian and earliest Westphalian intervals, and occur only within the most rapidly subsiding basins. They are distinguished from multistorey incised valley fills by their lateral extent (in excess of 80 km) and absence of marginal interfluves. The Yeadonian Rough Rock of the Pennine basin is a coarse-grained fluvial sand sheet interpreted as deposited in a laterally extensive, braided river system. Its sedimentological characteristics are essentially the same as those of fluvial-filled palaeovalleys.

Mouth-bar sandstones

Mouth bars within the Clare basin can be traced laterally for 4–6 km (Pulham 1989, Elliott & Davies 1994). In the Appalachian basin, mouth bars with 10–20 cm sandstone beds separated by 1–2 cm siltstone beds are exposed continuously for up to 5 km. They are traceable laterally for up to 25 km (Aitken & Flint 1995). Mouth-bar facies have a moderate proven reservoir potential. Proportions of sandstone within the facies are laterally variable, being higher close to the distributary mouth. Porosity–permeability in proximal mouth bars is related to the occurrence of fines, such as micas, deposited as high suspended load during flood events (e.g. Bristow & Myers 1989). Away from the sediment source, proportions of sandstone decrease markedly and finer facies become predominant. Reservoir potential in mouth-bar settings may be improved by concentration of sandstones into active growth faults associated with unstable delta front conditions. Some of the thickest and most sandstone-rich mouth-bar facies in the Clare basin are found associated with growth faults (Pulham 1989). Syn-depositional faulting is observed through the thickening of successive sandstone beds into the downthrown side of the fault, and sandstone-rich fills are at least 20 m thick. Less significant concentrations of sandy mouth bars may be located at the margins of active mud diapirs.

Sandstone-rich turbidite systems

Late Carboniferous turbidite systems are not currently exploited for hydrocarbons, although significant gas reserves are identified in the Morrowan Jackfork Formation of Arkansas (Shirley 1997). Major sandstone-prone turbidite successions in the Pennine and Clare basins reach several hundred of metres in thickness. The Alportian–Kinderscoutian Ross Formation of Ireland reaches 380 m in thickness and consists predominantly of sheet-like turbidites, although these are lenticular over several hundreds of metres (Chapin *et al.* 1994). Submarine channels occur within the sheet facies and are generally less than 300 m wide with limited basal relief (Chapin *et al.* 1994). Thickly bedded, massive sandstones generally form more than 80% of the channel. Sandstone-filled channels and sheet-form architectural elements are also recognized within the Kinderscoutian Shale Grit of the Pennine basin (Clark & Pickering 1996). Although estimates of the lateral continuity are restricted by the length of exposure, the channels here are at least 80–100 m wide.

Sand-prone intervals in turbidites are compartmentalized by several types of fine-grained interval. The most widespread, possibly basin-wide, fine-grained intervals are condensed shale horizons bearing thick-shelled goniatites (Elliott & Davies 1994). These intervals may be up to 9 m thick in the Clare basin, and would represent the most significant barriers to fluid flow. In the Shale Grit of the Pennine basin, goniatite-bearing intervals are not recognized although condensed fine-grained intervals containing *Promytilus* and *Sanguinolites* may also provide regional reservoir baffles/seals (Davies

& McLean 1996). Less significant fine-grained intervals such as interlaminated siltstones and claystones, and interbedded thin sandstones and siltstones, are described as separating sand-rich packages at outcrop. These units are of variable lateral development ranging from tens to hundreds of metres, and may effectively reduce sandstone connectivity (Chapin *et al.* 1994; Elliott & Davies 1994; Davies & Elliott 1996).

Relationship of palynology to sandstones

Arenaceous strata are generally barren of palynomorphs. Palynological correlation of them therefore relies upon data from associated coal seams (Peppers 1993) and/or mudstones (McLean & Murray 1996). The ease with which inter- and intrareservoir stratal units may be identified and differentiated using palynology determines the applicability of palynology to development projects. In general, fine-grained or carbonaceous strata which have formed under different sedimentological settings will contain different palynological assemblages. Such differences may not be sufficient to allow consistent differentiation of all sedimentological types. However, it is generally possible to readily differentiate the following groups of strata based upon microfloral characteristics: marine mudstones; lacustrine mudstones; overbank fines; palaeosols; coals with different mire development histories; and intrasandstone coals and mudstones. Consequently, the distribution of these facies in and around reservoir units (Figs 4 and 6) is important in understanding the limitations of palynological applications. In addition, data from lower delta-plain palaeoenvironments in the Namurian of Britain indicate that sequence stratigraphic systems tracts may contain diagnostic and correlatable palynological assemblages (Davies & McLean 1996). The application of this to palaeoenvironmental areas (principally the upper delta-plain areas of the Westphalian) has yet to be explored.

The distribution of palynomorph-bearing horizons in and around sandstone units is defined by the sandstone-body type. Fluvial components of incised valley fills may contain little in the way of fine-grained sediment. However, any intrafill mudstones and coals (related to short-lived abandonment episodes on the braid plain) will contain highly distinctive palynological assemblages dominated by the spores and pollen of colonizing vegetation. The lacustrine facies commonly found overlying the palaeovalley fill will also have a distinctive palynological signature. Similarly, the fine-grained components to estuarine- or marine-influenced facies in the upper part of North American incised valley fills will contain diagnostic microfloras. They may also contain facies-restricted, organic-walled fossils such as scolecodonts (annelid jaws) and foraminiferal test linings, which provide highly distinctive components to the palynological assemblages.

Mouth-bar sandstone bodies usually contain large amounts of fine-grained material. Palynological material from this is often indistinguishable from assemblages derived from over- and underlying strata. One exception is the Yeadonian Haslingden Flags of the Pennine basin which contain reworked Lower Palaeozoic and Devonian palynomorphs while adjacent strata do not (McLean & Chisholm 1996). This difference is related to a different provenance for the Haslingden Flags and their associated strata. It indicates that sandstones with different derivations may be palynologically discriminated if reworking of palynomorphs has occurred.

Although turbidite sandstone bodies are associated with large amounts of fine-grained strata, the palynological discrimination of these can be difficult. This is generally because all of the fine-grained material is deposited by the same sedimentological processes. Mudstone rip-up clasts may be common, but these are usually derived from more or less contemporary deposits on the shelf (see Owens in Spears and Amin 1981) and so may be difficult to discriminate from the turbidite mudstones. However, as in most depositional settings, condensed mudstone horizons, which represent the most significant barriers to fluid flow, may remain distinguishable from background sedimentation. A study of the Kinderscoutian Shale Grit of the Pennine basin (Davies & McLean 1996) identified palynological criteria for the recognition of such horizons, even where goniatite-bearing marine band facies are not developed.

A fundamental difficulty in the consistent recognition of subsandstone facies relates to the degree of incision of sandstone bodies, whether in a single-storey channel, an incised valley or a submarine channel. This is illustrated in Figs 4 and 6 in which different sandstone bodies are seen to incise to different degrees. Erosion of palynologically distinctive intra- or subsandstone horizons at the channel (or palaeovalley) base can preclude the successful palynological identification of the base of a sandstone unit. Basal erosion is an almost ubiquitous feature of sandstone bodies and so its degree must be estimated and considered as a limiting factor in the application of palynology to correlation of sandstone-related strata.

Case study: the Murdoch gas field, southern North Sea

The Murdoch gas field (southern North Sea, Block 44/22; Fig. 7) has a major early Duckmantian (early Westphalian B) reservoir sandstone unit. This sandstone, informally termed the 'Murdoch reservoir sandstone', correlates directly with the informal 'Caister sandstone unit' (Cameron 1993) in the adjacent Caister Field, Block 44/23 (Ritchie & Pratsides 1993). It consists of multistorey, stacked, tabular, cross-bedded, pebbly and conglomeratic sandstones (Green & Slatt 1992) within a coal-bearing, aggradational fluvial facies association typical of the late Langsettian (late Westphalian A) and early Duckmantian of the southern North Sea succession (Fig. 3). The reservoir unit is underlain by a high-gamma mudstone unit identified by palynology and cored macrofaunas (Turner *et al.* 1991) as the horizon of the Vanderbeckei Marine Band (= basal Duckmantian). The reservoir unit is overlain by another high-gamma mudstone unit which contains the non-marine bivalve *Anthracosia ovum*, and is interpreted as a lacustrine deposit. The Murdoch reservoir sandstone is interpreted as the deposits of a braided, fluvial system (Green & Slatt 1992). Within and beyond the Murdoch Field the base of the sandstone unit occurs at varying stratigraphical levels above the biostratigraphically distinct horizon of the Vanderbeckei Marine Band. Strata immediately beneath the reservoir unit are commonly barren of palynomorphs. This may indicate oxidative weathering of the subreservoir strata prior to sandstone deposition, or it may be due to post-depositional oxidative action of groundwaters percolating through the sandstone.

The Murdoch Field was discovered by the Conoco (UK) Ltd well 44/22-1. Five of the six subsequent wells penetrated the reservoir unit (well 44/22-2 reached terminal depth (TD) above base Permian) and were routinely analysed for palynology. During this exploration phase the palynological data were interpreted in terms of the regional palynostratigraphical schemes of Smith & Butterworth (1967) and Clayton *et al.* (1977). Prior to the onset of the development phase of drilling all six exploration wells were re-analysed in order to develop a palynostratigraphical development for the field. Because the exploration analysis had identified a major episode of reworking of Namurian–early Westphalian palynomorphs in the stratigraphical

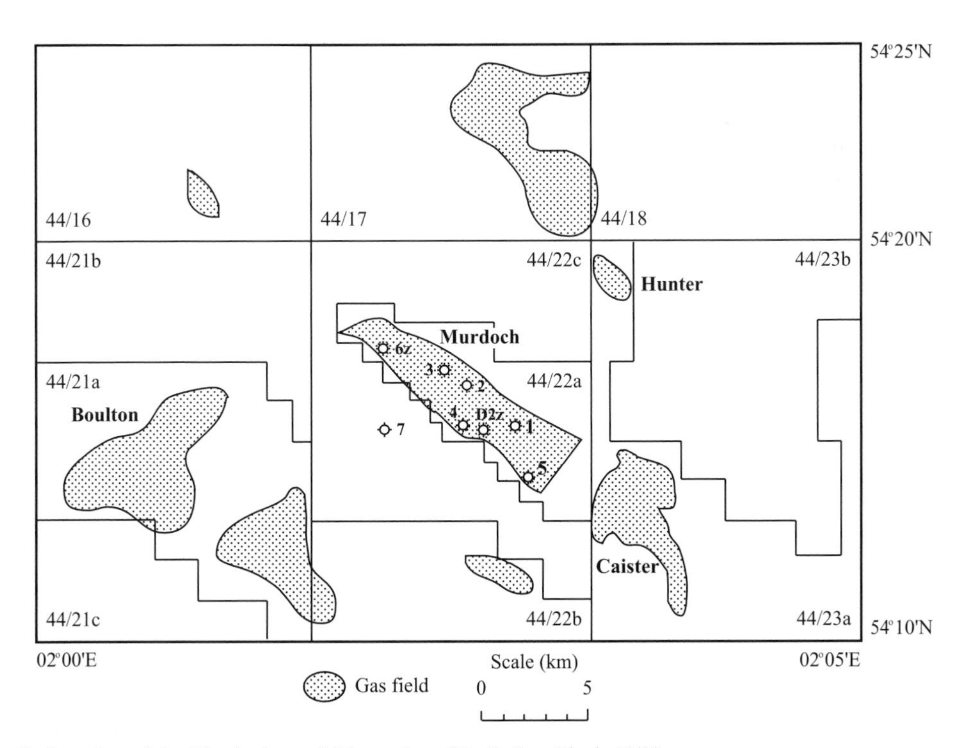

Fig. 7. Location of the Murdoch gas field, southern North Sea, Block 44/22a.

vicinity of the reservoir unit, it was decided to concentrate the development model on correlation of coal seams. The rationale for this being that the autochthonous nature of coal-seam assemblages would preclude contamination by reworking. It was also subsequently found that the relatively extreme oxidations applied to coal samples (Smith & Butterworth 1967) would destroy palynomorphs derived from mudstones which were caved down into ditch cuttings samples selected for analysis as coals. Coals were identified on electric logs, lithological logs and core descriptions, and systematically sampled from all wells. Comparisons of detailed analysis of cored coals with sonic logs indicated a general under-representation of coal seams on the electric logs. This proved particularly so in instances in which coals were stratigraphically very close or compounded, but provided only a single (undifferentiated) sonic signature. Cored coals were sampled by spot samples and by channel samples, the latter intended to mimic the representation of the coals in ditch cuttings material. Laboratory preparations for all samples were identical. Ditch cuttings samples proved to be commonly contaminated with siliciclastic mineral matter which required removal by density separation. As it is recognized that heavy liquid separation can bias palynological assemblages (Wolfram 1954), this technique was applied to all samples, including core material, irrespective of the occurrence or otherwise of mineral matter. Standardized counts of 250 palynomorphs were made from each prepared sample. Subjective (i.e. non-statistical) correlations of the sample counts were made and allowed the recognition of four palynologically distinctive coal seams. The remaining seams could not be characterized individually, but were divided into groups of palynologically similar seams.

As well as re-analysing the coal seams in the reservoir interval, it was decided to inspect the original data from several mudstone units closely associated with the reservoir. This procedure involved re-analysis of previously sample material prepared during the exploration phase. No attempt at standardization of preparation techniques could be made. However, standard counting and recording procedures were used. This allowed the recognition of three palynologically distinctive mudstone units: the Vanderbeckei Marine Band, an intrareservoir mudstone unit and the *A. ovum* lacustrine horizon. The association of different palynological assemblages with each of these relates to their fundamentally different depositional settings.

The development model (Fig. 3) was applied to seven development wells. In each case samples were prepared and analysed onshore. Rates of penetration in the Carboniferous section of the wells proved sufficiently slow to allow such practices to provide answers to stratigraphical questions within a realistic time frame. Nevertheless, upon receipt of material, samples were prepared, analysed and reported upon as rapidly

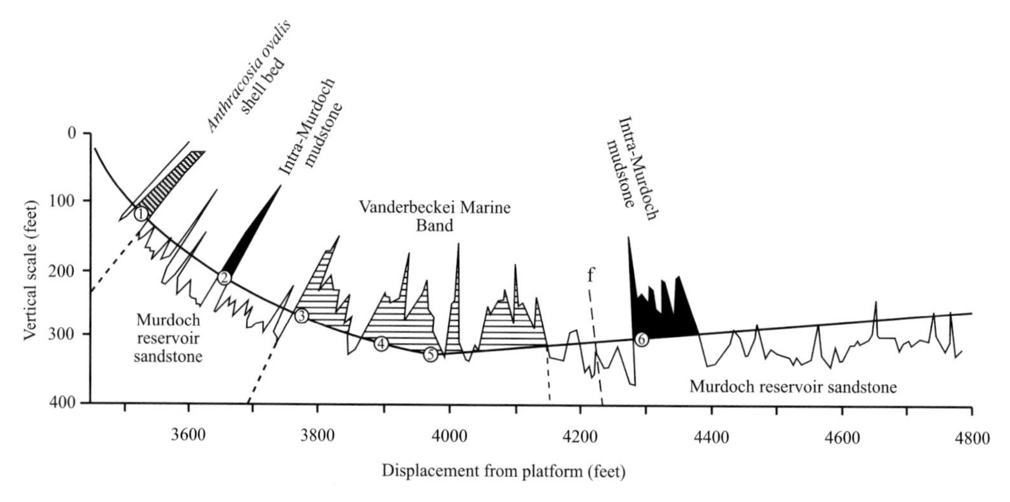

Fig. 8. Schematic well-track and gamma-ray log through the lower, subhorizontal section of well 44/22a-D2z (modified after Gunn *et al.* 1993). Numbered points along the well-track indicate palynologically significant horizons as described in the text. Note the upturn in the drilling direction at *c.* 4000-foot displacement (point 5). Identification and discrimination of the palynologically distinctive mudstone units during drilling provided stratigraphical control and provided interpretations relevant to progress of directional drilling.

as possible (while adhering to the standard procedures). The consequent degree of urgency for results provided a potential source of error, particularly when slow counts, unsociable hours of work and the high expectations of the development geologists were combined. Subsequent to the termination of drilling of each development well the data from all samples were reappraised under more relaxed conditions. Any variations in assemblage populations between correlated coals or mudstone units were identified and included into the development model where appropriate.

The initial benefits of applying the palynological development model were in identifying the stratigraphical level of the drill bit, and in predicting depths to the reservoir in new wells. In practice, determinations were required immediately the well penetrated beneath the Permian unconformity. However, secondary oxidation of the Carboniferous strata beneath the unconformity has often resulted in the loss of nearly all organic carbon, and so palynological assemblages are often not present until at least 25 feet beneath this level. Once palynomorph-bearing strata were reached by the bit, they were analysed and interpreted in terms of the development model. In particular, the identification of the D seam (see Fig. 3) identified a stratigraphical position approximately 250–300 feet above the top of the reservoir unit. This was confirmed by subsequent coal-seam analyses which identified Group 3 seams. Identification of the *A. ovum* band and the C seam were used to confirm penetration of the reservoir unit. This proved particularly useful in well sections in which the upper part of the reservoir unit consists of a heterolithic or fine-grained dominated section. Palynological identification of the horizon of the Vanderbeckei Marine Band and underlying palynologically distinctive coals identified that the bit had penetrated beneath the reservoir and was used to determine total depth. Development well 44/22a-D2z was drilled as a subhorizontal penetration of the Murdoch reservoir sandstone. As with other development wells, the palynological development model was used to monitor drilling and to predict stratigraphical positions above the reservoir unit. Once drilling reached the reservoir, on-shore palynological monitoring was used to identify the distinctive mudstone units (Fig. 8). This was made difficult by common caving of the *A. ovum* band (Fig. 8, point 1) and associated overlying mudstone units into the reservoir section. However, the highly distinctive nature of the palynological assemblages from the intra-Murdoch mudstone unit (Fig. 8, point 2) allow its successful recognition. The next mudstone unit to be penetrated proved difficult to identify initially. Palynological assemblages from a *c.* 3800-foot deviation (Fig. 8, point 3) consisted entirely of components associated with the supra-reservoir strata. However, subsequent sample material from a *c.* 3850-foot deviation (Fig. 8, point 4) provided assemblages distinctive of the horizon of the Vanderbeckei Marine Band. It was assumed that the first of these sets of data represented the immediate subreservoir strata which are commonly oxidized and devoid of palynomorphs, and that the assemblages here were entirely caved. The latter set of data identified the true stratigraphical position of the bit and this interpretation indicated the necessity of raising the section in order to re-enter the reservoir (Fig. 8, point 5 at *c.* 4000-foot displacement). Confirmation that the well had re-entered the reservoir was given by the identification of the intra-Murdoch mudstones at *c.* 4250-foot displacement (Fig. 8, point 6).

Conclusions

Successful application of palynology in the development of Late Carboniferous hydrocarbon reservoirs is contingent upon a large number of factors. The constraints which apply may be grouped into two types: objective constraints which are essentially geological and are not amenable to changes in methodology; and subjective constraints which are largely methodological, and relate to practical scientific and economic limitations. These are summarized in Fig. 9. Successful development applications require that these constraints are recognized, and that levels of confidence attached to predictions and interpretations made are affected by their recognition. It may appear to be a truism to state that reservoir models are only as good as the data upon which they are based. Most, if not all, hydrocarbon reservoirs are complex natural systems. In understanding and modelling them the tendency is to attempt to extract as much information from the data as possible. Unfortunately, the nature of biostratigraphical datasets is such that they are never complete, and are prone to a large number of biases. Consequently models based upon them are open to misinterpretation, particularly if basic assumptions and constraints are ignored. Sedimentological and palynological data derived from outcrop analogues can aid interpretation of subsurface reservoirs through the recognition and characterization of important inter- and intrareservoir intervals, and also by providing

Constraints	Effects	Recommendations	
Geological nature of the reservoir interval			Appreciate how acting constraints influence palynostratigraphy and define explicit levels of confidence on interpretations accordingly
Vertical and lateral distribution of inter- and intra-reservoir fine-grained sediments and coals Complexity of intra-reservoir units Degree of sandstone incision	Determine distribution of palynomorph-bearing strata	Use outcroop analogues to understand the stratigraphical and geographical distribution of palynomorphs in relation to the reservoir type	
Rate of evolution of the microflora Existence of stratigraphically restricted distinctive palynological assemblages Lateral continuity of palynological assemblages Degree of distinctiveness of different assemblages Oxidation beneath unconformities/sequence boundaries Occurrence of reworking events	Determine where stratigraphically restricted palynological assemblages occur, and how they may be identified		
Practical constraints: production of the development model			
Stratigraphy represented by well sections Availablitiy of exploration data	Determine prediction of distribution of palynomorph-bearing strata		
Quality of exploration data Availability of sample types (cores, swc, cuttings) Variable standards of analysis	Determine ability to consistently recognise and correlate distinctive assemblages	Appreciate deficiencies in the data	
Preparation of exploration material Aquisition of exploration data		Re-prepare and re-analyse. Standardise all techniques	
Practical constraints: application of the development model			
Turbine drilling	Limits recovery of palynomorphs	Appreciate technical limitations	
Use of oil-based drilling mud	Slows palynological preparation	Assess urgency of results, relate to R.O.P.s, etc. Move preparation and analysis to the wellsite if necessary	
Caving in cuttings Economic pressures Cost and time limitations on preparation and analysis Time limitations on relatively difficult biostratigraphy	Limit ability to recognise distinctive palynological assemblages		
Inaccuracies in the development model	Allow incorrect interpretation of data	Continually update with new data	

Fig. 9. Summary of constraints acting upon the effective application of palynology to hydrocarbon development situations.

information on their depostional and stratigraphic significance. Development biostratigraphy operates within a teamwork situation. While the development geologists may attempt to achieve very-fine-resolution biostratigraphical correlations it is important that biostratigraphers (who understand the fossil data best) remain objective. They need to define the limitations of their biostratigraphical models, and should provide levels of confidence on correlations, reasons why they may be incorrect or need modification and should be prepared to provide realistic alternative interpretations.

The authors wish to thank Conoco (UK) Ltd, Total Oil Marine plc and Arco British Ltd for permission to publish data from the Murdoch Field. Views expressed here are those of the authors and should in no way be taken to reflect those of the Murdoch Field partners. S. J. Davies wishes to acknowledge Conoco (UK) Ltd and ARCO British Ltd for sponsorship of post-doctoral research.

References

AITKEN, J. F. & FLINT, S. S. 1994. High frequency sequences and the nature of incised valley fills in fluvial systems of the Breathitt Group (Pennsylvanian) Appalachian Foreland Basin, Eastern Kentucky. *In*: BOYD, R., DALRYMPLE, R. W. & ZAITLIN, B. (eds) *Incised Valley Systems: Origin and Sedimentary Sequences*. Society of Economic Palaeontologists and Mineralogists, Special Publication, **51**, 353–368.

—— & ——1995. The application of high-resolution sequence stratigraphy to fluvial systems: a case study from the Breathitt Group, eastern Kentucky, USA. *Sedimentology*, **42**, 3–30.

ALPERN, B. 1963. Méthode d'extraction des spores des roches du Houiller. *Pollen et Spores*, **5**, 169–177.

ARCHER, A. W. & GREB, S. F. 1995. An Amazon-scale drainage system in the Early Pennsylvanian of Central North America. *Journal of Geology*, **103**, 611–628.

BAILEY, J. B., ARBIN, P., DAFFINOTTI, O., GIBSON, P. W. & RITCHIE, J. S. 1993. Permo-Carboniferous plays in the Silver Pit Basin. *In*: PARKER, J. (ed.) *Petroleum Geology of Northwest Europe: Proceedings of the 4th Conference*. Geological Society, London, 707–715.

BOWEN, D. W., WEIMER, P. & SCOTT, A. J. 1993. The relative success of siliciclastic sequence stratigraphic concepts in exploration: examples from incised valley fill and turbidite systems reservoirs. *In*: WEIMER, P. & POSMANTIER, H. W. (eds) *Siliciclastic Sequence Stratigraphy: Recent Developments and Applications*. American Association of Petroleum Geologists Memoir, **58**, 15–42.

BOWLER, M. M., RILEY, N. J. & CHRISTOPHER, R. A. 1995. Stratigraphy of the Trent Field [Abstract]. *In*: *Stratigraphic Advances in the Offshore Devonian and Carboniferous Rocks, UKCS and Adjacent Areas, Joint Meeting of the Geological Society and the Petroleum Exploration Society of Great Britain, London, January, 1995*, 5–6.

BRADLEY, W. H. 1956. Use of series subdivisions of the Mississippian and Pennsylvanian System. *American Association of Petroleum Geology Bulletin*, **40**, 2284–2285.

BRAGG, J. R., GALE, W. W., MCELHANNON, W. A., JR, DAVENPORT, O. W., PETRICHUK, M. D. & ASHCRAFT, T. L. 1982. Loudon surfactant flood pilot test. *In*: *Society of Petroleum Engineers – US Department of Energy, 3rd Joint Symposium on Enhanced Oil Recovery, Tulsa* (SPE Paper 20465), 933–952.

BRISTOW, C. S. 1988. Controls of the sedimentation of the Rough Rock Group (Namurian) from the Pennine Basin of northern England. *In*: BESLEY, B. M. & KELLING, G. (eds) *Sedimentation in a Synorogenic Basin Complex: The Upper Carboniferous of Northwest Europe*. Blackie, Glasgow, 114–131.

——1993. Sedimentology of the Rough Rock: a Carboniferous braided river sheet sandstone in northern England. *In*: BEST, J. L. & BRISTOW, C. S. (eds) *Braided Rivers*. Geological Society, London, Special Publications, **75**, 291–304.

—— & MYERS, K. J. 1989. Detailed sedimentology and gamma-ray log characteristics of a Namurian deltaic succession I: sedimentology and facies analysis. *In*: WHATELEY, M. K. G. & PICKERING, K. T. (eds) *Deltas: Sites and Traps for Fossil Fuels*. Geological Society, London, Special Publications, **41**, 75–80.

BROWN, D. P., GRASMICK, M. K. & NORTHCUTT, J. C. 1996. Cherokee Group (Pennsylvanian) production in Oklahoma: data from fluvial-dominated deltaic reservoirs. *Oklahoma Geological Survey Circular*, **98**, 249.

CALDER, J. H. 1994. The impact of climate change, tectonism and hydrology on the formation of Carboniferous tropical intermontane mires: the Springhill coalfield, Cumberland Basin, Nova Scotia. *Palaeogeography, Palaeoclimatology, Palaeoecology*, **106**, 323–351.

—— & GIBLING, M. R. 1994. The euramerican Coal Province: controls on Late Paleozoic peat accumulation. *Palaeogeography, Palaeoclimatology, Palaeoecology*, **106**, 1–21.

CALVER, M. A. 1969. Westphalian of Britain. *6ème Congrès Internationale de Stratigraphie et de Géologie du Carbonifère, Compte rendu, Sheffield, 1967*, **1**, 233–245.

CAMERON, N. & ZIEGLER, T. 1997. Probing the lower limits of a fairway: further pre-Permian potential in the southern North Sea. *In*: ZIEGLER, K., TURNER, P. & DAINES, S. (eds) *Petroleum Geology of the Southern North Sea: Future Potential*. Geological Society, London, Special Publications, **123**, 123–141.

CAMERON, T. D. J. 1993. Carboniferous and Devonian (Southern North Sea). *In*: KNOX, R. W. O'B. & CORDEY, W. G. (eds) *Lithostratigraphic Nomenclature of the UK North Sea*, Vol. 5. British Geological Survey, Nottingham.

CHAPIN, M. A., DAVIES, P., GIBSON, J. L. & PETTINGILL, H. S. 1994. Reservoir architecture of turbidite sheet sandstones in laterally extensive outcrops, Ross Formation, Western Ireland. *In*: WEIMER, P., BOUMA, A. H. & PERKINS, R. F. (eds) *Submarine Fans and Turbidite Systems, Sequence Stratigraphy, Reservoir Architecture and Production Characteristics. Edited GCSSEPM Foundation 15th Annual Research Conference, Submarine Fans and Turbidite Systems*, 53–68.

CHESTNUT, D. R. 1988. *Stratigraphic Analysis of the Carboniferous Rocks of the Central Appalachian Basin*. PhD thesis, University of Kentucky.

CLARK, J. D. & PICKERING, K. T. 1996. Namurian Shale Grit Formation slope channel fills, Alport Castles, northern England. *In*: CLARK, J. D. & PICKERING, K. T (eds) *Submarine Channels: Processes and Architecture*. Vallis Press, 155–158.

CLAYTON, G., COQUEL, R., DOUBINGER, J., LOBOZIAK, S., OWENS, B. & STREEL, M. 1977. Carboniferous miospores of western Europe: illustration and zonation. *Mededelingen Rijks Geologische Dienst*, **29**, 1–72.

CLEAL, C. J. 1991. Carboniferous and Permian biostratigraphy. *In*: CLEAL, C. J. (ed) *Plant Fossils in Geological Investigations. The Palaeozoic*. Ellis Horwood, Chichester, 182–215.

COLLINSON, J. D. & BANKS, N. L. 1975. The Haslingden Flags: bar finger sands in the Pennine Basin. *Proceedings of the Yorkshire Geological Society*, **40**, 431–458.

CORNFORD, C. 1990. Source rocks and hydrocarbons in the North Sea. *In*: GLENNIE, K. W. (ed.) *Introduction to the Petroleum Geology of the North Sea* (3rd end). Blackwell, Oxford, 294–361.

COWAN, G. & SHAW, H. 1991. Diagenesis of Namurian fluvio-deltaic sandstones from the Trumfleet Field, South Yorkshire. *Marine and Petroleum Geology*, **8**, 212–224.

DAVIES, S. J. & ELLIOTT, T. 1996. Spectral gamma ray characterisation of high resolution sequence stratigraphy: examples from Upper Carboniferous fluvio-deltaic systems, County Clare, Ireland. *In*: HOWELL, J. A. & AITKEN, J. F. (eds) *High Resolution Sequence Stratigraphy: Innovations and Applications*. Geological Society, London, Special Publications, **104**, 25–35.

—— & MCLEAN, D. 1996. Spectral gamma ray and palynological characterisation of marine bands in the Kinderscoutian (Namurian, late Carboniferous) of the Pennine Basin. *Proceedings of the Yorkshire Geological Society*, **51**, 103–114.

ELLIOTT, T. & DAVIES, S. J. 1994. *High Resolution Sequence Stratigraphy of an Upper Carboniferous Basin-fill Succession, Co. Clare, Western Ireland: Fieldtrip Guide B2*. University of Liverpool.

FELDMAN, H. R., GIBLING, M. R., ARCHER, A. W., WIGHTMAN, W. G. & LANIER, W. P. 1995. Stratigraphic architecture of the Tonganoxie palaeovalley fill (Lower Virgillian) in Northeastern Kansas. *American Association of Petroleum Geologists Bulletin*, **79**, 1019–1043.

GIBLING, M. R. & WIGHTMAN, W. G. 1994. Paleovalleys and protozoan assemblages in a Late Carboniferous cyclothem, Sydney Basin, Nova Scotia. *Sedimentology*, **41**, 699–719.

GOGARTY, W. B. & SURKALO, H. 1971. A field test of micellar solution flooding. *In*: *Society of Petroleum Engineers, 46th Annual Technical Conference and Exhibition, New Orleans*, 1–12.

GREB, S. F. & ARCHER, A. W. 1995. Rhythmic sedimentation in a mixed tide and wave deposit, Hazel patch sandstone (Lower Pennsylvanian), eastern Kentucky coal field. *Journal of Sedimentary Research*, **B65**, 96–106.

—— & CHESTNUT, D. R. 1994. Palaeoecology of an estuarine sequence in the Breathitt Formation (Pennsylvanian), Central Appalachian Basin. *Palaois*, **9**, 388–402.

GREEN, C. & SLATT, R. M. 1992. Complex braided stream depositional model for the Murdoch Field Block 44/22 UK Southern North Sea (abstract). *In*: *Proceedings of the Conference Braided Rivers: Form Process and Economic Applications*. Geological Society, London, 16.

GUION, P. D., BANKS, N. L. & RIPPON, J. H. 1995. The Silkstone Rock (Westphalian A) from the east Pennines, England: implications for sand body genesis. *Journal of the Geological Society, London*, **152**, 819–832.

GUNN, R., WANG, J., MURRAY, I. & REZIGH, A. 1993. Teamwork brings success in Quad 44 – the Murdoch Field development. *Petroleum Exploration Society of Great Britain Newsletter*, February, 4–7.

HAMPSON, G. J. 1997. A sequence stratigraphic model for deposition of the Lower Kinderscout delta, an Upper Carboniferous turbidite fronted delta. *Proceedings of the Yorkshire Geological Society*, **51**, x.

——, ELLIOTT, T. & DAVIES, S. J. 1997. The application of sequence stratigraphy to Upper Carboniferous fluvio-deltaic strata of the onshore UK and Ireland: implications for the southern North Sea. *Journal of the Geological Society, London*, **154**, 719–733.

——, —— & FLINT, S. S. 1996. Critical application of sequence stratigraphic concepts to the Rough Rock Group (Upper Carboniferous) of northern England. *In*: HOWELL, J. A. & AITKEN, J. F. (eds) *High Resolution Sequence Stratigraphy: Innovations and Applications*. Geological Society, London, Special Publications, **104**, 221–246.

HEDBERG, H. D. 1976. *International Stratigraphic Guide*. Wiley, London, 1–200.

HOWARD, R. H. & WHITTAKER, S. T. 1988. Hydrocarbon accumulation in a paleovalley at Mississippian–Pennsylvanian unconformity near Hardinville, Crawford County, Illinois: a model paleogeomorphic trap. *Illinois Petroleum*, **129**, 1–26.

HUGHES, N. F., DE JEKHOWSKY, B. & SMITH, A. H. V. 1964. Extraction of spores and other organic microfossils from Palaeozoic clastic sediments and coals. *Compte rendu 5ème Congrès Internationale de Stratigraphie et de Géologie du Carbonifère, Paris 1963*, **3**, 1095–1109.

JONES, C. M. & CHISHOLM, J. I. 1997. The Roaches and Ashover Grits: sequence stratigraphic interpretation of a 'turbidite-fronted' delta. *Geological Journal*, **32**, 45–68.

MACCARTHY, F. J., TISDALE, R. M. & AYERS, W. B., JR 1996. Geological controls on coalbed productivity in part of the North Staffordshire Coalfield, UK. *In*: GAYER, R. & HARRIS, I. (eds) *Coalbed Methane and Coal Geology*. Geological Society, London, Special Publications, **109**, 27–42.

MAYNARD, J. R., HOFMANN, W., DUNAY, R. E., BENTHAM, P. N., DEAN, K. P. & WATSON, I. 1997. The Carboniferous of western Europe: the development of a petroleum system. *Petroleum Geoscience*, **3**, 97–115.

MCCABE, P. J. 1977. Deep distributary channels and giant bedforms in the Upper Carboniferous of the Central Pennines, northern England. *Sedimentology*, **24**, 271–290.

MCLEAN, D. 1995. Palynostratigraphic classification of the Westphalian of the Southern North Sea Carboniferous Basin [Abstract]. *In*: *Stratigraphic Advances in the Offshore Devonian and Carboniferous Rocks, UKCS and Adjacent Areas*. Geological Society, London, 20–21.

—— & CHISHOLM, J. I. 1996. Reworked palynomorphs as provenance indicators in the Yeadonian of the Pennine Basin. *Proceedings of the Yorkshire Geological Society*, **51**, 141–151.

—— & MURRAY, I. 1996. Subsurface correlation of Carboniferous coal seams and inter-seam sediments using palynology: application to exploration for coalbed methane. *In*: GAYER, R. & HARRIS, I. (eds) *Coalbed Methane and Coal Geology*. Geological Society, London, Special Publications, **109**, 315–325.

MIJNSSEN, F. C. J. 1997. Modelling of sandbody connectivity in the Schooner Field. *In*: ZIEGLER, K., TURNER, P. & DAINES, S. (eds) *Petroleum Geology of the Southern North Sea: Future Potential*. Geological Society, London, Special Publications, **123**, 169–180.

MURRAY, D. K. 1996. Coalbed methane in the USA: analogues for worldwide development. *In*: GAYER, R. & HARRIS, I. (eds) *Coalbed Methane and Coal Geology*. Geological Society, London, Special Publications, **109**, 1–12.

OLIVEIRA, J. T. 1990. Part VI: South Portuguese Zone. Chapter 2, Stratigraphy and synsedimentary tectonism. *In*: DALLMEYER, R. G. & MARTINEZ GARCIA, E. (eds) *Pre-Mesozoic Geology of Iberia*. Springer, New York, 334–347.

O'MARA, P., MERRYWEATHER, M. & WANN, D. 1997. The Trent Gas Field [Abstract]. *In*: *5th Conference on Petroleum Geology of NW Europe, London, October 1997, Abstracts*, 143.

PEPPERS, R. A. 1993. Correlation of the 'Boskydell Sandstone' and other sandstones containing marine fossils in southern Illinois using palynology of adjacent coal beds. *Illinois State Geological Survey Circular*, **553**, 1–18.

——1996. Palynological correlation of major Pennsylvanian (Middle and Upper Carboniferous) chronostratigraphic boundaries in the Illinois and other coal basins. *Geological Society of America Memoir*, **188**, 1–111.

PUCKETTE, J., ABDALLA, A., RICE, A. & KELKAR, M. G. 1996. The Upper Morrow reservoirs: complex fluvio-deltaic depositional systems. *Oklahoma Geological Survey Circular*, **98**, 47–84.

PULHAM, A. J. 1987. *Depositional and Syn-sedimentary Deformation Processes in Namurian Deltaic Sequences of West County Clare, Ireland*. PhD thesis, University of Wales, Swansea.

——1989. Controls on internal structure and architecture of sandstone bodies within Upper Carboniferous fluvial-dominated deltas, County Clare, western Ireland. *In*: WHATELEY, M. K. G. & PICKERING, K. T. (eds) *Deltas: Sites and Traps for Fossil Fuels*. Geological Society, London, Special Publications, **41**, 179–203.

QUIRK, D. G. 1997. Sequence stratigraphy of the Westphalian in the northern part of the Southern North Sea. *In*: ZIEGLER, K., TURNER, P. & DAINES, S. (eds) *Petroleum Geology of the Southern North Sea: Future Potential*. Geological Society, London, Special Publications, **123**, 153–168.

RAISTRICK, A. 1935. The correlation of coal-seams by microspore-content. Part I. – The seams of Northumberland. *Transactions of the Institution of Mining Engineers*, **88**, 142–153, 259–264.

RAMSBOTTOM, W. H. C. 1977. Major cycles of transgression and regression (mesothems) in the Namurian. *Proceedings of the Yorkshire Geological Society*, **41**, 261–291.

—— & SAUNDERS, W. B. 1985. Evolution and evolutionary biostratigraphy of Carboniferous ammonoids. *Journal of Paleontology*, **59**, 123–139.

——, CALVER, M. A., EAGAR, R. M. C., HODSON, F., HOLLIDAY, D. W., STUBBLEFIELD, C. J. & WILSON, R. B. 1978. *A Correlation of Silesian Rocks in the British Isles*. Special report of the Geological Society of London No. 10, 1–81.

RITCHIE, J. S. & PRATSIDES, P. 1993. The Caister Fields, Block 44/23a. *In*: PARKER, J. R. (ed.) *Petroleum Geology of Northwest Europe: Proceedings of the 4th Conference*. Geological Society, London, 759–769.

ROSS, C. A. & ROSS, J. R. P. 1985. Late Paleozoic depositional sequences are synchronous and worldwide. *Geology*, **13**, 194–197.

RUST, B. R., GIBLING, M. R. & LEGUN, A. S. 1984. *Coal Deposition in an Anastomosing Fluvial System: the Pennsylvanian Cumberland Group South of Joggins, Nova Scotia, Canada*. International Association of Sedimentologists, Special Publication, **7**, 105–120.

SHIRLEY, K. 1997. Headlines gone, but activity goes on; Oklahoma's Arkoma Basin remains busy. *American Association of Petroleum Geologists, Explorer*, May, 14–15.

SMITH, A. H. V. & BUTTERWORTH, M. A. 1967. Miospores in the coal seams of the Carboniferous of Great Britain. *Special Papers in Palaeontology*, **1**, 1–324.

SPEARS, D. A. & AMIN, M. A. 1981. A mineralogical and geochemical study of turbidite sandstones and interbedded shales, Mam Tor, Derbyshire, UK. *Clay Minerals*, **16**, 333–345.

STAUB, J. R. 1994. Mine level analysis of planar and raised peat deposition in a wave- and tide-influenced deltaic shoreline setting; Beckley bed (Westphalian A), southern West Virginia. *Palaeogeography, Palaeoclimatology, Palaeoecology*, **106**, 203–221.

STOREY, M. W. & NASH, D. F. 1993. The Eakring–Dukeswood oil field: an unconventional technique for describing a field's geology. *In*: PARKER, J.

(ed.) *Petroleum Geology of Northwest Europe: Proceedings of the 4th Conference*. Geological Society, London, 1527–1537.

TURNER, N., MCLEAN, D., NEVES, R., MASON, M. & WATSON, H. K. 1991. The Vanderbeckei Marine Band of the Murdoch Field, Block 44/22a, southern North Sea: biostratigraphy and geology [Abstract]. *In*: *Programme of the Joint Meeting of the Geological Society Coal Geology Group and the Yorkshire Geological Society, Carboniferous of the Southern North Sea Basin and Onshore Flanks, Sheffield, September 1991*, 5.

WAGNER, R. H. 1974. The chronostratigraphic units of the Upper Carboniferous in Europe. *Bulletin de la Societé belge de Géologie*, **83**, 235–253.

——1984. Megafloral zones in the Carboniferous. *9ème Congrès Internationale de Stratigraphie et de Géologie du Carbonifère, Compte rendu, Washington & Champaign-Urbana 1979*, **2**, 109–134.

WHITTAKER, A., COPE, J. C. W., COWIE, J. W., GIBBONS, W., HAILWOOD, E. A., HOUSE, M. R., JENKINS, D. G., RAWSON, P. F., RUSHTON, A. W. A., SMITH, D. G., THOMAS, A. T. & WIMBLEDON, W. A. 1991. A guide to stratigraphical procedure. *Journal of the Geological Society, London*, **148**, 813–824.

WOLFRAM, A. 1954. Versuche zur Trennung der Sporenmorphen von organischen und anorganischen Beimengungen unter Berücksichtigung der Wirkung des Ultraschalls auf Kohlenmazerate. *Geologie*, **3**, 655–659.

ZIEGLER, P. A. 1989. *Evolution of Laurussia: A Study in Late Palaeozoic Plate Tectonics*. Kluwer, Dordrecht.

Microfossil assemblages as proxies for precise palaeoenvironmental determination – an example from Miocene sediments of northwest Borneo

M. D. SIMMONS,[1] M. D. BIDGOOD,[2] P. BRENAC,[3] P. D. CREVELLO,[4,5] J. J. LAMBIASE[4] & C. K. MORLEY[4]

[1] *Department of Geology and Petroleum Geology, University of Aberdeen, Aberdeen AB24 3UE, UK*

[2] *Grampian Stratigraphic Services, Unit 39, Howe Moss Avenue, Kirkhill Industrial Estate, Dyce, Aberdeen AB21 0GP, UK*

[3] *Brenac Stratigraphy International, Rhosili, 133 Peulwys Lane, Old Colwyn, Clwyd LL29 8YF, UK*

[4] *Department of Petroleum Geoscience, Universiti Brunei Darussalam, Tunku 2028, Brunei Darussalam*

[5] *Current address: Petrex-PetroGeos Petroleum Geoscience Consultants, PO Box 2905 Bander Seri Begawan, Brunei Darussalam*

Abstract: Hydrocarbon reservoirs in northwest Borneo are often developed in 'paralic' depositional settings, although current exploration is evaluating relatively deep-water turbiditic plays. In the absence of conventional core, and with only ambiguous wireline log and seismic signatures being available, the use of microfossil data is considered to determine precise depositional setting. This is important because different depositional settings imply different reservoir qualities in terms of architecture, connectivity, heterogeneity and poroperm characteristics.

Equivalents of the reservoir succession are well exposed in northwest Brunei, and contain well-preserved sedimentary features and ichnofossils to determine precise depositional setting. Microfossil assemblages (both palynomorphs and foraminifera) have been sampled from each depositional environment identified at outcrop and by using an iterative approach, and incorporating data on modern distributions, diagnostic microfossil assemblages and taxa have been identified which can be used as precise palaeoenvironmental proxies. By using this approach distal turbidite, proximal turbidite, open shelf with slumping, open shelf, lower shoreface, upper shoreface, tidal flat with tidal channels, lower distributary channel, lagoon–distributary channel margin and upper distributary channel depositional environments can be recognized.

The discovery of oil at Miri in northeast Sarawak in 1910, followed by the discovery of the Seria Oilfield in Brunei in 1929 (Harper 1975), signalled the establishment of northwest Borneo as a world-class hydrocarbon province. In Brunei, exploration moved offshore in the 1960s leading to the discovery of numerous fields such as Champion, Magpie and Iron Duke (Sandal 1996). All these onshore and offshore discoveries were in 'paralic' (i.e. marginal marine) sediments of Miocene age. At the time of writing, exploration is moving into the deeper water offshore, searching for hydrocarbons in the turbidite equivalents of the existing paralic plays. In offshore Sarawak, the Miocene-aged Luconia carbonates also provide exploration and production opportunities, although will not be referred to further herein.

Biostratigraphy has always played an important role in exploration studies in the region. The reservoir and adjacent sediments are normally rich in spores and pollen, allowing for exploration-scale biozonation using pollen lineages such as the mangrove *Florschuetzia* (Morley 1991; Sandal 1996). Broad-scale chronostratigraphic calibration is achieved using the

SIMMONS, M. D., BIDGOOD, M. D., BRENAC, P., CREVELLO, P. D., LAMBIASE, J. J. & MORLEY, C. K. 1999. Microfossil assemblages as proxies for precise palaeoenvironmental determination – an example from Miocene sediments of northwest Borneo. *In:* JONES, R. W. & SIMMONS, M. D. (eds) *Biostratigraphy in Production and Development Geology*. Geological Society, London, Special Publications, **152**, 219–241.

occasional occurrence of calcareous nannofossils and planktonic foraminifera in sediments representing more open marine settings. Benthonic foraminifera assemblages have been used to establish broad palaeoenvironmental interpretations (James 1984; Sandal 1996).

As well as exploration in the deeper water offshore, the industry in northwest Borneo, as in many other parts of the world, is interested in maximizing recovery from existing fields. To do so requires a precise understanding of the depositional setting of the reservoir, so as to assess reservoir architecture, connectivity and compartmentalization, in order to locate horizontal and multilateral production wells efficiently. In the absence of cores, depositional setting is often interpreted from log signatures and seismic facies. However, in northwest Borneo, these are often ambiguous, it being difficult, for example, to distinguish shoreface sands from distributary channel sands, which have differing reservoir properties. It is our contention that biostratigraphy could hold the key, in combination with other subsurface methods, for determining precise depositional environment, which is so important for effective production strategies. Whilst biostratigraphy has always been used to determine depositional setting of subsurface samples, these interpretations have often been too broad for use in detailed reservoir interpretation. Samples have often been described as 'fluvio-marine', 'paralic' or 'deltaic'. These limited interpretations result from the following uncertainties.

- There is little information on the environmental distribution of the modern-day equivalents of the microfossils occurring in the reservoir succession.
- In terms of benthonic foraminifera, the assemblages are dominated by agglutinating forms, which are often difficult to interpret in terms of depositional environment. Such assemblages are often interpreted as indicating either low-oxygen, deep-marine environments, or freshwater-influenced, marginal marine environments. To hedge their bets, some biostratigraphic contractors have given interpretations of such assemblages as 'marginal marine to bathyal marine' – obviously an interpretation of limited value! The problem is further compounded in that because of the pressure of time constraints on the reporting of subsurface studies, many agglutinating foraminifera are identified to generic level only. For example, '*Trochammina* sp.' will have a (palaeo)environmental range of marginal marine to bathyal marine, but it may be possible to identify species of *Trochammina* which are restricted to more precise depositional environments. R. W. Jones (pers. comm.) reports that in the Neogene basins of eastern Venezuela, speciation of *Trochammina* is indeed useful for such precise palaeoenvironmental definition.

In order to overcome these uncertainties we have conducted a study on the outcrop analogues of the subsurface reservoirs in northwest Borneo. There are extensive outcrops of the equivalents of the subsurface reservoirs in Brunei, in northeast Sarawak and on the island of Labuan (Fig. 1). These outcrops are broadly age equivalent to the reservoir successions (see the section on 'Chronostratigraphic results' below) and encompass the range of depositional environments known to occur in the subsurface. Sedimentary structures and trace fossils are often very evident at outcrop, permitting a precise and accurate interpretation of depositional environment (Lambiase *et al.* in press). A range of depositional environments from fluvial, through tidally influenced distributary channel, lagoon, estuarine, upper and lower shoreface, open inner shelf, deep shelf and proximal and distal turbidite are all recognized. We have sampled a representative set of outcrops encompassing the full range of depositional environments noted above. Each sample has been processed and prepared for the study of its content of foraminifera and palynomorphs. By undertaking an iterative approach, whereby the interpretations of depositional environment from the microfossils, sedimentology and ichnology are combined, we have been able to recognize microfossil assemblages which characterize each of the depositional environments recognizable from sedimentology. We believe that this will have great practical value in the subsurface where, in the absence of core, detailed sedimentology is not possible. Microfossils, as recovered from cuttings and sidewall cores, can now be used as precise palaeoenvironmental proxies for interpreting the reservoir succession, thus permitting biostratigraphy to be a valuable tool when applied to production and development geology. A very similar approach has been adopted in the palaeoenvironmental interpretations of the deltaic sediments of Miocene–Pliocene age of the East Venezuela basin (Jones *et al.* 1999).

Geological background

Excellent descriptions of the geology of Brunei and northwest Borneo can be found in the recent publication by Brunei Shell Petroleum (Sandal

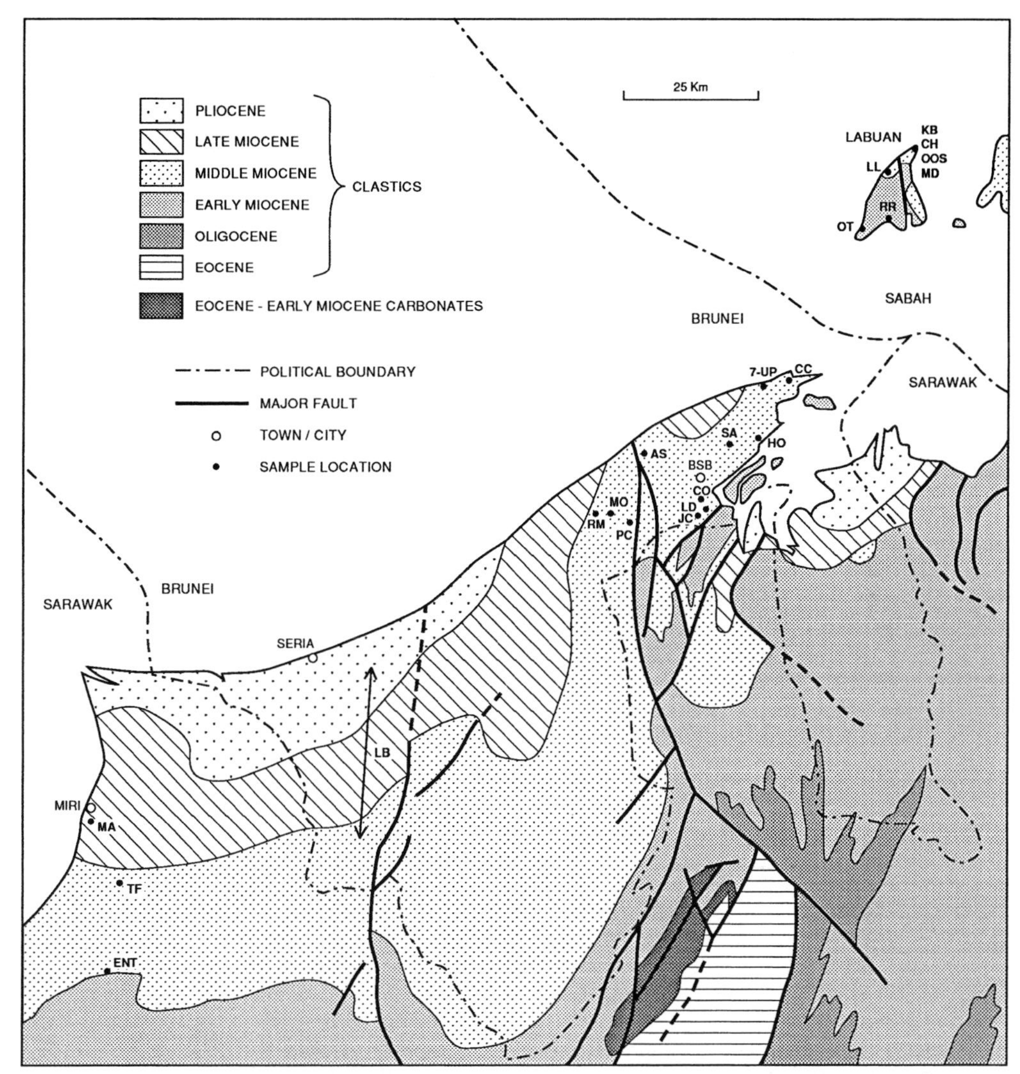

Fig. 1. Geological map of northwest Borneo (modified after Sandal 1996), showing the location of outcrops sampled during this study and referred to in the text. Political boundaries are approximate.

1996) and its predecessor (James 1984). As noted above, the main reservoirs of northwest Borneo were developed in a 'paralic' setting (see also Prosser & Carter 1997). The collision of the Sundaland plate and South China Sea portion of the Eurasian plate in the Late Palaeogene–Neogene led to the development of the Crocker–Rajang mountain belt. This uplifted sediments of the former Crocker–Rajang accretionary complex, subsequent erosion of which caused several rapidly prograding clastic depositional systems to develop. These are usually referred to as 'deltas', but we prefer to refer to them as prograding clastic systems as they represent a more complex set of depositional environments than can be envisioned within a deltaic model. Initially, an Early Miocene Melingan prograding clastic system was deposited, followed by a Middle–Late Miocene Champion system and a Late Miocene–Quaternary Baram system. Sediments of the Champion prograding clastic system are the focus of this study as they form the sediments at outcrop and the majority of subsurface reservoirs.

As noted above, the Neogene depositional systems are much more complex than can be characterized by envisaging three progradational delta complexes. First, a complex interaction between tectonics and sedimentation resulted in the progressive partitioning of the Neogene deltaic depocentres into sub-basins. A number of diapir-cored structures grew episodically and simultaneous with deposition (Morley *et al.* 1998), which profoundly altered depositional settings on a local scale. Secondly, high-frequency changes in sea level caused not only progradation, aggradation and retrogradation, but also marked changes in the profile of the coastline. In terms of a general depositional model for the studied sediments, analogy with parts of the coastline of modern-day northwest Borneo (e.g. Brunei Bay) is likely, as is analogy with the sub-Recent Baram delta, which was flooded *c.* 5400 years BP (Caline & Huong 1992). A coastline with variable wave, tidal and fluvial dominance (as in modern Brunei Bay) seems likely, into which major river systems were feeding. In times of falling relative sea level these formed progradational clastic systems ('deltas'). At times of relative sea-level rise, large estuaries were formed. Our studies of the outcrops (see Lambiase *et al.* in press) suggest that classical deltaic sequences are rarely observed. Instead, much of the Miocene sedimentation falls more readily into a model dominated by large estuaries such as modern-day Brunei Bay. The palaeogeographies of north-west Sabah presented by Rice-Oxley (1991) support this contention.

In terms of lithostratigraphy (Fig. 2) these sediments are assigned to the Belait Formation in Brunei and Labuan, which progrades over the older, basinal Setap and Temburong formations. In northeast Sarawak and western Brunei, the progradational complex includes the Sibuti, Lambir and Miri formations. This lithostratigraphic separation of northeast Brunei and northeast Sarawak reflects a further complication of the Neogene depositional systems in that the Lambir and Berekas sub-basins had independent controls on deposition, much like at present where in the west the Baram delta is being deposited, whilst in the east the large estuary of Brunei Bay occurs. In the Lambir sub-basin sedimentation tends to be wave-dominated, whilst in the Berekas sub-basin sedimentation is more tidally influenced. Geological maps published by James (1984) and Sandal (1996) indicate the occurrence of the Miri Formation in northeast Brunei, suggesting the need for a fundamental revision of the lithostratigraphy of the region.

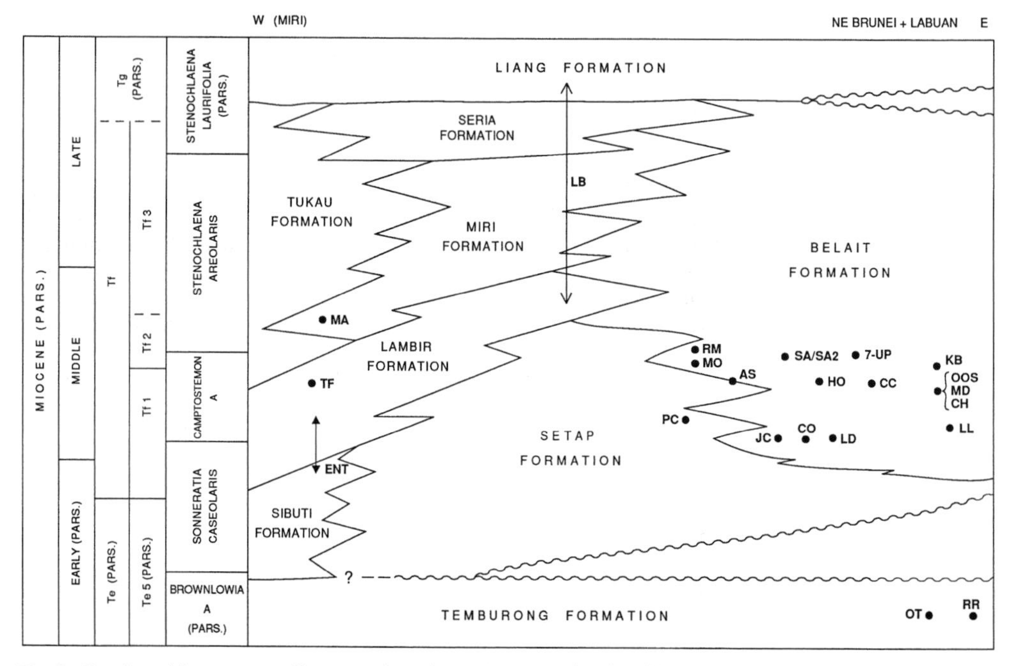

Fig. 2. Stratigraphic summary diagram of northwest Borneo, showing lithostratigraphy v. local biostratigraphy and chronostratigraphy (modified after Sandal 1996), and the stratigraphic position of outcrops sampled during this study and located in Fig. 1.

Sample locations

Figure 1 shows the geographic locations of the outcrops sampled during this study and Fig. 2 their stratigraphic position (see also discussion of chronostratigraphy below). Over 200 samples from 28 different localities have been studied in the course of this project.

Three localities were studied south of Miri in Sarawak. The ENT section in the Lambir Hills represents the transition from the Sibuti facies of the Setap basinal succession, into the wave-dominated prograding shoreface sediments of the overlying Lambir Formation. The lower part of the succession represents open shelf sedimentation (Sibuti Formation/facies), whilst the upper part (Lambir Formation) represents storm-influenced lower and upper shoreface deposition. The TF section, also in the Lambir Hills, represents upper shoreface deposition of the Lambir Formation cut by major distributary channels. The MA section represents a sand-rich tidal flat–beach succession in the Miri Formation, exposed in the crest of the Miri anticline, just south of the town of Miri.

Eighteen different outcrops were visited in Brunei, the most significant of which are as follows. CC is from an outcrop near the abandoned Brookerton Colliery where a tidally influenced estuarine succession is cut by a distributary channel, which is subsequently infilled by tidal flat deposits, culminating in coal deposition. 7-UP is from the 7-UP beach succession, which exposes a muddy embayment sequence punctuated by sand-filled tidal channels and tidal flats. AS, near Jerudong, represents the transition from the basinal Setap Formation to the prograding deposits of the Belait Formation. Open shelf sediments are well exposed. The nearby PC outcrop also represents open shelf deposition, but where syn-depositional tectonics have led to slumped blocks being redeposited on the margins of an open shelf environment. Lower shoreface deposits of the Belait Formation are well exposed at the MO locality to the west of Jerudong, whilst at RM there are good exposures of upper shoreface deposits. The SA2 succession along the Sungai Akar road presents a good succession of upper shoreface sediments, whilst the underlying SA succession represents a tidal flat–tidal channel complex. The HO succession on the Kota Batu road is also a tidal flat–tidal channel succession, cut by distributary channels and with an embayment succession culminating in coal deposition.

Seven outcrops were studied on the island of Labuan which represent a variety of depositional settings. OT is from the oil terminal at the southwest corner of the island, and is in proximal turbidite facies of the Temburong Formation. RR is from a temporary outcrop at Rancha-Rancha on the southern coast of the island and is the distal turbidite equivalent of OT. LL is Lyang-Lyang Beach on the west coast of the island were the basinal Setap Formation (in distal delta front/lower shoreface facies) is abruptly overlain by fluvial sediments of the Belait Formation equivalent. A similar situation occurs at Kubong Bluff (KB) on the northern tip of the island, although there the Belait Formation equivalent includes fluvial, tidal flat and lower shoreface facies. Further lower shoreface deposits are exposed at the MD outcrop close to KB, along with upper shoreface, tidal flat and tidally influenced distributary channel deposits. CH and OOS are further small outcrops of the lower shoreface facies of the Setap Formation at the northern end of the island.

Chronostratigraphic results

The paralic Belait, Lambir and Miri formations, which are the focus of this study, generally lack microfossils suitable for high-resolution chronostratigraphic determinations. Planktonic foraminifera and calcareous nannofossils are almost absent, whilst most benthonic foraminifera species tend to be long ranging, or are new species of uncertain age range. Age determinations thus tend to be based on spore and pollen palynology, as originally defined by Muller and co-workers (Muller 1964, 1969, 1972; Germeraad *et al.* 1968) and refined by Morley (e.g. Morley 1991). Age-significant forms include the lineage of mangrove pollen of the genus *Florschuetzia*, and the pollen *Camptostemon*, *Dacrydium*, *Praedapollis* and *Piceapollenites*. Within the open marine Setap Formation (and equivalents), beneath the Belait and Lambir formations, and in the offshore facies equivalents of the proximal facies of the Belait and Lambir formations, in addition to age significant spores and pollen, some age-significant planktonic foraminifera (Eckert 1970) and dinoflagellates (Besems 1993) occur.

At all the outcrops studied the Setap Formation and its equivalents, directly beneath the Lambir and Belait formations, is of Early Miocene age, based on the presence of *Florschuetzia trilobata* and the absence of *Camptostemon* sp. At the ENT section in the Lambir Hills a transitional section between the Setap Formation ('Sibuti facies') and the Lambir Formation is exposed. Here the Sibuti Formation contains planktonic foraminifera indicative of an Early

Miocene age (presence of *Globigerinoides obliquus*; absence of *Orbulina* spp.), whilst the presence of *Florschuetzia levipoli* and *Florschuetzia meridionalis* suggests a late Early Miocene age (the association of the calcareous nannofossils *Sphenolithus heteromorphus* with *Reticulofenestra haqii* and *Reticulofenestra minuta* also supports this age assignment). The overlying mudstones and sandstones of the Lambir Formation are no older than intra-Middle Miocene based on the co-occurrence of *F. trilobata* and *Camptostemon* sp. This suggests that there is an unconformity between the Setap and Lambir formations at this locality, with the earliest Middle Miocene being absent. The presence of slight angular disconformity in the outcrop and abrupt change in depositional environments supports this possibility. On the island of Labuan a similar situation exists where the uppermost Setap present (outcrop LL) is Early Miocene in age based on the abundance of *F. trilobata*, presence of *Praedapollis* sp. and absence of *Camptostemon* sp., whilst the lowest paralic sands (outcrop KB) are intra-Middle Miocene based on the presence of *Camptostemon* sp.

It appears that at all the localities studied, in the Lambir, Belait, Berakas and Labuan sub-basins (outcrops AS, ENT, KB, MO, PC and RM), the onset of paralic sedimentation (i.e. progradation of the various delta systems) is approximately synchronous within the resolution of biostratigraphy. These initial deltaic sands are all intra-Middle Miocene in age, corresponding to the stratigraphic inception of *Camptostemon* sp. *Florschuetzia meridionalis* also occurs in relatively large numbers, as opposed to *Florschuetzia trilobata* or *Florschuetzia levipoli.* It seems likely that an intra-Middle Miocene tectonic event caused uplift of the Crocker Range in central Borneo, and caused simultaneous progradation of clastic depositional systems into the South China Sea from both southeasterly and southwesterly directions.

Stratigraphic differentiation within the Belait, Lambir and Miri formations is hampered by the lack of chronostratigraphically significant microfossils. However, outcrops known to be in the upper part of the Belait Formation (at the SA–SA2 sections in Brunei) are regarded as being distinctively younger, no older than late Middle Miocene, based on the co-occurrence of *F. trilobata* and *Dacrydium* sp.

Turbiditic sediments of the Temburong Formation exposed on the southern part of Labuan Island (outcrops OT and RR) form the oldest sediments studied herein, being of early Early Miocene age based on the presence and abundance of *F. trilobata* and *Piceapollenites* sp. One clast from the OT locality demonstrates the reworking of Middle Eocene pelagic sediments with abundant Middle Eocene planktonic foraminifera, dinocysts and calcareous nannofossils being present. In other samples from these localities reworking of Oligocene sediments is evident.

It is clear that high-resolution correlation within the Belait, Lambir and Miri formations will not be possible using microfossil inception and extinction events. Quantitative studies such as those proposed by Chow (Shell Sarawak internal reports) for offshore Sarawak may be of value, using ratios of climatically sensitive pollen taxa to detect a high-resolution climatostratigraphic signal. However, the strong palaeoenvironmental control on microfossil assemblages as described below needs to be borne in mind, as this may result in correlations of similar depositional environment being confused with chronostratigraphically significant correlations. Studies of whole-rock geochemistry to recognize geochemical correlations may prove useful, although, once again, palaeoenvironmental (and also diagenetic) controls will have to discounted before correlations can be regarded as truly chronostratigraphic.

Palaeoenvironmental proxies

As noted above, the Miocene sediments of northwest Borneo are reasonably rich in palynomorphs and foraminifera. Foraminiferal assemblages are typically dominated by agglutinated forms, but in open marine shelfal settings calcareous forms are relatively common, including sparse planktonics. Comparison with the coeval faunas of Sabah (Whittaker & Hodgkinson 1979), Indonesia (Van Marle 1991) and the Sunda Shelf (Biswas 1976) is possible. The seminal review of the Recent foraminifera collected on the Challenger Expedition of the last century by Jones (1994) provides an invaluable guide for identification of taxa encountered in this study.

The taxonomy and palaeoenvironmental significance of the agglutinating foraminifera present in the studied sediments has recently been assessed by Bidgood *et al.* (in press), in a study allied to that presented here. Knowledge of the agglutinated foraminifera occurring in the Neogene–Recent sediments of northwest Borneo is relatively limited (although see Bronnimann & Whittaker (1993) for a discussion of Recent agglutinated foraminifera from the Malay Archipelago). It is clear from the work of Bidgood *et al.* (in press) that several undescribed taxa

ENVIRONMENT OF DEPOSITION	PALYNOLOGY										
	MANGROVE	BACK-MANGROVE	COMPETACEAE	FRESHWATER SWAMP	DIPTEROCAPACEAE	PALMS	GYMNOSPERMS	FRESHWATER ALGAE	DINOCYSTS	SPORES	REWORKING
DISTAL TURBIDITE	VR	O		O		O	O		VR	C-A	
PROXIMAL TURBIDITE	C-A									C-A	Abt
OPEN SHELF WITH SLUMPS	O	O		O-C		O-C			VR	C	
OPEN SHELF	R-O	R		R-O		R-O			R-A	C-A	
LOWER SHOREFACE	R-O	R		R		O-C			VR	C-A	
UPPER SHOREFACE	R	R		R		O-C				C	
TIDAL FLAT+ TIDAL CHANNEL	C	R	(C)	R	(C)	C		O	VR	C	
LAGOON OR DISTRIBUTARY CHANNEL MARGIN	Abt									R	
LOWER DISTRIBUTARY CHANNEL	C	C				C			VR	O-C	
UPPER DISTRIBUTARY CHANNEL + FLUVIAL	VR	C		C		VR		O		Abt	

ENVIRONMENT OF DEPOSITION	FORAMINIFERA													
	PLANKTONICS	CALCAREOUS BENTHONICS	CALCAREOUS AGGLUTS	AGGLUT DIVERSITY	FINE 'TROCHAMMINA'	WHITE SUGARY 'TROCHAMMINA'	COARSE 'TROCHAMMINA'	'TEXTUKURNUBIA'	'PSEUDOEPISTOMINA'	'CYCLAMMINA'	KARRERIELLA	TREMATOPHRAGMOIDES	MILIAMMINA	HAPLOPHRAGMOIDES CF. EGGERI
DISTAL TURBIDITE				LOW	C-A									
PROXIMAL TURBIDITE				LOW	C-A					R				R
OPEN SHELF WITH SLUMPS	R	Abt	C	MOD	C			C						
OPEN SHELF	R-C	C-A	O	MOD		C					O			
LOWER SHOREFACE				MOD		C			R-C					
UPPER SHOREFACE	VR	VR		MOD		C					O			
TIDAL FLAT+ TIDAL CHANNEL				MOD			C							
LAGOON OR DISTRIBUTARY CHANNEL MARGIN				LOW			R					C	C	
LOWER DISTRIBUTARY CHANNEL				MOD			C		R					
UPPER DISTRIBUTARY CHANNEL + FLUVIAL				NIL										

Fig. 3. Distribution of groups of microfossils recognized within this study v. deposition environments recognized at outcrop and thought to occur in the subsurface. This chart is the basis for the discussion in the palaeoenvironmental proxies part of this paper.

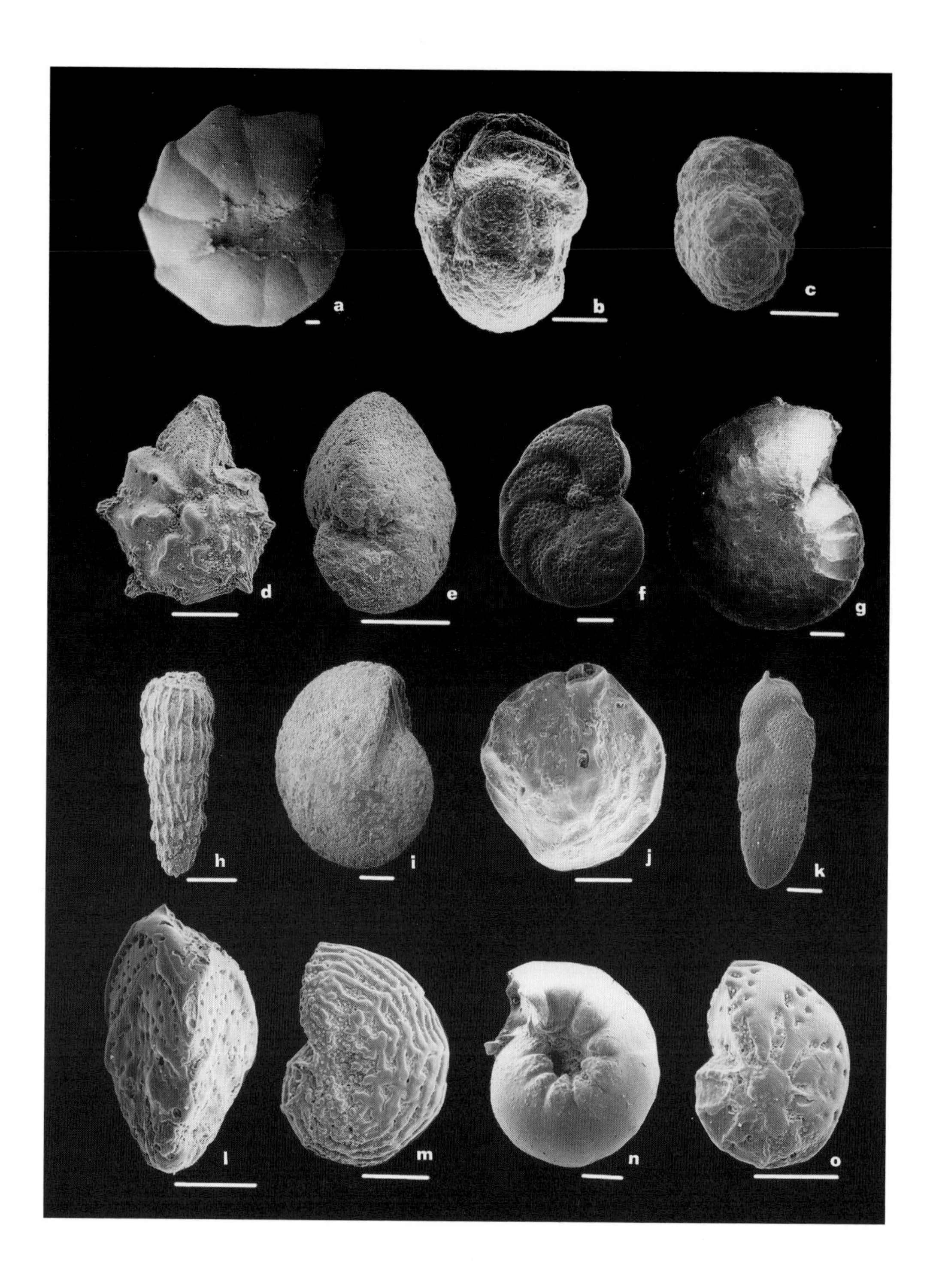
a
b
c
d
e
f
g
h
i
j
k
l
m
n
o

occur. This observation, together with the differentiation of forms of common genera such as *Trochammina*, *Recurvoides* and *Cyclammina*, permits the use of agglutinating foraminifera as reasonably precise palaeoenvironmental indicators. This is further supported by the work of Bronniman & Keij (1986) in reviewing the occurrence of modern agglutinated foraminifera in the estuaries and bays of Brunei, whilst the work of Dhillon (1968) and Ho (1971) provides useful information on the modern distribution of all groups of foraminifera in northwest Borneo.

Palynomorphs are often abundant in the studied material and can be used for palaeoenvironmental interpretation (see also Haseldonckx 1974). However, most samples contain material redeposited from hinterland environments, thus initial interpretations of palynological assemblages can result in interpretations which are significantly more proximal than when the whole microfossil assemblage and sedimentology are considered.

In the discussion of palaeoenvironmental proxies below, palynomorphs are grouped into categories based on their general environmental preferences and taxonomy.

Mangrove belt – including *Zonocostites* (=modern *Rhizophora*), *Avicennia* and *Florschuetzia* (= modern *Sonneratia*).

Back-mangrove – including *Acrostrichum* and Combretaceae.

Coastal gymnosperms – including *Casuarina*.

Freshwater swamp – including *Blumeodendron*, *Calophyllum*, *Dicolpopollis*, *Lanagiopollis*, Dipterocarpaceae, Pentace, *Camptostemon* and *Marginopollis concinnus* (=modern *Barringtonia*).

Coastal and mangrove palms – including *Oncosperma* and *Spinizonocolpites echinatus* (= modern *Nypa*).

Hinterland gymnosperms – including *Dacrydium* and *Pinus*.

Freshwater algae – including *Pediastrum* and *Botryococcus*, the later also being present in saline waters such as those present in lagoons and intertidal regions.

Marine dinocysts – including *Spiniferites*, *Systematophora* and *Operculodinium*.

Hinterland spores – including *Laevigatosporites*, *Verrucatosporites*, *Leiotriletes*, *Selaginella* and *Lycopodium*.

This gives the broad environmentally controlled elements of any palynological assemblage within the studied samples. Obviously, most pollen and spores can be dispersed from the location of the plant they originate from. In our attempt to nominate palaeoenvironmental proxies, we consider the relative abundance of these elements in the context of the production rate of pollen or spores from the parent plant. Thus, for example, we consider the abundance of freshwater swamp elements and back-mangrove elements even in sedimentary environments we know to be relatively distal.

Anderson & Muller (1975) concluded that palynological assemblages from a Miocene coal within the Belait Formation were similar to those from a Holocene peat. Although the Miocene coal sampled by these authors may be from a different locality to those sampled by us, we cannot concur with their conclusions relating to the origin of the Belait Formation coals. All the coals we have sampled are extremely rich in mangrove pollen (to the exclusion of other types) (see further discussion below), and we suggest that these coals formed by infilling of distributary channels by mangrove swamps during late highstand, or by accumulation of mangrove swamp vegetation on windward or current protected edges of distributary channel mouths as happens in modern Brunei Bay.

From the outcrops noted earlier, 10 distinct depositional environments can be recognized on the basis of their integrated sedimentology,

Fig. 4. Some representative foraminiferal taxa from various palaeoenvironments from the northwest Borneo area. All scale bars = 100 μm. **(a)** '*Cyclammina*' sp. Characteristic of proximal turbidite settings. Recorded so far from offshore borehole material only. Generic character (alveolar wall) unclear. **(b)** '*Recurvoides*' sp. Form characteristic of proximal and distal turbidite settings. Note the fine-grained agglutinated wall structure. Generic character (streptospiral coiling) unclear. **(c)** '*Trochammina*' sp. Form characteristic of proximal and distal turbidite settings. Note the fine-grained agglutinated wall structure. Apertural position unclear. **(d)** *Asterorotalia* sp. Characteristic of 'normal' open shelf and open shelf with slumps settings. **(e)** *Nonion* sp. Characteristic of 'normal' open shelf settings. **(f)** *Planulina* cf. *wuellerstorfi* (Schwager). Characteristic of 'normal' open shelf and particularly open shelf with slumps settings. **(g)** *Operculina* sp. Characteristic of 'normal' open shelf settings, probably deeper parts of the shelf. **(h)** *Rectuvigerina ?striata* (Schwager). Characteristic of 'normal' open shelf settings. **(i)**: *Lenticulina* sp. Characteristic of 'normal' open shelf settings. **(j)** *Quinqueloculina parkeri* (Brady). Characteristic of 'normal' open shelf settings, probably shallower parts of the shelf. **(k)** *Bolivinita spathulata* (Williamson). Characteristic of 'normal' open shelf and open shelf with slumps settings. **(l)** *Reusella simplex* (Cushman). Characteristic of 'normal' open shelf settings. **(m)** *Lenticulina* sp. nov. Characteristic of 'normal' open shelf settings. **(n)** *Hansenisca neosoldanii* Brotzen. Characteristic of 'normal' open shelf and open shelf with slumps settings. **(o)** *Elphidium* sp. Characteristic of 'normal' open shelf and open shelf with slumps settings.

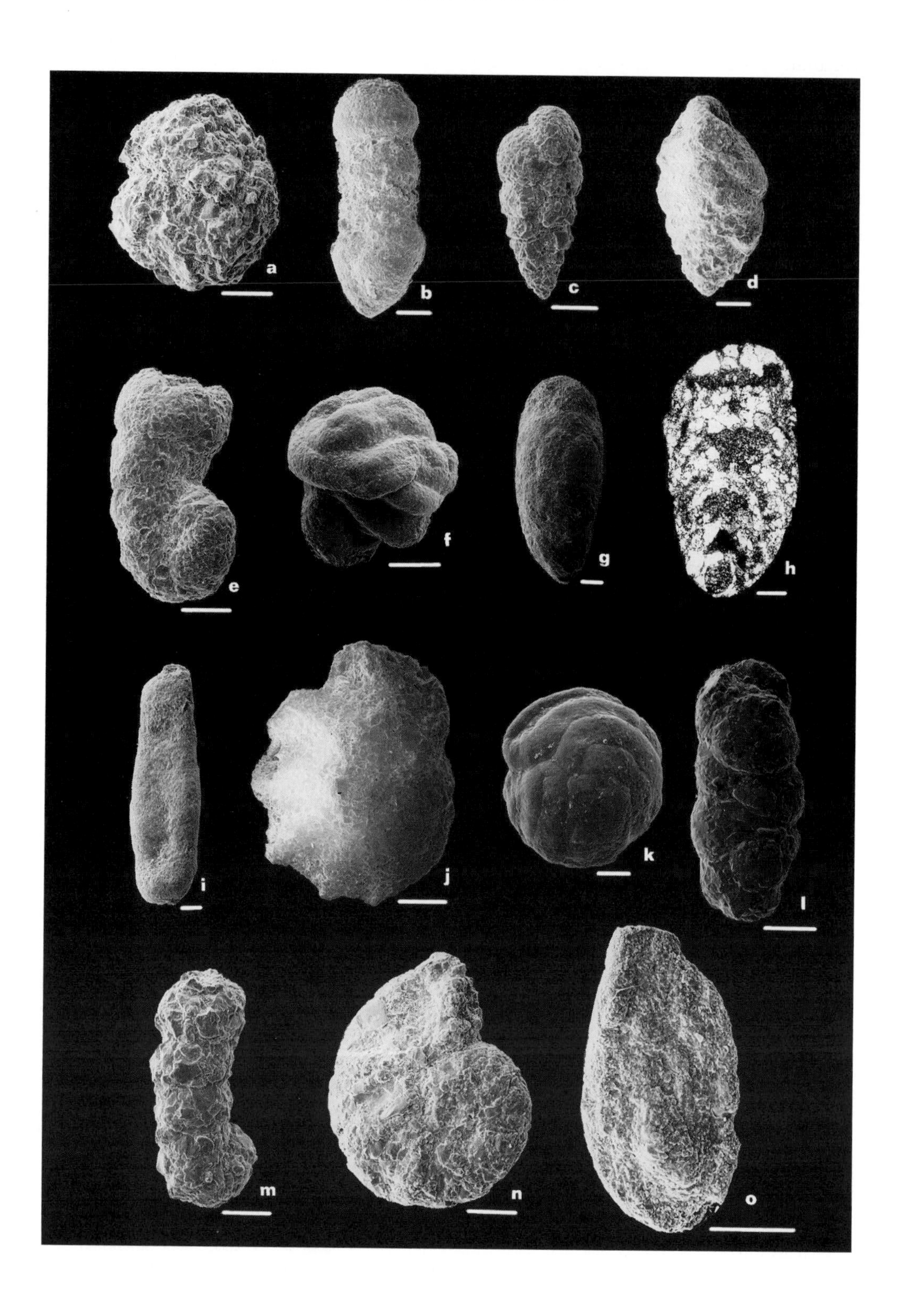
a
b
c
d
e
f
g
h
i
j
k
l
m
n
o

ichnology, micropalaeontology and palynology. These are: distal turbidite; proximal turbidite; open shelf with slumping; open shelf; lower shoreface; upper shoreface; tidal flat with tidal channels; lower distributary channel; lagoon–distributary channel margin; and upper distributary channel. We believe that each of these depositional environments has its own distinctive microfossil assemblages based on our empirical observations of the outcrops and our (limited) knowledge of the modern-day distributions of the equivalents of the microfossils occurring in these assemblages. An important point to stress is that it is essential to study palynology in conjunction with foraminifera. Neither discipline alone will give unequivocal results, whilst the combination of palynological data with foraminiferal data is a powerful tool for interpreting depositional environments. These results are summarized in Fig. 3 and considered in more detail below. Figures 4–13 illustrate the key microfossils and microfossil assemblages referred to in the text.

Distal turbidite

Distal turbidite deposits of northwest Borneo contain microfossil assemblages which differ markedly from what might be expected from similar-aged deposits from other parts of the world. Contrary to expectations, planktonic foraminifera and dinocysts are very rare or absent. The reason for this is that in the modern day the Baram, Temburong and Brunei rivers, and others, input immense amounts of freshwater onto the northwest Borneo shelf, causing a freshwater plume to be developed which extends far out into the South China Sea. Such freshwater plumes are thought to have been present during the Miocene and would have severely impacted upon the occurrence of stenohaline marine plankton such as planktonic foraminifera and most dinocysts. Other examples of this phenomenon have been described by Brenac & Richards (1998) and Dunay *et al.* (1998). Palynological assemblages are thus rather impoverished, containing very rare or occasional back-mangrove elements, gymnosperms, palms and freshwater swamp species. Mangrove pollen and dinocysts are extremely rare. However, spores can be common–abundant, testifying to their marked abilities for dispersion, and leading to possible misinterpretation of such assemblages as representing more proximal, marginal marine environments.

Foraminiferal assemblages are rather distinctive, containing abundant agglutinating forms especially fine-grained and often deformed specimens of '*Trochammina*/*Recurvoides*'. The fine-grained nature of the wall (individual agglutinated particles typically no more than 2–3 μm in

Fig. 5. Some representative foraminiferal taxa from various palaeoenvironments from the northwest Borneo area. All scale bars = 100 μm. (**a**) '*Trochammina*' sp. Note the coarse agglutinating wall structure which is often white and 'sugary' in texture. This is a characteristic of this taxon in open shelf and, particularly, shoreface assemblages. (**b**) *Martinoltiella*? sp. 1 (*sensu* Bidgood *et al.* in press). Characteristic of open shelf and upper shoreface settings. (**c**) *Textularia agglutinans*. Characteristic of open shelf and open shelf with slumps settings where Textulariidae are fairly diverse. This particular form is also found, rarely, in tidal channel and shoreface settings. (**d**) *Textularia* sp. 2 *sensu* Bidgood *et al.* (in press). Characteristic of open shelf and open shelf with slumps settings. (**e**) *Ammobaculites* sp. 1a (*sensu* Bidgood *et al.* in press). Characteristic of lagoonal depositional settings. (**f**) *Glomospira gordialis* (Jones & Parker). Characteristic of fully marine environments. (**g**) '*Textukurnubia*' sp. *sensu* Bidgood *et al.* (in press). This large highly distinctive cigar-shaped form has been found commonly in marine shelf with slumps settings. (**h**) '*Textukurnubia*' sp. *sensu* Bidgood *et al.* (in press). Thin-section photograph showing predominantly biserial chamber arrangement. (**i**) Uncertain genus No. 1 (*sensu* Bidgood *et al.* in press). This distinctive form of as yet unproven affinity is found in association with '*Textukurnubia*' and other agglutinants in open shelf with slumps settings. (**j**) '*Pseudoepistomina*' sp. *sensu* Bidgood *et al.* (in press). Another distinctive agglutinant which displays characteristic sunken chamber walls. It is found in shoreface and lower distributary channel–estuarine settings. (**k**) '*Glomospira*' *glomerata* (Hoeglund). This glomospirid is recorded from open marine (both shallow and deep) settings, as well as being found in lower shoreface environments. (**l**) *Karreriella* sp. *Karreriella* is recorded from most marine environments from deep waters up to shoreface settings. It displays changes in wall structure with fine-grained and coarse-grained forms (such as that illustrated here) recorded from offshore and nearshore waters, respectively. (**m**) *Ammobaculites exiguus* Cushman & Bronnimann. This species appears to have a preference for shallow-marine and shoreface settings but has also been recorded from lagoonal samples. It is commonly regarded throughout the area as a general 'fluvio-marine' indicator. (**n**) *Trematophragmoides bruneiensis* Bronnimann & Keij?. This form has been recorded from brackish (lagoonal, mangrove swamp or overbank) settings. (**o**) *Miliammina fusca* (Brady). This form has a wide environmental preference (brackish hypersaline marshes to upper bathyal) according to Murray (1991). However, in the northwest Borneo study area it is a commonly recorded component of marsh and lower distributary channel environments where it is often associated with abundant mangrove pollen.

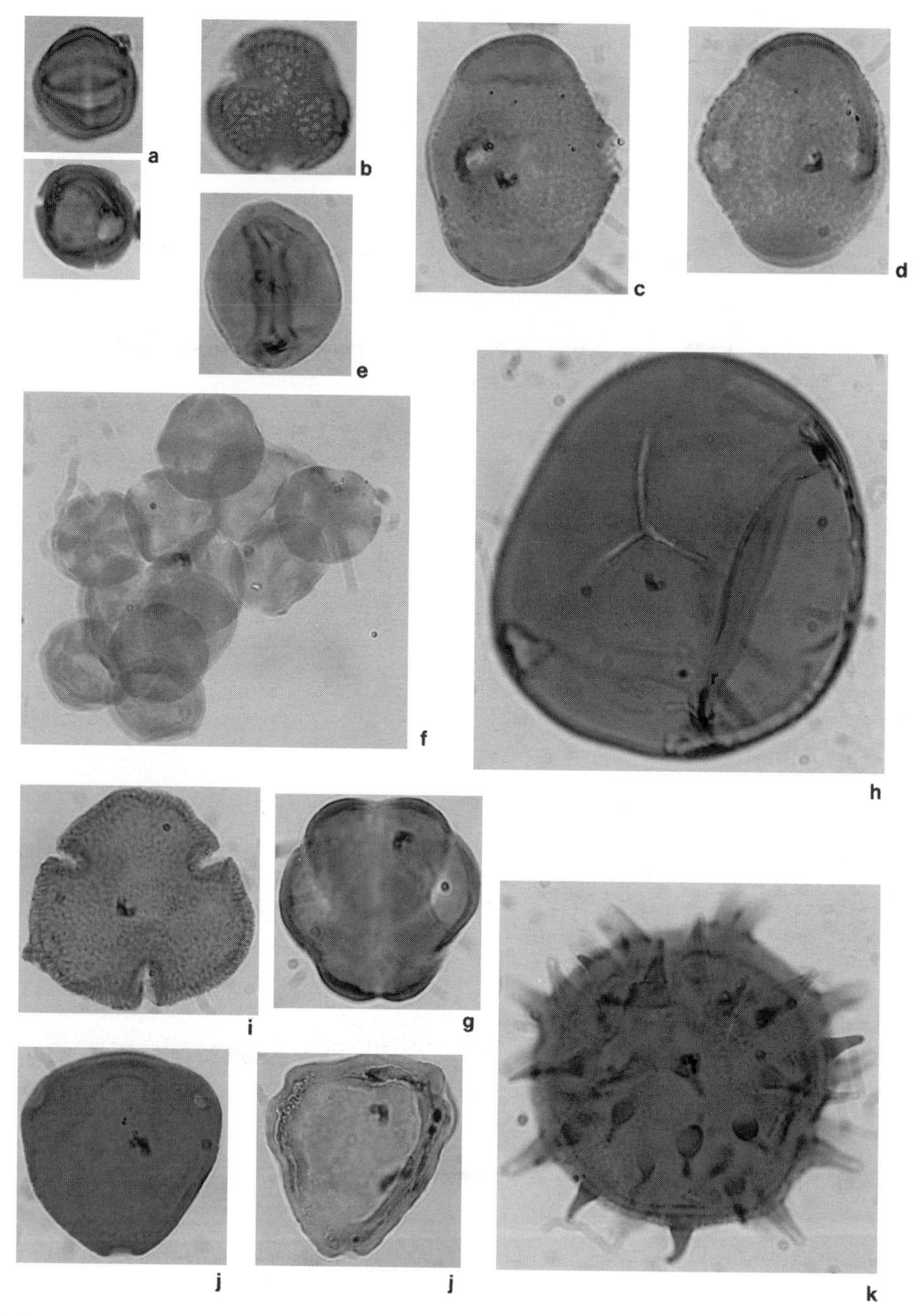

Fig. 6. Some representative palynological taxa and palynofacies from various palaeoenvironments from the northwest Borneo area. *Mangrove pollen assemblage*: (**a**) *Rhizophora* type (*Zonocostites ramonae*) – equatorial and polar views ×1000; (**b**) *Avicennia* sp. – polar view ×1000; (**c**) *Sonneratia caseolaris* (*Florschuetzia levipoli*) – equatorial view ×1000; (**d**) *Sonneratia alba* (*Florschuetzia meridionalis*) – equatorial view ×1000; (**e**) *Florschuetzia trilobata* – equatorial view ×1000. *Backmangrove miospore assemblage*: (**f**) Combretaceae–pollen massulae – different views ×1000; (**g**) *Lumnitzera* type–Combretaceae – subequatorial view ×1000; (**h**) *Acrostichum aureum* – proximal view ×1000; (**i**) *Brownlowia* type (*Discoidites bomeensis*) – polar view ×1000. *Coastal gymnosperm taxa*: (**j**) *Casuarnia* type *equisetifolia* – polar view ×1000. *Freshwater swamp pollen assemblage*: (**k**) *Camptostemon* sp. – subpolar view ×1000.

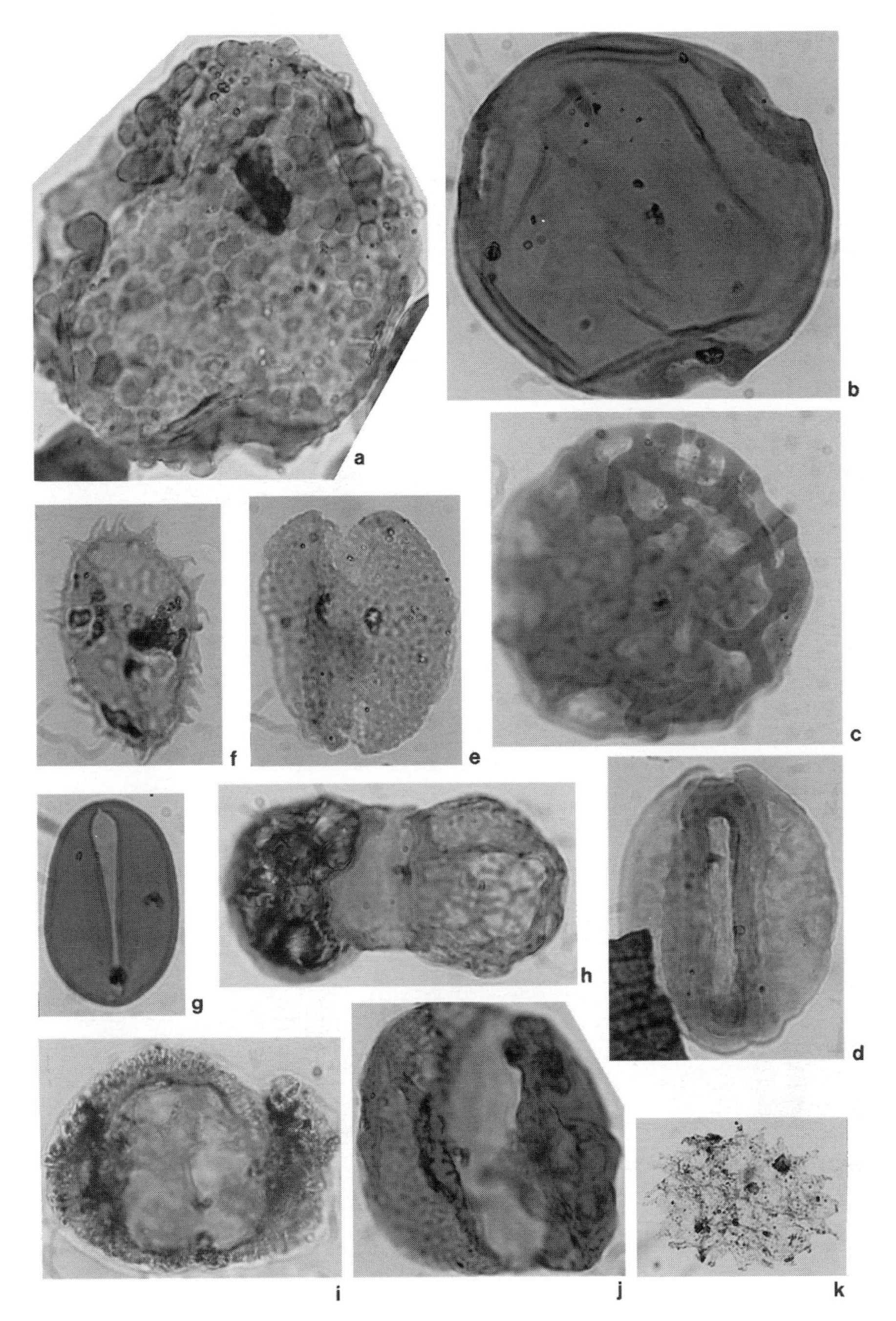

Fig. 7. Some representative palynological taxa and palynofacies from various palaeoenvironments from the northwest Borneo area. *Freshwater swamp pollen assemblage*: (**a**) *Alangium* sp. (*Lanagiopolis* spp.) – polar view ×1000; (**b**) *Durio* type – polar view ×1000; (**c**) *Intsia* type – subequatorial view ×1000; (**d**) *Barringtonia* sp. (*Marginipolis concinnus*) – equatorial view ×1000; (**e**) *Calamus* type (*Dicolpopolis* spp.) – distal view ×1000. *Coastal and mangrove palm pollen assemblage*: (**f**) *Nypa* sp. (*Spinizonocolpites echinatus*) – proximal view ×1000; (**g**) palmae undifferentiated – proximal view ×1000. *Hinterland gymnosperm pollen assemblage*: (**h**) *Podocarpus polystachius* ×1000; (**i**) *Dacrydium* sp. ×1000; (**j**) *Pinus* sp. ×1000. *Freshwater algae*: (**k**) *Pediastrum* sp. ×400.

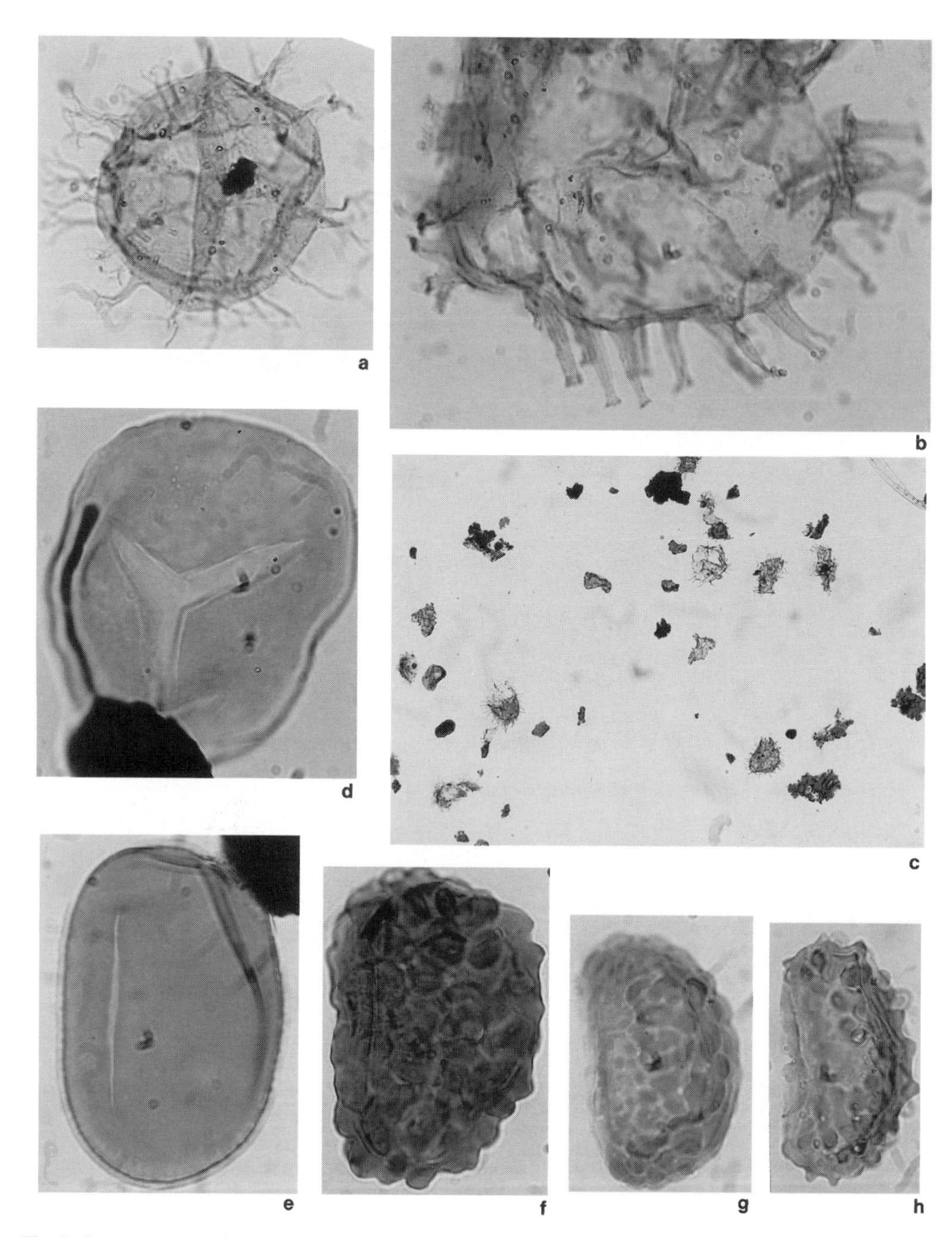

Fig. 8. Some representative palynological taxa and palynofacies from various palaeoenvironments from the northwest Borneo area. *Marine dinocyst assemblage*: (**a**) *Spiniferites ramosus* – proximal view ×1000; (**b**) *Polysphaeridium subtile* – half cyst ×1000; (**c**) palynofacies with numerous *Polysphaeridium subtile* (outcrop ENT sample 4) ×200. Open shelf environment characterized by this almost monospecific dinocyst assemblage. *Hinterland spore assemblage*: (**d**) *Leiotriletes* spp. (also called *Deltoidospora* spp.) – proximal view ×1000; (**e**) *Laevigatosporites* spp. ×1000; (**f**) *Verrucatosporites* spp. (robust form) ×1000; (**g**) *Verrucatosporites* spp. ×1000; (**h**) *Stenochlanea palustris* (*Verrucatosporites usmensis*) ×1000. Note: spores are generally very resistant to decay, oxidation and transport, generally more resistant than pollen. They are produced in large quantity and for these reasons are found in all types of environments, even relatively deep marine where they are transported.

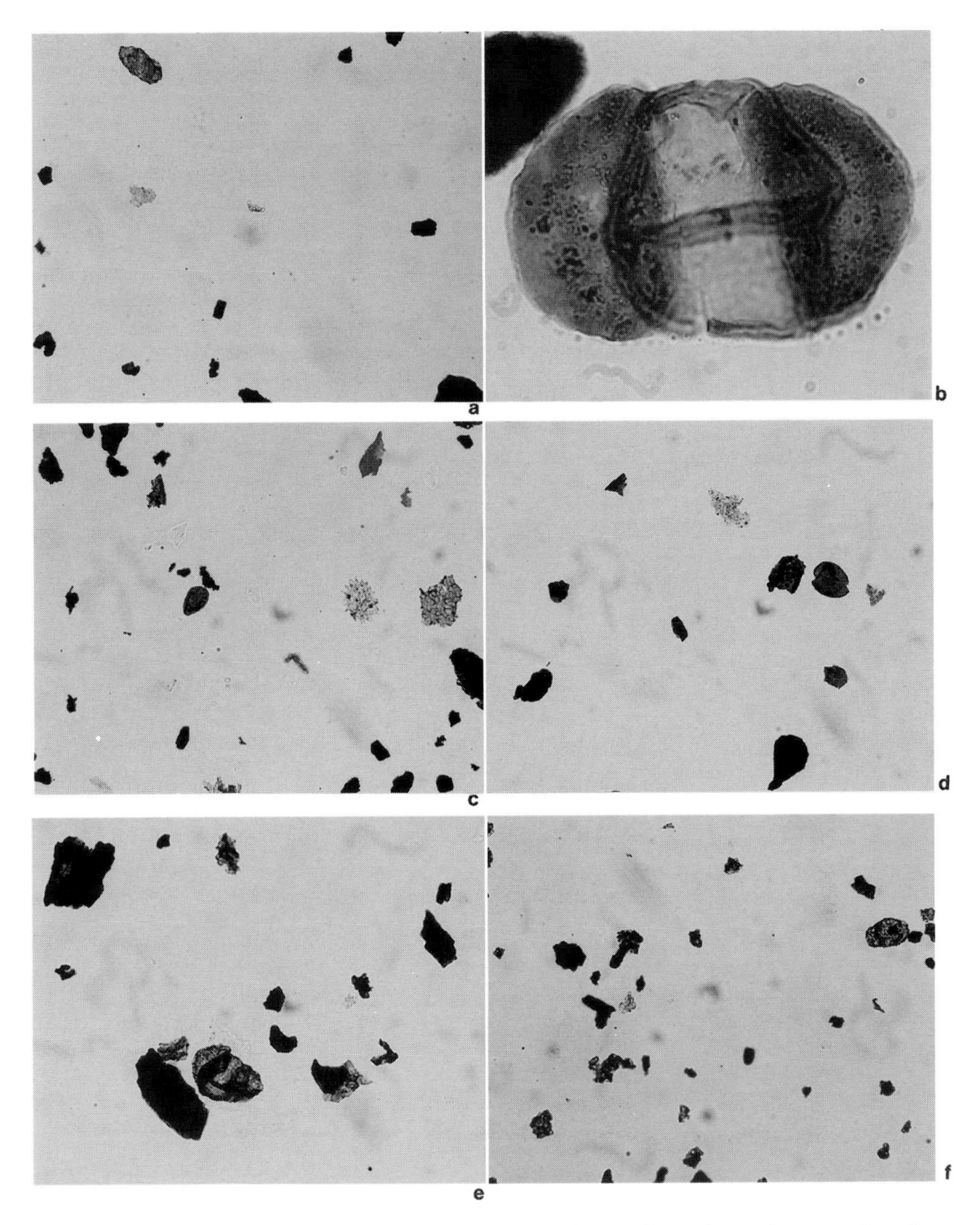

Fig. 9. Some representative palynological taxa and palynofacies from various palaeoenvironments from the northwest Borneo area. *Distal turbidite*: (**a**) palynofacies with altitude conifer *Pinus* sp. (outcrop RR sample 1) ×200; (**b**) *Pinus* sp. (outcrop RR sample 1) ×1000; (**c**) palynofacies with freshwater algae *Pediastrum* sp. (outcrop RR sample 4). Note: the palynofacies is characterized by rare kerogen particles, mainly dark (vitrinite and inertinite). The presence of *Pinus* sp. and other bisaccates reflects a deposition relatively far offshore. This taxon is generally easily transported in comparison to other palynomorphs. *Proximal turbidite*: (**d**) palynofacies with freshwater swamp *Calamus* type (*Dicolpopolis* spp.) and *Stenochlanea palustris* (*Verrucatosporites usmensis*) (outcrop OT sample 1) ×200; (**e**) palynofacies with *Pinus* sp. (outcrop OT sample 3) ×100; (**f**) palynofacies with *Pinus* sp. (outcrop OT sample 3) ×200. Note: the palynofacies is characterized by dark kerogen particles and the presence of bisaccates. The relative increase in particles or richer residue reflects a more proximal depositional setting.

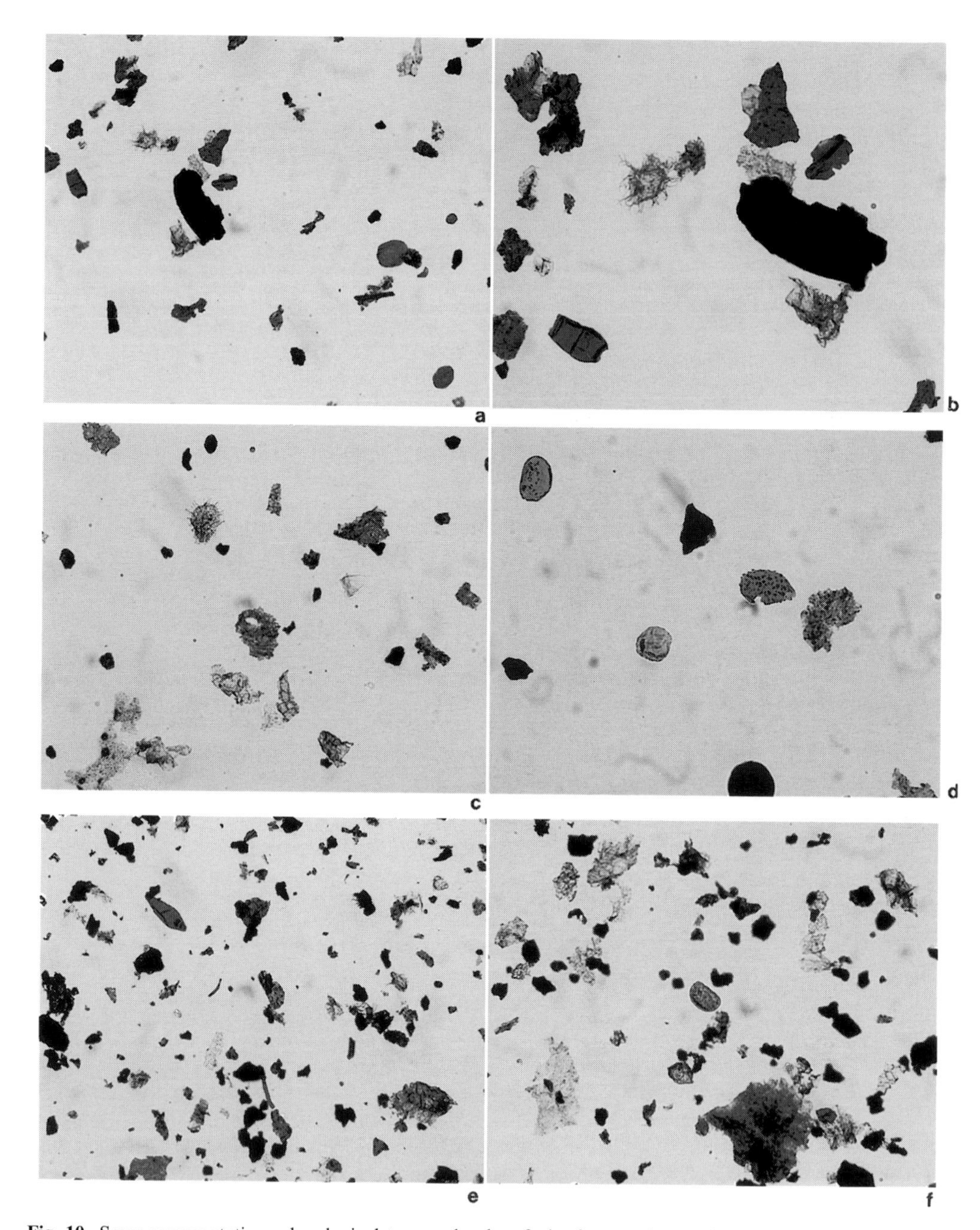

Fig. 10. Some representative palynological taxa and palynofacies from various palaeoenvironments from the northwest Borneo area. *Open shelf*: (**a**) palynofacies with rare dinocyst *Spiniferites ramosus* (outcrop ENT sample 14) ×100; (**b**) palynofacies with rare dinocyst *Spiniferites ramosus* (outcrop ENT sample 14) ×200; (**c**) palynofacies with dinocyst *Polysphaeridium subtile* (outcrop ENT sample 24) ×100; (**d**) palynofacies with spores (outcrop ENT sample 24) ×200. Note: the palynofacies is characterized by less numerous dark particles and the presence of more common dinocysts, although rare. Spores are common but not as abundant as in terrestrial environments located behind the mangrove belt. *Open shelf with slumps*: (**e**) palynofacies (outcrop PC sample 2) ×100; (**f**) palynofacies (outcrop PC sample 1) ×100. Note: the palynofacies is characterized by a mixing of numerous particles of different spore coloration and size. This probably reflects the sediment mixing itself while the slumping is active and the importance of transport of allochthonous material.

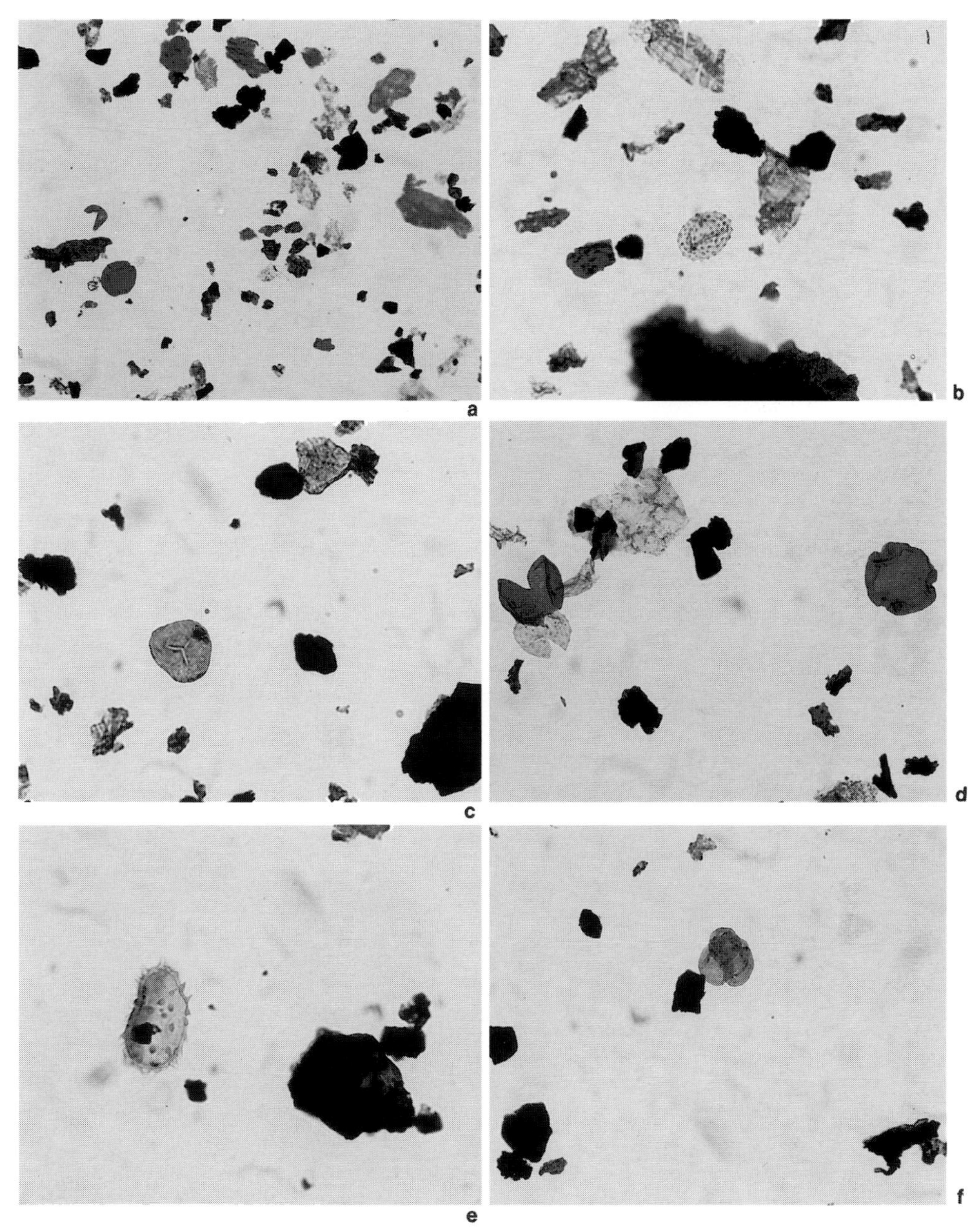

Fig. 11. Some representative palynological taxa and palynofacies from various palaeoenvironments from the northwest Borneo area. *Lower shoreface*: (**a**) palynofacies with freshwater swamp *Alangium* sp. (*Lanagiopolis* sp.) (outcrop OOS sample 1) ×100; (**b**) palynofacies with freshwater swamp *Stenochlanea palustris* (*Verrucatosporites usmensis*) (outcrop OOS sample 2) ×200; (**c**) palynofacies with backmangrove *Acrostichum aureum* (outcrop MO sample 3.5) ×200; (**d**) palynofacies with freshwater swamp *Alangium* sp. and spores (outcrop CH sample 2); ×200; (**e**) palynofacies with *Nypa* sp. (*Spinozonocolpites echinatus*) (outcrop MO sample 5) ×400. Note: the palynofacies is characterized by light coloured, as well as dark coloured, particles. There is also increased evidence of relatively large pollen taxa such as *Alangium* sp., *Acrostichum aureum* and other spores, *Spinozonocolpites echinatus*, which source from diverse terrestrial environments connected to the coastline. *Upper shoreface*: (**f**) palynofacies with freshwater swamp *Lophopetalum multinervum* (outcrop RM sample 3.9) ×200. Note: the palynofacies contains freshwater swamp taxa and common spores. The major palynological difference between lower shoreface and upper shoreface environments is the virtual absence of dinocysts in the latter.

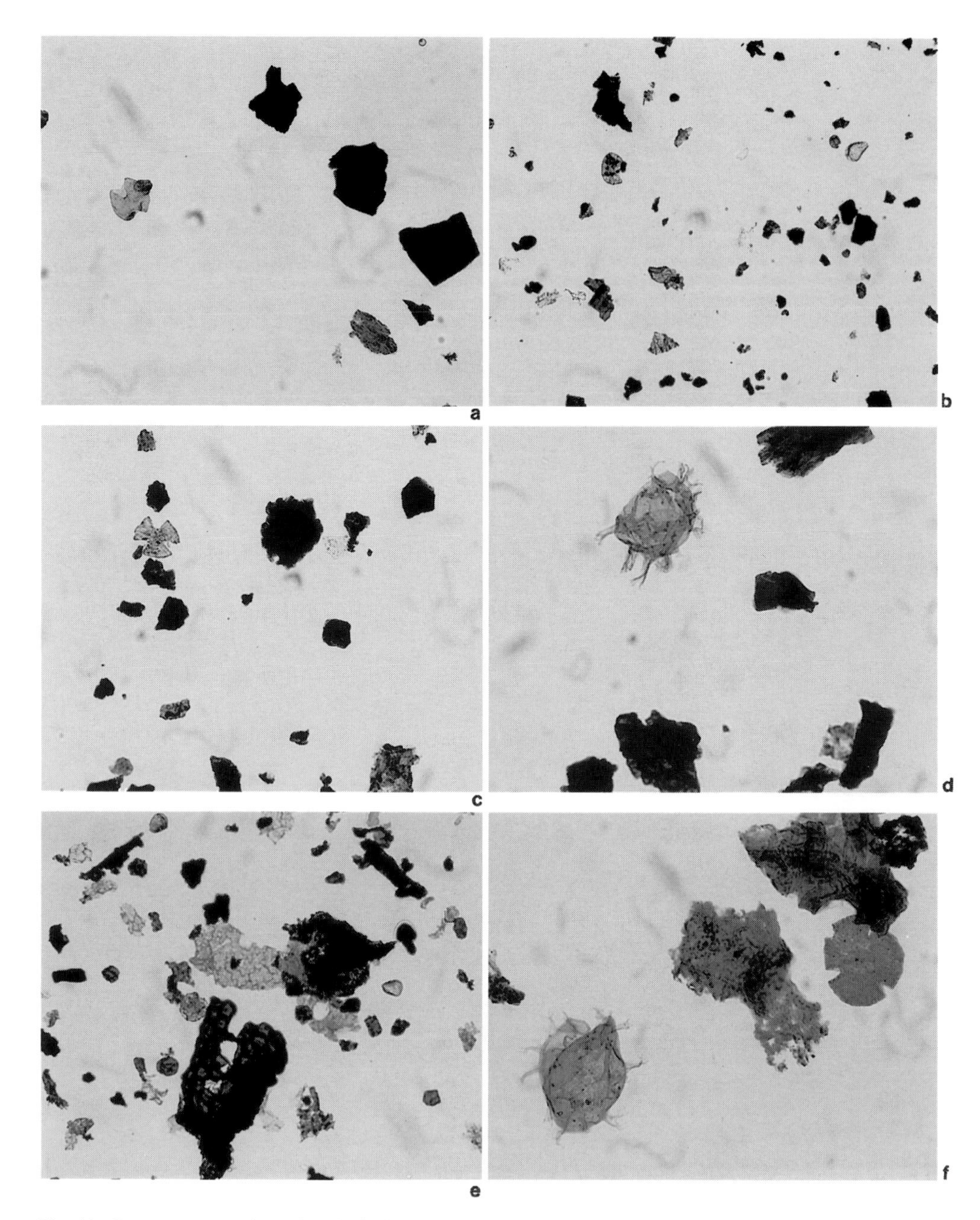

Fig. 12. Some representative palynological taxa and palynofacies from various palaeoenvironments from the northwest Borneo area. *Upper shoreface*: (**a**) palynofacies with *Calamus* type (*Dicolpopolis* sp.) (outcrop RM sample 15.3) ×200; (**b**) palynofacies with spores (outcrop RM sample 17.7) ×100. *Upper shoreface, tidally influenced*: (**c**) palynofacies with freshwater swamp Dipterocarpaceae (outcrop MD sample 4) ×200; (**d**) palynofacies with dinocyst *Hystricholpoma* sp. (outcrop MD sample 14) ×400. Note: the combined presence of dinocysts and taxa of freshwater environments indicates environmental zones where terrestrial and marine influences are well matched. *Sandy beach – tidal flat*: (**e**) palynofacies with numerous palynomorphs and fungal remains (outcrop MA sample 6) ×200; (**f**) palynofacies with dinocyst *Spiniferites ramosus* and backmangrove *Brownlowia* type (outcrop MA sample 9) ×400. Note: the presence of numerous and diverse palynomorphs indicates the proximity of terrestrial sources and vegetational environments close to the environment of deposition.

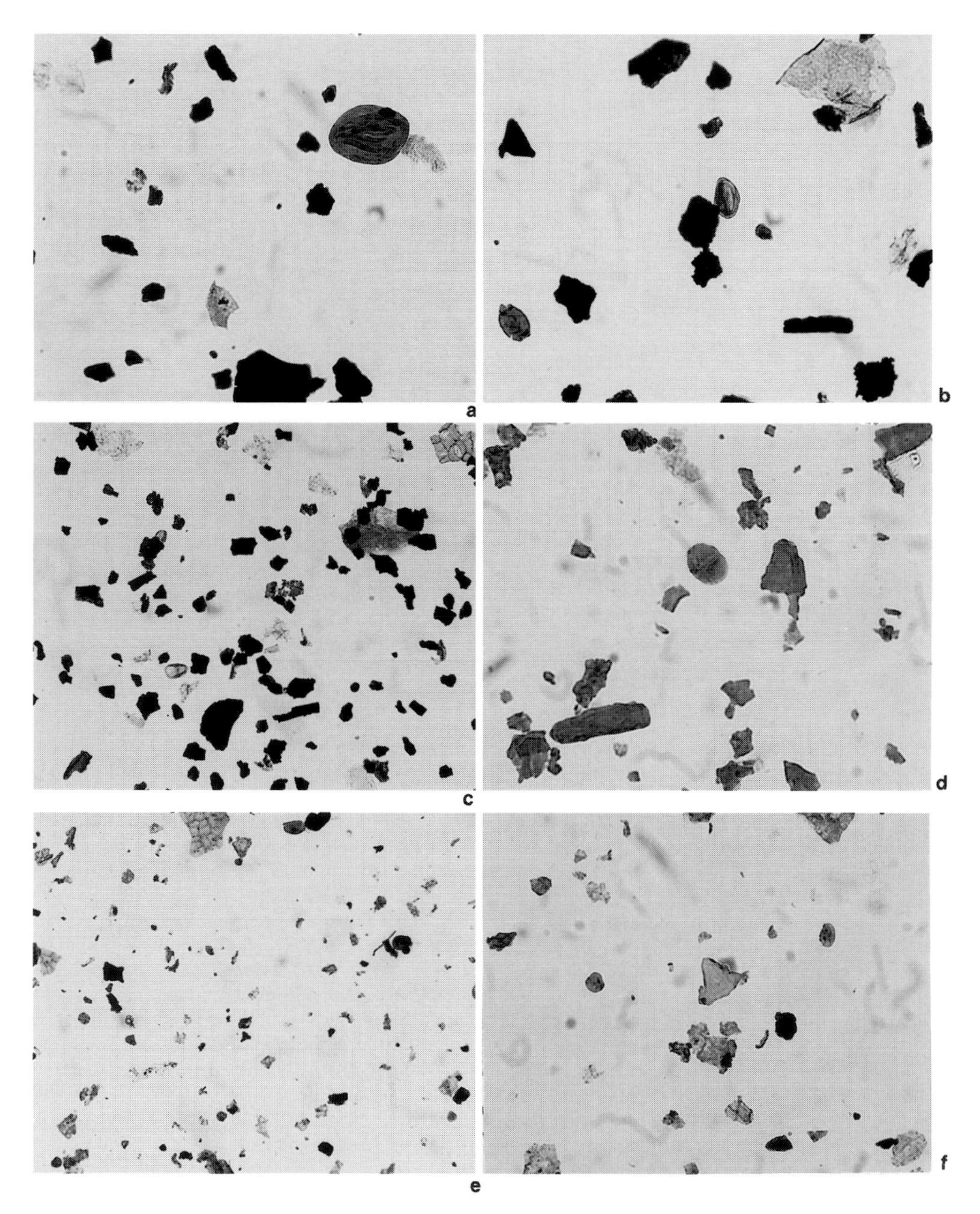

Fig. 13. Some representative palynological taxa and palynofacies from various palaeoenvironments from the northwest Borneo area. *Tidal flat–tidal channel complex*: (**a**) palynofacies with freshwater swamp *Crudia* sp. (*Striatricolpites catatumbus*) (outcrop SA sample 10.5) ×200; (**b**) palynofacies with mangrove *Florschuetzia* spp. (outcrop SA sample 58) ×200; (**c**) palynofacies (outcrop SA sample 58) ×100. Note: the palynofacies is characterized by numerous dark particles, most probably micropieces of peat which generates in freshwater swamps. Mangrove taxa can be locally abundant between channels. *Lagoon–distributary channel margin*: (**d**) palynofacies with *Rhizophora* sp. (*Zonocostites ramonae*) (outcrop HO sample 92.8) ×400; (**e**) palynofacies with *Rhizophora* sp. (outcrop 7UP sample 208) ×100; (**f**) palynofacies with *Rhizophora* sp. (outcrop 7UP sample 208) ×200. Note: the palynofacies is characterized by the absence or paucity of dark kerogen particles which indicates dominance of *in situ* sedimentation within the mangrove belt and lesser influence of transport from other environments, mostly terrestrial but also marine.

diameter) appears to be particularly distinctive of deposition in a distal turbidite setting. The assemblages lack the diversity of more shallow-water settings.

Proximal turbidite

Palynological assemblages from proximal turbidite settings are distinctive because of the abundant reworking of Oligocene and older palynomorphs which occur in these settings. Spores and mangrove pollen are also abundant (if not reworked). Foraminiferal assemblages are similar to those from a distal turbidite setting, containing abundant agglutinating foraminifera, especially fine-grained, deformed '*Trochammina*/*Recurvoides*'. Fine-grained and thin-walled '*Cyclammina*' and *Haplophragmoides* may also be present. Unlike the palynological assemblages, reworking is not obvious, thus the use of both palynology and foraminiferal micropalaeontology is essential to recognize this depositional environment.

Open shelf

Palynological assemblages from open shelf deposits are reasonably rich and diverse. Spores are common–abundant, whilst back-mangrove, freshwater swamp and palm pollen are rare or occasionally common. Mangroves are variable in their abundance but are generally rare with *Zonocostites* more common than *Florschuetzia*. Dinocysts are typically rare, but at some localities can be locally abundant, although monospecific. For example, at the ENT locality, abundant *P. subtile* occurs.

Foraminiferal assemblages are rich and diverse. Calcareous benthonic foraminifera are common or abundant and include *Asterorotalia*, *Nonion*, *Planulina*, *Operculina*, *Cibicidoides*, *Rectuvigerina*, *Lenticulina*, *Quinqueloculina* and *Bolivinita*. Agglutinating forms are also common and '*Recurvoides*/*Trochammina*' (with a white, sugary wall), *Karreriella*, *Textularia* and *Ammobaculites* sp. 1/1a. In some sections, for example ENT, planktonic foraminifera are also common.

The open shelf environment may be further subdivided on the basis on the relative proportion of calcareous benthonic taxa. Associations of common *Rectuvigerina*/*Quinqueloculina* suggest inner shelf palaeobathymetry, whilst those with dominant *Operculina* and with common *Operculina* and *Cibicidoides* represent middle and outer shelf settings, respectively.

Open shelf with slumps

As might be expected, there are similarities between the microfossil assemblages recovered from sediments representing this depositional setting and those from 'normal' open shelf settings. Palynological assemblages are very similar consisting of a mixed assemblage of moderately rich spores, occasional mangrove, back-mangrove, freshwater swamp and palm pollen, and very rare dinocysts. The foraminiferal assemblage is, however, reasonably distinctive. The background fauna is dominated by agglutinating foraminifera including large and deformed '*Recurvoides*', '*Trochammina*/*Recurvoides*' with a fine wall, *Bathysiphon*, '*Glomospira*', and a distinctive, probably new, form of agglutinating foraminifera we refer to as '*Textukurnubia*' (see Bidgood *et al.* in press). Overall diversity of agglutinating forms is moderate. In addition to the agglutinating foraminifera, the sections with slumps contain abundant calcareous benthonic foraminifera, especially the genus *Asterorotalia*, *Hansenisca*, *Planulina* cf. *wuellstorfii*, *Bolivinita* and *Cibicides* are also present in reasonable numbers, as are calcareous agglutinating forms such as *Textularia*. Planktonic foraminifera are rare.

Lower shoreface

Lower shoreface deposits can be distinguished from open shelf deposits by the absence of calcareous benthonics. They contain agglutinating foraminifera assemblages of moderate diversity and abundance, especially *Haplophragmoides*, '*Recurvoides*' and '*Trochammina*', which have a white, sugary wall. *Ammobaculites* may also be present together with a distinctive form we refer to as '*Pseudoepistomina*' (see Bidgood *et al.* in press). '*Glomospira*' *glomerata* is also a distinctive component of this depositional setting, although it also occurs in open shelf sediments.

Palynological assemblages are not especially rich, and tend to be dominated by spores and palm pollen. Mangroves, back-mangrove elements and freshwater swamp pollen are typically rare. Dinocysts are a very rare component of these assemblages.

Upper shoreface

Upper shoreface deposits are difficult to distinguish from the lower shoreface deposits, as described above, in terms of their microfossil assemblages. Palynologically, the assemblages are very similar, being dominated by spores and occasional palm pollen with mangroves, back-mangrove and freshwater swamp pollen all rare. In contrast to lower shoreface deposits, dinocysts are completely absent.

Foraminiferal assemblages are also similar to those from lower shoreface deposits, consisting mainly of common and moderately diverse assemblages of agglutinating foraminifera, especially '*Trochammina*', '*Recurvoides*' and *Haplophragmoides* with a white, sugary wall. A distinctive component is the presence of moderately common *Karreriella*.

Tidal flat–tidal channel

Tidal flat–tidal channel deposits can be distinguished from many of the more open marine and more distal environments by their diverse palynological assemblages which include common mangrove and palm pollen, and common spores. A distinctive element is the presence of moderately common *Casuarina*. Dipterocapaceae are also occasionally common, as are algae such as *Botryococcus* which can tolerate low salinity. Dinocysts are recorded in this setting but are very rare.

Foraminiferal assemblages tend to be of low abundance but of moderate diversity, and include coarse-grained '*Trochammina*' and '*Recurvoides*' (wall being coarser than in more distal settings), and occasional *Trematophragmoides* and *Ammobaculites* sp.

Lower distributary channel

Sediments from lower distributary channel deposits contain a distinctive palynological assemblage consisting of common mangrove, back-mangrove and palm pollen. Spores, as in many other depositional settings, are also common. Very rare dinocysts may also occur. Foraminiferal assemblages are of moderate diversity, consisting mainly of coarse-grained agglutinating forms such as '*Trochammina*' and *Trematophragmoides*. '*Pseudoepistomina*', which is common in lower shoreface deposits, also occurs as rare specimens in sediments from this depositional setting.

Lagoon–distributary channel margin

These settings occur where mangrove swamp vegetation is accumulating on windward or current protected edges of distributary channel mouths. They also occur where distributary channels are being infilled by mangrove swamps during falling sea level or due to autocyclic channel shifting. As such they contain palynological assemblages completely dominated by mangrove pollen. Foraminiferal assemblages are also monospecific, being dominated by either *Trematophragmoides* or *Miliammina fusca*.

Upper distributary channel

Upper distributary channel deposits represent near-freshwater conditions and as such are typically barren of foraminifera. Palynologically, they are distinctive, containing superabundant spores, common and diverse freshwater swamp and back-mangrove pollen, and very rare mangrove and palm pollen. Freshwater algae can also be common in this setting.

Discussion

A number of previous studies have noted that foraminifera and/or palynomorphs could be used as palaeoenvironmental proxies, largely based on the observation of the distribution of Recent microfauna/flora, in work which complements our own, outcrop-based, research.

Dhillon (1968) studied the distribution of modern foraminifera in the Lupar and Labuk estuaries of eastern Malaysia, in settings similar to what we describe as lower distributary channel. The Lupar Estuary, which has marked tidal influence, is dominated by what are assumed to be transported stenohaline forms such as *Pseudorotalia*, *Asterorotalia*, *Nonion*, *Lagena*, *Oolina*, *Discorbis*, *Cibicides*, *Amphistegina*, *Triloculina* and other taxa, especially in the distal part of the estuary. Indigenous forms are rare but include *Arenoparella*, *Haplophragmoides*, *Trochammina* and *Ammonia*. The redeposition of stenohaline forms has not been noted in our outcrop material, but the fact that it occurs in Recent sediments suggests that it may also occur in ancient sediments and needs to be taken into account in subsurface studies. In the Labuk Estuary indigenous agglutinating forms are much more common and include *Ammobaculites exiguus*, *Haplophragmoides* spp., *Miliammina* spp., *Arenoparrella*, *Jadammina* and *Trochammina* spp. *Ammonia* is common at the mouth of the estuary. Redeposited stenohaline forms are rare. In comparison to our lower distributary channel outcrop analogues, these assemblages are surprisingly rich and diverse. *Arenoparrella* has often been recorded from modern lower distributary channel sediments, but is difficult to recognize in our ancient material because of poor preservation of apertural characteristics. Jones *et al.* (1999) have noted *Arenoparrella*, *Haplophragmoides*, *Miliammina* and *Trochammina* in mangrove palaeoenvironments from Neogene sediments of Venezuela.

The depositional setting of the Labuk Estuary more closely fits that envisaged for the Miocene sediments reported upon herein. The estuary is relatively brackish and with a wide mouth opening out into Labuk Bay. Tidal influence is not as strong as in the Lupar Estuary. The dominance of agglutinating foraminifera matches our own results, and also compares well with the work of Ho (1971) who studied the distribution of Recent benthonic foraminifera in inner Brunei Bay. Ho recognized a *Trochammina* cf. *lobata* assemblage in the lower distributary channels, an *Ammobaculites* assemblage over much of Brunei Bay proper, corresponding to our lagoon setting, and an *Asterorotalia trispinosa* assemblage at the mouth of Brunei Bay corresponding to our open shelf setting. The *Trochammina* cf. *lobata* assemblage consists only of coarse-grained agglutinating forms such as *Trochammina* spp., *Miliammina fusca*, *Arenoparella*, *Haplophragmoides* and *Ammobaculites*. This matches well with the lower distributary channel assemblage described herein. Ho noted that the abundance of *Trochammina* may be related to the abundance of organic matter in the distributary channels.

The *Ammobaculites* assemblage of Ho (1971) has some similarities with the lagoon assemblages described herein. Abundance and diversity are both low and consist of a few species of coarse-grained *Ammobaculites* (including *A. exiguus*), *Haplophragmoides* and *Trochammina*. The *Asterorotalia trispinosa* assemblage is similar in character to elements of the open shelf assemblage described herein. *Asterorotalia* and *Ammonia* are very common, whilst *Elphidium*, *Florilus*, *Ammobaculites*, *Cellanthus* and *Ozawia* also occur alongside transported specimens of more typical outer shelf taxa such as *Operculina*, *Calcarina* and *Triloculina*.

In conclusion, studies of modern foraminiferal distributions appear to confirm our contention that microfossils can be used as precise palaeoenvironmental proxies. Ten precise depositional settings can be identified from microfossil (palynomorph and foraminiferal) assemblages, ranging from distal turbidite through to upper distributary channel. We believe that this opens the possibility to more widespread use of biostratigraphy as a tool in development and production geology in northwest Borneo, by helping to fingerprint the depositional settings of reservoir intervals and so better predict their architecture, connectivity, heterogeneity and poroperm characteristics.

This work has been financed and supported by Brunei Shell Petroleum (BSP), to whom the authors wish to express their gratitude. In particular, Dr Shirley van Heck, senior stratigrapher with BSP, and Dave Watters, former Head of Reservoir Characterization with BSP, have been extremely supportive. Christian Thomas, former MSc student at the University of Aberdeen, carried out micropalaeontological analysis of subsurface material from offshore Brunei which proved relevant to this study. Palynological analysis on a trial set of samples was carried out by Dr Bob Morley of Palynova. Bob Morley is also thanked for his insights into the palaeoenvironmental interpretation of palynological data. A small number of calcareous nannofossil analyses were carried out by Liam Gallagher of Network Stratigraphic Consulting Ltd. Palynological preparation was carried out at the University of Sheffield. Figures and plates were produced with the help of Barry Fulton and Walter Richie at the University of Aberdeen. Dr Bob Jones of BP is thanked for his critical review of the manuscript.

References

ANDERSON, J. A. R. & MULLER, J. 1975. Palynological study of a Holocene peat and Miocene coal deposit from NW Borneo. *Review of Palaeobotany and Palynology*, **19**, 291–351.

BESEMS, R. E. 1993. Dinoflagellate cyst biostratigraphy of Tertiary and Quaternary deposits of offshore northwest Borneo. *Geological Society of Malaysia Bulletin*, **33**, 65–93.

BIDGOOD, M. D., SIMMONS, M. D. & THOMAS, C. G. in press. Agglutinating foraminifera from Miocene sediments of north-west Borneo. *In*: HART, M. B. (ed.) *Proceedings of the 5th Workshop on Agglutinating Foraminifera, Plymouth 1997*.

BISWAS, B. 1976. Bathymetry of Holocene foraminifera and Quaternary sea-level changes on the Sunda Shelf. *Journal of Foraminiferal Research*, **6**, 107–133.

BRENAC, P. A. & RICHARDS, K. 1998. *Pediastrum* as a fossil guide in sequence stratigraphy. *In*: *Proceedings of the 9th International Palynological Congress, Houston 1996*.

BRONNIMANN, P. & KEIJ, A. J. 1986. Agglutinated foraminifera (Lituolacea and Trochamminacea) from brackish waters of the State of Brunei and of Sabah, Malaysia, Northwest Borneo. *Revue de Palaeobiologie*, **5**(1), 11–31.

—— & WHITTAKER, J. E. 1993. Taxonomic revision of some Recent agglutinated foraminifera from the Malay Archipelago in the Millett Collection, British Museum (Natural History). *Bulletin of the British Museum (Natural History), Zoology*, **59**, 107–124.

CALINE, B. & HUONG, J. 1992. New insights into the recent evolution of the Baram Delta from satellite imagery. *Geological Society of Malaysia Bulletin*, **32**, 1–13.

DHILLON, D. S. 1968. Notes on the foraminiferal sediments from the Lupas and Labuk estuaries, East Malaysia Borneo Region. *Malaysia Geological Survey Bulletin*, **9**, 56–73.

DUNAY, R. E., BRENAC, P. A. & EDWARDS, P. G. 1998. Palynology of the Lower Pliocene sequence offshore Equatorial Guinea. *In*: *Proceedings of the 3rd Symposium of African Palynology, Johannesburg 1997.*

ECKERT, H. R. 1970. Planktonic foraminifera and time – stratigraphy in well Ampa-2. *Brunei Museum Journal*, **2**, 320–326.

GERMERAAD, J. H., HOPPING, C. A. & MULLER, J. 1968. Palynology of Tertiary sediments from tropical areas. *Review of Palaeobotany and Palynology*, **6**, 189–348.

HARPER, G. C. 1975. *The Discovery and Development of the Seria Oilfield.* Muzium, Brunei.

HASELDONCKX, P. 1974. A palynological interpretation of palaeo-environments in Southeast Asia. *Sains Malaysiana*, **3**, 119–127.

HO, K. F. 1971. Distribution of Recent Benthonic foraminifera in the 'inner' Brunei Bay. *Brunei Museum Journal*, **2**, 124–137.

JAMES, D. M. D. (ed.) 1984. *Hydrocarbon Resources of Negara Brunei Darussalam.* Musium Brunei and Brunei Shell Petroleum, Special Publications. Kok Wah Press.

JONES, R. W. 1994. *The Challenger Foraminifera.* Oxford University Press/Natural History Museum, London.

——, JONES, N. E., KING, A. D. & SHAW, D. 1999. Reservoir biostratigraphy of the Pedernales Field, Venezuela. *This volume.*

LAMBIASE, J. J., SIMMONS, M. D., CREVELLO, P. D. & MORLEY, C. K. In press. Depositional settings, facies distribution and stratigraphic architecture of Neogene sediments in Northwestern Borneo: implications for characterization of petroleum reservoirs. *Sedimentology.*

MORLEY, C. K., CREVELLO, P. & AHMAD, Z. H. 1998. Shale tectonics – deformation associated with active diapirism: the Jerudong Anticline, Brunei Darussalam. *Journal of the Geological Society, London*, **155**.

MORLEY, R. J. 1991. Tertiary stratigraphic palynology in Southeast Asia; current status and new directions. *Bulletin of the Geological Society of Malaysia*, **28**, 1–36.

MULLER, J. 1964. A palynological contribution to the history of the mangrove vegetation in Borneo. *In*: *Ancient Pacific Floras.* University of Hawaii Press, 33–42.

——1969. A palynological study of the genus *Sonneratia* (Sonneratiaceae). *Pollen et Spores*, **11**, 223–298.

——1972. Palynological evidence for change in geomorphology, climate and vegetation in the Mio-Pliocene of Malaysia. *In*: ASHTON, M. & ASHTON, P. S. (eds) *The Quaternary Era in Malaysia.* University of Hull, Department of Geography, Miscellaneous Series, **13**, 6–16.

MURRAY, J. W. 1991. *Ecology and Palaeoecology of Benthic Foraminifera.* Longman, London.

PROSSER, D. J. & CARTER, R. R. 1997. Permeability heterogeneity within the Jerudong Formation: an outcrop analogue for subsurface Miocene reservoirs in Brunei. *In*: FRASER, A. J., MATTHEWS, S. J. & MURPHY, R. W. (eds) *Petroleum Geology of Southeast Asia.* Geological Society, London, Special Publications, **126**, 195–235.

RICE-OXLEY, E. D. 1991. Palaeo-environments of the Lower Miocene to Pliocene sediments in offshore NW Sabah area. *Geological Society Malaysia Bulletin*, **28**, 165–194.

SANDAL, S. T. (ed.) 1996. *The Geology and Hydrocarbon Resources of Negara Brunei, Darussalam.* Sybas, Bandar Seri Begawan, Brunei, Darussalam.

VAN MARLE, L. J. 1991. *Eastern Indonesia, Late Cenozoic Smaller Benthic Foraminifera.* Royal Dutch Academy of Sciences.

WHITTAKER, J. E. & HODGKINSON, R. L. 1979. Foraminifera of the Togopi Formation, eastern Sabah, Malaysia. *Bulletin of the British Museum (Natural History), Geology Series*, **31**, 1–120.

Reservoir biostratigraphy of the Pedernales Field, Venezuela

R. W. JONES,[1] N. E. JONES,[2] A. D. KING[3] & D. SHAW[3]

[1] *BP Exploration Operating Company Ltd, Chertsey Road, Sunbury-on-Thames, Middlesex TW16 7LN, UK*

[2] *BP Exploracion de Venezuela S.A., Edificio Centro Seguros Sud America, Avenida Francisco de Miranda (cruce con Avenida Mohenado), El Rosal-Chacao, Caracas, Venezuela*

[3] *EPOCA, Centro Empresarial Uniformes Miranda, Avenida Juan de Urpin, Barcelona, Venezuela*

Abstract: The geographic and geological settings of the Pedernales Field, and its history from discovery to development, are briefly outlined. Biostratigraphy and palaeoenvironmental interpretation, and their impact on integrated reservoir description (IRD) and production geology, are discussed in detail.

Historical and modern vintage micropalaeontological data enable the recognition of a number of 'zones' over the reservoir interval. These 'zones' are diachronous, although they do have correlative value when integrated into the framework of isochronous tectono-stratigraphic sequence boundaries and flooding surfaces identified on seismic sections and well logs. The 'zones' and their containing sediments can be tentatively calibrated against global standard biozones and hence sequence boundaries and flooding surfaces on the eustatic sea-level cycle chart.

Micropalaeontological and palynological data also enable the recognition of a number of biofacies interpreted on the basis of modern analogy as indicating a range of delta-top, delta-front and prodelta environments. The principal reservoir is interpreted as having been deposited in a range of delta-top to delta-front subenvironments, the seal in a range of distal prodelta subenvironments.

The ongoing high-resolution biostratigraphic work programme has revealed the existence of a large number of correlatable events over the reservoir interval. Integrated high-resolution biostratigraphic, well log and seismic correlation can be achieved both within and between the compartments of the field.

High-resolution biostratigraphic/biofacies data demonstrate the existence of a bathymetric gradient at the time the reservoir were being deposited. Gravity gliding along the observed gradient provides a possible mechanism for the locally observed section repetition and enhanced reservoir development, and also for the up-dip trapping mechanism.

The Pedernales Field is located in the wetlands of the Orinoco delta some 100 km north-northeast of Maturin in the northeasternmost part of the Eastern Venezuelan Basin, sometimes referred to as the Maturin sub-basin (Figs 1 and 2).

The geological history of the Maturin sub-basin is long and complex (see, for instance, Parnaud *et al.* 1995, and additional references therein). Most of our knowledge of its stratigraphy comes from the extensive surface outcrops in the Serrania del Interior. Some published (and much proprietary) information is also available on the subsurface. Unfortunately, no truthed data on the pre-Tertiary stratigraphy of the subsurface are available from anywhere east of the Quiriquire Field. The deepest wells on the Pedernales structure terminate in condensed Palaeocene–Eocene pelagic limestones assigned by Creole geologists to the essentially Trinidadian Navet Formation (an age-equivalent of the Caratas Formation).

The Barranquin, Garcia, El Cantil, Chimana, Querecual, San Antonio, San Juan, Vidono, Caratas, Tinajitas, Los Jabillos and Areo formations of the northern part of the Maturin sub-basin record a passive margin phase, which lasted until within the Late Oligocene (Parnaud *et al.* 1995) (Fig. 3). Associated clastic sediments, derived from the Guyana Shield to the south,

JONES, R. W., JONES, N. E., KING, A. D. & SHAW, D. 1999. Reservoir biostratigraphy of the Pedernales Field, Venezuela. *In:* JONES, R. W. & SIMMONS, M. D. (eds) *Biostratigraphy in Production and Development Geology*. Geological Society, London, Special Publications, **152**, 243–257.

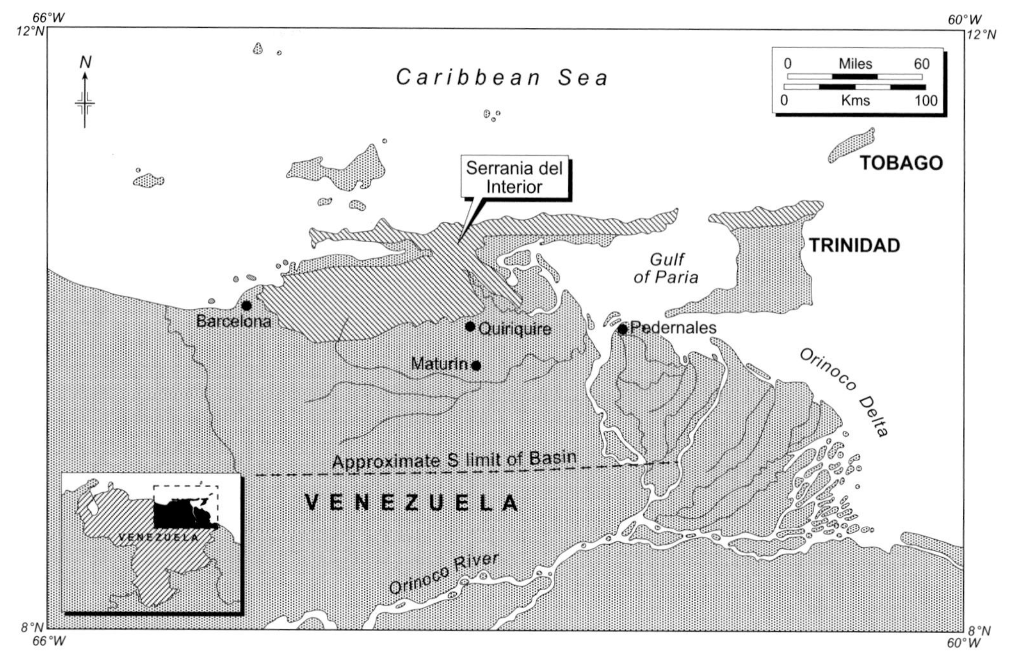

Fig. 1. Location map, Eastern Venezuelan basin. Hachure indicates mountainous regions (Serrania del Interior in eastern Venezuela, Northern Ranges in Trinidad). Pedernales Field is located on the Orinoco delta 100 km east-northeast of Maturin. Quiriquire Field, operated by Maxus of Dallas in partnership with BP, is located 25 km north of Maturin. Much of the intervening area falls within the Guarapiche Block, awarded to BP and partners in the PDVSA exploration licensing round (colloquially known as the 'aperture' in 1996).

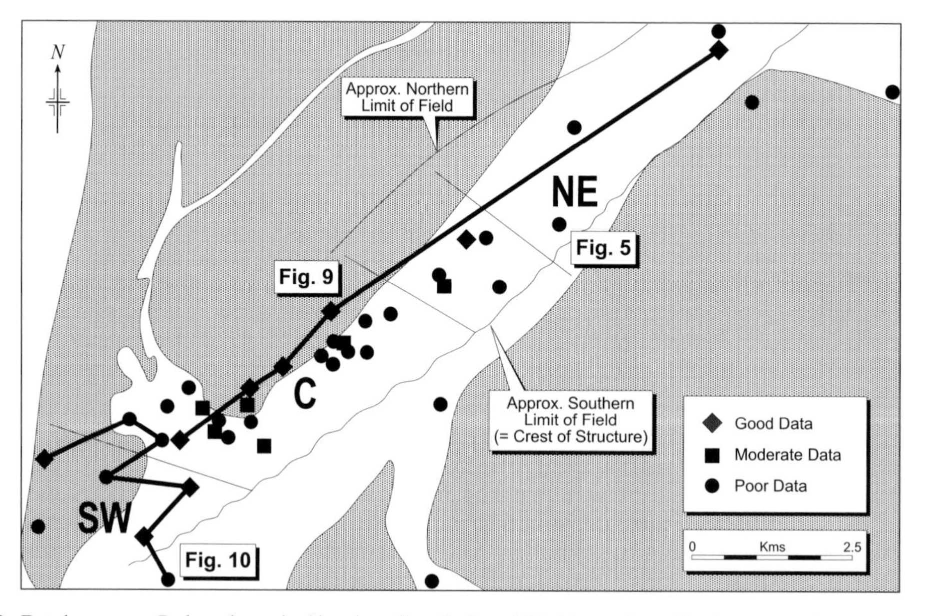

Fig. 2. Database map, Pedernales unit. Showing all wells for which biostratigraphic data are available, lines of the wireline log correlation panels (Figs 9 and 10) and line of seismic panel (Fig. 5). Also showing approximate outlines of the Pedernales structure, Pedernales Field (some areas still under appraisal), and principal stratigraphic–structural compartments thereof.

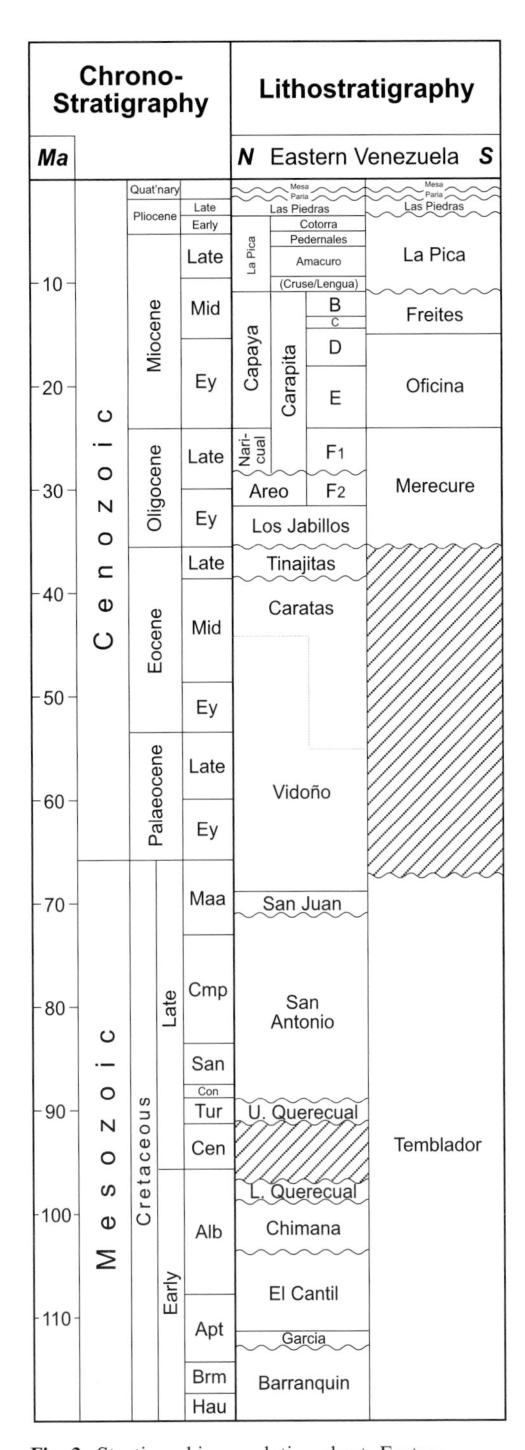

Fig. 3. Stratigraphic correlation chart, Eastern Venezuelan basin. Chronostratigraphy after Haq *et al.* (1988); lithostratigraphy after BP.

range from marginal marine sandstones through shallow-marine sandstones and shales to deep-marine shales. Lithology varies with eustatically mediated sea level, with the sandstones associated with regressions and the shales with transgressions. Carbonate sediments (platform and pelagic limestones) are only locally volumetrically significant.

The seismically demonstrably syn-tectonic Naricual, Carapita, La Pica, Las Piedras and Mesa/Paria formations, and their correlatives, record a foreland-basin phase, which has lasted from Late Oligocene to the present day (Parnaud *et al.* 1995) (Figs 3 and 4). The foreland-basin phase was initiated as a result of the oblique collision between the Caribbean and South American plates, which ultimately led to the formation of the Serrania del Interior. Because of the obliquity of the collision, the foreland-basin phase and the deposition associated with it was initiated earlier in the west than in the east, and migrated from west to east through time (see, for instance, Rohr 1991; Hoorn *et al.* 1995; Diaz de Gamero 1996). Associated clastic sediments, derived not only from the Guyana Shield to the south but also from the rising Serrania del Interior to the north, range from alluvial and fluvial sandstones and conglomerates through peri-deltaic sandstones and coals and prodeltaic shales to turbiditic sandstones and basinal shales. On a basinal scale, lithology varies with tectonically-mediated basin evolution and fill, with the turbiditic sandstones ('flysch') associated with early stages and the prodeltaic, peri-deltaic, fluvial and alluvial sandstones ('molasse') with late stages. On a smaller scale, lithology varies with eustatically mediated sea level, with peri-deltaic sandstones associated with regressions and prodeltaic shales with transgressions.

In terms of petroleum geology, the restricted basinal Querecual Formation of the passive-margin phase is the most important regional source rock. The marginal to shallow-marine San Juan and Los Jabillos formations, and their correlatives, are the principal reservoir rocks in the Quiriquire Deep Field and in the adjacent Furrial Trend. The Pedernales Member of the La Pica Formation of the foreland basin is the principal reservoir rock in the Pedernales Field (a smaller, lighter oil, accumulation also occurs in the underlying Amacuro Member). The non-marine Las Piedras Formation is the principal reservoir in the Quiriquire Shallow Field. The Cotorra Member of the La Pica Formation is the seal in the Pedernales Field. The Quiriquire Shallow Field is sealed by a tar-mat. A range of structural and stratigraphic trap types are known

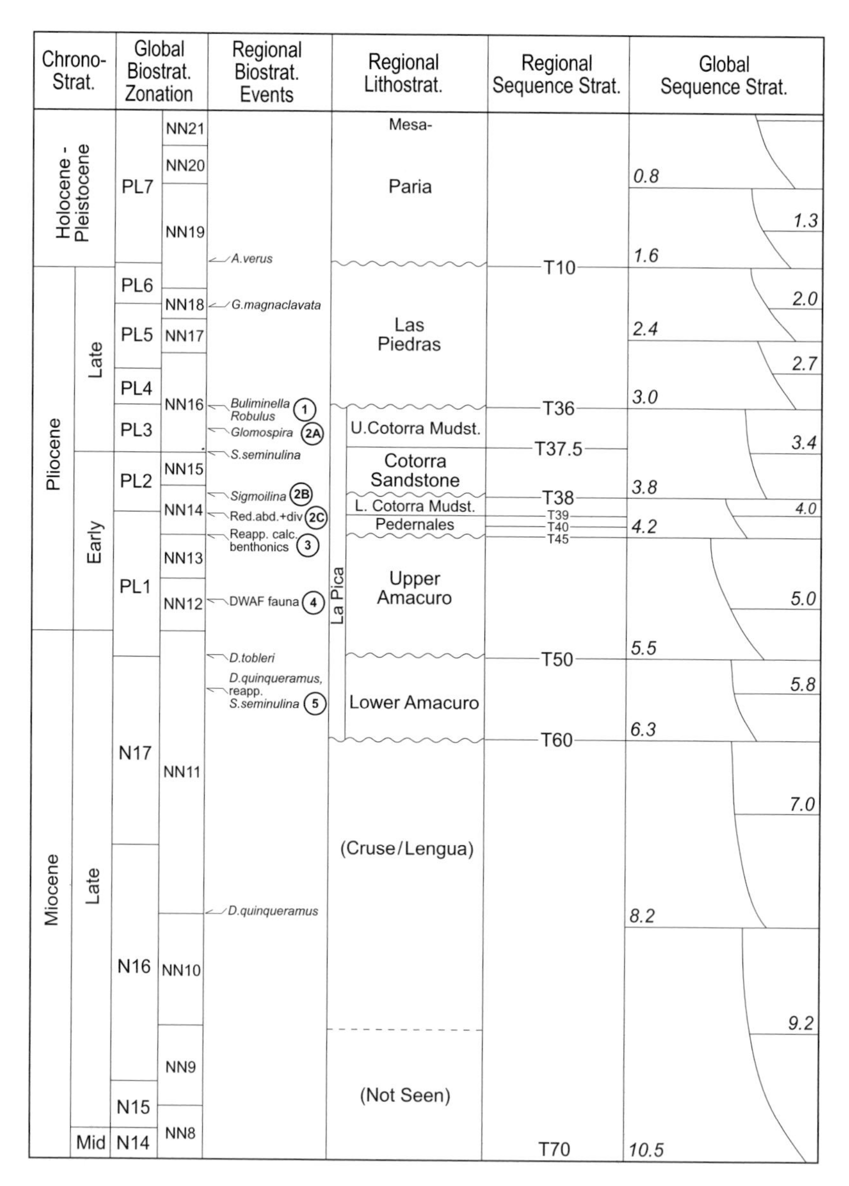

Fig. 4. Late Miocene–Holocene stratigraphic correlation chart, Eastern Venezuelan basin. Chronostratigraphy, calcareous nannoplankton biostratigraphy and sequence stratigraphy after Haq *et al.* (1988); planktonic foraminiferal biostratigraphy after Berggren *et al.* (1995), benthonic foraminiferal event biostratigraphy (1–5) after EPOCA; lithostratigraphy after BP. Calibration best fit.

in the basin. The trap to the Pedernales Field is currently interpreted as essentially structural, although there may be a stratigraphic component (see Fig. 5; see also below). Recent mapping has led to the recognition of a significantly higher degree of structural and stratigraphic compartmentalization than was previously envisaged.

History of Pedernales Field from discovery to development

Indications of hydrocarbons in the vicinity of the field were first noted over 150 years ago (Martinez 1989). On 3 October 1839, Jose Maria Vargas reported the presence of asphalt at

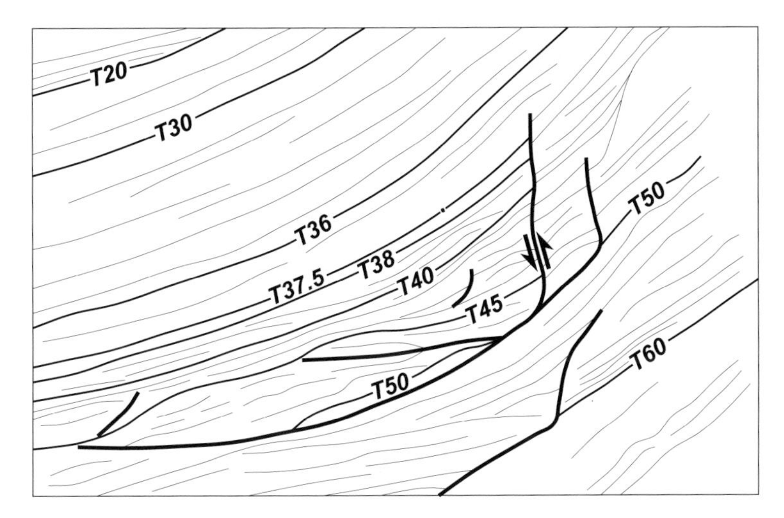

Fig. 5. Portion of seismic line showing structural interpretation. For line of section see Fig. 2. Principal reservoir located between T40 and T45.

Pedernales in the Lower Orinoco Canton to the Ministry of Finance, and recommended investigations to determine the form and extent of the resource, and the granting of concessions for its exploitation. Graham Company of Trinidad drilled four wells at La Brea, Capure Island, near Pedernales in 1890, and the following year the first issue of the *Boletin de la Riqueza Publica* (*Bulletin of Public Resources*) referred to the exploitation of the asphalt resource there (later described by L. V. Dalton in 1912 as 'desultory and ill-advised'). In 1903 L. Hirzel referred to the asphalt and petroleum resources of the Orinoco River (including Pedernales and Isla Plata).

Pedernales Field was discovered by Creole (Standard Oil of Venezuela) on completion of the wildcat well Pedernales-2 on 20 June 1933. Production began later that year, and was halted in 1986 (operatorship passing from Creole to Lagoven, a subsidiary of the state-owned oil company Petroleos de Venezuela S.A., on nationalization in 1976). During that 60-odd year period (interrupted by World War II and the end of the Creole–Texaco refining contract), cumulative production from about 60 wells totalled about 60 mmstb (Bennett *et al.* 1994; Gluyas *et al.* 1996; Jones & Stewart 1997).

The contract to reactivate the field on behalf of Lagoven was awarded to BP Venezuela in the licencing round in 1993 (Bennett *et al.* 1994; Gluyas *et al.* 1996; Jones & Stewart 1997). Production from the earliest two BP-operated wells was sufficiently encouraging for commerciality to be declared in March 1995. Results from later wells indicated that, despite depleted reservoir pressures, the production rate required to maintain materiality (11 500 barrels of oil per day) would be achieved. Phase II of the reactivation project was sanctioned in November 1996. At the time of writing (January 1997) a total of 17 oil-producer, water-injector and gas-disposal wells (PCA-49–PCA-65) have been drilled by BP. These wells have been sited on a combination of historical well-performance data, and 1994 vintage two-dimensional (2D) and 1996 vintage (fast-track) 3D seismic mapping.

This paper focuses on biostratigraphy and palaeoenvironmental interpretation, and their impact on integrated reservoir description (IRD) and production geology.

Historical biostratigraphy and palaeoenvironmental interpretation

Biostratigraphy

Biostratigraphic control on age-dating of the Late Miocene–Holocene of the Eastern Venezuelan basin is generally poor because of the widespread development at this time of marginal marine to non-marine conditions inimical to the open oceanic calcareous nannoplankton and planktonic foraminifera used in global standard biozonations (see Fig. 4).

Biostratigraphic data of both historical (Creole) and modern (EPOCA) vintage are available on a total of 43 wells in the Pedernales Unit (Fig. 2). Micropalaeontological (chiefly marine benthonic foraminiferal) data are available on all Creole and EPOCA wells, palynological (freshwater algal, terrestrially derived spore and pollen, and marine dinoflagellate cyst) data on some Creole and all EPOCA wells, and nannopalaeontological (marine calcareous nannoplankton) data on some EPOCA wells only. The palynological data on all Creole and some EPOCA wells (all those operated by Creole and Lagoven) are of questionable quality owing to contamination of samples by drilling fluids formulated from local river-mud and river-water containing modern mangrove pollen and other palynomorphs, effectively indistinguishable from their ancient counterparts. The quality of the data are also affected by sample type and spacing. Overall, good-quality data (core and cuttings, mean sample spacing of the order of 20 feet) are available on 10 wells, moderate data (cuttings only, mean sample spacing 50 feet) on six wells and poor data (cuttings only, mean sample spacing 100 feet) on 27 wells.

Creole (see also Barnola 1960) recognized five micropalaeontological 'zones' and 'subzones' over the Amacuro to Paria interval in wells in the Pedernales Field, with a mean duration (resolution) of the order of 2.1 Ma. The best-fit correlation of these 'zones' against established lithostratigraphic units (complicated by diachronous development) is as follows:

- B0 'zone' = Paria/Las Piedras Formations;
- B1a 'zone' (characterized by *Buliminella* and *Glomospira*) = La Pica Formation, Cotorra Member;
- B1b 'zone' (*Sigmoilina*) = La Pica Formation, Pedernales–Amacuro members;
- B1b 'zone', AFQ (agglutinating foraminifera/*Quinqueloculina*) 'subzone' = La Pica Formation, Pedernales Member, seal section;
- B1b 'zone', Lam (laminated) 'subzone' = La Pica Formation, Pedernales and Amacuro members, reservoir sections.

EPOCA recognized seven micropalaeontological 'zones' and 'subzones' over the Amacuro–Cotorra interval in wells in the Pedernales Field, with a mean duration (resolution) of the order of 0.4 Ma. The best-fit correlation of which against established lithostratigraphic units is as follows (see also Fig. 4):

- PED1 = La Pica Formation, Cotorra Member (Upper Cotorra Mudstone);
- PED2A = La Pica Formation, Cotorra Member (Upper Cotorra Mudstone);
- PED2B = La Pica Formation, Cotorra Member (Lower Cotorra Mudstone)/Pedernales Member, seal section;
- PED2C = La Pica Formation, Pedernales Member, reservoir section;
- PED3 = La Pica Formation, Upper Amacuro Member;
- PED4 = La Pica Formation, 'Middle' Amacuro Member;
- PED5 = La Pica Formation, Lower Amacuro Member ('Cruse/Lengua [of Trinidad] Equivalent' of deep-penetration (crestal) wells).

EPOCA recognize only one palynological zone (zone VIII of Lorente 1986).

All of Creole's and EPOCA's micropalaeontological 'zones' are characterized in terms of facies-controlled benthonic foraminifera, and are 'ecozones' rather than 'chronozones'. As such, they are inherently, and indeed demonstrably, diachronous on both the regional and reservoir scale. For example, calibration against seismically defined time lines clearly demonstrates that the youngest occurrence of the deep-water indicator *Sigmoilina* [*Silicosigmoilina*], also the index fossil for the B1b or PED2B 'zones', is younger down-dip than up-dip, whether on a regional scale or on the scale of individual structures within the region (see also below).

However, the 'zones' do have correlative value when integrated into the framework of isochronous tectono-stratigraphic sequence boundaries and flooding surfaces identified on seismic sections and well logs (on a workstation).

Moreover, the occurrence of the calcareous nannoplankton species *Discoaster quinqueramus* within EPOCA 'zone' PED5 has chronostratigraphic, as well as correlative, value, and indicates calibration against Late Miocene global standard calcareous nannoplankton zone NN11. A tentative calibration is also indicated between the containing sediments of the Lower Amacuro Member and the 5.8 Ma maximum flooding surface (MFS) on the eustatic sea-level cycle chart (Haq *et al.* 1988).

On the basis of cycle counting, further tentative calibrations are indicated between the base of the Upper Amacuro Member and the 5.5 Ma sequence boundary (SB), between the base of the Pedernales Member (reservoir section) and the 4.2 Ma SB, and between the top of the Pedernales Member/base of the Lower Cotorra Member (seal section) and the 4.0 Ma MFS. Independent corroborative evidence for the calibration of the Lower Cotorra Member with the 4.0 Ma MFS comes from Trinidad, where the lithostratigraphically equivalent Lower Forest Clay contains *Globigerina nepenthes*, the index species for planktonic foraminiferal zone PL1 of Berggren *et al.* (1995) (approximately 4.0–5.3 Ma) (unpublished observations).

Palaeoenvironmental interpretation

Palaeoenvironmental interpretation is based on integration of micropalaeontological and palynological (biofacies) and sedimentological data. The criteria for the characterization of subenvironments by means of micropalaeontological and palynology are summarized in Table 1, and the criteria for the characterization of micropalaeontological biofacies in Table 2 (see also Figs 6A and B).

Micropalaeontological biofacies. EPOCA recognize 13 micropalaeontological biofacies (*Centropyxis*, *Miliammina*, *Ammonia*, *Trochammina*, *Hanzawaia*, *Buliminella*, *Eggerella*, *Uvigerina*, *Bolivina*, *Glomospira*, *Uvigerina*/planktonic, *Alveovalvulina*/*Silicosigmoilina* and *Cyclammina*) in wells in the Pedernales Unit, which are essentially equivalent to those recognized by Batjes (1968) and Jones (1998) in the Late Miocene and Pliocene of Trinidad.

The palaeoenvironmental interpretation of the micropalaeontological biofacies is based on comparison between ancient and analogous modern biofacies (van Andel & Postma 1954; Phleger 1955, 1966; Saunders 1957, 1958; Todd & Bronnimann 1957; Drooger & Kaasschieter 1958; Lankford 1959; Bermudez 1966; Batjes 1968; Carr-Brown 1972; Pflum & Frerichs 1976; Radford 1981; Haman 1983; Boltovskoy, 1984; Hiltermann & Haman 1985; Culver, 1990; van der Zwaan & Jorissen, 1991; Scott *et al.* 1991; Plaziat 1995; Jones 1996, 1998).

The *Centropyxis* biofacies is interpreted as indicating a fluvially dominated delta-top environment, the *Miliammina* and *Trochammina* biofacies tidally dominated delta-top subenvironments, the *Buliminella* biofacies a delta-front environment, the *Eggerella* biofacies a proximal (shelf) prodelta environment, and the *Glomospira*, *Alveovalvulina*/*Silicosigmoilina* and *Cyclammina* biofacies distal (shelf–slope) prodelta environments. Certain species that occur in the *Glomospira*, *Alveovalvulina*/*Silicosigmoilina*, *Cyclammina* and *Uvigerina*/planktonic biofacies range through to the Recent, and have known bathymetric distributions that, assuming uniformitarianism, can be used to infer palaeobathymetries. The inferences are of palaeobathymetries of the order of 50–100 m for the *Glomospira* biofacies, 150–300 m for the *Alveovalvulina*/*Silicosigmoilina* biofacies, and 500 m for the *Cyclammina* and *Uvigerina*/planktonic biofacies (Jones in press).

The *Ammonia* and *Hanzawaia* biofacies are interpreted as indicating inner shelf, the *Bolivina* and *Uvigerina* biofacies middle–outer shelf, and the *Uvigerina*/planktonic biofacies outer shelf–basinal environments outside the area of deltaic influence.

The preponderance of agglutinating foraminifera over their calcareous counterparts in the peri-deltaic biofacies is noteworthy. Jones

Table 1. *Characterization of sub-environments by means of micropalaeontology and palynology.* Zonocostites ramonae *and other mangrove pollen also widely recorded in submarine fans (e.g. Pleistocene, Amazon Fan)*

Subenvironment	Micropalaeontological biofacies	Palynological biofacies
Fluvially dominated delta-top	*Centropyxis*	*Pediastrum*; palm/gymnosperm/grass/*Bombax*/spore; fungal spore; pteridophyte spore
Tidally dominated delta-top	*Miliammina* and *Trochammina*	*Zonocostites ramonae* (mangrove swamp); palm (palm swamp); palm/*Z. ramonae* (mixed swamp)
Delta-front	*Buliminella*	Dynocyst
Prodelta	*Eggerella* (proximal (shelf)); *Glomospira* (distal (shelf)); *Alveovalvulina*/*Silicosigmoilina* and *Cyclammina* (distal (slope))	Dinocyst
Inner shelf (outside deltaic influence)	*Ammonia* and *Hanzawaia*	Dinocyst
Middle–outer shelf (outside deltaic influence)	*Bolivina* and *Uvigerina*	Dinocyst
Outer shelf–basinal (outside deltaic influence)	*Uvigerina*/planktonic	Dinocyst

Table 2. *Characterization of micropalaeontological biofacies.* Trochammina *biofacies further characterized by* Arenoparrella mexicana, Ammobaculites salsus, Haplophragmoides bonplandi, H. hancocki, H. manilaensis, H. wilberti *and* Saccammina sphaerica

Biofacies	Dominant taxa	Accessory taxa
Centropyxis	*Centropyxis* spp.	
Miliammina	*Miliammina telemaqensis*	*Arenoparrella* sp.
Trochammina	*Trochammina* cf. *quadriloba* [*T.* cf. *pusilla*]	*Ammonia beccarii*, *Alveolophragmium* sp.
Buliminella	*Buliminella* spp.	*Nonionella* spp., *Nonion* spp., *Suggrunda porosa*, *Bolivina* sp., *Fursenkoina* [*Virgulina*] sp., *Gyroidina parva*, *Epistominella* sp.
Eggerella	*Eggerella forestensis*	Astrorhizidae [*Psammosiponella*] sp., *Hippocrepinella hirudinea*, *Saccammina* sp., *Glomospira gordialis*, *Haplophragmoides* sp., *Textularia* sp., *Trochammina* sp.
Glomospira	*Glomospira gordialis*	*Saccammina* sp., *Ammodiscus* sp., *Haplophragmoides carinatus*, *?Spirosigmoilina tenuis* [*Silicosifmoilina* sp.], *Sigmoilopsis schlumbergeri*, *Cyclammina cancellata*
Alveovalvulina/*Silicosigmoilina*	*Alveovalvulina suteri*, *Spirosigmoilina tenuis* [*Silicosigmoilina* sp.]	*Saccammina* sp., Astrorhizidae [*Psammosiphonella*] sp., *Glomospira gordialis*, *Cyclammina cancellata*, *Martinottiella pallida*, *Sigmoilopsis schlumbergeri*
Cyclammina	*Cyclammina cancellata*	Astrorhizidae [*Psammosiphonella*] sp., *Ammodiscus* sp., *Usbekistania* [*Glomospira*] *charoides*, *Glomospira gordialis*, *Conglophragmium* [*Haplophragmoides*] *coronatum*, *Haplophragmoides carinatus*, *Recurvoides higginsi*, *R. obsoletum*, *Trochammina* sp., *Valvulina flexilis*, *Martinottiella pallida*, *Alveovalvulina suteri*, *Guppyella miocenica*, *?Spirosigmoilina tenuis* [*Silicosigmoilina* sp.], *Sigmoilopsis schlumbergeri*, *Discamminoides tobleri*, *Texularia* sp.
Ammonia	*Ammonia beccarii*	
Hanzawaia	*Hanzawaia strattoni*	*?Alveolophragmium* sp., *Lenticulina calcar*, *Cassidulina* sp., *Nonionella* sp., *Cancris oblongus*, *Ammonia beccarii*
Bolivina	*Bolivina floridana*	*Sigmoilopsis schlumbergeri*
Uvigerina	*Uvigerina* spp.	*Bolivina* spp., *Bulimina* sp., *Buliminella* sp., Miliolidae, Nodosariidae, *Cassidulina* sp.
Uvigerina/planktonic	*Uvigerina* spp., planktonics	Abundant and diverse calcareous benthonic and planktonic species

(1996) noted a correlation between the development of (deep-water) agglutinating foraminiferal (DWAF) faunas and, among other things, tectono-eustatic sea-level lowstands and lowstand systems tracts. Jones (1998) proposed two mechanisms for the development of DWAF faunas and the effective exclusion of their calcareous benthonic counterparts, one feeding back into the other. The first mechanism involves rapid sedimentation associated with a lowstand, resulting in dissolution of calcareous benthonic species in early diagenetic reactions involving the decomposition of organic material.

Palynological Biofacies. EPOCA also recognize eight palynological biofacies (*Pediastrum*, palm/gymnosperm/grass/*Bombax*/spore, fungal spore, pteridophyte spore, *Zonocostites ramonae*, palm, palm/*Z. ramonae* and dinocyst). The palaeoenvironmental interpretation of the palynological biofacies is based on comparison between ancient and analogous modern biofacies (Muller

(a)

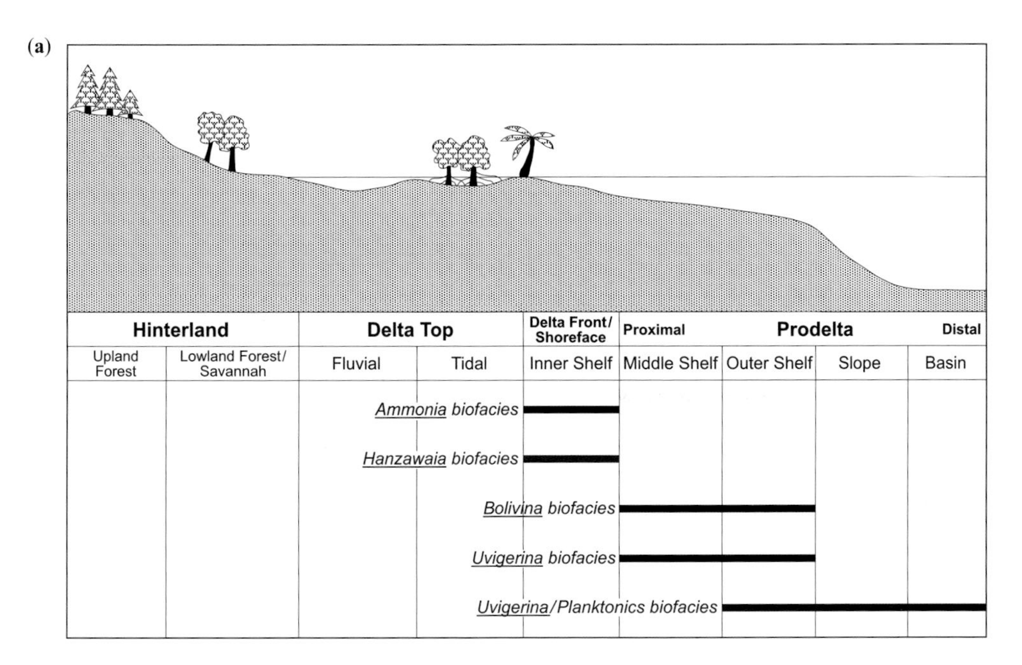

(b)

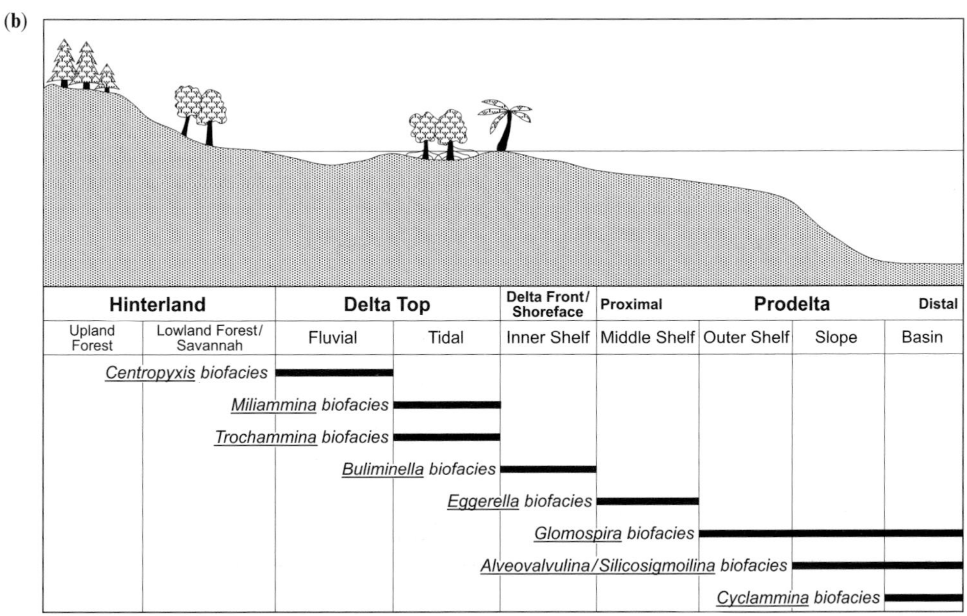

Fig. 6. Palaeoenvironmental interpretation of micropalaeontological biofacies. Showing interpreted distributions of micropalaeontologiacal biofacies in relation to an idealized profile. A = 'clear water'; B = 'turbid water'.

1959; Williams & Sarjeant 1967; Lorente 1986; Traverse 1988; Plaziat 1995). The *Pediastrum* biofacies is interpreted as indicating a fresh-water origin, the palm/gymnosperm/grass/*Bombax*/spore biofacies a hinterland, riverside or mixed-swamp origin, the fungal spore biofacies a back-swamp origin, the pteridophyte spore biofacies a fern-swamp origin, the *Zonocostites ramonae* biofacies a coastal mangrove-swamp origin, the palm biofacies a palm-swamp origin, the palm/*Z. ramonae* biofacies a mixed coastal swamp origin, and the dinocyst biofacies a marine origin. Importantly, mangrove swamps are best developed in sheltered parts of the tidally

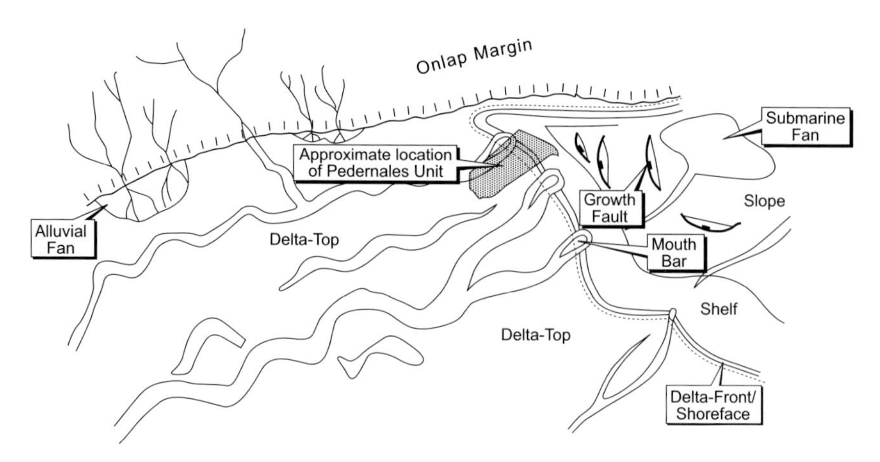

Fig. 7. Regional distribution of facies at time of Pedernales Member reservoir deposition. Time slice defined on seismic, wireline log and biostratigraphic evidence.

influenced area of the modern Orinoco Delta (Lorente 1986). They are rare in storm-influenced sedimentary regimes.

Integration of micropalaeontological, palynological and sedimentological data. Integration of micropalaeontological, palynological and sedimentological data indicate that the reservoir section of the Amacuro Member was deposited in a prodelta environment characterized by turbiditic sedimentation. The reservoir section of the Pedernales Member (Figs 7 and 8a) was deposited in a range of delta-top to delta-front subenvironments, locally modified by shoreline processes (e.g. distributary channels in southwestern and central parts of the field, shorefaces (locally slipped) in the northeastern part of the field). The seal section of the Pedernales Member (Fig. 8b) was deposited in a range of prodelta subenvironments (see also below). Incidentally, outcrop analogues for the Amacuro and Pedernales Member reservoir and seal sections are to be found in Trinidad. Here, there is a strong correlation between (micro)biofacies and ichnofacies. Upper–middle shoreface sands are characterized by *Buliminella* and associated (micro)biofacies and *Skolithos* ichnofacies with *Ophiomorpha* (including *O. nodosa*, probably produced by the shrimp *Callianassa major*, still living in the moderate- to high-energy littoral environments of the eastern seaboard of the United States (Bromley 1990)) and *Thalassinoides*. Offshore sands and shales are characterized by *Uvigerina*/planktonic and associated (micro)biofacies and *Cruziana* ichnofacies with *Chondrites*, cf. *Gyrolithes*, *Schaubcylindrichnus*, *Rhizocorallium* and *Teichichnus*.

The observed close vertical proximity of the delta-top and delta-front sediments of the Pedernales Member reservoir section and the distal prodelta sediments of the seal section in wells indicates a close proximity of delta-top, delta-front and prodelta environments at the time of deposition, and hence a lowstand shelf-edge-delta rather than a highstand shelf-delta (cf. van Wagoner *et al.* 1990). Evidence of the extensive growth-faulting typically found in this type of setting has been observed on subsurface seismic sections from Pedernales, and also in surface outcrop sections along the adjacent Southern (Moruga) Coast of Trinidad.

Ongoing biostratigraphic work programme and its impact on integrated reservoir description (IRD)

The principal objective of the ongoing high-resolution biostratigraphic data acquisition/work programme is to facilitate correlation over reservoir sections both within and between sequences. The impact of results to date on key areas of IRD is discussed below. IRD will likely also be significantly impacted by a planned pilot strontium-isotope stratigraphic study.

Field-wide correlation

At the time of writing, integrated high-resolution biostratigraphic–well log–seismic correlation (on a workstation) over the Pedernales Member reservoir section has be achieved both *within* and *between* the compartments of the field

(a)

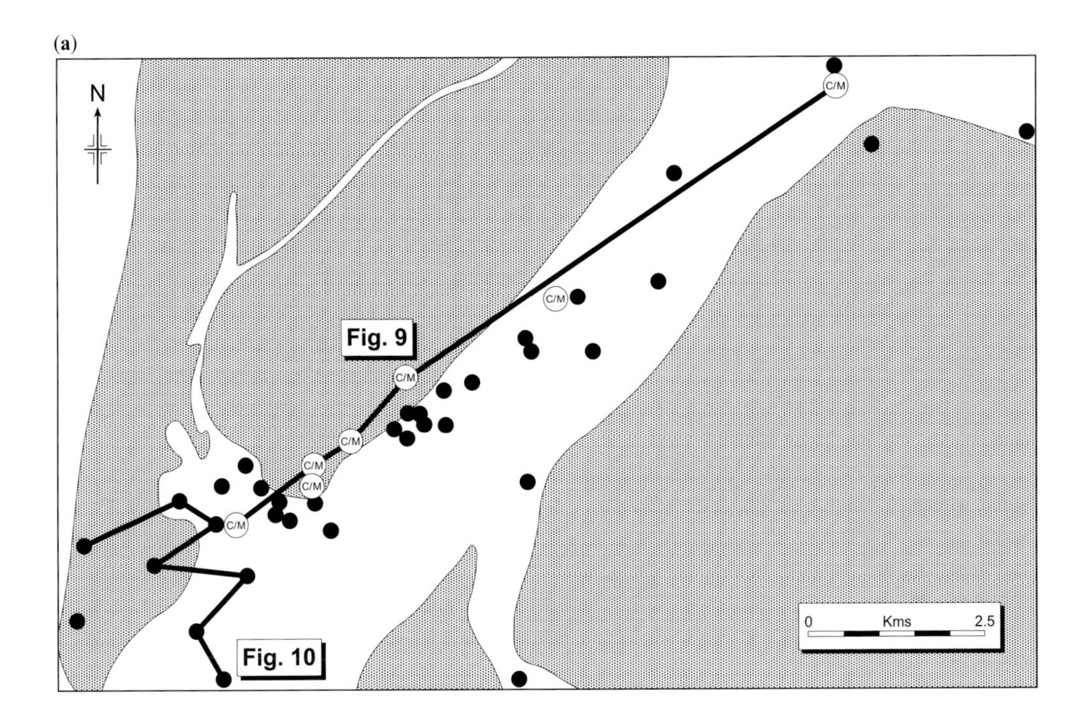

(b)

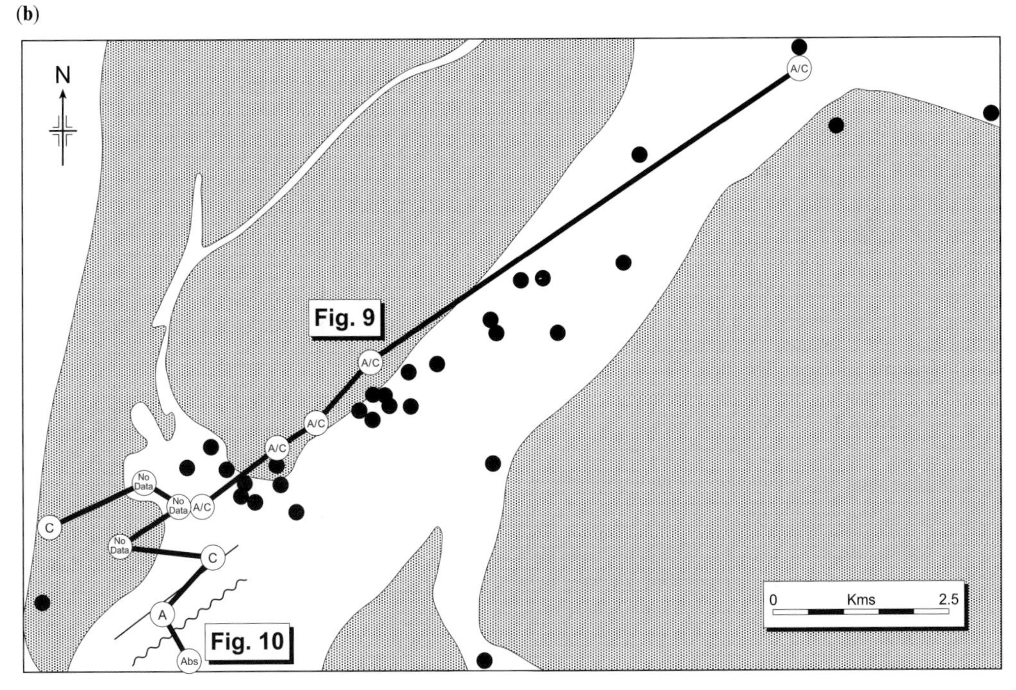

Fig. 8. (**a**) Field-wide distribution of facies at time of Pedernales Member, reservoir unit deposition. Time slice defined on seismic, wireline log and biostratigraphic evidence. C/M, *Centropyxis/Miliammina* biofacies (NB: reservoir sandstones in PCA-58 (to the northeast) encased in shales developed in *Alveovalvulina/ Cyclammina* biofacies, and interpreted as olistolithic). (**b**) Field-wide distribution of facies at time of Pedernales Member, seal unit deposition. Time slice defined on seismic, wireline log and biostratigraphic evidence. A, *Alveovalvulina* biofacies; C, *Cyclammina* biofacies.

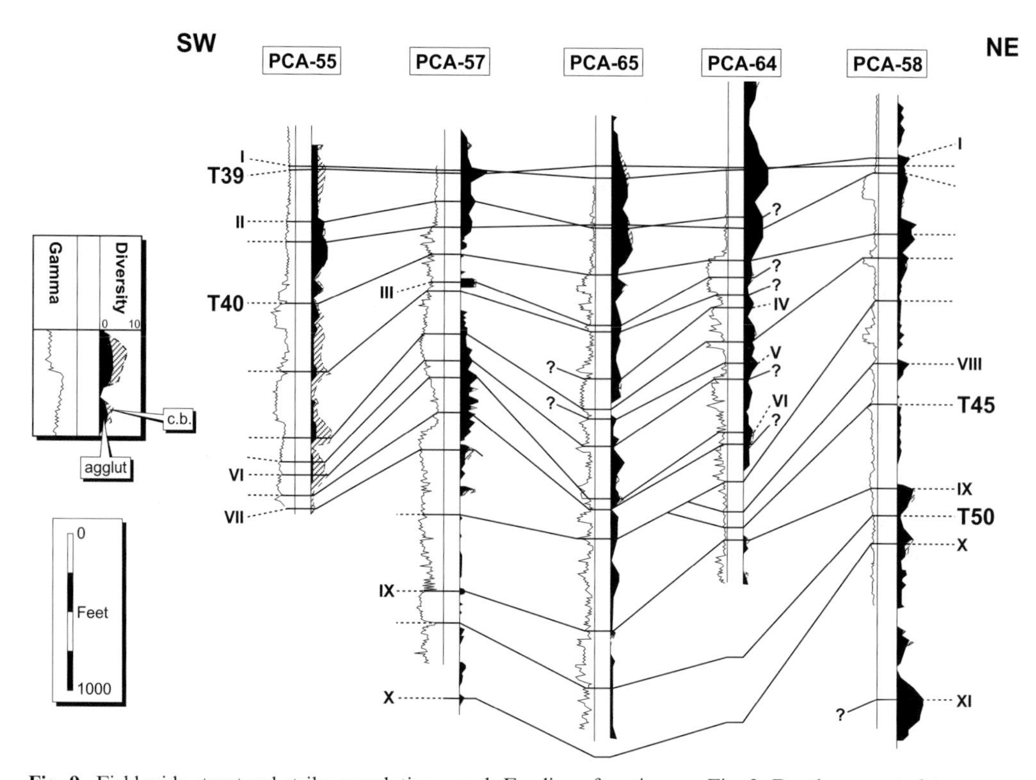

Fig. 9. Field-wide structural strike correlation panel. For line of section see Fig. 2. Depths corrected to true vertical depth (TVD). T39–T50 = time lines (flooding surfaces and sequence boundaries) identified interactively (on sun work-station) on seismic and wireline log evidence. T39 = top Pedernales; T50 = base Pedernales/top Amacuro; I–XI = biostratigraphic events (I = diversity maximum associated with seal unit (confidence limit 100%); II = reappearance of agglutinating foraminifera (80–100%); III = appearance of calcareous benthonic foraminifera (60%); IV = log event (40%); V = log event (40%); VI = reappearance of calcareous benthonic foraminifera (60–80%); VII = second reappearance of calcareous benthonic foraminifera (40%); VIII = second reappearance of agglutinating foraminifera (?40%); IX = third reappearance of calcareous benthonic foraminifera (?80%); X = reappearance of agglutinating and calcareous benthonic foraminifera (?40%); XI = third reappearance of agglutinating foraminifera including *Cyclammina* (?20%)). Micropalaeontological biofacies used in constraining correlation (on workstation, cf. Westcott *et al.* 1998), but omitted for clarity. Micropalaeontological and sedimentological evidence indicates that the sandstones in the southwestern and central areas are distributary channel sandstones and that those in the northeast are shoreface sandstones (locally displaced downslope).

(Fig. 9), albeit in a limited number of wells. Eleven correlatable events have been recognized, with a mean resolution thought to be of the order of 20 000 years. The events include abundance–diversity peaks and troughs, and biofacies–bathymetric trends. Abundance and diversity trends have also been used in the identification of condensed sections and in the stratigraphic correlation of coeval sediments in the Palo Seco Field in neighbouring Southwest Trinidad (Hudson *et al.* 1993).

Depositional model

High-resolution biostratigraphic–biofacies data demonstrate the existence of a bathymetric gradient at the time the Amacuro and Pedernales reservoirs were being deposited, with, in the case of the Amacuro, deep-marine biofacies on the flank and shallow-marine biofacies on the crest, and in the case of the Pedernales, marginal marine biofacies (associated with low-stand wedge reservoir sands) on the flank and a hiatus on the crest of the (past and present) Pedernales structure (Fig. 10). This probably indicates structural, as well as stratigraphic, control on reservoir deposition. Gravity gliding along the observed gradient provides a possible mechanism for the locally observed section repetition and enhanced reservoir development (in 'sweet spots'), and also for the up-dip trapping mechanism.

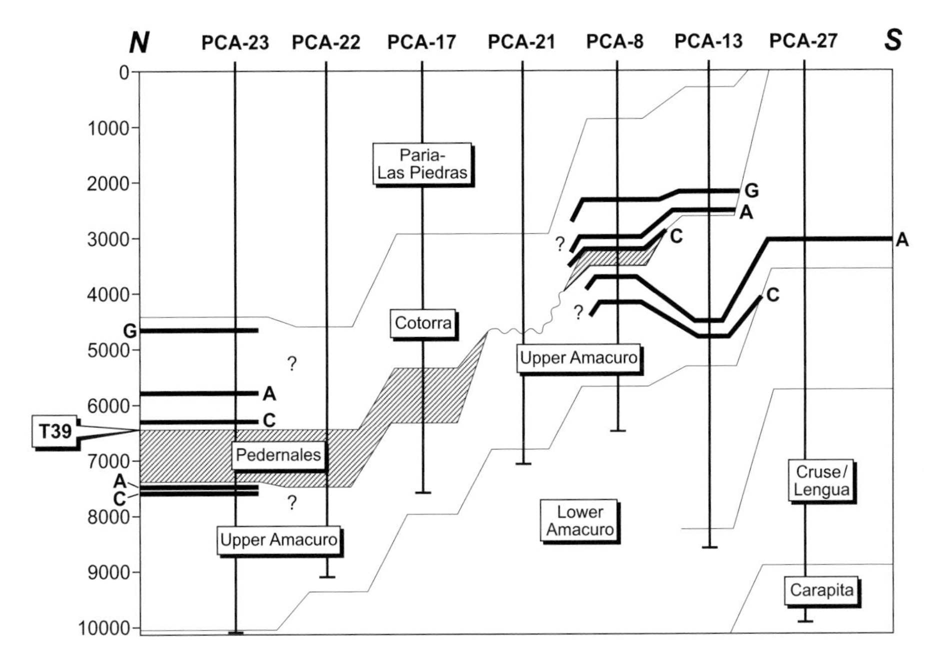

Fig. 10. Structural dip correlation panel, southwest Pedernales. For line of section see Fig. 2. Depths as drilled. Paria–Carapita = lithostratigraphic units adjusted to sequence stratigraphy. G = *Glomospira* biofacies; A = *Alveovalvulina* biofacies; C = *Cyclammina* biofacies.

Potential future applications

Potential future applications of biostratigraphy include picking top reservoir and base reservoir (true depth (TD)) and biosteering multilateral wells at rig-site.

Conclusions

Biostratigraphy and palaeoenvironmental interpretation

Historical and modern vintage micropalaeontological data enable the recognition of a number of 'zones' over the reservoir interval in wells in the Pedernales Field. These 'zones' are characterized in terms of facies-controlled benthonic foraminifera, and are diachronous, although they do have correlative value when integrated into the framework of isochronous tectono-stratigraphic sequence boundaries and flooding surfaces identified on seismic sections and well logs (on a workstation). The 'zones' and their containing sediments can be tentatively calibrated against global standard biozones, and hence sequence boundaries and flooding surfaces on the eustatic sea-level cycle chart. The principal reservoir appears to be associated with the 4.2 Ma SB, the seal with the 4.0 Ma MFS. The ongoing high-resolution biostratigraphic work programme has revealed the existence of 11 correlatable events (abundance and diver-sity peaks and troughs, and biofacies–bathymetric trends) over the reservoir interval, with a mean resolution thought to be of the order of 20 000 years.

Micropalaeontological and palynological data also enable the recognition of a number of biofacies over the reservoir interval. These biofacies are interpreted on the basis of modern analogy as indicating a range of delta-top, delta-front and prodelta environments. Integration of micropalaeontological, palynological and sedimentological data indicate that the principal reservoir was deposited in a range of delta-top to delta-front subenvironments, and the seal in a range of prodelta subenvironments. The observed close vertical proximity of the delta-top and delta-front sediments of the Pedernales Member reservoir section and the distal prodelta sediments of the seal section in wells indicates a close proximity of delta-top, delta-front and prodelta environments at the time of deposition, and hence a lowstand shelf-edge-delta rather than a highstand shelf-delta.

Impact on integrated reservoir description (IRD)

Integrated high-resolution biostratigraphic–well log–seismic correlation over reservoir sections can be achieved both within and between the structural–stratigraphic compartments of the Pedernales Field.

High-resolution biostratigraphic–biofacies data demonstrate the existence of a bathymetric gradient at the time the reservoir was being deposited, and hence suggest structural as well as stratigraphic control on reservoir deposition. Gravity gliding along the observed gradient provides a possible mechanism for the locally observed section repetition and enhanced reservoir development, and also for the up-dip trapping mechanism.

BP are thanked for permission to publish, and for providing word-processing and drafting facilities.

References

BARNOLA, A. 1960. Historia del Campo de Pedernales. *Memoria Tercier Congreso Geologico Venezolano*, **2**, 552–573.

BATJES, D. A. J. 1968. Paleoecology of foraminiferal assemblages in the Late Miocene Cruse and Forest Formations of Trinidad, Antilles. *In*: SAUNDERS, J. B. (ed.) *Transactions of the Fourth Caribbean Geological Conference, Trinidad and Tobago*, 141–156.

BENNETT, J. *et al.* 1994. The Pedernales Field: unravelling reservoir complexity. *In*: *VII Congreso Venezolano de Geofisica*, 486–493.

BERGGREN, W. A., KENT, D. V., SWISHER, C. C., III & AUBRY, M.-P. 1995. A revised Cenozoic geochronology and chroniostratigraphy. *In*: BERGGREN, W. A. *et al.* (eds) *Geochronology, Time Scales and Global Stratigraphic Correlation*. Society of Economic Paleontologists and Mineralogists, Special Publication, **54**, 129–212.

BERMUDEZ, P. J. 1966. Consideraciones sobre los Sedimentos del Miocene Medio al Reciente las Costas Central y Oriental de Venezuela, Primera Parte. *Boletin de Geologia*, **14**, 333–412.

BOLTOVSKOY, E. 1984. Foraminifera of Mangrove Swamps. *Physis (Buenos Aires), Secc. A*, **42**(109), 1–9.

BROMLEY, R. G. 1990. *Trace Fossils: Biology and Taphonomy*. Chapman & Hall, London.

CARR-BROWN, B. 1972. The Holocene/Pleistocene contact in the offshore area east of Galeota Point, Trinidad, West Indies. *In*: *Transactions of the Sixth Caribbean Geological Conference, Margarita, Venezuela*, 381–397.

CULVER, S. J. 1990. Benthic foraminifera of Puerto Rican mangrove–lagoon systems: potential for palaeoenvironmental interpretation. *Palaios*, **5**, 34–51.

DIAZ DE GAMERO, M. 1996. The changing course of the Orinoco during the Neogene: a review. *Palaeogeography, Palaeoclimatology, Palaeoecology*, **123**, 385–402.

DROOGER, C. W. & KAASSCHIETER, J. P. H. 1958. Foraminifera of the Orinoco–Trinidad–Paria shelf. *Verhandelingen der Koninklijke Nederlandse Akademie van Wetenschappen, Afdeeling Natuurkunde, Eerste Reeks*, **22**.

GLUYAS, J., OLIVER, J., WILSON, W. W. & TINEO, M. 1996. Pedernales oilfield, Eastern Venezuela: the first 100 years. *American Association of Petroleum Geologists Bulletin*, **80**, 1294.

HAMAN, D. 1983. Modern Textulariina (foraminifera) from the Balize Delta. Louisiana. *In*: VERDENIUS, J. G., VAN HINTE, J. E. & FORTUIN, A. R. (eds) *Proceedings of the First Workshop on Arenaceous Foraminifera*. Continental Shelf Institute, Trondheim, 59–88.

HAQ, B. U., HARDENBOL, J. & VAIL, P. R. 1988. Mesozoic and Cenozoic chronostratigraphy and eustatic cycles. *In*: WILGUS, C. K., HASTINGS, B. S., KENDALL, C. G. ST C, POSAMENTIER, H., ROSS, C. A. & VAN WAGONER, J. C. (eds) *Sea-level Changes: An Integrated Approach*. Society of Economic Paleontologists and Mineralogists, Special Publications, **42**, 71–108.

HILTERMANN, H. & HAMAN, D. 1985. Sociology and synecology of brackish-water foraminifera and thecamoebinids of the Balize Delta, Louisiana. *Facies*, **13**, 287–294.

HOORN, C., GUERRERO, J., SARMIENTO, G. A. & LORENTE, M. A. 1995. Andean tectonics as a cause for changing drainage patterns in Miocene northern South America. *Geology*, **23**, 237–240.

HUDSON, D., KEENS-DUMAS, J., GOBERDHAN, H., DEOKIE, C., LAKMAN, C. & ARCHIE, C. 1993. Applied sequence stratigraphic analysis of well logs, Cruse Formation, Palo Seco Field, Trinidad: a technique for detailed reservoir description. *In*: *Society of Petroleum Engineers 11th Technical Conference, Trinidad, 23–25 June 1993 (TT93005)*.

JONES, N. E. & STEWART, R. C. S. 1997. Reactivation of the Pedernales Field, Venezuela: a sleeping giant awakes. *Abstracts, Geological Society Field Reactivation for the 21st Century Conference, Bath*. Geological Society, London.

JONES, R. W. 1996. *Micropalaeontology in Petroleum Exploration*. Oxford University Press, Oxford.

—— in press. Palaeoenvironmental interpretation of the Late Miocene and Pliocene of Trinidad based on micropalaeontological data. *In*: *Transactions of the Fourteenth Caribbean Geological Conference, Trinidad and Tobago*.

LANKFORD, R. R. 1959. Distribution and ecology of foraminifera from East Mississippi delta margin. *American Association of Petroleum Geologists Bulletin*, **53**, 2068–2099.

LORENTE, M. A. 1986. *Palynology and Palynofacies of the Upper Tertiary in Venezuela*. (Dissertaciones Botanicae, Bd. 99) J. Cramer, Berlin.

MARTINEZ, A. R. 1989. *Venezuelan Oil: Development and Chronology*. Elsevier Applied Science, London.

MULLER, J. 1959. Palynology of Recent Orinoco delta and shelf sediments. *Micropaleontology*, **5**, 1–32.

PARNAUD, F., GOU, Y., PASCUAL, J.-C., CAPELLO, M. A., TRUSKOWSKI, I. & PASSALACQUA, H. 1995. Petroleum geology of the central part of the Eastern Venezuelan basin. *In*: TANKARD, A. J., SUAREZSORUCO, R. & WELSINK, H. J. (eds) *Petroleum Basins of South America*. American Association of Petroleum Geologists, Memoir, **62**, 741–756.

PFLUM, C. E. & FRERICHS, W. E. 1976. *Gulf of Mexico Deep-water Foraminifera*. Cushman Foundation for Foraminiferal Research, Special Publication, **14**.

PHLEGER, F. B. 1955. Ecology of foraminifera in South-Eastern Mississippi delta area. *American Association of Petroleum Geologists Bulletin*, **39**, 712–752.

——1966. Patterns of living marsh foraminifera in South Texas coastal lagoons. *Boletin Societa Geologica Mexicana*, **28**(1), 1–44.

PLAZIAT, J.-C. 1995. Modern and fossil mangroves and mangals: their climatic and biogeographic variability. *In*: BOSENCE, D. W. J. & ALLISON, P. A. (eds) *Marine Palaeoenvironmental Analysis from Fossils*. Geological Society, London, Special Publications, **83**, 73–96.

RADFORD, S. S. 1981. Depth distributions of Recent foraminifera in selected bays, Tobago Island. *Revista Espanola de Micropaleontologia*, **13**(2), 219–238.

ROHR, G. M. 1991. Paleogeographic maps, Maturin basin of E. Venezuela and Trinidad. *In*: GILLEZEAU, K. A. (ed.) *Transactions of the Second Geological Conference of the Geological Society of Trinidad and Tobago*, 88–105.

SAUNDERS, J. B. 1957. Trochamminidae and certain Lituolidae (Foraminifera) from the Recent brackish-water sediments of Trinidad, British West Indies. *Smithsonian Miscellaneous Collections*, **134**(5), 1–16.

——1958. Recent foraminifera of mangrove swamps and river estuaries and their fossil counterparts in Trinidad. *Micropaleontology*, **4**(1), 79–92.

SCOTT, D. B., SUTER, J. R. & KOSTERS, E. C. 1991. Marsh Foraminifera and Arcelleans of the Lower Mississippi Delta: controls on spatial distribution. *Micropaleontology*, **37**(4), 373–392.

TODD, R. & BRONNIMANN, P. 1957. *Recent Foraminifera and Thecamoebina from the Eastern Gulf of Paria, Trinidad*. Cushman Foundation for Foraminiferal Research, Special Publication, **3**, 1–43.

TRAVERSE, A. 1988. *Paleopalynology*. Unwin Hyman, Boston, Massachusetts.

VAN ANDEL, TJ. & POSTMA, H. 1954. Recent Sediments of the Gulf of Paria. *Verhandelingen der Koninklijke Nederlandse Akademie van Wetenschappen, Afdeeling Natuurkunde, Eerste Reeks*, **20**(5), 1–245.

VAN DER ZWAAN, G. J. & JORISSEN, F. J. 1991. Biofacial patterns in river-induced shelf anoxia. *In*: TYSON, R. V. & PEARSON, T. H. (eds) *Modern and Ancient Continental Shelf Anoxia*. Geological Society, London, Special Publications, **58**, 65–82.

VAN WAGONER, J. C., MITCHUM, R. M., CAMPION, K. M. & RAHMANIAN, V. D. 1990. *Siliciclastic Sequence Stratigraphy in Well Logs, Cores and Outcrops: Concepts for High-resolution Correlation of Time and Facies*. American Association of Petroleum Geologists (Methods in Exploration Series, **7**).

WESCOTT, W. A., KREBS, W. N., SIKORA, P. J., BOUCHER, P. J. & STEIN, J. A. 1998. Modern applications of biostratigraphy in exploration and production. *The Leading Edge*, September 1998, 1204–1210.

WILLIAMS, D. B. & SARJEANT, W. A. S. 1967. Organic-walled microfossils as depth and shoreline indicators. *Marine Geology*, **5**, 389–412.

High-resolution sequence biostratigraphy of a lowstand prograding deltaic wedge: Oso Field (late Miocene), Nigeria

J. M. ARMENTROUT,[1] L. B. FEARN,[1] K. RODGERS,[1] S. ROOT,[1]
W. D. LYLE,[1] D. C. HERRICK,[1] R. B. BLOCH,[1]
J. W. SNEDDEN,[1] & B. NWANKWO[2]

[1] *Mobil Exploration and Producing Technical Center, PO Box 650232, Dallas, TX 75265-0232, USA*
[2] *Mobil Producing Nigeria Limited, Lekki Expressway, Victoria Island PMB 12054, Lagos, Nigeria*

Abstract: High-resolution biostratigraphic analyses provide calibration of both depositional systems and specific depositional environments. Such studies facilitate prediction of reservoir geometry and the lateral continuity of both reservoir sandstones and sealing mudstones. Data from the offshore Nigeria Oso Field are presented as a case study of one late Miocene deltaic depositional system. Within the Oso Field, marine mudstones are recognized as effective top seals across the entire field in contrast to marginal marine to non-marine mudstones which have more restricted distribution and act as intrafield baffles. Three sequence stratigraphic models have been considered: (1) the first is based on biostratigraphic data integrated with core sedimentology; (2) a second model constructed using core sedimentology; and (3) a third based on interpreted regional seismic reflection profiles and well-log data. Integration of all data shows that the lower Oso Field producing interval is interpreted as a lowstand prograding wedge, and the upper interval as a prograding distal transgressive or alternatively distal highstand system tract.

The Oso Field was discovered in 1967 by Mobil Producing Nigeria Limited in the offshore Niger Delta Joint Venture Area (Fig. 1). Approximately one billion barrels of condensate are recoverable from the field which is being produced at the rate of approximately 110000 barrels per day. The decline in reservoir pressure has resulted in a series of studies to better define the reservoir architecture and pressure compartments in anticipation of a gas-injection programme. This paper focuses on the biostratigraphic calibration of the Oso Field, both regionally and within the context of reservoir architecture.

Oso Field occurs within a roll-over anticline downthrown to a regional growth fault, Fault 1 (Fig. 2). The field produces from four zones; only the 1Y1 and the 2Y2 units of the Biafra Member of the Agbada Formation are discussed here (Fig. 3). Seismic reflection profiles clearly display the anticlinal structure and the aggradational stacking pattern of both prograding reservoir intervals (Fig. 4).

Snedden *et al.* (1992) and Thompson & Snedden (1996) have described the reservoir as a tide-influenced fluvial deltaic system based on detailed core sedimentology, Formation MicroScanner logs and dipmeter data (Fig. 5). High-resolution biostratigraphy was studied using fossils recovered from samples collected from each mudstone in the conventional cores from eight wells and from cuttings through the entire late Miocene–earliest Pliocene section from one well. The biostratigraphic data, combined with the core sedimentology and log analysis, were integrated into a regional sequence stratigraphic framework permitting biostratigraphic characterization of both the sandstone reservoirs and mudstone seals.

Multiple sequence stratigraphic models have been considered: (1) one based on biostratigraphic data integrated with core sedimentology; (2) a second model constructed using core sedimentology; and (3) a third based on interpreted regional seismic reflection profiles and well-log data (Fig. 6). Integrating all the data results in the following interpretation of the Oso Field producing intervals: the lower interval (1Y1) was deposited as a lowstand prograding

ARMENTROUT, J. M., FEARN, L. B., RODGERS, K., ROOT, S., LYLE, W. D., HERRICK, D. C., BLOCH, R. B., SNEDDEN, J. W. & NWANKWO, B. 1999. High-resolution sequence biostratigraphy of a lowstand prograding deltaic wedge: Oso Field (late Miocene), Nigeria. *In:* JONES, R. W. & SIMMONS, M. D. (eds) *Biostratigraphy in Production and Development Geology*. Geological Society, London, Special Publications, **152**, 259–290.

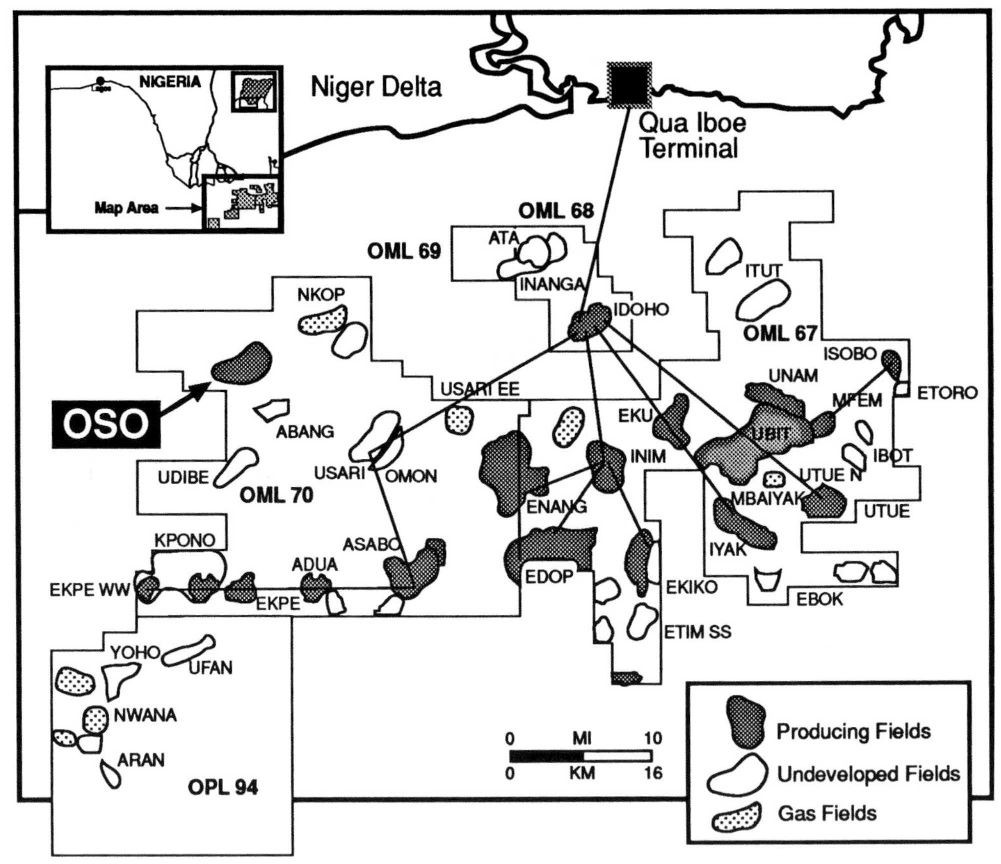

Fig. 1. Field distribution map of the Nigerian offshore Joint Venture producing area. Oso Field is in the northwestern portion of OML 70, approximately 12 miles offshore.

wedge; the upper interval (2Y2) is interpreted as either a prograding distal transgressive deposit or alternatively as a distal highstand systems tract.

Depositional model

Snedden *et al.* (1992) analysed nearly 4000 feet of conventional core, dipmeter, biostratigraphic and wireline-log data to develop a highly constrained depositional model for the Oso Field 1Y1 interval. That reservoir interval, with typical thicknesses of 500–800 feet (true vertical thickness), formed in a deltaic system intermediate in terms of wave, tide and fluvial processes. Characteristic depositional facies include distributary channel-fill, tidal creek, tidal flat, lagoon, mouth bar, delta front and prodelta (Fig. 5). Highest reservoir quality was measured within distributary sandstones, where pebbly massive lithologies had measured porosities of 28% and permeability as high as 50+ D. Channel-fills are described as having sharp erosional bases, and sharp to gradational tops. Channels trend NW–SE and NE–SW, reflecting coastal plain gradients and channel splitting. Snedden *et al.* (1992) interpreted non-amalgamated channel widths as ranging from 1000 to 10 000 feet, and thicknesses between 30 and 90 feet. They interpreted these dimensional ranges as similar to moderately sinuous channel systems of the modern Niger delta (Weber 1971).

The tidally influenced sandstones exhibited slightly higher porosities but lower permeabilities than the distributary channel sandstones (Snedden *et al.* 1992). Non-reservoir lagoonal and interchannel mudstones interbedded within the reservoir interval exhibit soft-sediment deformation, interpreted to reflect rapid deposition and topographic relief related to oversteeping and undercutting of the channel-margin sides. These mudstones have horizontal permeabilities

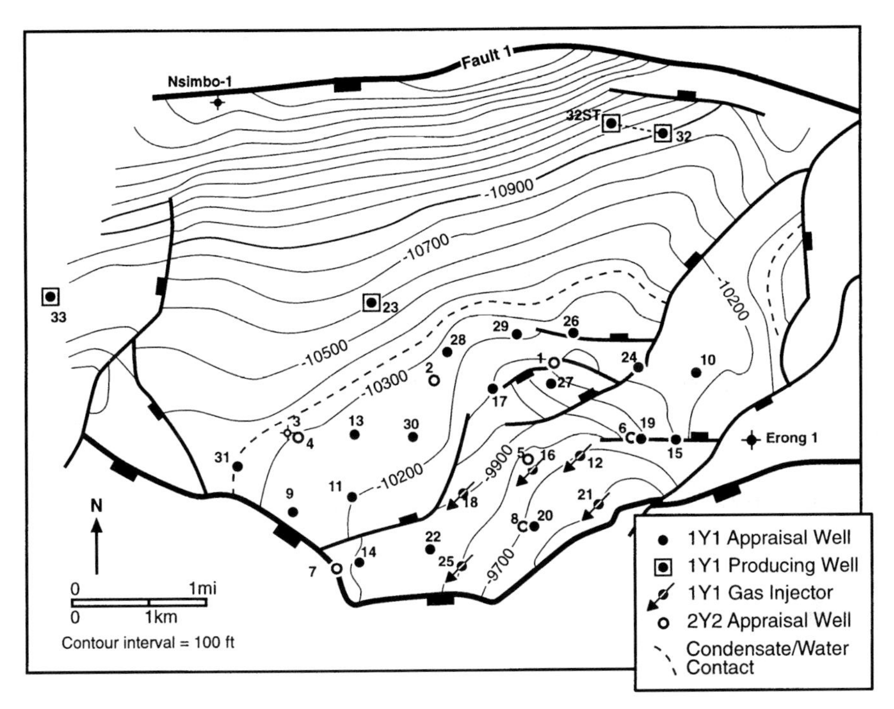

Fig. 2. Structural map of Oso Field at the top of the 1Y1 reservoir (see Fig. 4). Distribution of wells reflects the development of the field as of December 1997. High-resolution biostratigraphy was done on conventional core samples from wells 8, 11, 12, 16, 24, 28, 31 and 32.

in the range of 10–100 mD, but are effectively impermeable in a vertical dimension (Snedden *et al.* 1992).

The top of the 1Y1 reservoir sandstone is a transgressive surface overlain by marine mudstones (Snedden *et al.* 1992; Fearn *et al.* 1997). This marine mudstone overlies a variety of reservoir facies (Fig. 5), suggesting very rapid transgression, probably with little erosion at the ravinement surface (Snedden *et al.* 1992).

Snedden *et al.* (1992) cited three lines of evidence for the Oso Field 1Y1 reservoir interval being a lowstand prograding complex (Posamentier & Vail 1988). First, many of the spontaneous potential and gamma-ray log-motif patterns of the 1Y1 reservoir show funnel-shaped uniform coarsening-upwards patterns, but the log-motif of Oso well 23 has a sharp base. This sharp base suggests an unconformity over marine shales for the 1Y1 sandstone in this more northern, more proximal well (Fig. 2). This fits the pattern for a lowstand prograding wedge deposit as defined by Van Wagoner *et al.* (1990), Posamentier *et al.* (1992) and Posamentier & Allen (1993). Second, the 1Y1 sandstone is very coarse grained relative to most other sandstones within the Oso Field. This suggests up-dip valley incision and erosion during a rapid fall in sea level with progradation of the sand-prone depositional wedge downlapping onto the underlying sequence boundary, a pattern well imaged in Fig. 4. Third, the transgressive marine mudstone over the 1Y1 reservoir interval indicates a rapid rise in sea level indicative of an initial transgression.

Preliminary data on the 2Y2 reservoir, assembled for this study, suggests that the 2Y2 is more tidally influenced than the 1Y1 reservoir interval, and that the 2Y2 is less proximal relative to the 1Y1 depositional facies within the area of the Oso Field.

Palaeontological data

Two scales of biostratigraphic analysis have been used in the study of the Oso Field, one regional in scope and a second within the field itself. The regional study was undertaken integrating well-log, seismic and biostratigraphic sequence stratigraphy in order to construct a chronostratigraphic framework. Using this framework, a series of palaeogeographic maps was

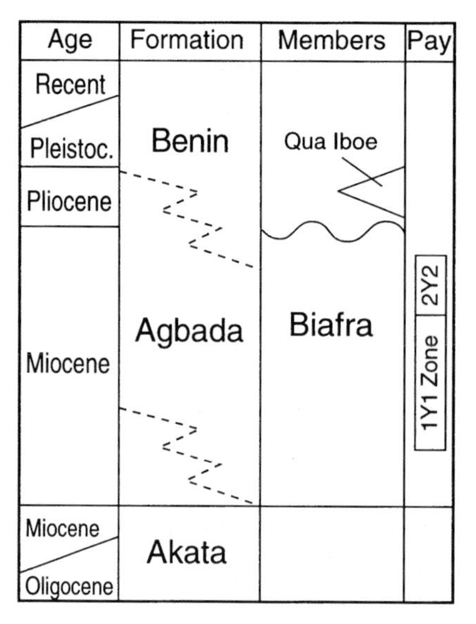

Fig. 3. Stratigraphic column for the Nigeria offshore showing the Oso Field 1Y1 and 2Y2 reservoir intervals of the Biafra Member of the Agbada Formation. Each of these formations consists of a single predominate lithofacies of a Neogene prograding deltaic system representing progressively younger sediments toward the south (offshore). The position of the 1Y1 and 2Y2 pay zones are show relative to the regional stratigraphy and not to scale. The age of each depositional sequence within this prograding system is presented in Fig. 7.

constructed to constrain the depositional setting of all fields in the Joint Venture Area. Biostratigraphic data are based on both well-cutting and conventional core mudstone samples processed using standard laboratory methods and examined by experienced palaeontologists working for several service companies. Because of the sandy character of the reservoir intervals most in-field biostratigraphy is based on analysis of core mudstone samples.

The high-resolution data consist of detailed analyses of extinction and first occurrence events, and the numerical abundances of foraminifera and palynomorphs. Because of variance between laboratories and contract biostratigraphers, considerable effort was undertaken by Mobil palaeontologists to standardize taxonomy and normalize statistical data.

Palynomorph counts were made using a fixed traverse method. All palynomorphs seen in five traverses across slides of both the $>20\,\mu$m sieved fraction and the $<20\,\mu$m unsieved fraction slides were counted. Original sample volumes were equal, as were residue concentrations. All palynomorphs contribute to the 'pollen sums'. Dinocyst and algal curves are expressed as per cent of total palynomorphs. Mangrove statistics are expressed as per cent of the total freshwater and brackish water pollen and spore assemblage. Other palynomorph statistics, including *Mannulutus*, are shown as per cent of the total freshwater pollen and spore assemblage.

Foraminiferal strew slides were prepared from uniform volumes of sediment. Calcareous nannoplankton slides were prepared from time-settled solutions of sediment. Examination of both types of slides consisted of numerous traverses across the slide identifying each specimen as to species and counting the number of each species observed.

The traditional stratigraphic nomenclature of the Niger delta Neogene section consists of three formations: (1) the deeper marine mudstones of the Oligocene–Miocene and younger Akata Formation; (2) the progradational prodelta mudstones of the Miocene and younger Agbada Formation; and (3) the sandstone-dominated coastal plain and shallow-marine deposits of the Plio-Pleistocene and younger Benin Formation (Fig. 3) (Weber 1990). This succession of prograding lithofacies is punctuated by a series of depositional sequences clearly defined on seismic reflection profiles and wireline logs. The methodology used is presented in Vail (1987) and Armentrout (1996), and the assembled data types and interpretations for one sequence are shown in Fig. 7. Thirteen depositional sequences, numbered younger to older 100–1300 in Fig. 8 were interpreted throughout the Joint Venture Area. Eight additional sequences between the sea floor and Sequence 100 have been mapped locally (unnumbered sequences above Sequence 100 in Fig. 8).

In addition, local studies were undertaken within each field. These studies integrated wireline-log motif and core biostratigraphic analysis with core sedimentology, where available, to define the specific depositional environment of the sandstone reservoirs and mudstone seals. Using both the integrated regional and local datasets, depositional sequences were subdivided into systems tracts (Posamentier & Vail 1988). Palaeogeographic maps of each systems tract were constructed as a predictive tool for both interpolation and extrapolation of reservoir and seal distribution.

Chronostratigraphic analysis

To understand both stratigraphic and environmental aspects of the depositional sequences of the offshore Niger delta Joint Venture Area,

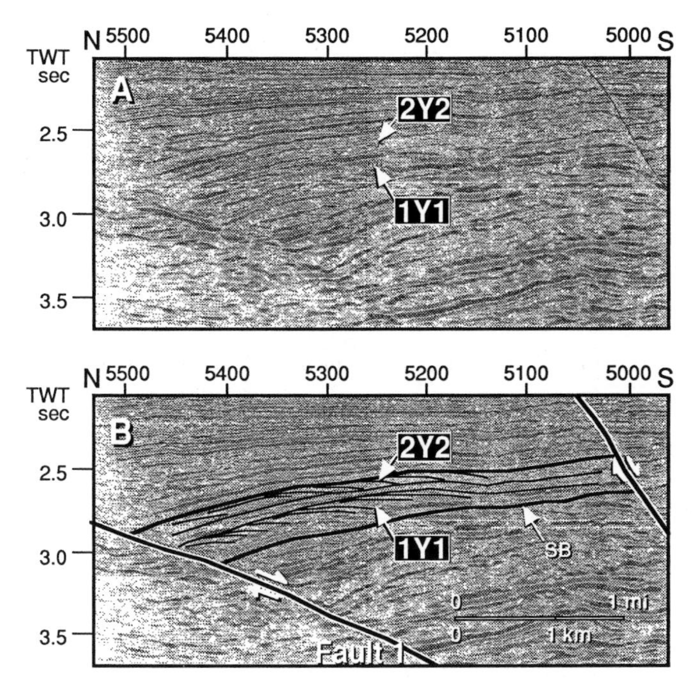

Fig. 4. N–S 3D seismic reflection profile showing the prograding clinoform geometry of the 1Y1 and 2Y2 reservoir intervals down thrown to Fault 1: A = uninterpreted; B = interpreted. The high-amplitude and continuous reflections above each prograding interval are interpreted to be a marine flooding surface based on association with an influx of planktonic microfossils in a laterally continuous mudstone. The black triangles are positioned pointing upward at the interpreted shelf-edge, chosen at the inflection between the relatively flat topset and inclined foreset of the most basinward clinoform.

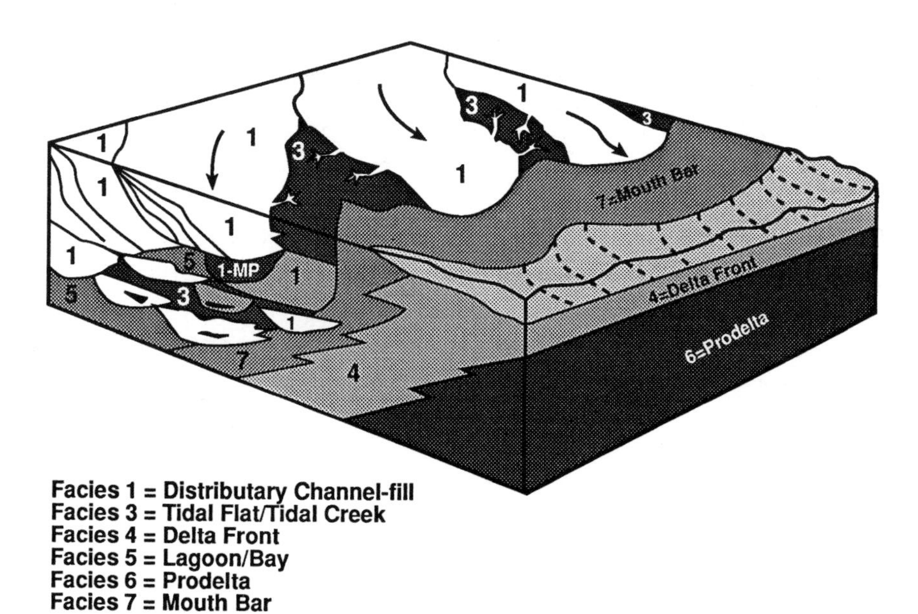

Fig 5. Depositional model for the 1Y1 reservoir interval of Oso Field, based on analysis on more than 4000 feet of conventional core (from Snedden *et al.* 1992; Thompson & Snedden 1996).

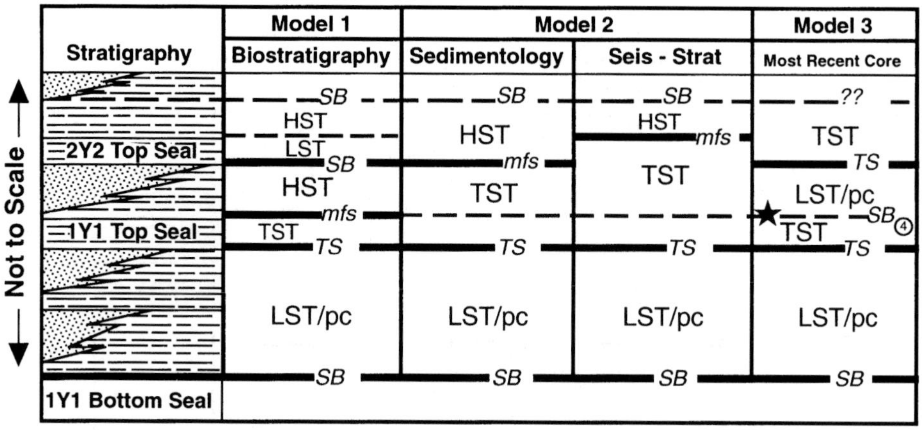

Stratigraphy	Model 1 Biostratigraphy	Model 2 Sedimentology	Model 2 Seis - Strat	Model 3 Most Recent Core
	SB	SB	SB	??
	HST LST SB	HST mfs	HST mfs	TST TS
2Y2 Top Seal	HST mfs	TST	TST	LST/pc SB ④ (star)
1Y1 Top Seal	TST TS	TS	TS	TST TS
Not to Scale	LST/pc SB	LST/pc SB	LST/pc SB	LST/pc SB
1Y1 Bottom Seal				

Fig. 6. Comparison of the sequence stratigraphic interpretations using different datasets. All four interpretations considered data from each discipline but based the final interpretation on the strengths of one specific set of observations. In particular, Model 3 places a high-frequency (? fourth-order) sequence boundary at the base of the 2Y2 sandstone (star) based on a sharp-based log-motif correlated with a sandstone channel-base observed in one conventional core.

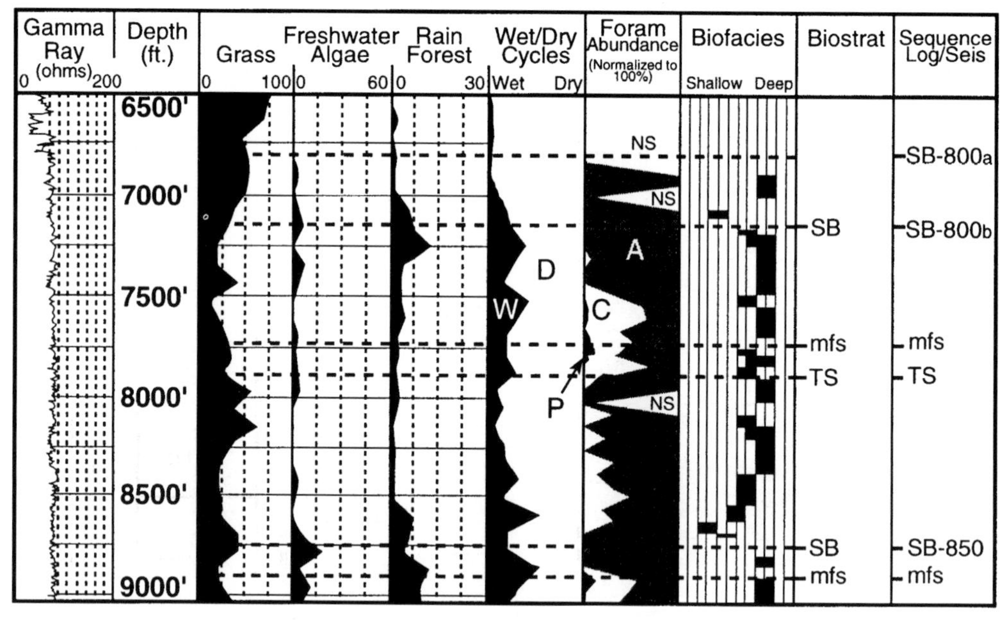

Fig. 7. Abridged version of the integrated log–biostratigraphic–seismic well panels used in the regional study. The well shown is the Kpono West 1 and was chosen for the quality of the biostratigraphy due to the lack of sandstone in the drilled section: the 1Y1 and 2Y2 sandstones of the Oso Field are not encountered in the displayed interval. Wet–dry cycles are interpreted from variations in grass and rain forest–mangrove palynomorphs; marine flooding events are recognized by increases in planktonic foraminiferal abundance and shallow-to-deepening events interpreted from foraminiferal biofacies analysis. These biostratigraphic patterns are interpreted as candidate sequence stratigraphic datums and compared to mapped seismic sequence stratigraphic datums. Abbreviations are: W = wet; D = dry; A = arenaceous foraminifera; C = calcareous foraminifera; P = planktonic foraminifera; NS = no sample; SB = sequence boundary; mfs = maximum flooding surface; TS = transgressive surface. The surface SB-800a is based on log-motif analysis and SB-800b is based on biostratigraphy.

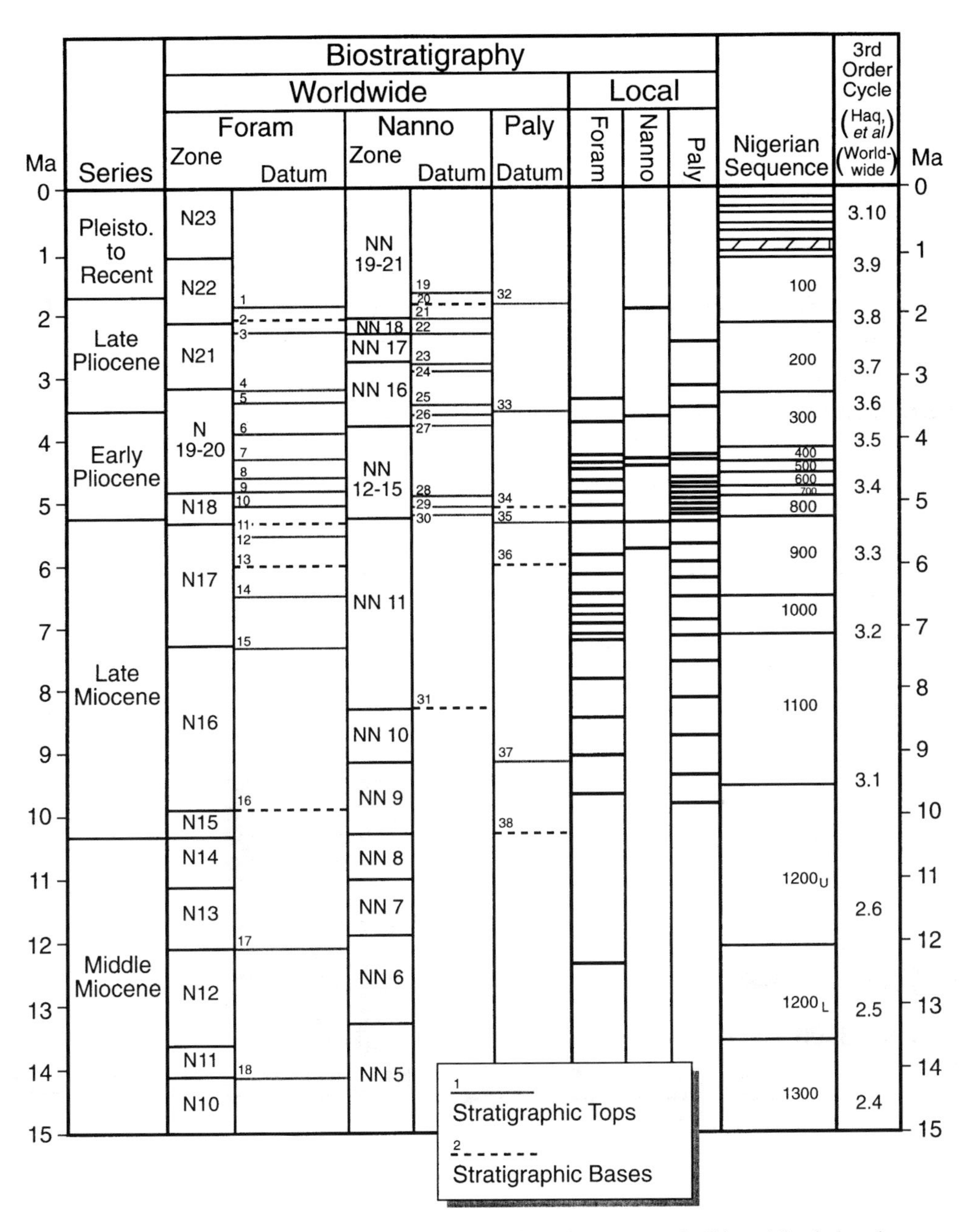

Fig. 8. Chronostratigraphic correlation chart for the depositional sequences of offshore Nigeria based on a synthesis of 33 wells in the Joint Venture Area. Observed chronostratigraphically significant planktonic foraminifera, calcareous nannofossil and palynomorphs are noted relative to their globally recognized bioevent position: species names for bioevents 1–38 are in Table 1. The local Nigerian depositional sequences are placed centered on the age range of key fossils recovered from each sequence and the superpositional order of the sequences. Time scale after Berggren *et al.* (1995).

Table 1. *Global bioevent species that occur locally in the Nigerian strata penetrated by the wells used in the synthesis study*

Taxonomic group	No.	Top/base	Genus/species
Foraminifera	1	Top	*Globorotalia tenuitheca*
	2	Base	*Globorotalia truncatulinoides*
	3	Top	*Globigerinoides extremus*
	4	Top	*Globoquadrina altispira*
	5	Top	*Globorotalia margaritae*
	6	Top	*Globoquadrina globosa*
	7	Top	*Globoquadrina venezuelana*
	8	Top	*Globorotalia plesiotumida*
	9	Top	*Globigerinoides seiglei*
	10	Top	*Globorotalia merotumida*
	11	Base	*Neogloboquadrina dutertrei*
	12	Top	*Globogerinoides mitra*
	13	Base	*Neogloboquadrina pseudopima*
	14	Top	*Globorotalia lenguaensis*
	15	Top	*Globorotalia juanai*
	16	Base	*Neogloboquadrina acostaensis*
	17	Top	*Globorotalia fohsi robusta*
Nannoplankton	18	Top	*Globorotalia peripheroacuta*
	19	Top	*Calcidiscus macintyrei*
	20	Base	*Gephyrocapsa oceanica*
	21	Top	*Discoaster brouweri*
	22	Top	*Discoaster pentaradiatus*
	23	Top	*Discoaster surculus*
	24	Top	*Discoaster asymmetricus*
	25	Top	*Discoaster challengeri*
	26	Top	*Discoaster variabilis*
	27	Top	*Sphenolithus abies*
	28	Top	*Ceratolithus acutus*
	29	Top	*Triquetrorhabdulus rugosus*
	30	Top	*Discoaster quinqueramus*
	31	Base	*Discoaster quiqueramus*
Palynomorph	32	Top	*Retibrevitricolporites obodoensis*
	33	Top	*Praedapollis flexibilis* (comm/consis)
	34	Base	*Echitricolporites mcneilyi*
	35	Top	*Verrutricolporites microporus*
	36	Base	*Fenestrites spinosus* (comm/consis)
	37	Top	*Verrutricolporites rotundiporus*
	38	Base	*Fenestrites spinosus*

a correlation framework was constructed. This framework consisted of correlated seismic sequences integrated with age significant events identified by fossil species first and last downwell occurrences. Observations were compiled from 62 wells, 33 with high-resolution biostratigraphy, and a seismic reflection profile grid with an average 1.0×1.0 mile spacing over most of the N–S 28 mile by W–E 60 mile area (Fig. 1). The data were integrated following standard sequence stratigraphic methodology through three successive iterative analyses.

Planktonic foraminifera provide the primary age diagnostic taxa and were available from 28 wells distributed throughout the Joint Venture Area (Fig. 1). Planktonic foraminifera provide 18 correlation events with 14 extinction data points, and four first (last downhole) occurrences for the Neogene interval studied (Fig. 8). Calcareous nannofossils were equally valuable for age dating but these taxa were studied in only eight wells. Twelve correlation events with 11 extinction data points and two first occurrences were noted for calcareous nannofossils. The planktonic foraminifera and calcareous nannofossils are tied to the world-wide biochronozones of Blow (1969) and Martini (1971), and are plotted after Berggren *et al.* (1995) in Fig. 8. The names for each of the bioevents are given in Table 1. These correlations provide calibration of the local Nigerian sequences to the world-wide time scale.

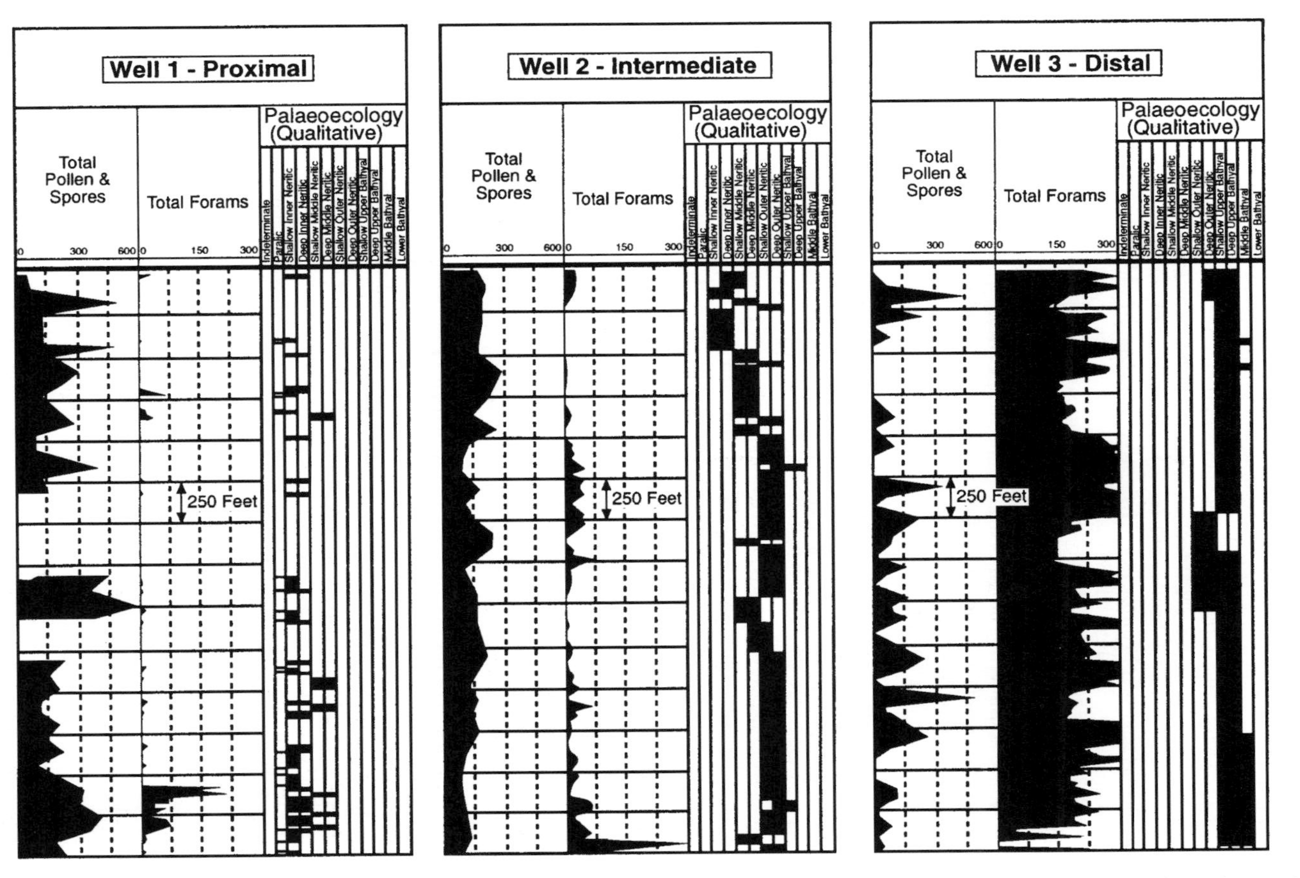

Fig. 9. Representative biostratigraphic datasets from wells located along a N–S transect illustrating the near-shore to off-shore decrease in pollen and spore (non-marine) recovery and the reciprocal increase in foraminiferal (marine) fossil recovery. The overall utility of each discipline is apparent from this distribution. The proximal well is Nkop 1, the intermediate well is Ibom 1 and the distal well is Okwok 2.

Palynology was studied in 37 wells. Regionally useful pollen and spore bioevents consist of four extinction events, and three first occurrences (Fig. 8). The name of each of the palynomorph bioevents is given in Table 1. Age assignment of these events follows Germeraad *et al.* (1968), Legoux (1978), Muller *et al.* (1987) and Salard-Cheboldaeff (1990). These pollen and spore bio-event age assignments should be regarded as approximations as none are specifically tied to the world-wide planktonic biochronozones. The palynological zonations of Poumot (1989) and Morley & Richards (1993), compiled for Nigeria, provide a preliminary tie of local events with the more global studies.

Numerous local foraminiferal, nannoplankton and palynomorph events are noted in the data assembled from the wells of the Nigerian Joint Venture Area. The stratigraphic position of each of these events allows for tentative correlation with the more global bioevents. Their position relative to the global events is shown in Fig. 8, but each local event remains unspecified in this report.

Wells from the northern part of the study area reflect shallower marine and non-marine facies with proportionately more abundant terrigenous palynomorph specimens than marine fossils. The more southerly, further offshore, wells reflect deeper marine deposition and have proportionately more marine fossils than terrigenous palynomorphs (Fig. 9). Thus, planktonic calibration is best in the southern wells and is projected northward into the less marine facies using wireline-log patterns, seismic reflection horizons and biostratigraphically defined climate cycles as correlation datums.

Climate cycles

Of particular note in the local bioevents are palynological assemblages diagnostic of vegetation cycles attributed to climate fluctuations. Climate cycles can be recognized as successive phases of humidity v. aridity expressed as a relative predominance of mangrove and rain forest taxa v. savannah taxa. In the Neogene, these climate cycles are interpreted to be linked with sea-level fluctuations driven by glacial–interglacial cycles. Poumot (1989) recognized six major cycles spanning the time equivalent to sequences 1100–200 of this study. Morley & Richards (1993) use similar aridity–humidity vegetational cycles as part of criteria for recognizing specific zones.

In this study, two climatic end-member assemblages can be consistently recognized, representing successive arid and humid phases of climatic cycles tied to periods of sea-level lowstand and transgressive-to-highstand, respectively. Observed species diversity is higher for the humid transgressive–highstand interval v. lower diversity associated with the more arid lowstand interval. This is due primarily to presence of dinocycts in the transgressive–highstand facies. As many as 15 cycles are present within the succession of sequences 100–1300.

The aridity events are characterized by relative maxima in the grass pollen *Monoporites annulatus*, the colonial freshwater algae *Pediastrum* spp., and in some cases elevated levels of other herbaceous pollen such as *Cyperaceaepollis* spp., *Echitricoloporites* spp. and *Fenestrites* spp. Concurrent with these maxima are minima in the primary humidity indicators. This includes mangrove taxa *Zonocostites ramonae*, *Psilatricoloporites crassus* and *Acrostichum* spp., the rain forest and back-mangrove species *Pachydermites diederixi*, and open freshwater swamp forest index species such as *Retitricolporites irregularis* and certain palm species. Conversely, the humidity events are characterized by maxima in the humid climatic indicators and minima in the aridity indicators listed above. Interpretations of arid–dry and humid–wet cycles are shown in Fig 7.

The grass pollen *M. annulatus* alone may be the most definitive taxon in delineating the climatic cycles (Morley & Richards 1993). In addition, in the more marine facies, mangrove taxa provide a stronger signature of the humid phase of the climate cycle in contrast to the stronger signal of rain forest taxa for humid phases within the more proximal non-marine facies (van der Zwan & Brugman 1997). Morley & Richards (1993) place a lot of emphasis on charred cuticle as an arid indicator in conjunction with *M. annulatus*, not just the pollen itself. In our study, charred graminaeae cuticle were not counted because we were very unclear as to the correct method for counting this palynomaceral. They are often fragmented and we questioned at what size does a fragment constitute a count. Grass pollen was easier to quantify and seems to be a good proxy to charred cuticle based on the published curves of Morley & Richards (1993).

Each of the events, in addition to defining either aridity or humidity maxima, can be further characterized and used as a correlation event. The characteristics of a climate phase providing correlation utility can consist of variation in magnitude, aerial extent of diagnostic and accessory species, and on faunal associations (these are the local bioevents of Fig. 8). The temporal context of these climate events must be carefully checked against independently derived

(a)

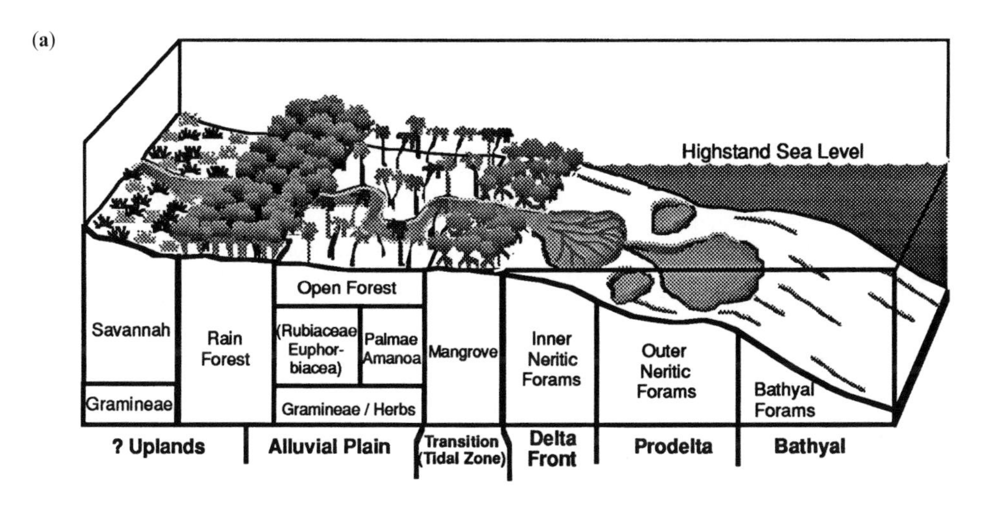

(b)

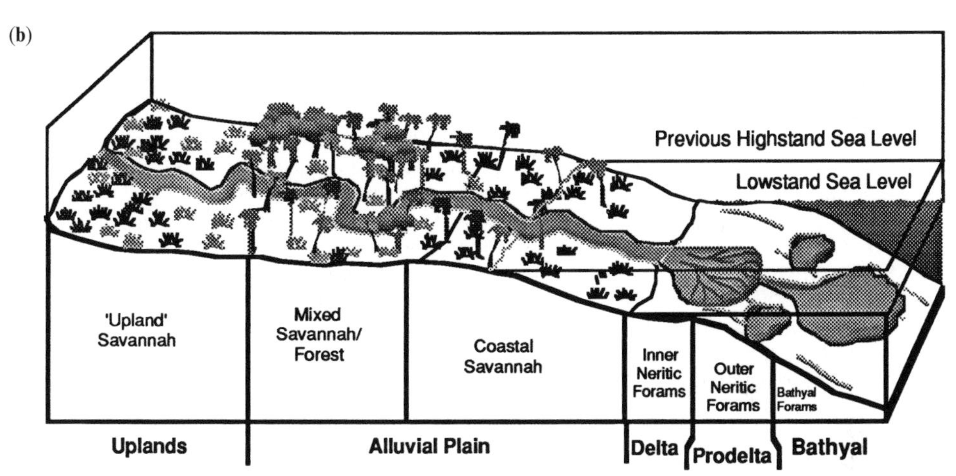

Fig. 10. Diagrams of Neogene subtropical floral and faunal communities for relative highstand and lowstand of sea level (modified from Poumot 1989). (**a**) The highstand, modelled from the present, is characterized by coastal plain areas of restricted savannah and expanded rain forest and coastal mangrove systems, and a relatively broad shelf with inner–outer neritic foraminiferal faunas. (**b**) The lowstand, modelled on the late Pleistocene, has expanded areas of savannah with restricted remnants of forest species, and a basinward shift of the outer neritic foraminiferal biofacies onto the physiographic slope.

chronostratigraphic correlations. This is essential to separate palynological assemblages related to fluvial systems draining different but contemporaneous vegetational ecosystems v. those reflecting global climate cycles. Use of the climate cycles as correlation tools is especially important in the more proximal facies of the northern Joint Venture Area where few marine fossils occur (Fig. 9).

The vegetationally defined climate cycles of this study show a correlation between aridity events and lowstand systems tracts, and humidity events with transgressive and highstand systems tracts. The systems tracts have been independently identified from seismic reflection profile analysis in accordance with criteria defined by Vail (1987) and Posamentier & Vail (1988). Figure 10 (after Poumot 1989) presents conceptual diagrams of the floral and faunal distribution patterns associated with highstand and lowstand systems tracts. The palynological assemblages can therefore be used as a tool to predict probable systems tract type in areas devoid of detailed seismic stratigraphic analysis, or in concert with seismic systems tract analysis to increase confidence in interpretations.

Reviewer Paul Ventris (written comm., March 1998) comments that the mangrove is a key indicator of marine intervals down-dip of the mangrove belt, with high abundance occurrences in two depositional intervals. These intervals are: (1) the late lowstand–early transgressive phase of slow relative sea-level rise when much of the coastal plain is aggrading and supporting a wide mangrove belt; and (2) the highstand phase of relative sea level associated with progradation. In Ventris' experience, the maximum flooding surface-condensed section interval has low mangrove abundance interpreted to be associated with maximum inundation of the coastal plain and restriction of the mangrove distribution.

In our data we do not see the increase in the mangrove assemblage associated with our interpretation of the late lowstand and early transgressive intervals at the top of the 1Y1 reservoir interval and the overlying transgressive mudstones of the 1Y1 top seal. This cannot be a sampling artifact as all conventional core mudstones from this interval have been sampled and examined using both the $>20\,\mu$m and $<20\,\mu$m sieved fractions.

Construction of an age model

On average, three uniquely chronostratigraphically significant, globally recognized bioevents per well can be recognized. Integration of biostratigraphic data from each well was achieved using seismic sequence boundaries and maximum flooding surfaces as a preliminary correlation framework. Because of growth fault deformation of the study area, seismic correlations are often not uniquely demonstrable across faults.

Chronostratigraphically significant bioevents provide a check of seismic correlations. All bioevent occurrences were tested against biofacies to assure that occurrences were independent of environmental control (see Armentrout 1996). Each cross-fault correlation was tested using local bioevent occurrences in wells on each side of each fault, where available.

Each biochronostratigraphically significant event occurs in only a few wells resulting in an age assignment for each sequence with some variance. This results in the appearance that most sequences overlap in age with adjacent sequences. Within the Nigerian Joint Venture Area, we have chosen as the proposed age model for the depositional sequences, a stacking order centred on the midpoint age of each sequence constrained by superposition of successive sequences (Fig. 8). This age model will evolve as additional data become available from the new exploration wells being drilled into deeper-water facies where abundant planktonic bioevents will be encountered, and as three-dimensional (3D) seismic grids facilitate more precise correlations.

The proposed age model for the sequences has undergone several revisions as a result of iterations both between biostratigraphic disciplines and between seismic stratigraphers and biostratigraphers. The sequence succession shown in Fig. 8 therefore represents our collective 'best fit' using all available 'good' data. The position of sequences 100, 300, 800 and 900 are reasonably well constrained. Sequences 100, and 400–700 are constrained by superposition between the better constrained sequences. The uppermost sequences 100 and 200, and lowermost sequences, 1100, 1200 and 1300, are less well constrained and their apparent longer duration may be in part due to poorly constrained calibration. The eight sequences between Sequence 100 and the present are well defined on seismic reflection profiles, but are not biostratigraphically calibrated as they occur above the sampled stratigraphy of the studied wells. The calcareous nannofossil *C. macintyre* was observed within the lower intervals of Sequence 100, indicating an age of 1.47 Ma (Berggren *et al.* 1995), and suggesting that the eight youngest sequences might correlate with the late Pleistocene–Holocene glacial–interglacial oxygen isotope stages (see Berggren *et al.* 1995).

Application of the age model

The resulting age model for depositional sequences constructed for this study facilitates numeric modelling. For instance, Fig. 11 compares two hydrocarbon maturation models using the burial history plot for the Oso Field area. Identical source horizon, thermal gradient (2.0°F per 100 feet), surface intercept (55°F) and sequence thicknesses are used in each model. The 'old timing model' is based on age assignments for depositional sequences prior to the rigorously integrated study resulting in Fig. 8. In this old model the onset of early oil generation is predicted to occur for speculated Oligocene source rocks (base of the illustrated stratigraphic section) at approximately 19 Ma BP. The 'new timing model' which uses the age relationships defined in Fig. 8, predicted early oil generation to occur for the speculated Oligocene source rocks at approximately 14.5 Ma BP. This difference of nearly 5 Ma between models for initial generation of hydrocarbons makes relatively later-forming traps more prospective than the old model suggested.

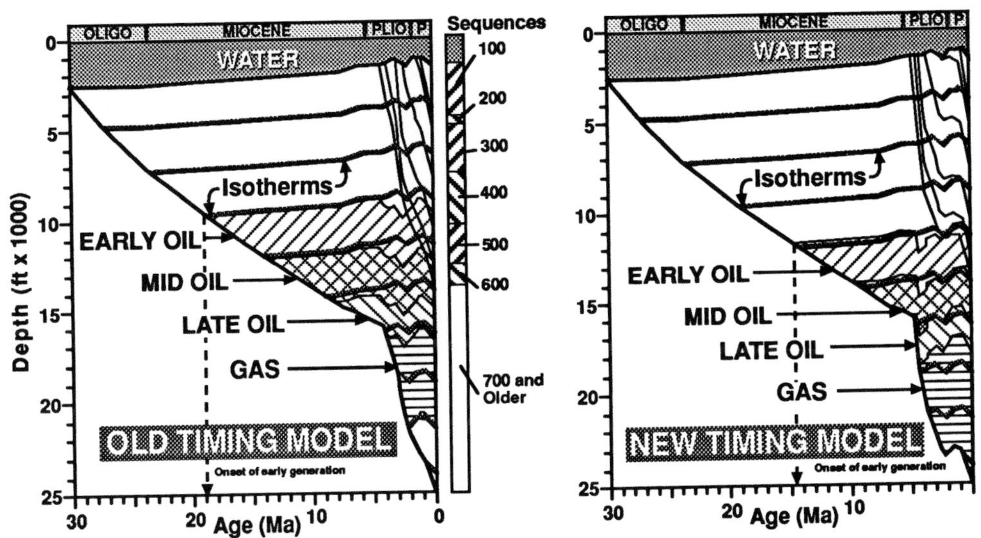

Fig. 11. Burial history and hydrocarbon maturation model for the Oso Field area, contrasting an old timing model with a new timing model based on the chronostratigraphy of Fig. 8. The only difference for these 'Basinmod' calculations is the age of the mapped horizons which bound the sequences. A thermal gradient of 2.0°F per 100 ft and a sea floor intercept of 55°F were used in both models. Based on these calculations, the onset of hydrocarbon generation is estimated as approximately 4.5 Ma later in the new timing model than in the old model.

Systems tract biofacies analysis

Using the integrated biostratigraphic and seismic sequence stratigraphic framework, a set of biofacies maps have been constructed for specific-time intervals. A biofacies is defined as a single species or group of species that inhabit a specific environment(s), and can therefore be used to identify that environment (see Fig. 10). Foraminiferal biofacies are discussed here, and palynological biofacies are presented under the following section on the Oso Field.

Biofacies interpretations of an individual well or biofacies maps based on multiple wells provide important information about the depositional environments of potential reservoir units and the distribution of seal types. For example, shallow-water biofacies can be associated with environments characterized by laterally extensive deltaic or marine sands deposited under shallow-water, wave or tidally dominated processes. In contrast, deep-water biofacies are most often associated with environments dominated by gravity-flow processes in which sand-prone channel-fed fan deposits may occur. Mudstones deposited within transgressive marine environments are likely to be geographically extensive and provide more effective seals than non-marine coastal plain mudstones or shallow-marine mudstones deposited in interdistributary bays between delta lobes.

A foraminiferal biofacies can include both planktonic and benthic taxa. Benthic taxa live either on the sea floor or within the upper few centimetres of marine sediments where physical–chemical conditions (e.g. substrate, pH, eh, etc.) can be highly variable. As many foraminiferal species have very specific physical and chemical requirements, their geographic distribution can be fairly limited. Many physical and chemical parameters vary with depth, making water depth inferences from benthic assemblages a useful environmental descriptor.

The highest variability of foraminiferal biofacies occurs in shallower (neritic–upper bathyal) marine environments. As water depth increases ecological conditions become less variable and more laterally extensive. This pattern is illustrated in comparing the three wells shown in Fig. 9. The proximal well has very rapid environmental changes of shallow-marine and paralic biofacies, whereas the distal well shows long intervals of the same deep-water environments.

Planktonic foraminifera inhabit the upper layers of oceanic water where physical and chemical conditions are less variable. Therefore, they have a much broader distribution, and do not provide the detailed environmental information that many benthic species provide. Some planktonic foraminifera have a specific depth zone that they inhabit. The impingement of these depth zones along continental margins is

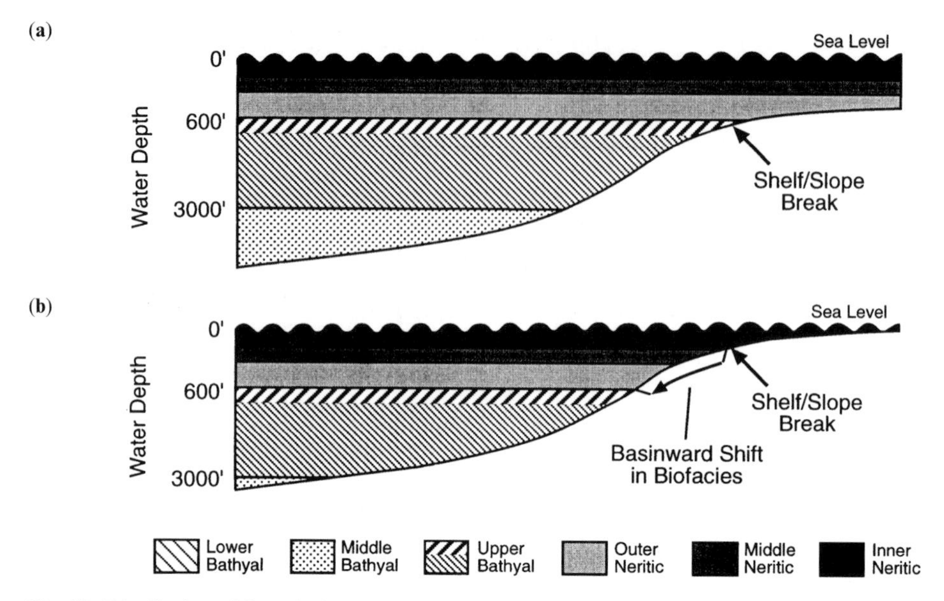

Fig. 12. Distribution of foraminiferal biofacies v. the shelf–slope break between highstand and lowstand of sea level. The traditional definition of foraminiferal biofacies is based on the present-day highstand of sea level with the neritic biofacies on the shelf and bathyal biofacies on the slope. During lowstand, the biofacies shift basinward with some neritic biofacies occurring on the physiographic slope.

the primary factor that limits their shoreward distribution. The impingement boundaries provide a minimum water depth which is used for biofacies interpretations.

As sea level rises and falls, the biofacies assemblages will shift their physiographic location in order to stay within their optimal physical and chemical environment. Consequently, the biofacies assemblages will have a different occurrence along the basin margin profile during lowstand than during highstand. Figure 12 illustrates the end-member distribution of biofacies relative to the shelf–slope break of a continental margin. Neritic biofacies occupy shelf positions today at a relative highstand of sea level, but occupied an outer shelf and upper slope position during the late Pleistocene relative lowstand of sea level.

The neritic–bathyal marine environment boundary is most often placed at 600 feet, coincident with the edge of the continental shelf (Hedgpeth 1958). While this may be a reasonable global average for the highstand of today's sea level, significant local variability exists. For example, in the Gulf of Mexico, the shelf–slope depositional inflection often occurs at a water depth of approximately 300 feet, which reflects the 300-foot eustatic rise in sea level since the late Pleistocene lowstand. Much of the Niger delta shelf–slope break is near 300 feet of water depth. Thus, biofacies assemblages are distributed relative to their optimal physical and chemical needs, and cannot be uniquely related to physiographic provinces, a mistake frequently made in palaeogeographic reconstructions.

The relationship of fossil biofacies assemblages to the palaeoshelf–slope break is an excellent approach to identifying relative position of sea level through time, but requires high-resolution analysis of both the foraminiferal assemblages and the seismic reflection profiles of each depositional cycle.

High-resolution biofacies analysis consists of making interpretations on each sample rather than 'smoothing' interpretations of multiple samples. Figure 13 illustrates such interpretations. The palaeoecology column of qualitative analyses is based on inspection of each and every sample by the palaeontologist with a biofacies interpretation unique to that sample. In contrast, the palaeoecology column under Palex Summary is a computer analysis using depth-calibrated species with the interpretation averaged over an interval of three–five samples. While the objectivity of a computer analysis is perhaps more reproducible, important detailed information can be lost. All analyses necessitate very careful sample evaluation to preclude misinterpretations

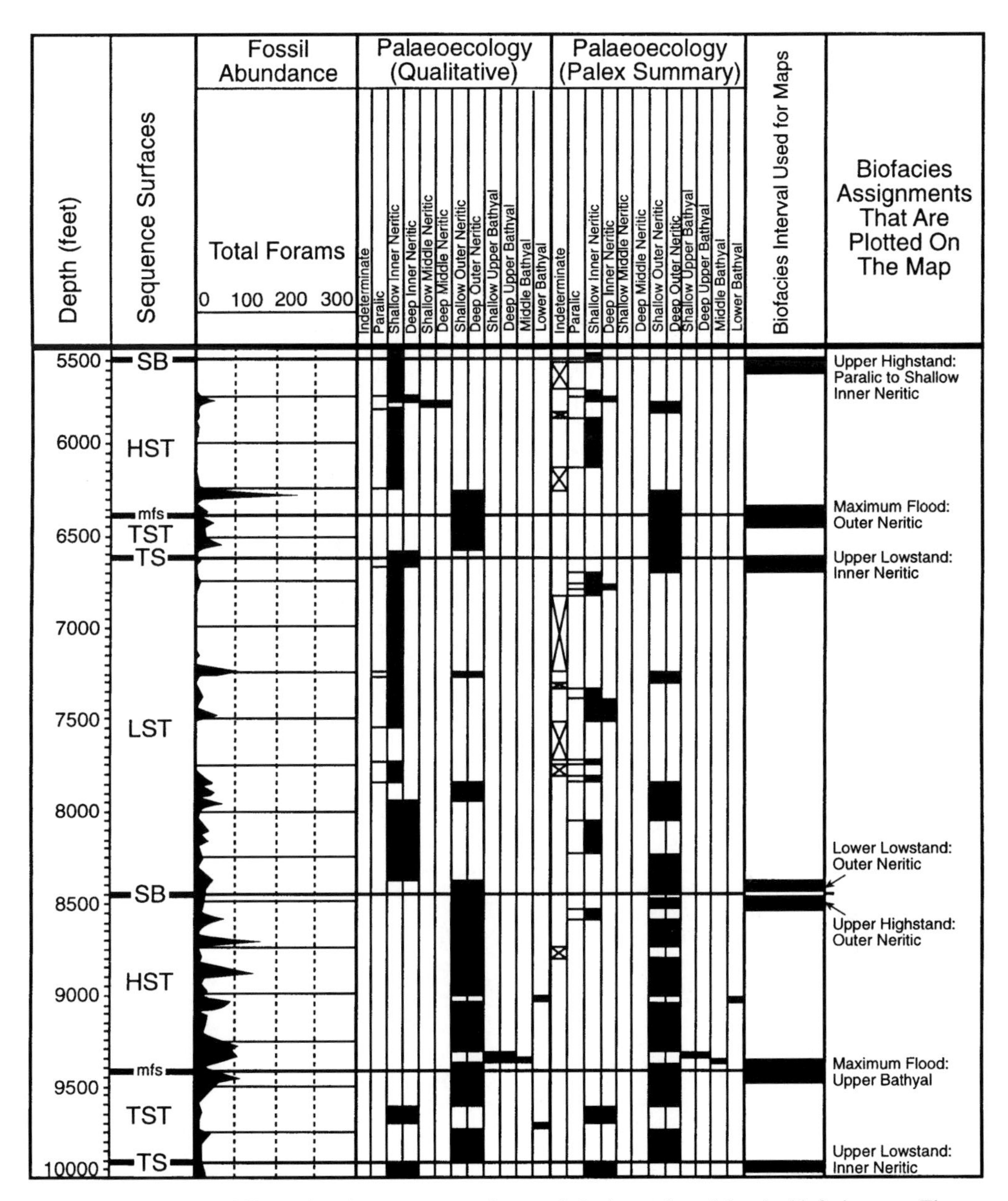

Fig 13. Well data panel illustrating the sequence surfaces and the interval used for the biofacies map. The qualitative palaeoecology is based on analysis of individual samples from the Utai 2 well; Palex summary palaeoecology is based on a computer program integrating data over a five-sample interval using specified water depth ranges for specific species. Specific criteria for selecting the mapped biofacies for each systems tract are discussed in the text. This well was chosen for the quality of the biofacies data and does not show the 1Y1 or 2Y2 interval of Oso Field.

due to samples modified by cavings, lost-circulation material, cement dilution from drilling-out a casing shoe, etc.

Five foraminiferal biofacies can be recognized in the eastern Niger delta Joint Venture Area. The biofacies assemblages range from paralic (marginal marine) to inner neritic (<150 foot water depth) biofacies to upper bathyal (600–1500 foot water depth) biofacies. Non-marine biofacies are typical of the onshore areas, and deeper water bathyal biofacies will be encountered in the exploration wells drilled further seaward, or in wells drilled to greater depths in which the deep-water facies will be found below the prograding shelf depositional systems.

Owing to limited data on the ecological distribution of modern foraminifera in the offshore areas of Nigeria, the local biofacies are not well defined. We have used species criteria defined by Adegoke *et al.* (1976) for the Niger delta, as well as criteria for the Gulf of Mexico by Tipsword *et al.* (1966) and Poag (1981) with modifications developed by Mobil (Fearn *et al.* 1997; Rodgers pers. comm. 1997). Ongoing statistical studies (Cosign Theta Cluster Analysis) of the foraminiferal biofacies assemblages suggest the occurrence of three bathyal associations: (1) dominantly planktonic forms interpreted to indicate well-oxygenated bottom water; (2) *Bolivina*/*Uvigerina* assemblage interpreted to indicate low-oxygenated bottom water; and (3) dominantly arenaceous forms possibly suggesting turbid bottom water (R. Echols, pers. comm., April 1998). Preliminary analysis suggests that the low-oxygen *Bolivina*/*Uvigerina* assembly does not uniquely occur within a specific systems tract but is better developed in the late Pliocene and Pleistocene interval than the late Miocene–early Pliocene.

Palaeogeographic maps

The biofacies maps have been constructed for four intervals within each of the regionally correlated depositional sequences (Fig. 8). These four intervals are: (1) the lower lowstand–prograding systems tract, based on the deepest biofacies within 100 feet immediately above the sequence boundary; (2) the upper lowstand–prograding complex systems tract, based on the shallowest biofacies within the upper 100 feet of the lowstand immediately below the transgressive surface; (3) the maximum flooding surface, based on the deepest water biofacies associated with the maximum flooding surface; and (4) the highstand systems tract, based on the shallowest biofacies within the uppermost 100 feet immediately below the overlying sequence boundary (Fig. 14). An example of the intervals used is shown in Fig. 13.

Each biofacies map is based on a relatively limited number of wells, especially for the older, more deeply buried sequences, such as Sequence 1100, the sequence of the primary reservoir interval in Oso Field. Contouring these limited data requires integration with the observations from the seismic reflection profile analysis and iteratively with the successively older and younger biofacies maps. The seismic

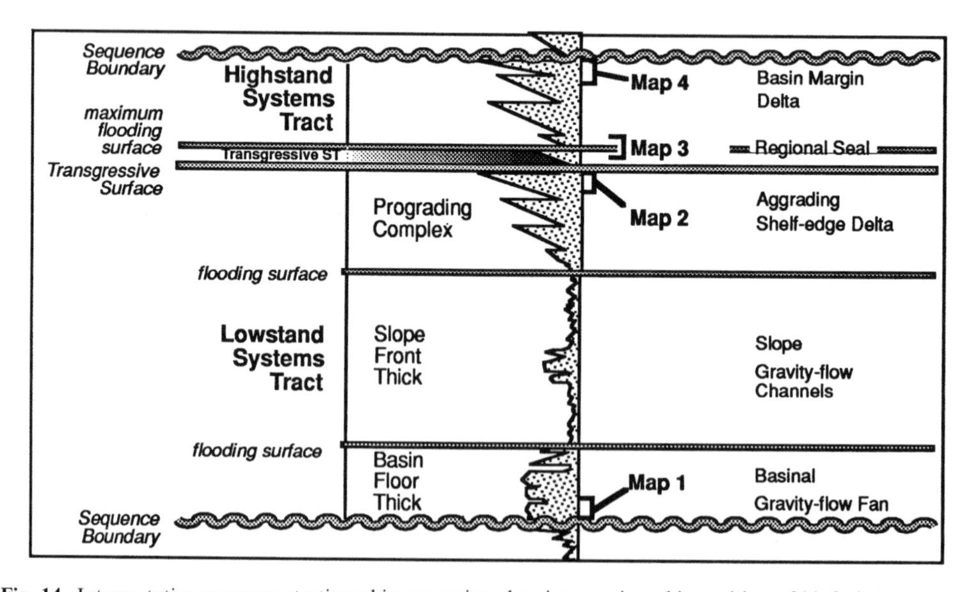

Fig. 14. Interpretative sequence stratigraphic succession showing stratigraphic position of biofacies maps presented as Figs 15–18. Note that the relatively proximal position of the Oso Field results in the stratigraphically lowest facies deposited being part of the upper slope front thick. The earliest deposits of the depositional sequence are interpreted to have bypassed the study area and to have been deposited further basinward as a basin floor (or slope basin) thick which was not penetrated in the wells studied.

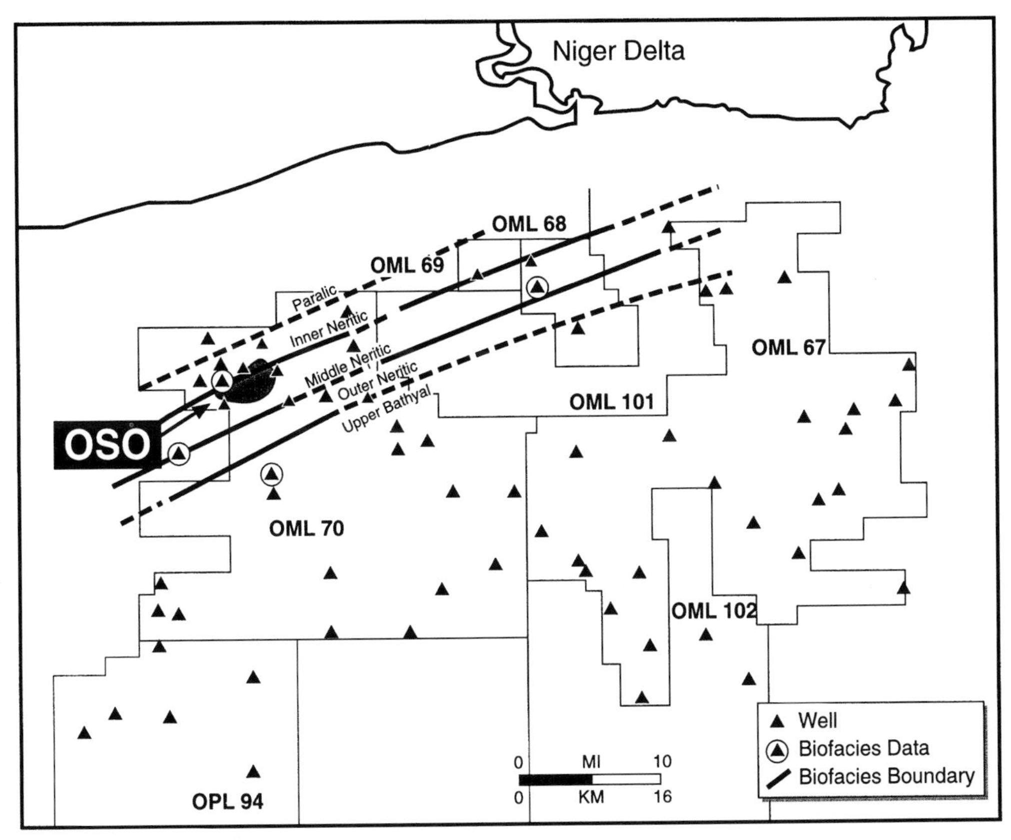

Fig. 15. Biofacies map for the deepest foraminiferal biofacies in the lowermost 100 feet of the Sequence 1100 lowstand. Data are available from only four wells penetrating this interval, requiring that the mapped pattern be integrated with the seismic facies analysis of the same systems tract and compared to the mapped pattern of the underlying and overlying systems tracts inorder to suggest possible biofacies trands. This map is for a very proximal area of the lowstand and thus represents biofacies associated with the lower part of the lowstand prograding complex, and neither the slope nor basin floor components.

observations consist of the shelf–slope break occurrence and associated seismic facies (see Fig. 4) (see Armentrout 1991, 1996 for additional examples). For each of the biofacies maps presented for Sequence 1100 (Figs 15–18), the biofacies are contoured between control points parallel to the interpreted shelf–slope break. The topset–foreset inflection on seismic reflection profiles is used as a proxy for the shelf–slope break (see Fig. 4). Biofacies for the Oso Field stratigraphic intervals are presented in Figs 19 and 20.

The prograding complex of the Oso Field 1Y1 reservoir is interpreted as a lowstand facies based on its position along a regionally extensive shelf–edge supported both by seismic facies mapping of the topset–foreset inflection, by the close proximity of deeper-water foraminiferal assemblages immediately south of the field, and by the predominance of grass pollen and relative absence of mangrove pollen associated with arid climates and relative lowstand of sea level (Morley & Richards 1993; Fearn *et al.* 1997).

The biofacies map (Fig. 15) for the lower lowstand–prograding complex systems tract, characterized by the bottom-seal mudstones of the 1Y1 reservoir, shows the Oso Field area occurring within the inner–middle neritic biofacies interval. This relationship reflects the shallow-marine foraminiferal assemblages associated with the initial marine deepening over the underlying sequence boundary in a marine offshore shelfal setting. The mapped interval is associated with the base of the lowstand prograding complex because of the relative proximal location of the Oso Field. The lowstand systems tract slope front thick consisting of channel–levée complexes, and the basin floor thick consisting of

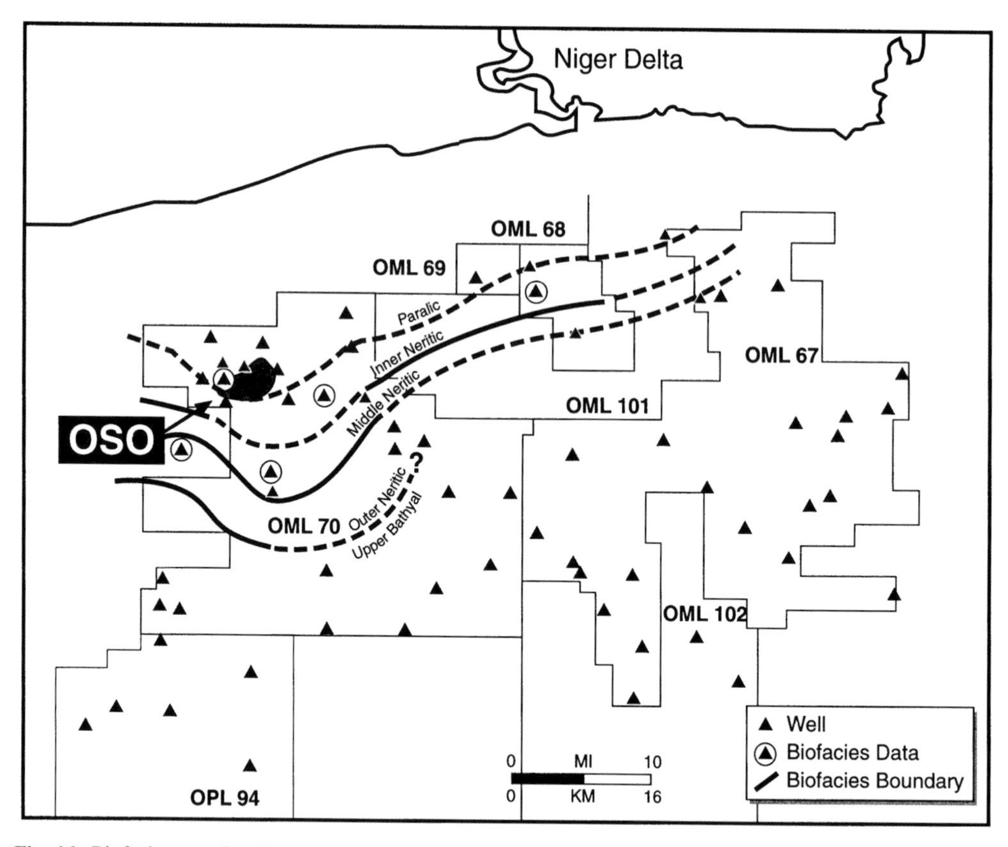

Fig. 16. Biofacies map for the uppermost portion of Sequence 1100 lowstand, coincident with the upper 100-feet interval of the prograding 1Y1 reservoir. Biofacies data are available from five exploration and seven production wells penetrating this interval.

gravity-flow transported deposits, are absent in the immediate area of the Oso Field. These lowstand deposits are interpreted to have probably bypassed the Oso Field area and may occur in undrilled areas furthei south. Note the rapid deepening south of Oso reflecting the relative proximity of the Oso Field lowstand prograding complex to the upper slope.

In Fig. 16, the upper lowstand–prograding complex systems tract biofacies map, the shallow-water biofacies in the area of the Oso Field shift southward relative to the biofacies of the lower lowstand–prograding complex map of Fig. 15. This is interpreted to reflect the basinward progradation of the Oso Field lowstand deltaic complex. The seismic reflection profile of Fig. 4 clearly shows the progradational nature of the depositional system. On the basis of four nearby exploration wells, Oso Field is mapped within the inner neritic biofacies reflecting the interpretation of the interdistributary mudstone foraminiferal assemblages within the upper lowstand–prograding complex systems tract. In fact, the reservoir sandstones interpreted from core as distributary channel and distributary mouth-bar deposits are deposited in paralic coastal plain settings which prograde into the inner neritic environment (Snedden *et al.* 1992; Thompson & Snedden 1996). Typically, in such interbedded paralic and inner neritic deposits, the quantitative analysis selects the deeper foraminiferal biofacies within each sample which is further averaged toward the deeper biofacies in the computer analysis, depending on how the analysis is programmed. Indeed, biofacies analysis derived from core in seven production wells within the Oso field are dominated by paralic biofacies.

On the biofacies map of Fig. 17 for the initial flooding surface or maximum flooding surface (depending on the preferred sequence stratigraphic interpretation – see Discussion), the biofacies in the Oso Field area shift northward reflecting the transgression across the underlying

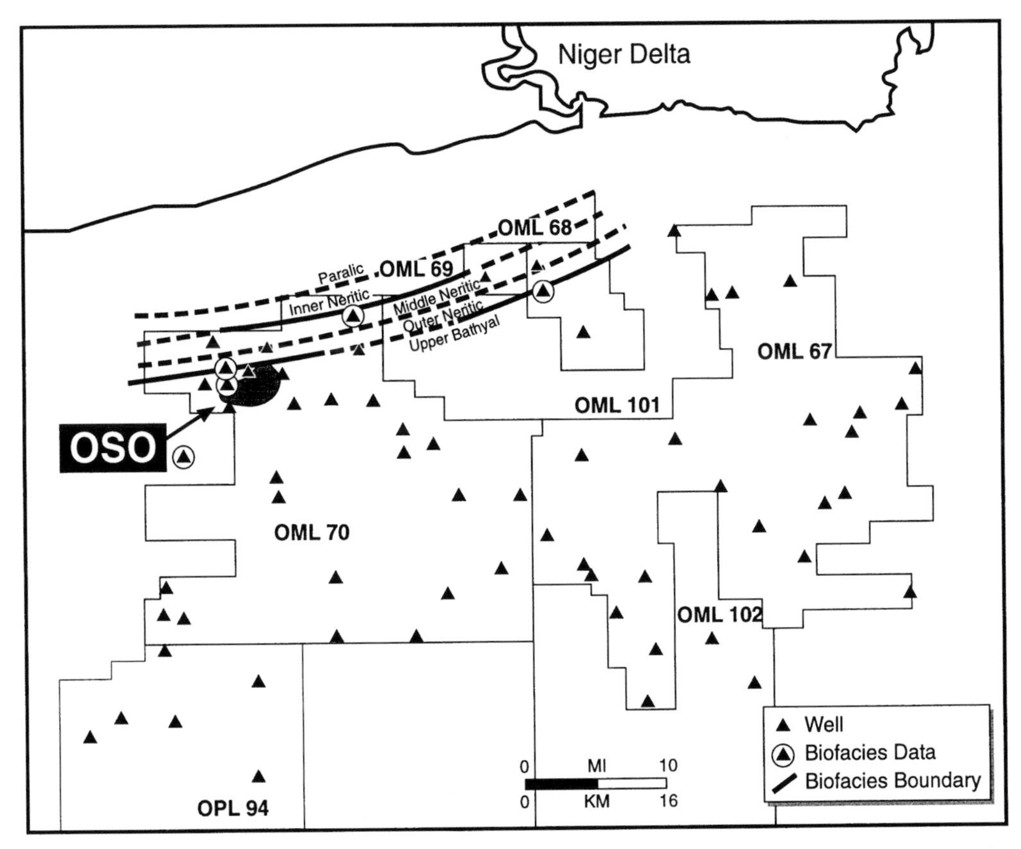

Fig. 17. Biofacies map of the deepest water biofacies associated with the wireline log and seismic stratigraphic defined maximum flooding surface of Sequence 1100. Biofacies data are available from six wells.

lowstand systems tract of Fig. 16. At Oso Field, the mudstones between the 1Y1 and 2Y2 reservoir correlate with this interval, and provide a regionally extensive top seal. The foraminiferal biofacies assemblages from these mudstones indicate open marine conditions with upper bathyal biofacies assemblages, suggesting water depths in excess of 600 feet (Fig. 20). This rapid and very large deepening could reflect a combination of absolute sea-level rise resulting in relative sediment starvation in the Oso Field area, combined with continued sea-floor subsidence due to both mudstone compaction and movement on Fault 1 (Fig. 2).

During deposition of the highstand, as shown in Fig. 18, the paralic and neritic biofacies shift basinward reflecting the progradation of the highstand systems tract relative to the maximum flooding surface transgressive facies of Fig. 17. The Oso Field occurs within a paralic–inner neritic setting suggesting a slightly more marine condition for the deposition of the 2Y2 reservoir than the 1Y1 reservoir below, which can be interpreted as a relative transgressive shift of depositional environments. A wave-dominated shoreline within the 2Y2 dominantly transgressive succession is indicated by abundant hummocky cross strata and wave ripples in 2Y2 cores along with a marine ichnofauna (Snedden pers. comm. 1997).

The biofacies maps of Sequence 1100 clearly illustrate the basinward progradation and landward transgressions associated with the reservoir and seal intervals of the Oso Field. These biofacies maps, part of a regional set of systems tract biofacies maps for sequences 1200–300 (Fig. 8), provide useful information on depositional environments. Combined with net-sand isopachs, isochron or isopach maps, and seismic facies maps for the same systems tracts, a high-resolution palaeogeographic map can be constructed. Such high-resolution maps can be used to predict the potential for reservoir and seal facies between and beyond data control points.

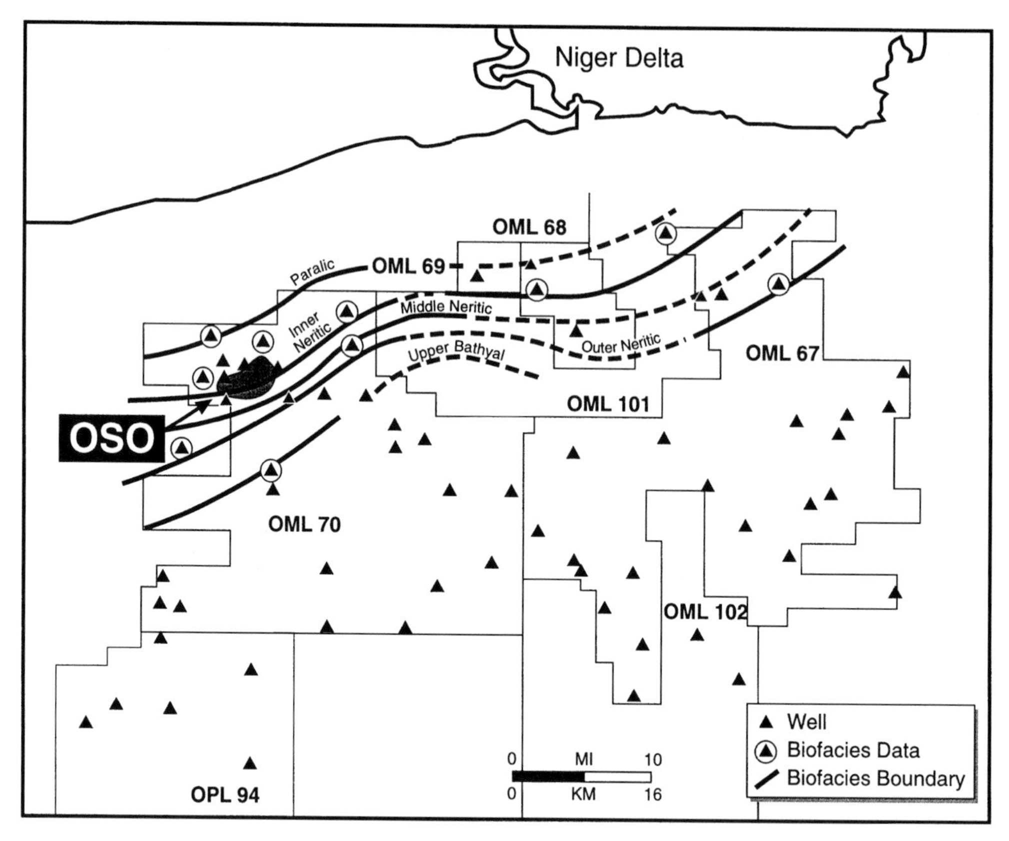

Fig. 18. Biofacies map for the uppermost portion of Sequence 1100 highstand systems tract, based on the shallowest biofacies within 100 feet below the 1000 sequence boundary. Data are available from 10 wells penetrating this interval.

Oso Field study

The primary reservoirs of Oso Field, the 1Y1 and 2Y2 sandstones, occur within Sequence 1100 dated as late Miocene (Fig. 8). This interval is part of the Biafra Member of the Agbada Formation (Fig. 3). Deposition of the reservoir intervals occurred during a dynamic phase of growth-fault movement as demonstrated by the increased thickness in the Sequence 1100 interval toward Fault 1 (Fig. 4).

Integration of seismic reflection profiles, wireline-log profiles, biostratigraphic data and conventional core sedimentology result in a reservoir framework of progradational cycles encased between marine mudstones which provide seals. The Oso Field 1Y1 and 2Y2 geometric framework is shown in Fig. 4; Figs 19 and 20 have the log profile and biostratigraphic data.

The generalized reservoir architecture shows the 1Y1 progradational cycle separated from the 2Y2 progradational cycle by the marine shale of the 1Y1 top seal, and the entire reservoir interval is capped by the marine mudstone 2Y2 top seal. Each of the prograding cycles consists of a set of individual coarsening-upwards parasequences each capped by a mudstone.

Nearly 4000 feet of conventional core was taken in eight of the 33 wells within the Oso Field area (Fig. 2). Palaeontological analysis of mudstone samples from these cores included examination of pollen and spores, dinocysts, foraminifera, calcareous nannoplankton and, in one core, diatoms. Palaeobathymetric analysis was carried out on the benthic foraminiferal assemblage following the methodologies of Tipsword *et al.* (1966), Adegoke *et al.* (1976) and Poag (1981). The high-resolution biostratigraphic data for one well are presented in Fig. 19. Data from all eight wells cored and studied have been distilled into a single chart shown as Fig. 20, which permits characterization of each reservoir sandstone and mudstone interval in the Oso Field.

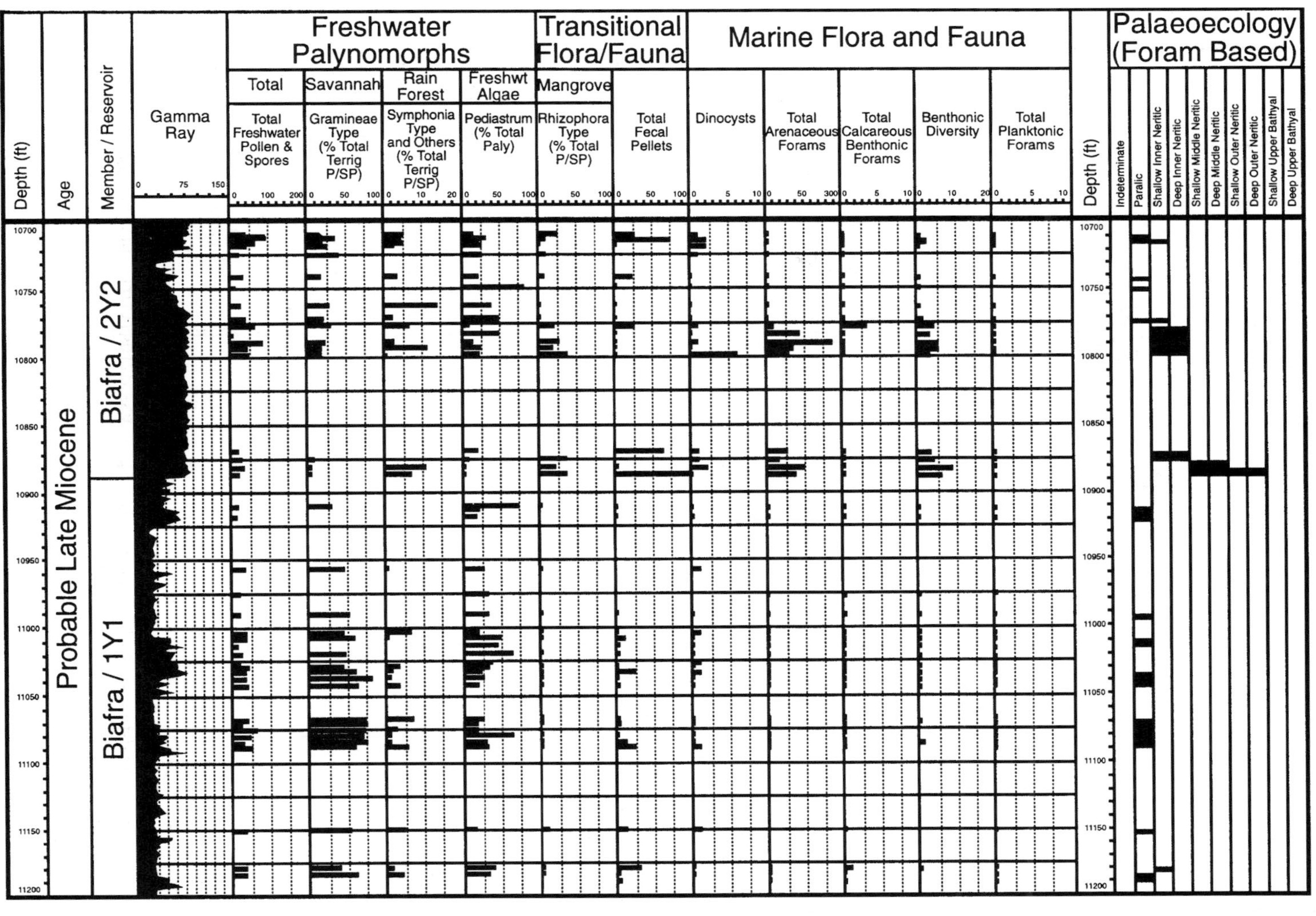

Fig. 19. Biostratigraphic profile of the high-resolution biostratigraphy for Oso 24-C well showing the detailed analysis of core samples. The absolute values for observed specimens varies from group to group but uniform relative abundances have been used in the summary diagram of Fig. 20.

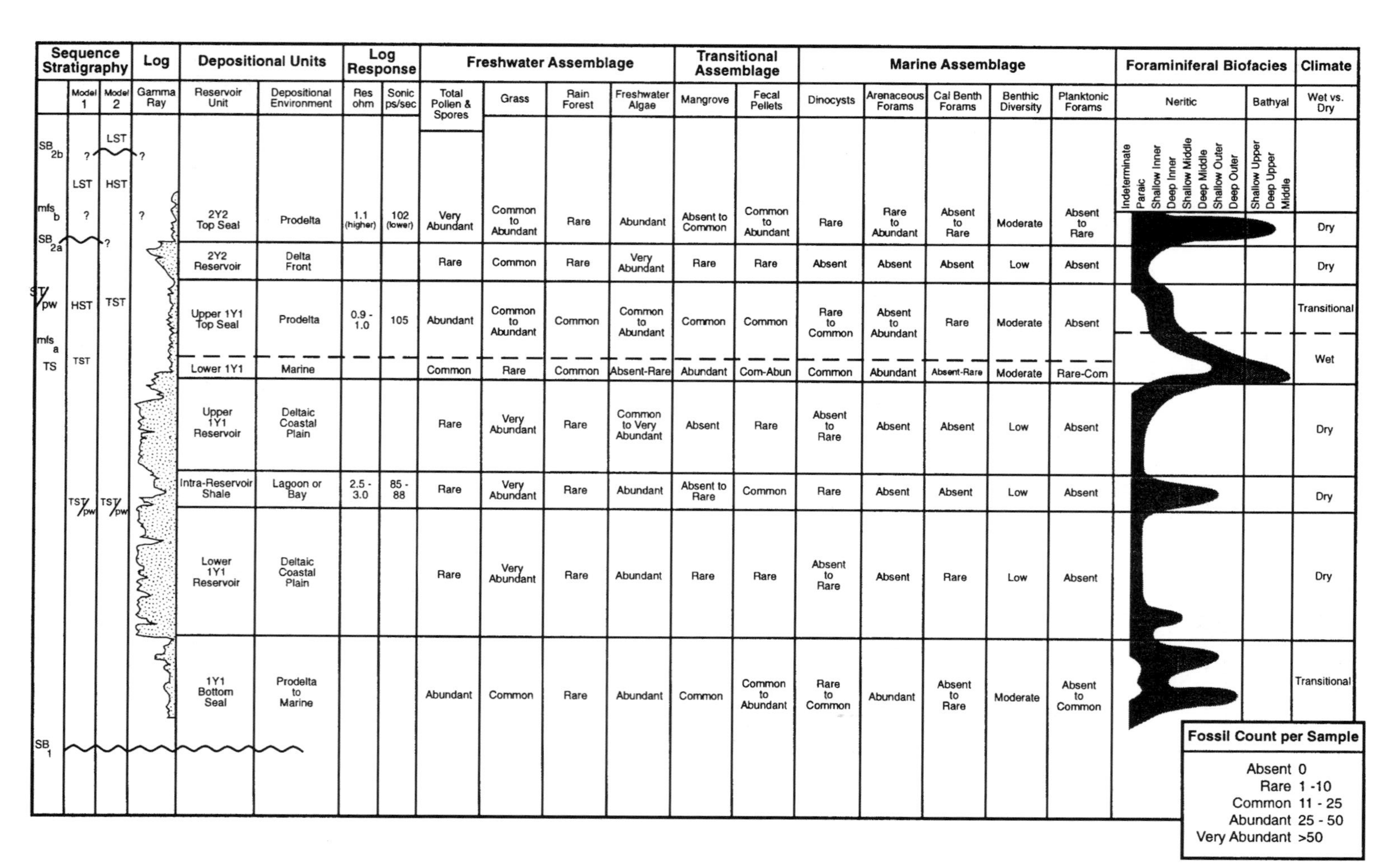

Fig. 20. Chart summarizing the biostratigraphic characterization of stratigraphic intervals within the Oso Field based on analysis of both cuttings and conventional core samples such as displayed in Fig. 19. 1Y1 bottom seal characteristics are derived from nearby exploration wells as this interval is not penetrated in production wells such as Fig. 19. All relative abundances of observed fossil taxa are based on the same numerical ranges. The alternative models for sequence stratigraphic interpretation are discussed in the text and in Fig. 6. The log profile is from Fig. 19.

Palynological analysis proved to be a major key to understanding the palaeoclimate and very useful in characterizing the depositional setting of the Biafra mudstones. The interpretations presented here draw on the work of Germeraad *et al.* (1968), Poumot (1989) and Morley & Richards (1993). All samples examined from the cored intervals were barren of calcareous nannofossils and diatoms.

Biostratigraphic data

The lower producing interval, the 1Y1 reservoir, consists of two or three coarsening-upwards to blocky sandstones separated by mudstones (log profile of Figs 19–21). Core sedimentology of reservoir intervals are interpreted as coastal plain, distributary channel, tidal channel, lagoon, delta-front and shoreface environments (Snedden *et al.* 1992; Thompson & Snedden 1996) (Fig. 5).

Most intra-1Y1 shales contain terrestrial palynomorph assemblages dominated by Gramineae (grass) pollen and *Pediastrum* spp., a freshwater algae, with essentially no marine floral or faunal components. These mudstones are interpreted as interdistributary coastal plain overbank and lagoonal deposits accumulated during a relative arid climate. The intra-1Y1 mudstones are vertically impermeable and appear to compartmentalize the field within the hydrocarbon leg (Snedden *et al.* 1992). Below the condensate–water contact these mudstones are not interrupting the flow units and are interpreted as discontinuous due to incision and cutouts by distributary channel-fills (Snedden *et al.* 1992; Thompson & Snedden 1996). A few intra-1Y1 mudstones are several feet thick, and locally contain some marine inner–middle neritic benthic foraminifera and rare dinocysts, suggesting local marine incursions possibly due to autocyclic lobe abandonment and development of bays.

The upper producing interval, the 2Y2 sandstone, is interpreted as forming in a coastal plain to strand plain to lower shoreface–inner shelf depositional setting (Thompson & Snedden 1996). The 2Y2 facies suggest a somewhat more distal marine facies overlying the more proximal coastal plain facies of the underlying 1Y1 sandstone within the Oso Field area. These interpretations suggest a northward shift in depositional environments from 1Y1 to 2Y2 reservoir sandstone time, indicating increased accommodation space. The increased accommodation space may be in response to either of two factors. A relative rise in sea level may have caused decreased rates of sediment supply to the Oso Field area resulting in apparent transgression. Alternatively, or additionally, increased rates of offset along Fault 1 with constant rate of sediment supply would have resulted in increased accommodation space and apparent transgression, especially as the underlying prodelta mudstones dewatered and compacted. This suggests that a regionally significant relative drop in sea level (?lowstand) would facilitate bypass or progradation of the sand across Fault 1 into the rapidly subsiding (locally apparent relative rise in sea level) area of 2Y2 sand deposition.

The palaeogeographic maps for the systems tracts of the 1Y1–2Y2 intervals of Sequence 1100 suggest regional transgression independent of movement on Fault 1, suggesting a regional transgression forced by eustasy rather than a uniquely local apparent transgression associated with Fault 1. However, the biofacies maps are based on limited data and have been contoured using the seismic shelf–slope inflection as a guide. The maps are therefore not a definitive tool to answer the question of which factor dominated the change in accommodation space.

The interpreted shift from inner neritic to upper bathyal biofacies during the transgression of the 1Y1 top seal mudstone over the 1Y1 reservoir suggests a change in nearly 450 feet of water depth (Fig. 20). The cycle chart of Haq *et al.* (1988) shows such a sea-level change associated with the TB 3.2 third-order cycle dated at 6.3–8.2 Ma, in the late Miocene, possibly correlative with the 1100 sequence of this study (see Fig. 8).

The thin, distal intra-2Y2 mudstones contain a similar spectrum of foraminiferal biofacies and palynofacies as the thin and distal intra-1Y1 mudstones, but with a slightly higher mangrove pollen count (Figs 19 and 20). Deposition during a dry, possibly glacial, climate phase is suggested for both reservoir intervals with the 2Y2 deposited in a slightly less arid conditions than the 1Y1 reservoir (Fig. 20) (Poumot 1989; Morley & Richards 1993).

The lowermost 10 feet of mudstone between the 1Y1 and 2Y2 sandstones, called the 1Y1 top seal (Fig. 20), is fully marine, initially middle neritic deepening to upper bathyal, with arenaceous benthic foraminifera, cosmopolitian marine dinoflagellates, rare planktonic and calcareous benthic foraminifera, a palynoflora characterized by abundant mangrove pollen and spores, and the complete absence of freshwater algae, such as *Pediastrum* spp. (Fig. 20). This mudstone is interpreted to be a marine transgressive shale deposited during a relatively wet, possibly interglacial climate.

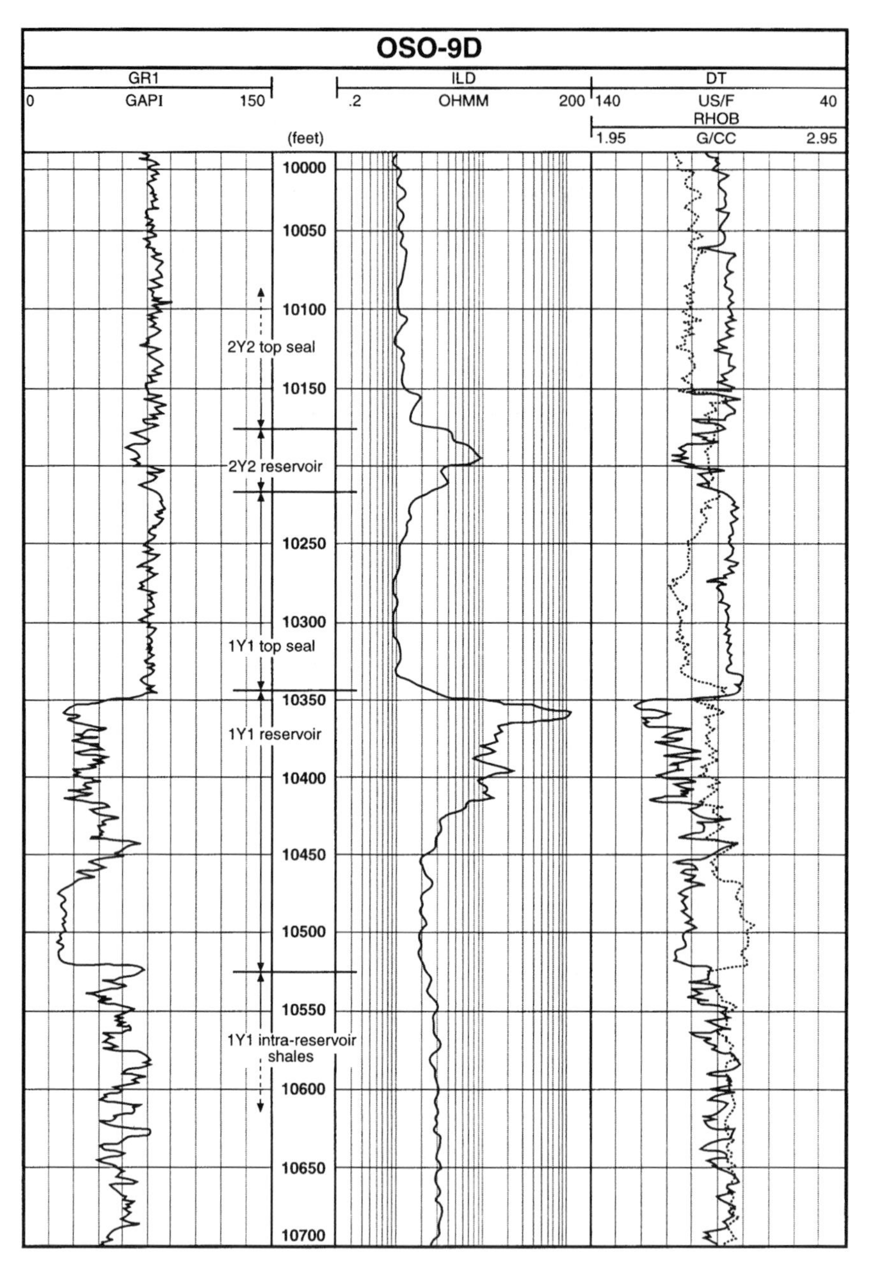

Fig. 21. Electric logs of the Oso 9D well annotated with the Oso Field reservoir and seal intervals. The petrophysical characteristics of each interval are inventoried in Table 2.

Ten feet above the base of this transgressive mudstone a maximum palaeowater depth is indicated by upper bathyal foraminiferal assemblages. Above this maximum water-depth biofacies interval, the upper 1Y1 top seal mudstone contains progressively shallower-water biofacies assemblages upward into the shallow inner neritic and paralic 2Y2 sandstone (Fig. 20). This shallowing-upwards mudstone contains relatively more pollen and spores, and significantly more grass pollen, relative to rainforest pollen, suggesting progressively drier conditions on land.

These patterns suggest that the 1Y1 top seal mudstone contains a maximum flooding surface (Fearn *et al.* 1997) (see Model 1 mfs/a of Fig. 20) of a transgressive, humid climate phase separating progradational deltaic depositional phases characterized by relatively more arid climates (see the section on 'Sequence stratigraphy').

The mudstone overlying the 2Y2 sandstone contains open marine, mainly arenaceous (middle neritic–upper bathyal) foraminifera, common Gramineae pollen, abundant freshwater algae and rare mangrove pollen (Fig. 20). This assemblage suggests marine deposition during a drier, possibly glacial climate (Poumot 1989; Morley & Richards 1993).

The biostratigraphic data clearly separate two types of mudstones within the Oso Field, and reinforce correlations based on log-motif analysis. The more laterally continuous mudstones are the fully marine 1Y1 and 2Y2 top seals, and the more local the 1Y1 and 2Y2 intrareservoir non-marine to paralic mudstones.

Petrophysical data

Wireline-log petrophysical data from seven Oso Field wells are presented in Table 2, and are used to characterize the marine and non-marine mudstones. Each log was 'blocked' for intervals of similar petrophysical response. Figure 21 shows a wireline-log suite of gamma ray, sonic travel time, density and deep induction from one Oso Field well. The depth interval shown extends from 10 000 to 10 650 feet, and contains a portion of the 2Y2 top seal, the 2Y2 reservoir, the 1Y1 top seal and the 1Y1 reservoir.

The 1Y1 intra-reservoir mudstones shown in Fig. 21, between 10 580 and 10 610 feet, have resistivity values of 2.5–3.0 ohm-metres (Ω-m) with corresponding sonic travel times of 85–88 μs foot^{-1}. This is in contrast with the 1Y1 top seal mudstones between 10 220 and 10 350 feet, which have resistivities of 0.9–1.1 Ω-m with sonic travel times of 105 μs foot^{-1}.

The marine top seal mudstones have lower resistivity and lower sonic velocity (high travel time) than the non-marine mudstones within the 1Y1 reservoir. The shallowing-upwards upper 1Y1 top seal mudstones have the same petrophysical characteristics as the 1Y1 top seal.

The lowermost 10–20 feet of the 2Y2 top seal mudstones have an average resistivity of approximately 1.1 Ω-m with an average sonic travel time of 102 μs foot^{-1}. These values are more like the 1Y1 top seal marine mudstone than the 1Y1 intra-reservoir non-marine mudstones, as is expected as both top seals are fully marine mudstones (Fig. 20).

Above the lower 20 feet of the 2Y2 top seal, the mudstones have lower resistivity and slower sonic travel times (lower velocity) than the lowest part of the 2Y2 top seal mudstones. The exact interpretation of this behaviour has not been established. Based on biostratigraphic analysis of core samples, it is possibile that this petrophysical difference is related to early less marine v. later more fully open marine phases of a transgression resulting in differences in the cementation style of the mudstones.

The petrophysical data inventoried in Table 2 for the 1Y1 and 2Y2 marine top seals mudstones, and the data discussed above for 1Y1 non-marine intra-reservoir mudstones, clearly distinguishes the marine and non-marine mudstones.

Seismic geometry

The mudstones of the 1Y1 and 2Y2 top seals correlate with high-amplitude seismic reflections which are continuous across the Oso Field, and occur above the prograding oblique clinoforms of both the 1Y1 and the 2Y2 reservoir intervals (Fig. 4). This reflects the laterally uniform deposition of the marine transgressive mudstones over the entire field area compared to the

Table 2. *Petrophysical data for the 1Y1 and 2Y2 reservoirs of the Oso Field (*log data missing or interpreted as unreliable)*

Well	2Y2 R_{sh} (Ω-m)	1Y1 R_{sh} (Ω-m)	2Y2 shΔt (μs foot^{-1})	1Y1 shΔt (μs foot^{-1})	2Y2 density (g cm^{-3})	1Y1 density (g cm^{-3})	1Y1 shale thickness (feet)
9D	1.1	0.9–1.1	102	105	2.48	2.50	132
13D	1.2	1.0	95	105	2.51	2.50	120
25B	1.0–1.3	1.0	*	*	2.45	2.44	150
26C	1.1–1.5	1.2	95	99	2.54	2.54	153
28E	1.1	1.0	100	102	2.50	2.48	107
29E	*	*	100	104	2.53	2.50	130
31D	1.5	1.1	95	100	2.52	2.51	190

R_{sh}, resistivity; shΔt, sonic travel time.

more locally developed lithofacies of the non-marine interval of the deltaic reservoirs. Within the non-marine intervals, the mudstones of the interdistributary areas are interpreted to be discontinuous due to restricted depositional extant and due to erosion by channels, and, therefore, are only local baffles (Fig. 5). These discontinuous mudstones do not appear to be uniquely identifiable on the seismic reflection profiles.

Intermediate between the marine and the non-marine mudstones are the intrareservoir paralic interdistributary bay mudstones and probable delta-front mudstones. These mudstones appear to correlate best with locally continuous clinoform foreset reflections (Fig. 4), which are probably caused by mudstone deposited over an abandoned prograding deltaic lobe after avulsion of the distributary channel supplying sediment to the lobe. Based on the correlation to seismic reflections, the marine mudstones extend over at least 10 miles, the paralic and prodelta mudstones for perhaps as much as 1500–2000 feet, and the non-marine mudstones somewhat less than 1000 feet.

RFT pressure data

Effectiveness of seal can be tested with reservoir pressure data. RFT (Repeat Formation Tester) pressure data for the 1Y1 and 2Y2 reservoirs was assembled for eight wells in Oso Field (Fig. 22). Analysis of the data revealed that in some wells the pressures in the 1Y1 and the 2Y2 lie on the same pressure–depth gradient indicating communication between the two reservoirs (e.g. well Pr-26C). In others cases, the reservoir pressures lie on different gradients indicating no communication (e.g. well Pr-29E).

The quantitative sealing capacity derived from the pressure data presented in Fig. 22 is summarized in Table 3. It should be noted that much of the spread in pressures at any given depth in this field is due to pressure depletion with production as a function of time. Hence, in those wells having discontinuous pressure gradients, the indicated pressure seal may be effective only over the relatively short time-span of production from the field. It is not possible to determine whether the seal was effective

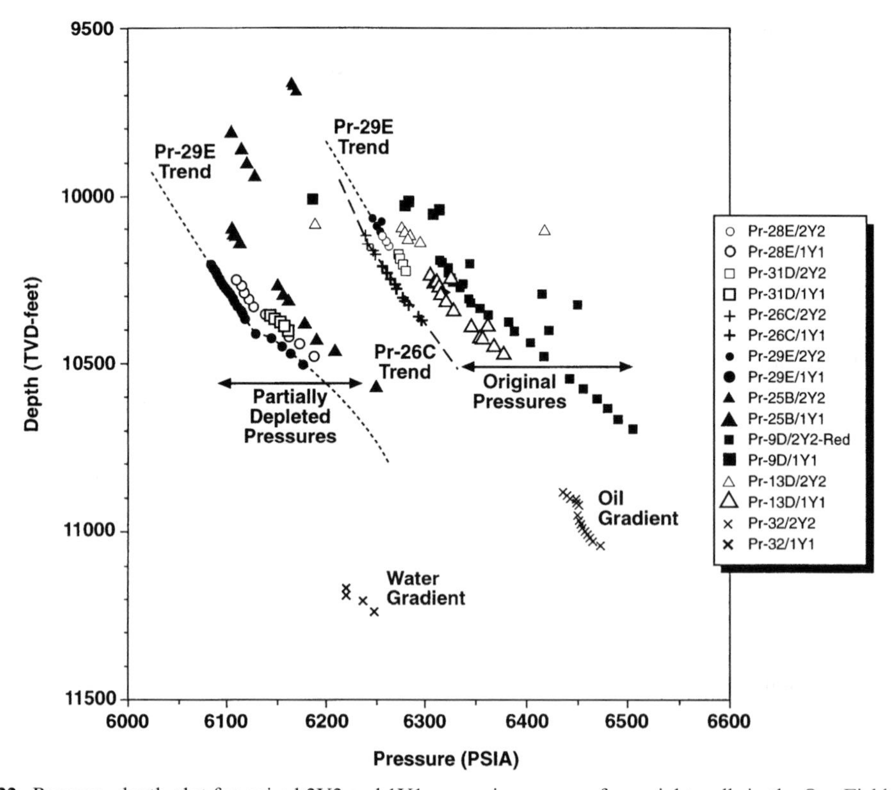

Fig. 22. Pressure–depth plot for paired 2Y2 and 1Y1 reservoir pressures from eight wells in the Oso Field. Note the uniform trend of pressure between the 2Y2–1Y1 reservoirs for the well Oso Pr26C suggesting no effective separation by the 1Y1 top seal. In contrast, the data for well Oso Pr-29E show different pressures suggesting effective separation by the 1Y1 top seal.

Table 3. *Inventory of seal integrity interpretations and thickness of the 1Y1 top seal based on measurements plotted in Fig. 22*

Well	Pressure seal between 1Y1 and 2Y2	Shale thickness (feet)
9D	No	132
13D	No	120
25B	Yes	150
26C	No	154
28E	Yes	107
29E	Yes	130
31D	Yes	190
32	Yes	150

in isolating the 1Y1 and 2Y2 reservoirs over previous geological time as no pre-production pressure data are available in the studied wells.

When mapped, the pressure data do not suggest any uniquely interpretable geographical pattern that accounts for the relative sealing capacity of the 1Y1 top seal mudstone. The thickness of the shale does not suggest a correlation with seal integrity and thickness. The most likely explanation for the apparent random arrangement of seal integrity is that the shale has the properties of a good seal, but local faulting and fracturing has ruptured the marine mudstone seal in a currently unpredictable manner.

The pressure test data suggest that the two marine mudstones provide a top seal across the entire Oso Field, an area of more than 6×10 miles, except where faulting disrupts the seal or offsets the mudstones and juxtaposes sandstones.

Mudstone bed continuity

One of the most important issues in developing the gas-injection–pressure maintenance programme at Oso Field concerns the lateral extent of the relatively thin (less than 20 feet thick) non-marine mudstone beds within the 1Y1 interval. Special core analysis from one Oso Field well demonstrated that permeabilities, particularly permeability perpendicular to bedding (kv), are low enough to prevent or severely restrict gas migration through these barriers over the immediate production life of the field (Snedden *et al.* 1992). For effective pressure maintenance and hydrocarbon sweep to occur, gas must migrate laterally around these barriers within the pressure compartment of the 1Y1 reservoir bounded by the marine mudstone top seal and bottom seal. Therefore, lateral continuity of the non-marine mudstones becomes a critical consideration in production of the 1Y1 reservoir, and ultimately the 2Y2 reservoir.

Snedden *et al.* (1992) assembled information on outcrop and subsurface mudstone continuity data demonstrating a strong function of depositional environment (Zeito 1965; Varrien *et al.* 1967; Geehan *et al.* 1986). Marine mudstones display the greatest mapped lateral extent, with 80% or more of the mudstones exceeding 2000 feet in length (Fig. 23). Non-marine and channel and point-bar mudstones typically exhibit the lowest continuity, with few exceeding 1200 feet in length. The degree of channelization and influence of wave action are thought to play a role in determination of shale length (Snedden *et al.* 1992).

The published curves presented in Fig. 23 are based largely on Cretaceous siliciclastic outcrops of the western United States (Zeito 1965). Applicability to the Neogene Niger delta was tested by plotting the separation distances for high confidence correlations of intrareservoir non-marine mudstones for closely spaced pairs of well bores, including both appraisal and production wells. Separation distances for the closely spaced well bores ranged from 52 to 1217 feet (Fig. 23). Snedden *et al.* (1992) provide a discussion of the correlation exercise.

Examination of 300 different mudstone beds penetrated by paired wells in the Oso Field suggest that less than 20% of the intrareservoir non-marine mudstones can be correlated over distances of 1000 feet or greater (Snedden *et al.* 1992). Snedden *et al.* (1992) concluded that the thin non-marine mudstone beds may in some places act as baffles to gas flow, but they are not likely to be barriers to gas dispersal on a reservoir scale.

In contrast, the fully marine transgressive mudstones provide effective top seals except where breached by faulting and juxtaposition of sandstones. The intermediate paralic–prodelta mudstones are potential baffles within the field, but appear to not significantly impede flow between existing gas-injection and condensate-production wells.

The biostratigraphic and petrophysical calibration of mudstones thus provides a rapid means of characterizing mudstone types and providing a means to predict the possible lateral extent and seal or baffle character of each mudstone.

Sequence stratigraphy

Alternative interpretations of depositional systems tracts have been made using the available data (Fig. 20 – see Fig. 6 for models). In interpretation Model 1 palaeontologists prefer the

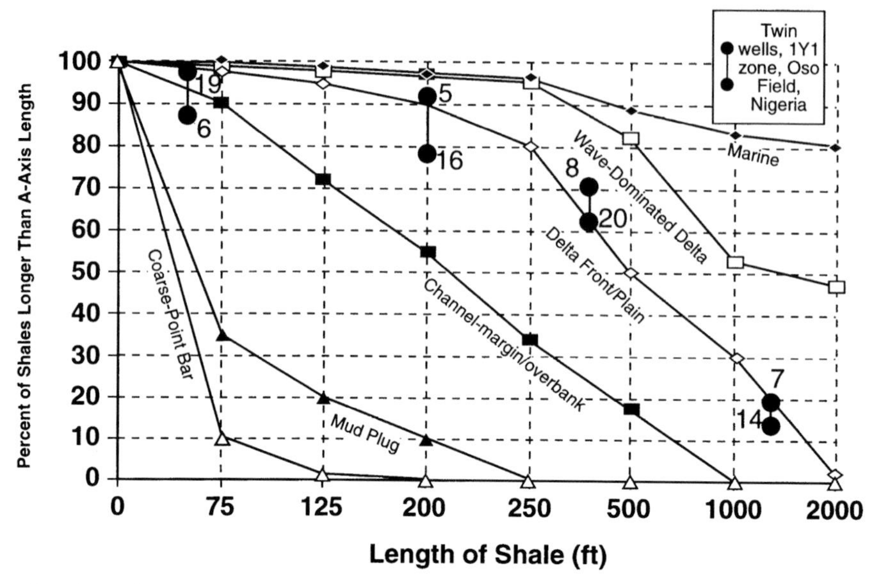

Fig. 23. Cumulative frequency distribution for shale lengths of different depositional environments (from Snedden *et al.* 1992). See text for discussion.

interpretation of the 1Y1–2Y2 mudstone as containing both a transgressive surface (TS) at the base and a stratigraphically higher condensed section (mfs/a of Fig. 20), characterized by the presence of planktonic and deeper-water benthic foraminifera, and dinocysts, the rare occurrence of grass and freshwater algal palynomorphs, and common rain forest and mangrove pollen, suggesting a wet climate in contrast to the mudstones within the 1Y1 and 2Y2 reservoir intervals which have dry climate fossil assemblages.

The above information has been interpreted to suggest that the 2Y2 interval was deposited during a drying climate phase following a wet climate, and therefore is a possible highstand systems tract overlying the relatively thin transgressive systems tract of the lower 1Y1 top seal mudstone. This interpretation of the 2Y2 reservoir as a highstand systems tract places a candidate sequence boundary at the top of the 2Y2 sandstone (SB2a of Fig. 20; also Model 1 of Fig. 6).

In contrast, sedimentologists prefer Model 2 (Figs 6 and 20), placing the transgressive surface (TS) at the top of the 1Y1 sandstone as do the palaeontologists, but interpreting the maximum flooding surface–condensed section above the 2Y2 sandstone (mfs/b of Fig. 20). Snedden *et al.* point out that within the 2Y2, five progradational parasequences occur in a retrogradational stacking pattern suggesting a transgressive systems tract. These thicken toward the north due to syn-depositional movement on Fault 1 (Figs 2 and 4), resulting in an increase in accommodation space due both to fault movement and regional rise in sea level. Snedden *et al.* also suggest that the absence of subaerial erosion or exposure at the top of the 2Y2 sandstone, based on core examination, does not suggest a sequence boundary and therefore the 2Y2 is a transgressive systems tract deltaic complex. This suggests that the 2Y2 is a back-stepped deltaic system deposited within the transgressive systems tract, and the overlying sequence boundary (SB2b of Fig. 20) is somewhat higher in the section. Regional seismic sequence stratigraphic analysis in general agrees with this model.

The most recent addition to the dataset is a core in Oso well 32. The base of the 2Y2 sandstone is very sharp-based suggesting the possibility of an erosional cycle boundary. John Snedden (pers. comm., December 1997) suggests that this sharp-based sandstone could be interpreted as a high-frequency sequence boundary related to a relative fall in sea level.

The regional seismic sequence stratigraphy study correlated sequences at a scale that does not address the details of the Oso Field systems tracts. The lower sequence boundary is interpreted below the base of the 1Y1 reservoir interval where it is noted by the downlapping reflections of the lowstand prograding complex (lower sequence boundary on Fig. 6). On the regional seismic grid, the overlying sequence boundary is interpreted far above the Oso Field producing interval. The seismic datum interpreted as the maximum flooding surface correlates somewhat above the top of the 2Y2 reservoir suggesting that the 2Y2 sandstone is part of a transgressive systems tract. Well-log and biostratigraphic patterns suggest that several high-frequency depositional cycles occur between the seismically correlated maximum flooding surface at the top of the 2Y2 reservoir and the regionally mapped sequence boundary at the top of seismic sequence 1100.

The sedimentologist's interpretation is in general agreement with that of the regional seismic stratigraphic analysis, and it is tempting to follow those two complimentary interpretations. However, the high-resolution biostratigraphic analysis of both foraminiferal biofacies and palynologically defined climate cycles cannot easily be dismissed. If the palynological assemblage indicated a humid–wet climate within the 2Y2 top seal, it would be possible to place the maximum flooding surface within this mudstone, and attribute the arid–dry palynomorph assemblage to local coastal plain environments. However, the arid–dry palynomorph assemblage persists stratigraphically upward through hundreds of feet of the post-2Y2 top seal mudstones suggesting deposition within a very mud-prone lowstand systems tract.

Model 3 is a modification of the earlier Models 1 and 2. It is based largely on observations of a single core near the northern (most landward) part of the field. The core has a very sharp base for the 2Y2 sandstone. That sharp base suggests the possibility of either an unconformity due to a rapid drop in relative sea level and a facies downshift basinward, or the base of a distributary channel. Elsewhere in the field, the basal 2Y2 grades upward in coarsening-upwards prograding patterns. This most landward sharp-based sandstone suggest the possibility of a high-frequency sequence lowstand systems tract, with the underlying highstand systems tract being entirely shale and very thin due the distal setting of the Oso Field area.

These different sequence stratigraphic interpretations reflect the integration of a spectrum of observations using different datasets. The palynomorph assemblages of both the 1Y1 and 2Y2 reservoirs suggest a dry, possibly glacial, climate phase. The sandstones are separated by a mudstone with an interval containing abundant planktonic foraminifera, suggesting a maximum flooding surface. Thus, Model 3 is perhaps a reasonable interpretation of the field-scale data. The regional seismic sequence stratigraphy analysis, Model 2 does not recognize the 2Y2 as a distinct sequence because it is below the resolution of the older vintage industry data. Resolution of such scale-dependent problems are typical in integrated studies.

Continued drilling within and near the Oso Field is providing additional data that is and will continue to constrain the interpretations of the depositional environment. Thus, our future interpretations are likely to differ from those presented here.

Summary

An age model has been proposed in which 21 depositional sequences recognized in the Joint Venture Area offshore of the Niger delta have been correlated with the middle Miocene–Pleistocene geological time scale. This age model results in improved correlation and modified numeric modelling, such as burial history and maturation analyses. Biostratigraphic correlations within a small area, such as a fault block, can attain a moderate level of precision. Regional correlations require integration of the biostratigraphy with a network of seismic reflection profiles and wireline-log based stratigraphic correlations.

A consistent means of characterizing sequence stratigraphic components (surfaces and systems tracts) has been developed on the relative abundance of palynological humidity v. aridity indicators, and on both subtle and major trends in benthic foraminiferal abundance and biofacies trends.

The mapped distribution of foraminiferal biofacies associated with sequence stratigraphic surfaces provides a completely independent means of defining palaeogeographic limits of wave-dominated v. gravity-flow dominated sandy reservoirs for a given time increment. Analysis of the Oso Field data currently available suggests the following interpretation, shown schematically in Fig. 24.

1Y1 sandstones were deposited as a lowstand shelf-edge prograding deltaic complex with autocyclic shifting of sandy lobes subsequently draped by laterally discontinuous interdistributary and lagoonal mudstones. Associated palynomorphs suggest a dry climate interpreted as a glacial interval associated with a relative lowstand of sea-level depositional phase.

Deposition of the 1Y1–2Y2 shale occurred when the rate of accommodation exceeded the sediment supply resulting in a marine transgression over the entire field area. The maximum flooding surface may occur within this interval. The associated palynomorph assemblage

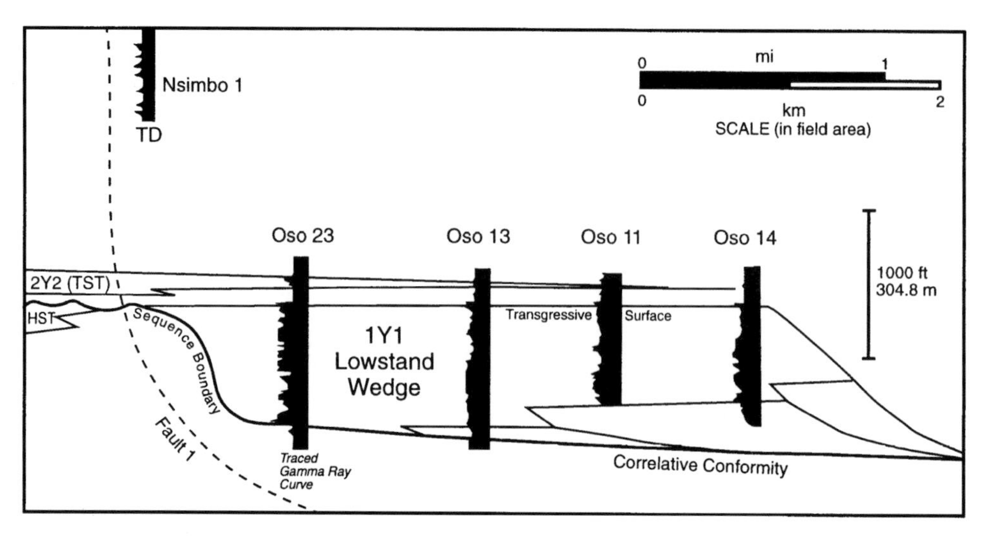

Fig. 24. Schematic sequence stratigraphic model for the 1Y1 and 2Y2 reservoirs of Oso Field, modified from Snedden *et al.* (1992) and Thompson & Snedden (1996), showing their interpretation of the 2Y2 reservoir as a transgressive systems tract. The alternative interpretation suggested by the biostratigraphic data places the 2Y2 reservoir in a highstand systems tract. The most recent interpretation suggests that the 2Y2 is a high-frequency sequence. See the text discussion for further comment. Note that the figure is not to scale north of Fault 1.

suggests a wet climate probably indicating an interglacial climate associated with a relative transgressive sea-level depositional phase.

2Y2 sandstone deposition occurred as a progradational event. The associated dry climate could reflect onset of a glacial climate with a relative lowering of sea level. The overall post-2Y2 deepening at Oso Field during a glacial interval must reflect increased accommodation due to decreased sediment flux with continued regional subsidence, perhaps in part created by growth-fault movement. Persistence of this dry climate during deposition of the post-2Y2 mudstone suggests that the 2Y2 sandstone may be a late highstand deposit followed by a mudstone-dominated lowstand systems tract due to a lateral shift of the sand supply. Alternatively, the 2Y2 reservoir interval may be an early transgressive systems tract delta, as discussed earlier, and strongly suggested by log-motif stacking patterns and core sedimentology.

The open marine 2Y2 mudstone is several hundred feet thick and is an effective top seal for the entire Oso Field. The thinner 50–100 foot thick 1Y1–2Y2 marine mudstone is also an effective seal except locally where faulting has juxtaposed the 1Y1 and 2Y2 sandstones. Most intrareservoir non-marine–paralic mudstones are local baffles interpreted as channel breached interdistributary mudstones and lobe-abandonment mudstone drapes.

High-resolution biostratigraphy, in concert with core sedimentology, log analysis and seismic sequence stratigraphy, permits construction of a much more complete characterization of the depositional environment of the deltaic sandstones and mudstones of the Oso Field. Palaeontological data on both terrestrial and marine fossils provide a definitive environmental analysis of mudstones, comparable to the detailed interpretation of sandstone depositional environments constructed from core sedimentology. Using this more complete environmental characterization of the Oso Field stratigraphy, plans have proceeded with both infill drilling and delineation of gas-injection support of production flow patterns for more cost-effective field development.

The information presented here comes from Mobil reports prepared by the authors, and relies heavily on technical input from many other Mobil employees. The authors gratefully acknowledge that support and the authorization to publish from Mobil Technology Corporation and Mobil Producing Nigeria Limited. Much of this work was sponsored through the support of Daniel Lambertaikhionbare, Babajide Agbabiaka and Victor Oyofo of Mobil Producing Nigeria Unlimited. Reviews of this manuscript were provided by Jerry Ragan, Bruce Kofran, Steve Lowe and Paul Ventris. In particular, Ventris' review caused the authors to reconsider several of their interpretations and significantly improve this paper. The final draft is the responsibility of the senior author.

References

ADEGOKE, O. S., OMATOSOLA, N. E. & SALAMI, N. B. 1976. *Benthic Foraminiferal Biofacies Off the Niger Delta.* Maritime Sediments, Special Publication, **1**, 279–292.

ARMENTROUT, J. M. 1991. Paleontologic constraints on depositional modeling: Examples of integration of biostratigraphy and seismic stratigraphy, Gulf of Mexico. *In:* WEIMER, P. & LINK, M. H. (eds) *Seismic Facies and Sedimentary Processes of Submarine Fans and Turbidite Systems.* Frontiers in Sedimentary Geology Series, Springer, New York, 137–170.

——1996. High resolution sequence biostratigraphy: examples from the Gulf of Mexico Plio-Pleistocene. *In*: HOWELL, J. A. & AITKEN, J. F. (eds) *High Resolution Sequence Stratigraphy: Innovations and Applications.* Geological Society, London, Special Publications, **104**, 65–86.

BERGGREN, W. A., KENT, D. V., SWISHER, C. C. & AUBRY, M. P. 1995. A revised Cenozoic geochronology and chronostratigraphy. *In*: BERGGREN, W. A., KENT, D. V. & HARDENBOL, J. (eds) *Geochronology, Time Scales and Global Stratigraphic Correlations: A Unified Temporal Framework for an Historical Geology.* Society of Economic Paleontologists and Mineralogists Special Publication, **54**, 129–212.

BLOW, W. H. 1969. Late Middle Eocene to Recent planktonic foraminiferal biostratigraphy. *In*: BRONNIMANN, P. & RENZ, H. H. (eds) *Proceedings of the First International Conference on Planktonic Microfossils (Geneva 1967)*, Vol. 1. E. J. Brill, Leiden, 199–421.

FEARN, L. B., RODGERS, B. K., ROOT, S. A., ARMENTROUT, J. M., COOKE, J. C. & SONUGA, M. S. 1997. Applications of high-resolution biostratigraphy for Niger Delta basin depositional sequence correlation, and Oso Field reservoir flow-unit analysis, Late Miocene through Pleistocene, Nigeria. *Nigerian Association of Petroleum Explorationists Bulletin*, in press.

GEEHAN, G. W. *et al.* 1986. Geologic prediction of shale continuity, Prudhoe Bay Field. *In*: LAKE, L. W. & CARROLL, H. B. (eds) *Reservoir Characterization.* Academic Press, New York, 63–82.

GERMERAAD, J. H., HOPPING, C. A. & MULLER, J. 1968. Palynology of Tertiary sediments from tropical areas. *Reviews of Paleobotany and Palynology*, **6**, 189–348.

HAQ, B. U., HARDENBOL, J. & VAIL, P. R. 1988. Mesozoic and Cenozoic chronostratigraphy and cycles of sea level change. *In*: WILGUS, C. K., POSAMENTIER, H., ROSS, C. A. & KENDALL, C. G. ST. C. (eds) *Sea Level Change: An Integrated Approach.* Society of Economic Paleontologists and Mineralogists, Special Publication, **42**, 71–108.

HEDGPETH, J. W. 1958. Classification of marine environments. *In:* HEDGPETH, J. W. (ed.) *Treatise on Marine Ecology and Paleoecology.* Geological Society of America, Memoir, **67**, 17–27.

LEGOUX, O. 1978. Quelques especes de pollen caracteristiques du Neogene du Nigeria. *Bulletin des Centres de Récherches Exploration–Production, Elf-Aquitaine*, **2**(2), 265–317.

MARTINI, E. 1971. Standard Tertiary and Quaternary calcareous nannoplankton zonation. *In*: FARINACCI, A. (ed.) *Proceedings of the Second Planktonic Conference (Roma 1970).* Tecnoscienza, Rome, 739–785.

MORLEY, R. J. & RICHARDS, K. 1993. Gramineae cuticle: a key indicator of late Cenozoic climatic change in the Niger Delta. *Review of Paleobotany and Palynology*, **77**, 119–127.

MULLER, J., DI GIACOMO, E. & VAN ERVE, A. W. 1987. *A Palynological Zonation for the Cretaceous, Tertiary and Quaternary of Northern South America.* American Association of Stratigraphic Palynologists, Contribution Series, **19**, 7–76.

POAG, C. W. 1981. *Ecologic Atlas of Benthic Foraminifera of the Gulf of Mexico.* Marine Science Institute, Woods Hole, Massachusetts.

POSAMENTIER, H. W. & ALLEN, G. P. 1993. Variability of the sequence stratigraphic model: effects of local basin factors. *Sedimentary Geology*, **86**, 91–109.

—— & VAIL, P. R. 1988. Eustatic controls on clastic deposition – sequence and systems tract models. *In*: WILGUS, C. K., POSAMENTIER, H., ROSS, C. A. & KENDALL, C. G. ST. C. (eds) *Sea Level Change: An Integrated Approach.* Society of Economic Paleontologists and Mineralogists, Special Publication, **42**, 125–154.

——, ALLEN, G. P., JAMES, D. P. & TESSON, M. 1992. Forced regressions in a sequence stratigraphic framework: concepts, examples, and exploration significance. *American Association of Petroleum Geologists Bulletin*, **76**, 1687–1709.

POUMOT, C. 1989. Palynological evidence for eustatic events in the tropical Neogene. *Bulletin des Centres de Récherches Exploration–Production, Elf Aquitaine*, **13**, 437–453.

SALARD-CHEBOLDAEFF, M. 1990. Intertropical African palynostratigraphy from Cretaceous to Late Quaternary times. *Journal of African Earth Sciences*, **11**, 1–24.

SNEDDEN, J. W., THOMPSON, L. B., FEARN, L. B., RODGERS, B. K., MAXWELL, G. S., NIETO, J. A., STEELE, L. E., WRIGHT, F. M., OKONKWO, A. A. O., BABAYEMI, T., FADASE, A. O. & SCHOENEWALD, D. 1992. Depositional and sequence stratigraphic model, 1Y1 reservoir, Biafra member, Oso Field, Nigeria. *Nigerian Association of Petroleum Explorationists Bulletin*, **7**, 9–23.

TIPSWORD, H. L. J., SETZER, F. M. & SMITH, F. L., JR 1966. Interpretation of depositional environment in Gulf Coast exploration from paleoecology and related stratigraphy. *Gulf Coast Association of Geological Societies Transactions*, **XVI**, 119–130.

THOMPSON, L. B. & SNEDDEN, J. W. 1996. Geology and reservoir description of 1Y1 reservoir, Oso Field, Nigeria using FMS and dipmeter. *In*: *Gulf Coast Section Society of Economic Paleontologists and Mineralogists Foundation Seventeenth Annual Research Conference Proceedings*, 315–323.

VAIL, P. R. 1987. Seismic stratigraphy interpretation procedure. *In*: BALLY, A. W. (ed.) *Atlas of Seismic*

Stratigraphy Volume 1. American Association of Petroleum Geologists, Studies in Geology **27**, 1–10.

VAN DER ZWAN, K. & BRUGMAN, P. 1997. *Biosignals From the Ea-Field, Nigeria*. Biostratigraphy in Production and Development Geology, Abstracts of Presentation, University of Aberdeen, 16–17 June.

VAN WAGONER, J. C., MITCHUM, R. M., JR, CAMPION, K. M. & RAHMANIAN, V. D. 1990. *Siliciclastic Sequence Stratigraphy in Well Logs, Cores, and* Outcrops. American Association of Petroleum Geologists, Methods in Exploration Series, **7**.

VERRIEN, J. P., COURNAD, G. & MONTADERT, L. 1967. Application of production geology methods to reservoir characteristics analysis from outcrop observation. *In*: *Proceedings of the Seventh World Petroleum Congress, Mexico*, 425–446.

WEBER, K. J. 1971. Sedimentologic aspects of oilfields in the Niger Delta. *Geologie en Mijnbouw*, **50**, 559–576.

——1990. Niger Delta reservoir geology; historical growth of the sedimentological model and its application to field development. *In*: BROOKS, J. (ed.) *Classic Petroleum Provinces*. Geological Society, London, Special Publications, **50**, 367–383.

ZEITO, G. A. 1965. Interbedding of shale breaks and reservoir heterogeneties. *Journal of Petroleum Technology*, **17**, 1223–1228.

Biosignals from the EA Field, Nigeria

C. J. VAN DER ZWAN[1] & W. A. BRUGMAN[2]

[1] *SIEP-RTS, EPT-HM, P.O. Box 60, 2280 AB Rijswijk, The Netherlands*
[2] *Parklaan 3, 3765 GH Soest, The Netherlands*

Abstract: The EA Field study was undertaken to determine reservoir connectivity across growth faults. In this high subsidence area, conventional biostratigraphy does not have sufficient resolution to provide a production-scale time frame. Consequently, a new tool had to be developed which would provide such accuracy. This new high-resolution correlation tool, Biosignals, is based on the relationship 'pollen–parent plant–ecology'. Pollen, thus related, have been grouped into 'vegetation' zones, which have been logged per well and summarized on saw blade diagrams. Sequences of vegetation zones could be recognized, showing an upward trend from spore dominated, via swamp, rain forest, savanna to montane.

A number of potential regional time lines could be recognized: (1) a climate curve; (2) a change from the freshwater alga *Pediastrum* to the brackish water alga *Botryococcus*; and (3) the top of frequent fungi.

Based on this study the following conclusions may be drawn: (i) vegetational stacking patterns reflect fourth-order sequence stratigraphy; (ii) Biosignals suggest the tidal channels in well EA-13 to be lowstand deposits, although many of them belong to truncated sequences (only lowstand systems tract (LST) and transgressive systems tract (TST) preserved); (iii) on a third-order scale, the sandy middle part of the sequence (EA-1, 6000–6500 feet) correlates with dry climatic conditions and to low eustatic sea level. The two regionally extensive maximum flooding surfaces correlate with humid climatic intervals and eustatic high sea level; and (iv) the new high-resolution correlation tool provides a detailed biostratigraphical subdivision, within the existing conventional biozonation, and thus enables correlation of reservoir bodies across growth faults.

The EA Field lies in the offshore of the Niger delta (Fig. 1). The hydrocarbon-bearing reservoirs are deltaic–shallow-marine clastics of Late Tertiary age. In this offshore setting subsidence rates are high, with deposition of thousands of feet of sediments in a few million years. Consequently, conventional biostratigraphy does not supply sufficient resolution to provide the time frame required to enable reliable correlation of reservoirs across growth faults. In the EA Field the main reservoirs, the D5–D7 sands, are all deposited in one palynozone: Shell Nigeria's P850 zone of late Miocene age. The present study was undertaken to develop a new biostratigraphical correlation tool which would have sufficient resolution to recognize additional time lines within the existing biozonation, and thus be capable of effecting accurate correlation across growth faults.

Work methods

The Biosignal technique used in this study is founded in Quaternary palynology (Faegri & Iversen 1975; Janssen 1974; Birks & Birks 1980). In Quaternary studies, ecological knowledge of the parent plant of the pollen is used to recognize ecological and climatological changes through time. The same principles have been applied to the late Miocene of the EA Field. As the parent plants are known for 75% of the pollen in this relatively young interval (Muller 1981; Sowumni 1981*a, b*), the pollen–parent plant–ecology relationship could be established (Brugman *et al.* pers. comm. 1996). Ecologically related pollen have been grouped into vegetation zones. These vegetation zones have been logged per well and summarized on sawblade diagrams. Vegetational stacking patterns can be recognized, which correlate with fourth-order eustatic sequences. In this area, regional time lines can be recognized based on these ecological changes.

Some 40 samples per well from three wells were studied: EA-1 (5600–7000 feet), EA-13 (5300–7200 feet) and EA-5 (5200–7200 feet) (Fig. 1). Samples were mainly cuttings, only in well EA-13 were sidewall and core samples available.

VAN DER ZWAN, C. J. & BRUGMAN, W. A. 1999. Biosignals from the EA Field, Nigeria. *In:* JONES, R. W. & SIMMONS, M. D. (eds) *Biostratigraphy in Production and Development Geology*. Geological Society, London, Special Publications, **152**, 291–301.

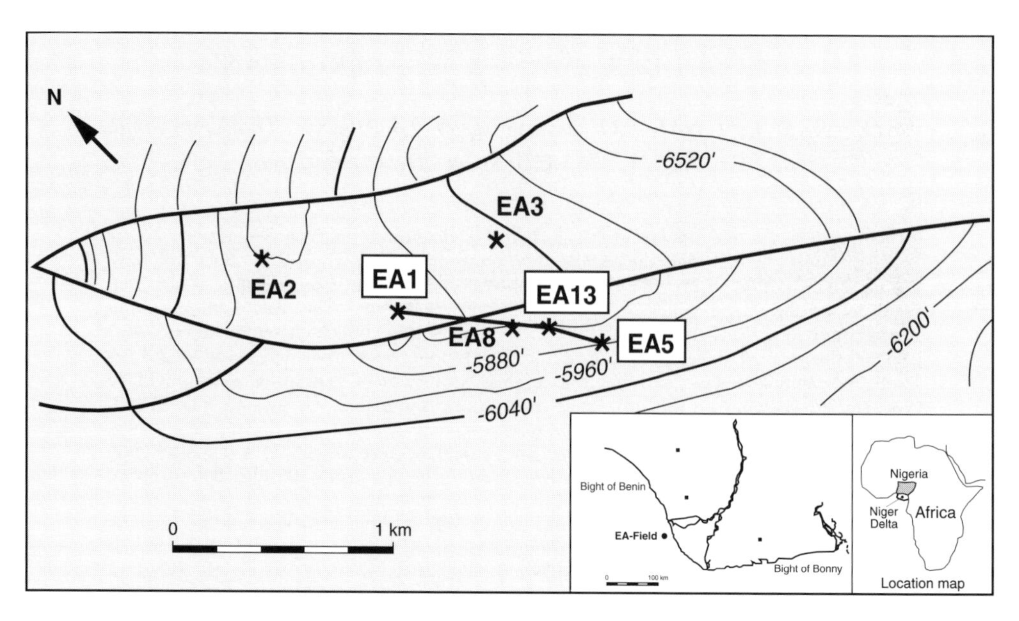

Fig. 1. Location map of the EA Field, Nigeria Top D6.0 sands.

Sample type is one of the main issues in this technique. Arguably, the Biosignal technique should be applied on core and sidewall samples only. Comparison between data from the well EA-13 (with cores and sidewall samples) and wells EA-1 and EA-5 (cuttings only) shows that in well EA-13 a high-frequency signal can be recognized. This signal, however, has a much higher frequency than required for the present study. A signal more suited for the present study is obtained from cuttings samples of wells EA-1 and EA-5. The better type of signal obtained could be explained by the smoothing effect of cuttings samples. Owing to the mixing of cuttings from a relatively large interval (10–30 feet) in one sample, the signal represents the whole interval (10–30 feet) rather than an individual layer. This contrasts with the signals obtained from core or sidewall samples; these should be considered spot observations, only representative for the 1 foot sampled. For the present study the resolution obtained from the cuttings is preferable. However, if in future studies bed-to-bed correlations have to be effected, dense spot sampling, using core and sidewall samples, may be required.

Standard palynological sample preparation techniques have been applied, using HF, HCl treatment and sieving. Use of a 10-μm sieve is recommended. Although use of a 15-μm sieve may affect the relative distribution between pollen, it is not thought to affect the relative frequency distribution of the pollen themselves. This despite the fact that, in particular, small mangrove pollen may be under-represented.

Standardized point counting techniques are used to acquire quantitative distribution data. Each time counting 100 palynomorphs. Starting with a count of 100 specimens of all palynomorphs, followed by a point count of 100 land-derived palynomorphs (including algae), 100 land-derived palynomorphs (excluding algae) and 100 marine palynomorphs. These data form the basis for calculating pollen-sums of all groups (Faegri & Iversen 1964; Janssen 1974). Data are manipulated in MS EXCELL spreadsheets and displayed in RAGWARE (Figs 3 and 4). The average value per group is calculated from all samples, and the above average value per group is shown as the shaded part of the curve (Figs 3, 4, 6, and 7).

Biosignals

Vegetational stacking patterns

Vegetation is the basis for climate recognition (Good 1974; Janssen 1974; Birks & Birks 1980). Simplified, oak trees are characteristic for a temperate climate, whereas palm trees are characteristic for the tropics. Climate varies through

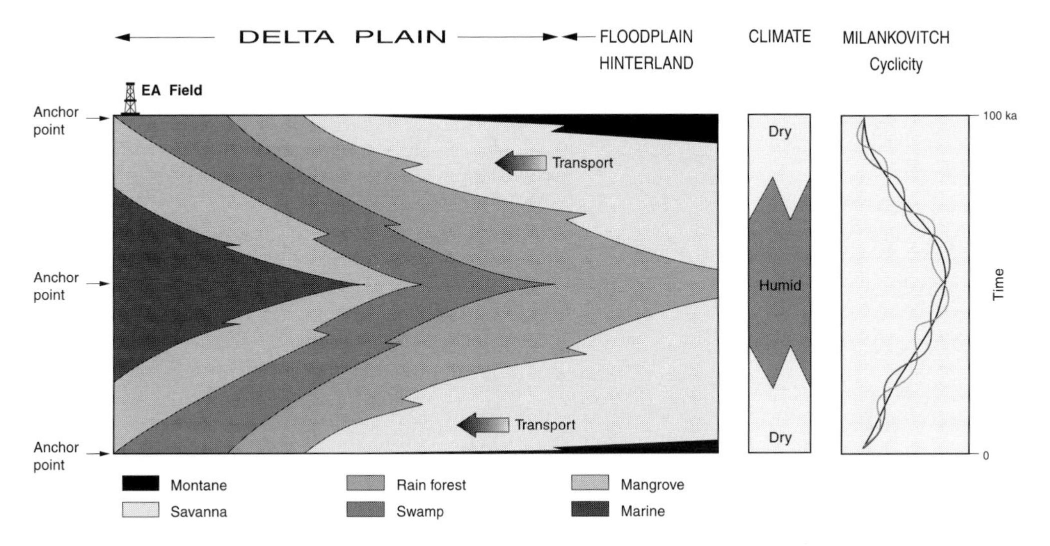

Fig. 2. Vegetational development in one climatic cycle.

time, which results in shifting of the climatic belts and of the vegetation belts (Birks & Birks 1980). During glaciations tundras expand southwards, whereas during interglacials the tree line moves northwards. In the equatorial region these climatic changes are expressed as variations between wetter and dryer climate. They can be recognized in the variation of the floodplain (hinterland) vegetation, from rain forest (wet) to savanna (dry).

In the pollen record these climatic changes are expressed in terms of quantitative changes in the palynofloras derived from the hinterland (Visscher & van der Zwan 1981; van der Zwan *et al.* 1985; Visscher *et al.* 1994). In the tropics these climatic changes are reflected in the variation in relative frequency between rainforest and savanna pollen (Morley & Richards 1993). However, the pollen record is usually not recorded on the flood plain, but in more coastal, deltaic or offshore settings in which sea-level fluctuations play an important role. Consequently, when logging intervals deposited in an offshore setting, not only is the entire hinterland pollen recorded but also pollen reflecting coastal vegetation belts, such as swamp, mangrove or marine elements. Theoretically, climatic changes in these settings would be reflected as follows (Fig. 2): during a dry climate, savanna and montane elements would be dominant and expand, partly at the expense of wetter flora elements such as rainforest, into the delta. When the climate becomes wetter, the rainforest expands far onto the flood plain, to be replaced in the delta by swamp, mangrove or marine elements. The next reversal in climate would have the opposite effect resulting in an increase in savanna and montane flora elements.

As a consequence of the different settings in which the pollen is recorded, the frequency of the marine, coastal or flood-plain elements will vary. This will result in a different local pollen record, however, the cyclicity contained within the regional pollen record will remain the same, which results in correlatable anchor points. It is tempting to assume that in the deltaic setting the main anchor points, maximum savanna–montane and maximum mangrove–marine, correspond in time to sequence boundaries and maximum flooding surfaces, respectively.

Transport of pollen will affect the pollen record (Muller 1959: Traverse & Ginsburg 1966; Poumot 1989; Dupont & Agwu 1991; van der Kaars 1991). This effect is most noticeable on bisaccate pollen, which have superb floating ability, but is less on non-saccate or angiosperm pollen of approximately the same size and weight. In the latter cases, it is assumed that in a marine depositional setting the pollen record gives a fair representation of the hinterland and local vegetation. Nevertheless, the abundance of spores during dry climatic conditions is attributed to increased erosion and run-off during periods with relatively less vegetation to hold the top-soil.

Brugman *et al.* (pers. comm. 1993) established the parent plant–pollen relationship of the Neogene of the EA Field. Five vegetation zones

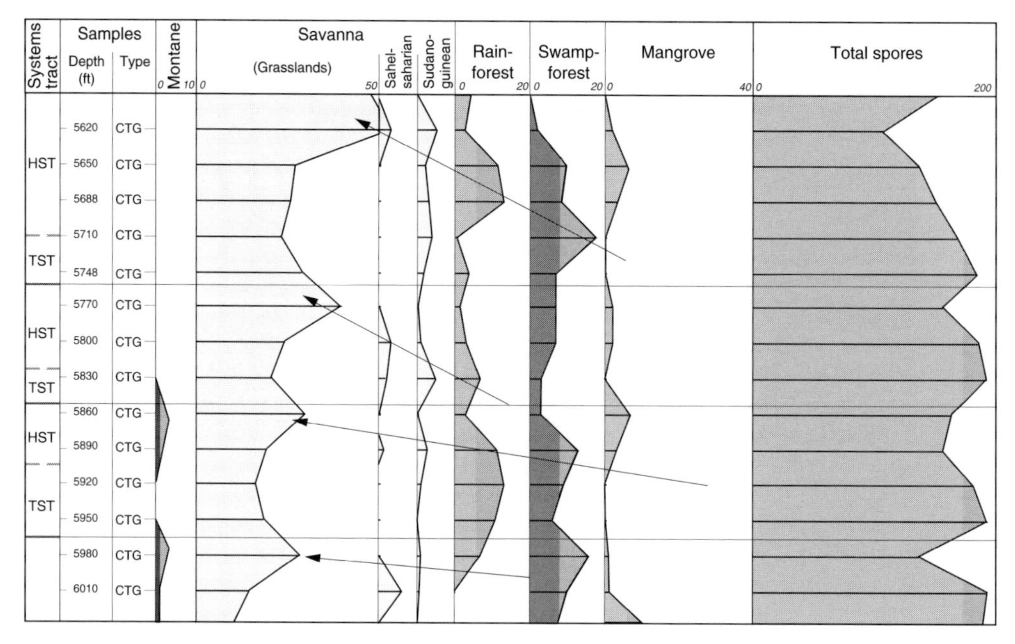

Fig. 3. Vegetational stacking patterns, well EA-1. Darker shaded areas in pollen curves indicate above average frequencies. Arrows (→) indicate trends in vegetational stacking patterns. Numbers above pollen curves indicate actual numbers of terrestrial pollen counted in 'land-derived pollen sum'.

could be recognized: montane, savanna, rain forest, swamp, mangrove, and with an additional category of 'undifferentiated spores' (Fig. 3).

- *Montane* – only occasionally developed. Characteristic taxa include *Ericaceae* and *Myricaceae*.
- *Savanna* – dominated by *Graminae*. Sahelo–Saharian and Sudanian taxa. The Sahelo–Saharian savanna represent a somewhat dryer savanna than the Sudanian–Guinean savanna. Although *Graminae* pollen dominate the savanna, they are also represented in the freshwater swamp (Keay 1959).
- *Rain forest* – this vegetation zone is represented by many different taxa, for example, *Acanthaceae*, *Rutaceae*, *Euphorbiaceae*, *Caesalpiniaceae*, *Mimosaceae*, *Ctenophonaceae* and a variety of Palmae (see remarks in the next paragraph on swamp forest). As many of these taxa are poor pollen producers, this phase is difficult to differentiate.
- *Swamp forest* – The main taxa are *Symphonia* and Palmae.

Remarks: (1) notable increases in the abundance of freshwater swamp taxa is probably related to periods of sea-level rise. During these periods the swamp forest is thought to expand, which is often accentuated by a proliferation of *Botryococcus* and/or *Pediastrum*.

(2) Poumot (1989) relates a palm-dominated phase to a rise in sea level, during which an extensive sandy coast would be formed. Palms, however, are extant in many different biotopes in West Africa, ranging from savanna to coastal swamp. Although palms do expand during sea-level rise, this may not be due to an expansion of a sandy coast as proposed by Poumot (1989), but to an overall expansion of the palms during a warmer climate.

- *Mangrove* – this vegetation zone is characterized by a dominance of *Rhizophora*.
- *Spore zone* – this zone is characterized by a dominance of spores (Pteridophytes, Lycopodiaphytes and Bryophytes).

Remarks: in near-shore sequences increases in the abundance of spores may be related to warmer, interglacial periods, resulting in periods of sea-level rise. However, the abundance of spores in the marine realm may also be observed in near-shore settings due to the fact that they are generally heavier and more resistent than most other palynomorphs, and therefore settle out earlier and/or are over-represented in higher-energy near-shore environments.

In a well with a complete sequence, vegetation zones commence with the spore-dominated zone and range through to the montane vegetation zone. Stacking pattern recognition is best effected by focusing on the anchor points: the presence of montane taxa or the maximum in savanna vegetation (Morley & Richards 1993).

Climate curve

Climatic variation is controlled by Milankovitch cyclic variations in the orbit of the Earth around the Sun (eccentricity, 100 and 400 ka), and in the obliquity (41 and 54 ka) and precession (23 and 19 ka) of the Earth's axis (Berger 1980; de Boer & Smith 1994). The 100- and 400-ka cycles result mainly in variations in temperature, and have been correlated with the Pleistocene glaciations. The obliquity and, particularly, the precession cycles are well expressed in mediterranean and equatorial regions, where they result in variation in precipitation (Versteegh 1995).

Studies on Neogene pollen in Colombia (Hooghiemstra & Ran 1994: Hooghiemstra & Cleef 1995) demonstrated the direct relationship between the abundance of arboreal pollen and the 100-ka (eccentricity), and the 19- and 23-ka (precession), cycles. Similar relationships could be established between dinoflagellates and the 23-, 41- and 100-ka cycles in the Neogene of the Mediterranean (Versteegh 1995).

The recognition of Milankovitch-order cyclicity studies in Colombia by Hooghiemstra & Ran (1994) suggests that such a signature may also be recorded from Nigerian data. Their recognition of the 100-ka (eccentricity) cycle is particularly important as this frequency has been correlated with glaciations, and thus provides a link to sea-level fluctuations. In the present study, the character and frequency of the vegetational stacking

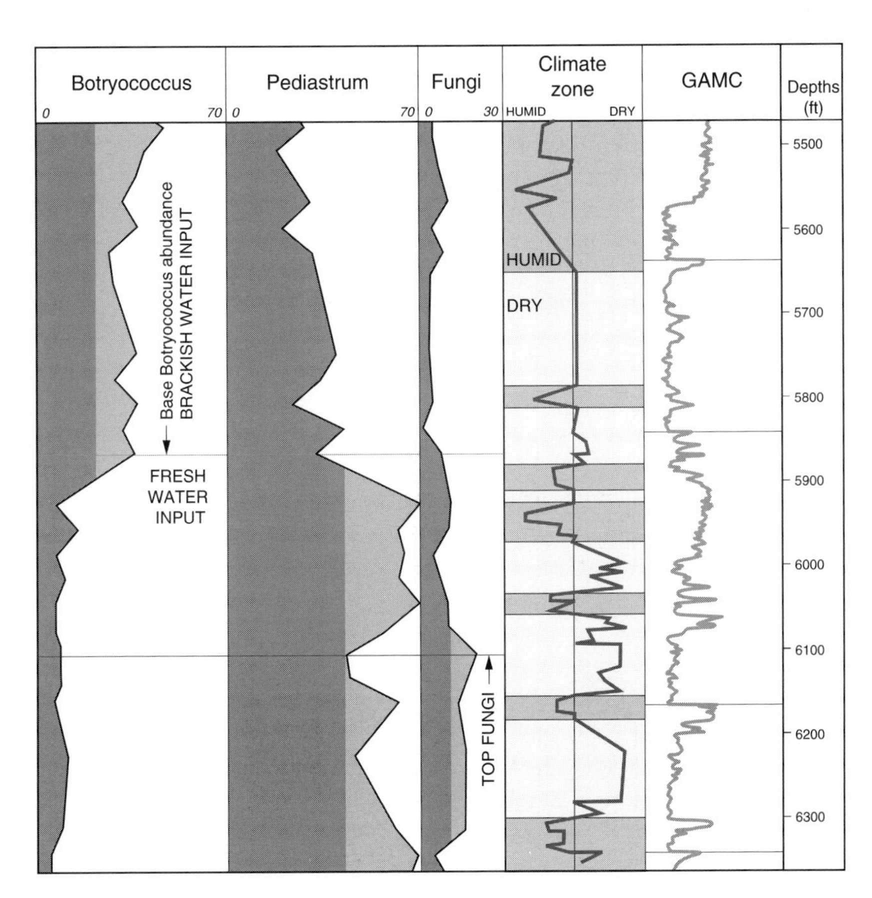

Fig. 4. Regional time lines, well EA-5. Darker shaded areas in pollen curves indicate above average frequencies. Numbers above pollen curves indicate actual numbers of fungal and algal palynomorphs counted in 'land-derived pollen sum'. Climate zone and curves based on the ratio savanna/rain forest pollen.

patterns would suggest a similar interpretation (compare Morley & Richards 1993). The recognition of the 100-ka (eccentricity) cycle would be in line with the number of cycles (six) recognized in the studied interval and the calculated duration of the late Miocene P850 Shell palynozone (approximately 600 ka).

Hinterland floras are thought not to be affected by changes in coastal environment, but only by regional climatic variation. This variation can be expressed in the composition of the hinterland vegetation: the ratio between the savanna and the rain forest. As changes in climate are synchronous and not affected by local environmental variation, this ratio expressed as a curve provides a means for reliable chronostratigraphic correlations for this area (Fig. 4). The climate curve obtained reflects the interaction of all Milankovitch frequences, rather than a direct link to the eccentricity cycle alone, and would require statistical manipulation to extract individual frequences. Even so, it provides a tool for chronostratigraphic correlation.

Time lines

In the wells studied, additional time lines could be recognized based on regional ecological changes (Fig. 4).

Botryococcus *and* Pediastrum. The two main fossilizing chlorococcale green algae are *Pediastrum* and *Botryococcus* (Traverse 1988; Zippi *et al.* 1991; Nielsen & Sorenson 1992; Tyson 1993). In comparison with *Botryoccoccus*, *Pediastrum* appears to be less tolerant of raised salinities, and to flourish in deeper, more permanent, better-mixed, harder-water and more nutrient-rich lake waters (Wake & Hillen 1980; Reynolds

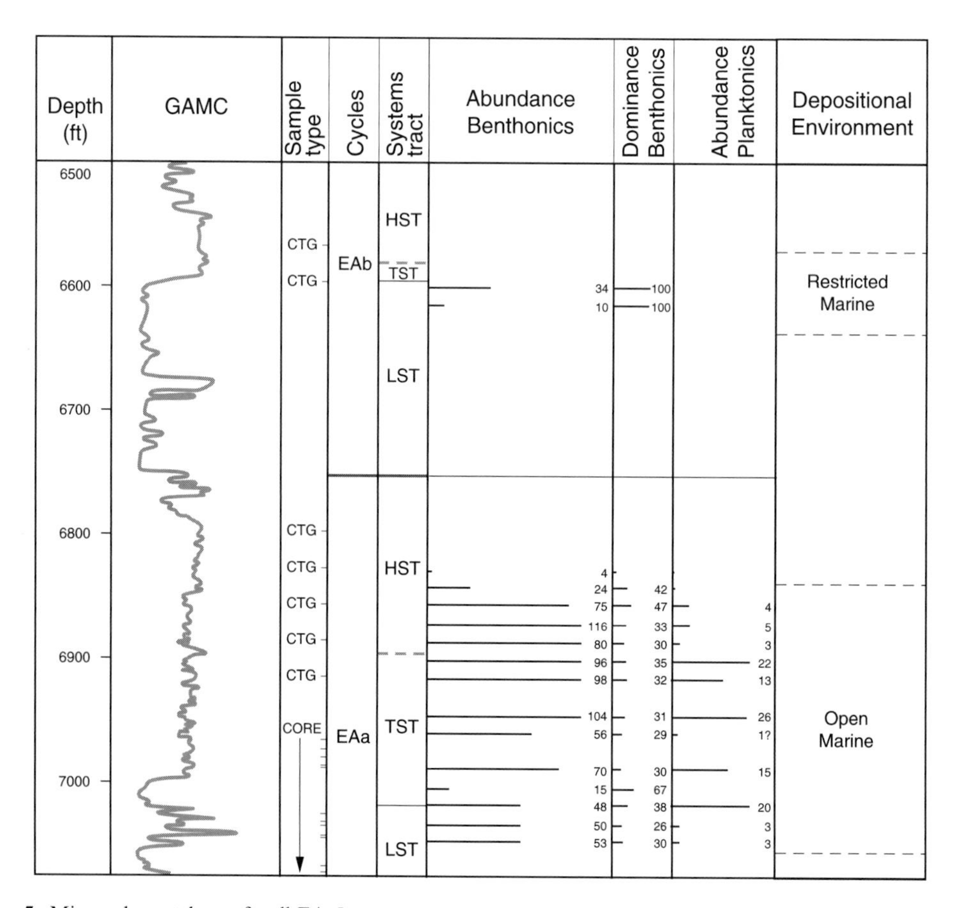

Fig. 5. Micropalaeontology of well EA-5.

1984; Talbot & Livingstone 1989). Where they do occur together, the much greater growth rate of *Pediastrum* generally results in its dominance. Most fossil records of *Botryococcus* are from lacustrine, fluvial, lagoonal and deltaic facies (Piasecki 1986; Batten & Lister 1988), but it is also recorded from facies that are at least temporarily hypersaline (Hunt 1987).

In all three wells studied a significant change occurs from dominance of the 'freshwater' alga *Pediastrum* to the 'brackish' water alga *Botryococcus*. Such a change in itself would hardly be considered of regional significance, but for the fact that they reflect a major change in the salinity of the water mass. This suggests a major change in either the basin shape or the source of the run-off from the hinterland. In both cases, they would reflect a significant basin-wide event.

Fungi. Fungi are efficient in rapid degradation of woody tissue under aerobic conditions. They respond quickly to environmental stress and disturbance, and as such are an indication for ecosystem destabilization (Visscher *et al.* 1996). Fungal remains, in comparison with pollen, are less subject to initial wind transport because they are generally associated with decaying plant material at the soil surface (with a very dense plant cover). Fungal remains are therefore extremely abundant in most delta deposits. They are probably sourced mainly from the backswamp soils during heavy rainfall and/or by erosion. In deltas they tend to be concentrated either within the delta or in near-shore environments (compare Muller 1959).

A significant uphole decrease in abundance occurs in the group of the fungi. Consequently, on a smaller scale, in the EA area their change in dominance is also considered to constitute a local correlative event.

The micropalaeontological signal

Foraminiferal abundance and dominance data are available from the upper and lower part of well EA-13 (Fig. 5). Variation in the abundance and dominance of benthonics and planktonics suggests variations in depositional environment. High abundance and low dominance of benthonics, together with the presence of planktonics, suggests an open marine environment, whereas low abundance and high dominance of benthonics together with the absence of planktonics suggests a restricted marine environment. They provide a key input in the determination of environment of deposition.

Sequence stratigraphy

An exploration-scale (third-order; Ten Hove pers. comm. 1992) and a production-scale (fourth order; Le Varlet pers. comm. 1991) sequence stratigraphy has been established for the EA Field area. In this shelf setting, the model predicts shoreface progradation during highstand (Posamentier *et al.* 1988; van Wagoner *et al.* 1990) and valley incision during lowstand, the latter are subsequently filled with estuarine or shoreface sands (compare Fig. 7).

Vegetation-based modifications

Vegetational stacking patterns in the wells studied show remarkable cyclicity. An ideal vegetational stacking pattern shows an upward trend from abundance of spores and occasionally savanna, to mangrove, swamp forest, rain forest, savanna and, ultimately, to montane (Figs. 3 and 6).

Comparison of the vegetational stacking patterns with sequence stratigraphy suggests the abundance of swamp forest and rain forest

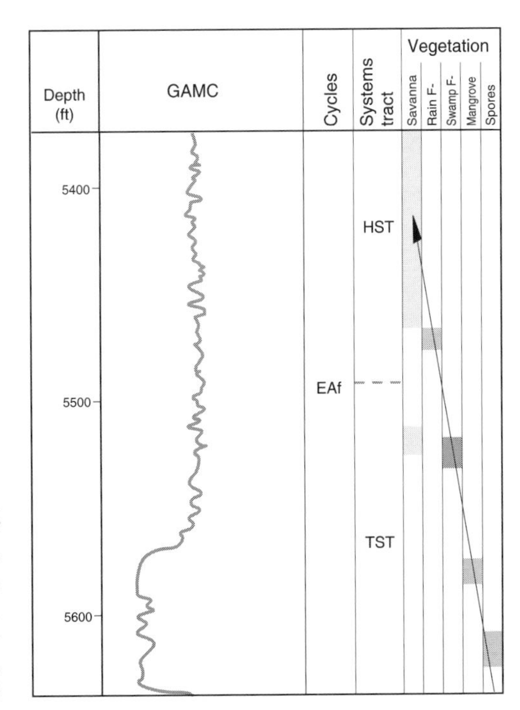

Fig. 6. A fourth-order sequence stratigraphy, well EA-5. Darker shaded areas in pollen curves indicate above average frequencies (compare also Figs 3 and 4). Arrows (→) indicate trends in vegetational stacking patterns.

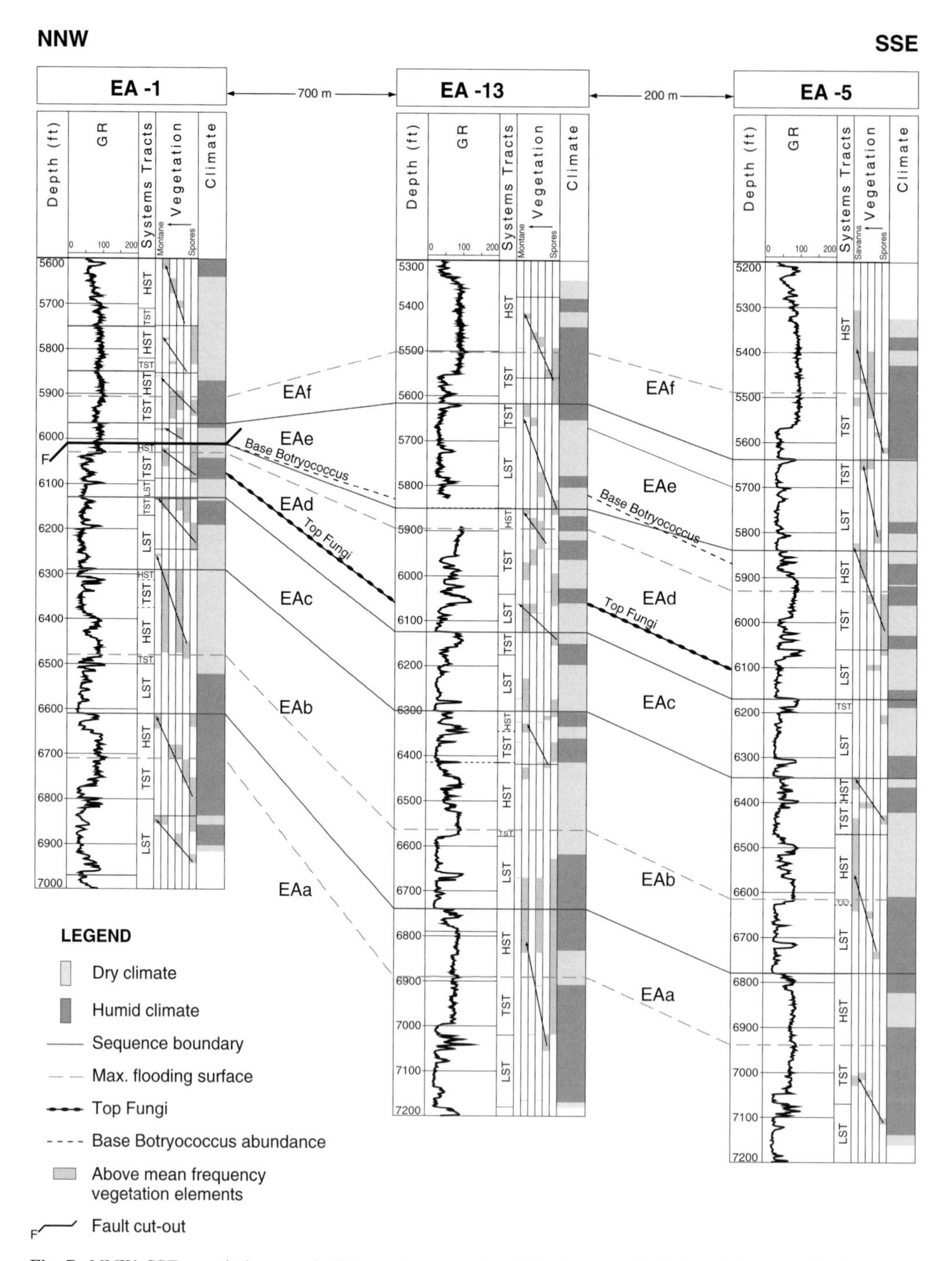

Fig. 7. NNW–SSE correlation panel of the main reservoirs (D5–D7 sands) in the late Miocene (Shell Nigeria's P850 zone) of wells EA-1, EA-13 and EA-5. In the vegetation column, darker shaded areas indicate above average frequencies in pollen curves (compare also Figs 3 and 4). Arrows (→) indicate trends in vegetational stacking patterns.

vegetation to approximate the maximum flooding surface (Figs 3 and 6). Abundance of mangrove pollen is related to stable and high sea level (Poumot 1989). Also in the EA Field, mangrove pollen occurs most frequently during transgressive and highstand systems tract (Figs 3, 6 and 7). The highstand up to the sequence boundary is characterized by the abundance of savanna and, ultimately, montane vegetation, although these rare flora elements are not always represented.

Lowstands often have a rather mixed association. The most consistent vegetational element is the abundance of spores, which are related to erosion and high run-off (Poumot 1989). Savanna vegetation, which is theoretically associated with the transition from highstand, via sequence boundary to lowstand, is frequently represented. It should be noted that spores and also savanna occur commonly not only on a fourth-order scale, but throughout the studied interval (Fig. 3). This could suggest a relationship with third-order lowstand (Ten Hove pers. comm. 1992) rather than with fourth-order cyclicity for most of the other flora elements.

The relationship between vegetational stacking patterns and sequence stratigraphy could be traced throughout the study area. The boundaries between the vegetational stacking patterns correlate very well with the sequence boundaries and maximum flooding surfaces in wells EA-1, EA-13 and EA-5. Occasionally, sequence boundaries have been shifted to better match the vegetational stacking pattern.

Correlations between EA-1, EA-13 and EA-5

Correlations between wells EA-1, EA-13 and EA-5 have been made, using regional time lines based on the change in abundance of *Botryococcus* and of fungi (Fig. 7). Using these time lines as a frame, further correlations are based on the climate zonation. Dry belts have to correlate with dry belts in other wells, and humid belts with humid belts, respectively.

The climatic zonation works both on a fourth- and third-order scale. In the latter case, the sand-dominated middle part of the sequence correlates with the third-order lowstand of Ten Hove (pers. comm. 1992), which would support the interpretation of increased run-off during lowstand. The two third-order maximum flooding surfaces in the upper (EAf mfs) and lower part (EAa mfs) of the studied interval correlate with humid climatic zones. On a bed-by-bed scale, there appears to be no direct relationship between climatic zone and lithology.

In well EA-5 (6344–6460 feet; Fig. 7) an additional vegetational stacking pattern could be recognized within one log-defined sequence (6344–6615 feet), which could be traced to well EA-13 (Fig. 7). If these stacking patterns are considered to be of the same hierarchical level, then this suggests that one log-defined sequence could have a duration of more than one vegetational stacking pattern. If these vegetational stacking patterns approximate to one eccentricity cycle of 100 ka, this suggests 200 ka or more years for the log-defined sequence. The alternative interpretation would suggest the second vegetational stacking pattern to be of a lower (fifth) order and to represent a modification on the higher (fourth) order cycle.

The absence of highstand vegetation in well EA-5 (5634 feet) has been interpreted as truncation of the sequence under a sequence boundary (Fig. 6). On the other hand, the relative absence of savanna vegetation in the lowstands of most wells could suggest the limited development of the lowstand in this area. This is not inconsistent with a shelf setting (Posamentier *et al.* 1988; van Wagoner *et al.* 1990). Ten Hove (pers. comm. 1992) suggested lowstand deposits in the EA Field area to be mainly developed as incised valley fills.

Based on the above time frame a fault cut-out, due to a growth fault, is suggested in well EA-1 (170 feet at 6010 feet) of a significant sand package present in wells EA-13 and EA-5 (Fig.7; compare also Fig. 1).

Conclusions

For this Late Miocene (P850 palynozone) interval, the 'Pollen–parent plant' relationship and its environmental relationship have been established. In this area, the vegetational stacking pattern swamp–rain forest–savanna correlates with transgressive–highstand systems tract, whereas the abundance of spores correlates with high run-off during lowstand.

Based on the floras studied a number of potential time lines could be recognized, such as:

- a climate curve, based on the ratio savanna/rain forest;
- the change from *Pediastrum*-dominated to *Botryococcus*-dominated algal floras indicating a regionally extensive change from a brackish to a freshwater mass;
- the top frequent fungi, indicating a regionally extensive change in the terrestrial ecosystem.

Foraminifera provide an excellent tool for the recognition of maximum flooding surfaces, which provide time lines.

Based on this interpretation the following conclusions may be drawn:

- vegetational stacking patterns reflect fourth-order sequence stratigraphy and are probably related to 100-ka Milankovitch eccentricity cycles;
- Biosignals suggest the tidal channels in well EA-13 (EAb and EAc cycles) to be lowstand deposits;
- on a third-order scale, the sandy middle part of the sequences studied correlates with an extensive dry climatic interval, Such climatic conditions would be related to a sea-level low with high run-off and, subsequently, extensive lowstand deposits. The two regionally extensive maximum flooding surfaces correlate with a humid climatic interval. These climatic zones can be correlated both on third- and on fourth-order cycle scale;
- application of the new high-resolution 'Biosignal' tool allows accurate correlation of reservoir bodies across growth faults and recognition of fault cut-outs.

Thanks are due to Shell Petroleum Development Company, Nigeria, for permission to publish this paper. The authors are indebted to the following colleagues for their valuable contributions, discussions and advice: Hans Roersma, Tim Potter, Hans Hageman, Marietta Vroon ten Hove, Peter Oriaifo, Le Varlet and Tom Faulkner.

References

BATTEN, D. J. & LISTER, J. K. 1988. Evidence of freshwater dinoflagellates and other algae in the English Wealden (Early Cretaceous). *Cretaceous Research*, **9**, 171–179.

BERGER, A. 1980. The Milankovitch astronomical theory of palaeoclimates: a modern review. *Vistas in Astronomy*, **24**, 103–122.

BIRKS, H. J. B. & BIRKS, H. H. 1980. *Quaternary Palaeoecology*. Edward Arnold, London.

DE BOER, P. L. & SMITH, D. G. (eds) 1994. *Orbital Forcing and Cyclic Sequences*. International Association of Sedimentology, Special Paper, **19**.

DUPONT, L. M. & AGWU, O. C. 1991. Environmental control of pollen grain distribution patterns in the Gulf of Guinea and offshore NW-Africa. *Geologische Rundschau*, **80**, 567–589.

FAEGRI, K. & IVERSEN, J. 1975. *Textbook of Pollen Analysis*. 3rd edition. Hafner, New York.

GOOD, R. 1974. *The Geography of Flowering Plants*. Longman, London.

HOOGHIEMSTRA, H. & CLEEF, A. M. 1995. Pleistocene climatic change and environmental and generic dynamics in the North Andean montane forest and parano. *In*: CHURCHILL, S. P. *et al.* (eds) *Biodiversity and Conservation of Neotropical Montane Forest*. New York Botanical Garden, 35–49.

—— & RAN, E. T. H. 1994. Late Pliocene–Pleistocene high resolution pollen sequence of Colombia: an overview of climatic change. *Quarternary International*, **21**, 63–80.

HUNT, C. O. 1987. Dinoflagellate cyst and acritarch assemblages in shallow-marine and marginal-marine carbonates: The Portland Sand, Portland Stone and Purbeck Formations (Upper Jurassic/Lower Cretaceous) of southern England and northern France. *In*: HART, M. B. (ed.) *Micropalaeontology of Carbonate Environments*. British Micropalaeontological Society Series. Ellis Horwood, Chichester, 208–225.

JANSSEN, C. R. 1974. *Verkenningen in de palynologie*. Utrecht.

KEAY, R. W. J. 1959. *An Outline of Nigerian Vegetation*. Government Printer, Lagos.

MORLEY, R. J. & RICHARDS, K. 1993. *Graminae* cuticle: a key indicator of Late Cenozoic climatic change in the Niger Delta. *Review of Palaeobotany and Palynology*, **77**, 119–127.

MULLER, J. 1959. Palynology of recent Orinoco delta and shelf sediments: reports of the Orinoco shelf expedition. *Micropaleontology*, **5**(1), 1–32

——1981. Fossil pollen records of extant angiosperms. *Botany Review*, **47**, 1–142.

NIELSEN, H. & SORENSON, I. 1992. Taxonomy and stratigraphy of late-glacial *Pediastrum* taxa from Lysmosen, Denmark – a preliminary study. *Review of Palaeobotany and Palynology*, **74**, 55–75.

PIASECKI, S. 1986. Palynological analysis of the organic debris in the Lower Cretaceous Jydegard Formation, Bornholm, Denmark. *Grana*, **25**, 119–129.

POSAMENTIER, H. W., JERVEY, M. T. & VAIL, P. R. 1988. Eustatic controls on clastic deposition I – Conceptual framework. *In*: WILGUS, C. K., POSAMENTIER, H., ROSS, C. A. & KENDALL, C. G. ST. C. (eds) *Sea Level Changes – An Integrated Approach*. Society of Economic Paleontologists and Mineralogists, Special Publication, **42**, 109–124.

POUMOT, C. 1989. Palynological evidence for eustatic events in the tropical Neogene. *Bulletin de Centres de Récherches Exploration – Production Elf Aquitaine*, **13**, 437–453.

REYNOLDS, C. S. 1984. *The Ecology of Freshwater Phytoplankton*. Cambridge University Press, Cambridge.

SOWUMNI, M. A. 1981*a*. Nigerian vegetational history from the Late Quarternary to the present day. *Palaeoecology of Africa*, **13**, 217–234.

——1981*b*. Aspects of Late Quarternary vegetational changes in West Africa. *Journal of Biogeography*, **8**, 457–474.

TALBOT, M. R. & LIVINGSTONE, D. A. 1989. Hydrogen Index and carbon isotopes of lacustrine organic matter as Lake Level indicators. *Paleogeography, Palaeoclimatology, Palaeoecology*, **70**, 121–137.

TRAVERSE, A. 1988. *Paleopalynology*. Allen and Unwin, New York, 1–600.

—— & GINSBURG, R. N. 1966. Palynology of the surface sediments of Great Bahama banks as related to water movements and sedimentation. *Marine Geology*, **4**, 417–459.

TYSON, R. V. 1993. Palynofacies analysis. *In*: JENKINS, D. G. (ed.) *Applied Micropaleontology*. Kluwer, Dordrecht, 153–191.

VAN DER KAARS, W. A. 1991. Palynology of marine piston cores: A Late Quartenary vegetational and climatic record for Australasia. *Paleogeography, Palaeoclimatology, Palaeocology*, **85**, 239–302.

VAN DER ZWAN, C. J., BOULTER, M. C. & HUBBARD, R. N. L. B. 1985. Climatic change during the Lower Carboniferous in Euramerica, based on multivariate statistical analysis of palynological data. *Palaeogeography, Palaeoclimatology, Palaeoecology*, **52**, 1–20.

VAN WAGONER, J. C., MITCHUM, R. M., CAMPION, K. M. & RAHMANIAN, V. D. 1990. *Siliciclastic Sequence Stratigraphy in Well Logs, Cores and Outcrops*. American Association of Petroleum Geologists, Methods in Exploration Series, **7**.

VERSTEEGH, G. J. M. 1995. *Palaeoenvironmental Changes in the Mediterranean and North Atlantic in Relation to the Onset of Northern Hemisphere Glaciations (2.5 Ma BP) – A Palynological Approach*. PhD thesis, University of Utrecht.

VISSCHER, H. & VAN DER ZWAN, C. J. 1981. Palynology of the circum-Mediterranian Triassic: phytogeographical and palaeoclimatological implications. *Geologische Rundschau*, **70**, 625–636.

——, BRINKHUIS, H., DILCHER, D. L., ELSIK, W. C., ESHET, Y., LOOY, C. V., RAMPINO, M. R. & TRAVERSE, A. 1996. The terminal Paleozoic fungal event: Evidence of terrestrial ecosystem destabilization and collapse. *Proceedings of the National Academy of Science, USA*, **93**, 2155–2158.

——, VAN HOUTE, M., BRUGMAN, W. A. & POORT, R. J. 1994. Rejection of a Carnian (Late triasic) 'pluvial event' in Europe. *Review of Palaeobotany and Palynology*, **83**, 217–226.

WAKE, L. V. & HILLEN, L. W. 1980. Study of a 'bloom' of the oil rich alga *Botryococcus braunii* in the Darwin River reservoir. *Biotechnology and Bioengineering*, **22**, 1637–1656

ZIPPI, P. A., WELBOURN, P. & NORRIS, G. 1991. *Peridinium* and *Pediastrum* of recent lake acidification. *In*: *24th Annual Meeting of the American Association of Stratigraphic Palynologists*, Abstracts.

Applied palaeontology: a critical stratigraphic tool in Gulf of Mexico exploration and exploitation

B. J. O'NEILL,[1] A. E. DuVERNAY[2] & R. A. GEORGE[3]

[1] *Shell Offshore Inc., P.O. Box 61933, New Orleans, LA 70161, USA*
[2] *Shell Deepwater Development Inc., P.O. Box 61933, New Orleans, LA 70161, USA*
[3] *Total Biostratigraphic Services Inc., 117 Evangeline Drive, Slidell, LA 70460, USA*

Abstract: Recent exploration has relied heavily on detailed biostratigraphic correlations using foraminifera and calcareous nannoplankton, particularly in salt flank positions poorly imaged by seismic. Shell's Bonnie, a deep near-salt well, was drilled at Eugene Island Block 95 to the depth of projected amplitude anomalies without encountering significant sand bodies. Detailed palaeontological correlation to nearby wells verified that the objective section had been penetrated. Objective sands were interpreted to be missing because of stratigraphic thinning on to the dome. A sidetrack well encountered significant hydrocarbon reserves. Appraisal drilling in the Mars basin (Mississippi Canyon blocks 763, 806 and 807) was aided by the recognition of regional condensed sections between gravity-flow units providing correlation of individual reservoir units within the field. Biostratigraphy showed stratigraphic pinch-out of reservoir sands rather than absence from erosional or structural truncation near vertical salt faces. The interpretation improved volume estimates and the planning of future wells.

In an era where technological advances in wireline logging and geophysical methods have received tremendous attention, applied palaeontology remains a critical borehole tool for stratigraphic analysis. Micropalaeontology using traditional methods as well as newer more quantitative methods has significant technical and monetary impact on exploration and exploitation projects. Application of palaeontology can range from the regional to the individual reservoir level. Traditional Gulf Coast methodologies have been documented elsewhere (LeRoy 1977; Poag 1977; Ventress 1991) as have more quantitative techniques (Barnette & Butler 1987; Armentrout & Clement 1990; Martin & Fletcher 1995; Armentrout 1996; Jones *et al.* 1996; Jones 1997). In this paper two examples of the application of palaeontology to Gulf Coast drilling operations and the impact of these applications are discussed.

Shell's 'Bonnie', a Gulf of Mexico shallow-water salt dome field, and 'Mars', a deep-water, salt-rimmed mini-basin field, will be used to illustrate the value of high-quality biostratigraphic correlations. At Bonnie, palaeontology played a critical role in an operational decision which led to the gas discovery. At Mars, biostratigraphy played a crucial role in unravelling structural and stratigraphic complexities in the appraisal wells before the decision to proceed with development, and Shell's second tension leg platform, was made.

Prospect Bonnie

Prospect Bonnie is located on the continental shelf, offshore of Louisiana in Eugene Island

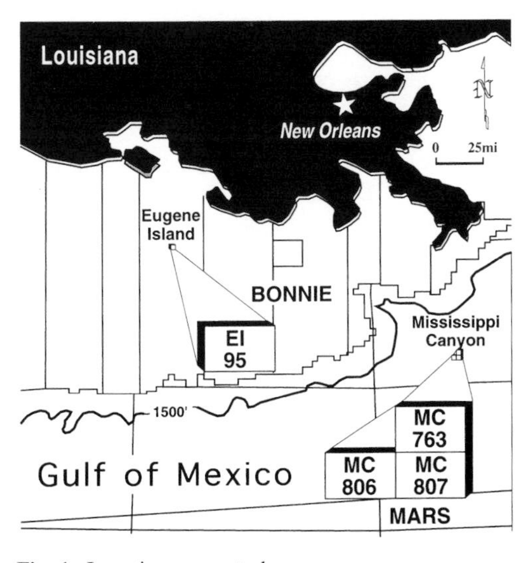

Fig. 1. Location map, study areas.

O'NEILL, B. J., DUVERNAY, A. E. & GEORGE, R. A. 1999. Applied palaeontology: a critical stratigraphic tool in Gulf of Mexico exploration and exploitation. *In*: UNDERHILL, J. R. (ed.) *Development and Evolution of the Wessex Basin*, Geological Society, London, Special Publications, **133**, 303–308.

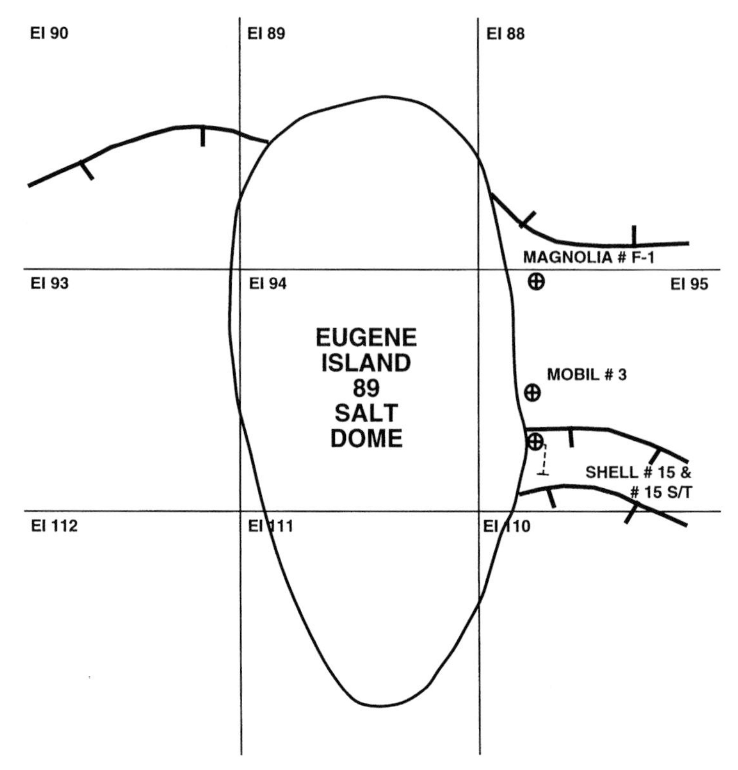

Fig. 2. Simplified map of Eugene Island 89 Salt Dome showing location of referenced wells and major faults.

Block 95 (Fig. 1). Bonnie was a 'bright spot' supported, late Miocene deltaic prospect on the eastern flank of Eugene Island 89 salt dome, located beneath existing shallow production (Fig. 2). Shell gained access to the prospect by shooting a three-dimensional (3D) seismic survey and drilling the initial discovery well.

The Bonnie wildcat location was picked using purchased 3D seismic, reprocessed at Shell (Fig. 3). The well was designed to test a downthrown fault block. The trap initially was defined by a salt face and the two bounding down to the south faults. Stratigraphic control was excellent: several nearby Mobil (Magnolia) wells within the block (Fig. 2) penetrated reservoir quality sands within the PM 8 (*Robulus* 'E'), M 2 (*Bigenerina* 'A') and M 2.2 (*Cristellaria* 'K') sections (Table 1). Styzen (1996) summarized the Shell biostratigraphic zonation for the Gulf of Mexico.

Although operational plans for the Bonnie test (EI 95 #15) called for foraminiferal biostratigraphy, it was considered low priority because of the strong geophysical and geological control for the project. Priorities changed suddenly when the well reached a depth of 9651 feet without encountering any significant sand bodies within the depth range of the anticipated objective section. Samples from the Bonnie well (Fig. 4) and from the believed equivalent section in the nearby Mobil S/L 685 #3 (Fig. 2) were examined for regional and, especially, local correlation markers while the wireline logs were being run. The logs confirmed that the section penetrated was essentially sand free, but were ambiguous as to whether: (1) the velocity model for the seismic was incorrect and the objective section had not yet been penetrated; (2) the well was too close to the salt stock and the reservoir units were missing because of lateral facies changes near the dome; or (3) unforeseen structural complexities had removed the reservoirs. Biostratigraphy was able to discriminate among these possibilities.

Because of the short interval sampled, only local palaeontological markers could be identified in the Shell #15. No regional condensed sections were sampled. However, the local markers identified in the wildcat and the Mobil #3 definitively showed that the #15 had indeed penetrated a substantial portion of the objective section in a sand-poor interval. A review of the samples from the Magnolia S/L 685 #F-1 (Fig. 2) further substantiated the new local field markers

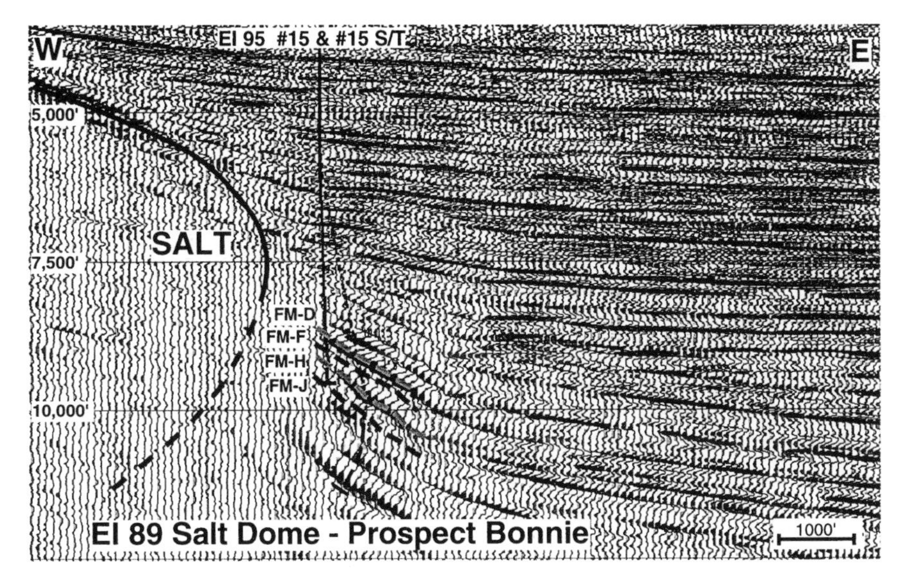

Fig. 3. E–W seismic profile through Bonnie location showing prospective amplitude anomalies.

bounding the reservoir sand units. Each of the field markers represented a downhole faunal increase and change. Field Markers (FM) 'D', 'F', 'H' and 'J' are defined by the first downhole occurrence (local LAD) of *Cyclammina* '3', *Uvigerina carapitana*, *Lenticulina americana* var. C-3 and *Uvigerina lirettensis*, respectively (Table 1).

As these data were being collected, discussions were underway as to whether to deepen the #15 or to drill a sidetrack hole further off-structure. Attempts were made to obtain a wireline tool in order to run a vertical seismic profile (VSP) to validate the seismic velocity model used. No such tool was available and it would have been several days before one could be brought to the location. The definitive biostratigraphic correlations between the three wells (Shell #15, Mobil #3 and Magnolia #F-1) prompted the decision to drill a sidetrack. Making this decision based on the biostratigraphy rather than waiting for the VSP saved 3–4 days rig time ($50 000–$70 000).

The #15ST was drilled to a measured depth of 12 985 feet and encountered numerous gas-bearing reservoirs from the M 2 through to the M 2.8 sections (Fig. 5). All four of the local biostratigraphic horizons established in the #15 were found in the #15ST (Fig. 6). The kick-off

Table 1. *Biostratigraphic zonation for Bonnie area. Ages from Styzen (1996)*

Age (Ma)	Shell marker name	Description of biostratigraphic events
7.55	PM 8	LAD *Lenticulina* sp. 'E'
7.75	M 2	LAD *Bigenerina floridana*
	FM B	*Textularia articulata* acme
7.77	M 2.2	LAD *Lenticulina cf. cristi* var. 'K'
	FM D	LAD *Cyclammina cancellata* var. '3'
	FM F	Local LAD *Uvigerina carapitana*
	FM H	LAD *Lenticulina americana* var. 'C-3'
	FM J	Local LAD *Uvigerina lirettensis*
8.85	NM B	LAD *Discoaster prepentaradiatus*
8.90	M 2.8	LAD *Bolivina thalmanni*

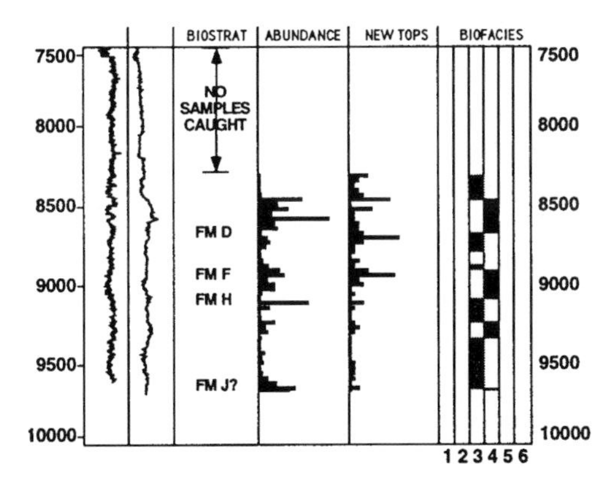

Fig. 4. Biostratigraphic interpretation of the EI 95 Shell #15 (biofacies code: 1 = littoral, 2 = inner neritic, 3 = middle neritic, 4 = outer neritic, 5 = upper bathyal, 6 = middle bathyal).

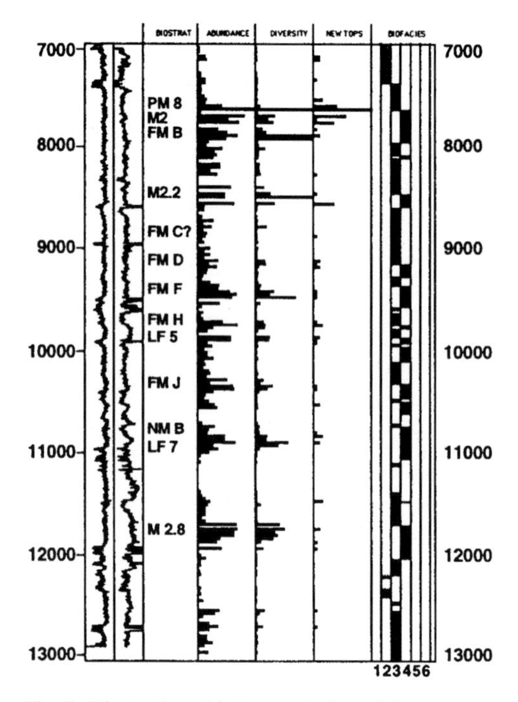

Fig. 5. Biostratigraphic interpretation of the EI 95 Shell #15ST (Biofacies code: 1 = littoral, 2 = inner neritic, 3 = middle neritic, 4 = outer neritic, 5 = upper bathyal, 6 = middle bathyal).

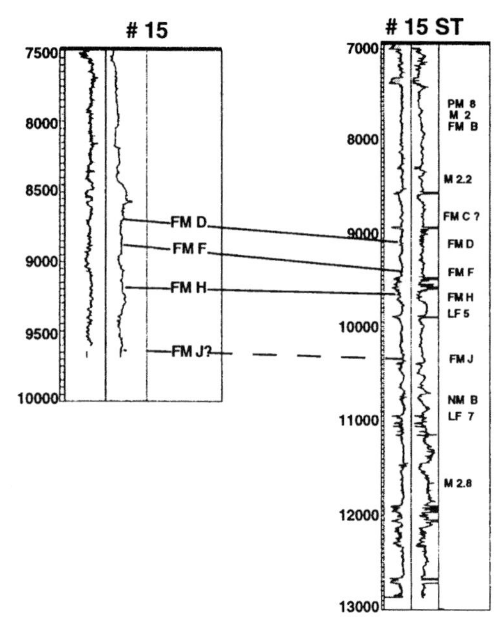

Fig 6. Wireline cross-section EI 95 Shell #15 and #15ST showing biostratigraphic correlations.

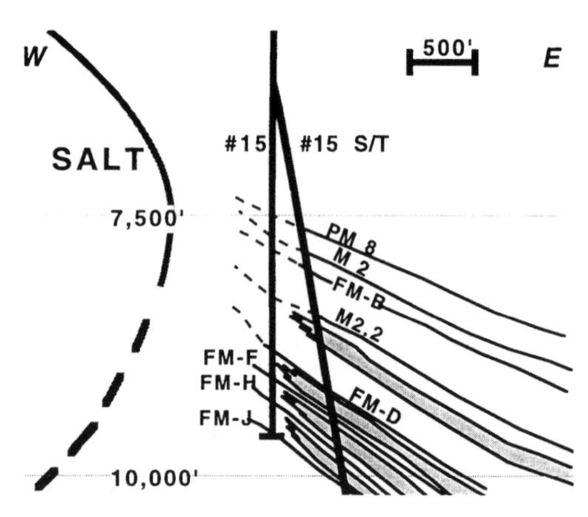

Fig 7. Schematic cross-section through EI 95 Shell #15 and #15ST, illustrating stratigraphic pinch-out of reservoir sands between the original hole and sidetrack.

point for the sidetrack was shallow enough to sample the PM 8 and M 2 regional shales (Fig. 5 and Table 1). The well was deepened and penetrated the regional NM B and M 2.8 shales. The regional transgressive shales provided age control above and below the reservoir section (Fig. 5 and Table 1). The model of rapid lateral facies changes and stratigraphic pinch-out of the reservoir sands was confirmed (Figs 6 and 7). Subsequent drilling in the area (Shell #17–#20) has proved reserves of 100–120 billion cubic feet (bcf) and further confirmed the reliability of the biostratigraphic field markers established in the initial investigation.

Mars development

The 'Mars' development is located in the Mississippi Canyon blocks 763, 806 and 807. It is approximately 50 miles south of the mouth of the Mississippi River in 3100 feet water depth (Fig. 1). The discovery well was drilled in 1989 in MC 763. The discovery well and subsequent delineation wells were drilled and evaluated using calcareous nannoplankton, foraminifera and palynology to understand the stratigraphic framework of the Mars mini-basin (Table 2). Mahaffie (1994) has described the geology of the Mars reservoirs.

In 1991 an appraisal well was drilled in MC 806 to test the existence and or extent of interpreted pay horizons imaged under salt (Fig. 8 left). The projected well path was expected to penetrate the chaotic unit just beneath the interpreted fault where amplitude anomalies terminate abruptly. The well encountered three bedded, sequential

Table 2. *Biostratigraphic zonation for Mars Basin. Ages from Styzen (1996)*

Age (Ma)	Shell marker name	Description of biostratigraphic events
1.15	P 1.6	GOM LAD *Hyalinea balthica*
3.09	P 3	LAD *Dentoglobigerina altispira*
5.15	PM 4	LAD *Globigerinoides mitra*
5.80	NPM 4.3	LAD *Discoaster berggrenii*
6.10	NPM BV	LAD *Discoaster bergenii*
7.80	NM AA	LAD *Minylitha convallis*
8.85	NM B	LAD *Discoaster prepentaradiatus*
9.10	NM C	LAD *Discoaster bollii*

biostratigraphic units in the chaotic seismic package near the shallow salt body (Fig. 8 right).

These three units are considerably older than the adjacent sediments in the basin and comparable in age to the deep reservoir section. This interval is interpreted to be a displaced block of sediment because further drilling encountered younger sediments, recognized by a key nannofossil assemblage, coeval to those in the basin centre. The interpretation has been aided by a strategically placed casing point, which prevented downhole contamination from the displaced block.

Below the displaced block, the well encountered a dramatically thinned equivalent to the deep reservoir section and finally penetrated massive salt. The interval from NM A (*Minylitha convalis*) to NM B (*Discoaster prepentaradiatus*) thinned from over 1700 feet. (200-foot net sand) at the basin centre to merely 300 feet (10-foot net sand) at its margin (Fig. 9; Table 2).

The significance of this interpretation is that the MC 806 well had, in fact, penetrated the entire objective section and that the reduced net sand was due to stratigraphic thinning as opposed to salt truncation. The biostratigraphic data from these wells enabled the Mars team to model the salt movement through time, controlling reservoir architecture and constraining volume estimates. These data were critical as the decision to proceed with the $1.2 billion development was made. The Mars tension leg platform (TLP) was installed in 1996 and current production exceeds 100 000 barrels per day.

Conclusions

Applied palaeontology played a critical operational role during drilling at prospect Bonnie. The data were gathered and interpreted quickly. The decision to drill the sidetrack well was made in a timely manner using firm biostratigraphic correlation to nearby wells. Detailed analysis of the foraminiferal faunas with key nannofossil control points, provided a high-resolution local zonation. This zonation has proved robust during Shell's ongoing drilling around the EI 89 dome.

At Mars, biostratigraphy was crucial to understanding the reservoir architecture near the salt bodies. Stratigraphic thinning was identified from biostratigraphic correlations. Recognizing this thinning improved the accuracy of reservoir models and the volume estimates based upon those models. The nature of the displaced sediment block penetrated by the MC 806 #1 was resolved only by palaeontological correlation.

Discussion

Applied palaeontology remains an important tool in hydrocarbon exploration and exploitation. By exploiting signals produced by dynamic biological communities, resolution finer than seismic loop level is attainable. Using multiple

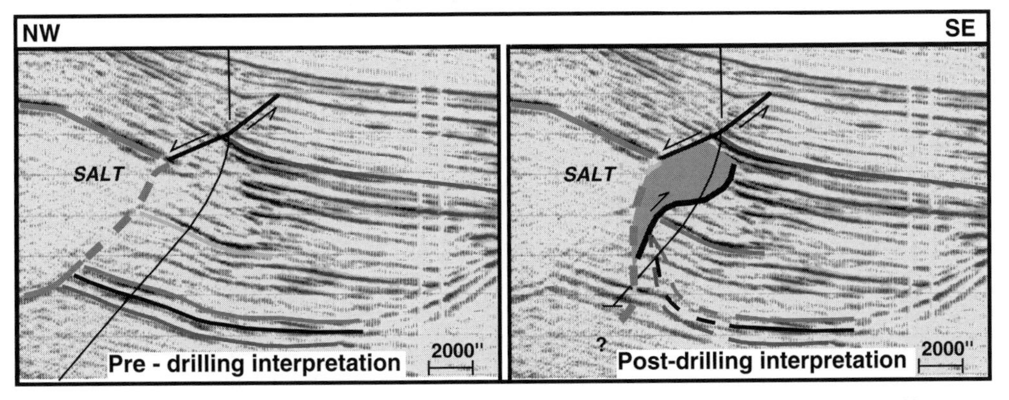

Fig. 8. Mars basin seismic profiles showing projected MC 806 #1 well path (left). Reinterpreted profile incorporating drilling results and biostratigraphic analysis (right).

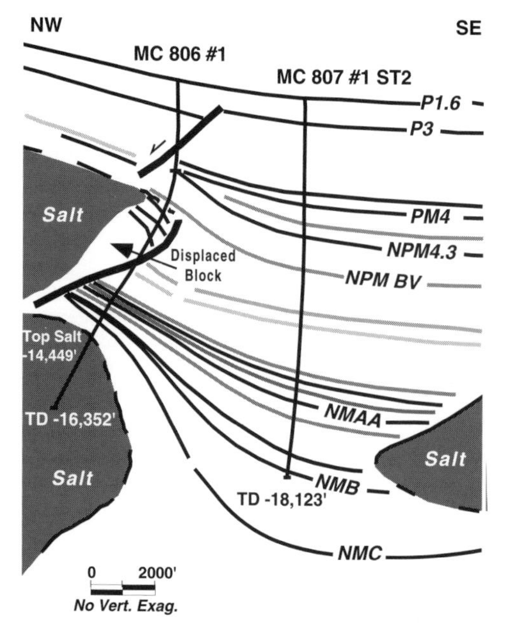

Fig. 9. Detailed cross-section illustrating stratigraphic thinning defined by palaeontological correlation.

fossil groups (e.g. foraminifera and calcareous nannofossils) greatly improves stratigraphic resolution. Resolution is also improved by examining each 30-foot interval rather than 60- or 90-foot intervals. Fundamental to this process is the integrity of the ditch cutting samples. Direct involvement by the biostratigraphers in planning and executing the sampling programme greatly improves the likelihood of high-quality samples being collected. Examination of palaeontological residues during drilling can provide timely data and effect operational decisions having significant monetary impact. High-quality samples accurately analysed provide critical information for hydrocarbon exploration and exploitation.

The authors would like to thank L. Reugger and M. Styzen for nannofossil stratigraphy at Mars, M. Rannik for palaeontological contributions at both Bonnie and Mars, M. Mahaffie for geological interpretations at Mars, and D. Zaengle and A. Rolph for Bonnie geology.

This article was modified and reprinted with permission from The Gulf Coast Section, Society of Economic Palaeontologists and Mineralogists Foundation.

References

ARMENTROUT, J. M. 1996. High resolution sequence biostratigraphy: examples from the Gulf of Mexico Plio-Pleistocene. *In*: HOWELL, J. A. & AITKEN, J. F. (eds) *High Resolution Sequence Stratigraphy: Innovations and Applications*. Geological Society, London, Special Publications, **104**, 65–86.

—— & CLEMENT, J. F. 1990. Biostratigraphic calibration of depositional cycles: A case study in High Island–Galveston–East Breaks areas, offshore Texas. *In*: ARMENTROUT, J. M. & PERKINS, B. F. (eds) *Sequence Stratigraphy as an Exploration Tool: Concepts and Practices in the Gulf Coast*. Gulf Coast Section SEPM Foundation Eleventh Annual Research Conference, 21–51.

BARNETTE, S. C. & BUTLER, D. M. (eds) 1987. *Innovative Biostratigraphic Approaches to Sequence Analysis: New Exploration Opportunities*. Gulf Coast Section SEPM Foundation Eighth Annual Research Conference.

JONES, G. D. 1997. Interpreting sequence stratigraphic architecture from biostratigraphic signatures: Case studies from the northern Gulf of Mexico, *GCAGS Transactions*, **47**, 654.

——, GARY, A. & WATERS, V. 1996. Applying the integrated paleontological system: Interpreting sequence stratigraphic architecture from microfossil signatures, Oligocene to Pleistocene section, Gulf of Mexico. *In*: REPETSKI, J. E. (ed.) *Sixth North American Paleontological Convention Abstracts of Papers*. Paleonotological Society, Special Publication, **3**, 202.

LEROY, D. O. 1977. Economic microbiostratigraphy. *In*: LEROY, L. W., LEROY, D. O. & RAESE, J. W. (eds) *Subsurface Geology*. Colorado School of Mines, Golden, Colorado, 212–233.

MAHAFFIE, M. J. 1994. Reservoir classification for turbidite intervals at the Mars discovery, Mississippi Canyon 807, Gulf of Mexico. *In*: WEIMER, P., BOUMA, A. H. & PERKINS, B. F. (eds) *Submarine Fans and Turbidite Systems; Sequence Stratigraphy, Reservoir Architecture and Production Characteristics, Gulf of Mexico and International*. Gulf Coast Section SEPM Foundation Fifteenth Annual Research Conference, 233–244.

MARTIN, R. E & FLETCHER, R. R. 1995. Graphic correlation of Plio-Pleistocene sequence boundaries, Gulf of Mexico: Oxygen isotopes, ice volume, and sea level. *In*: MANN, K. O. & LANE, H. R. (eds) *Graphic Correlation*. SEPM Special Publications, **53**, 235–248.

POAG, C. W. 1977. Biostratigraphy in Gulf Coast Petroleum Exploration. *In*: KAUFFMAN, E. G. & HAZEL, J. E. (eds) *Concepts and Methods of Biostratigraphy*. Dowden, Hutchinson & Ross, Stroudsburg, Pennsylvania, 213–233.

STYZEN, M. J. 1996. *Late Cenozoic Chronostratigraphy of the Gulf of Mexico*. Gulf Coast Section SEPM Foundation.

VENTRESS, W. P. S. 1991. Paleontology and its application in south Louisiana hydrocarbon exploration. *In*: GOLDTHWAITE, D. (ed.) *An Introduction to Central Gulf Coast Geology*. New Orleans Geological Society, New Orleans, 85–97.

Index

Page numbers in *italics* refer to illustrations; those in **bold** type to tables.